METHOD OF DISCRETE VORTICES

S. M. Belotserkovsky
I. K. Lifanov

Russian Academy of Sciences
Moscow, Russia

Translated by
V. A. Khokhryakov

English Edition Editor
G. P. Cherepanov
Florida International University
Miami, Florida

CRC Press
Boca Raton Ann Arbor London Tokyo

Library of Congress Cataloging-in-Publication Data

Belotserkovsky, Sergeĭ Mikhaĭlovich.
Method of discrete vortices / Sergei M. Belotserkovsky, Ivan K. Lifanov.
p. cm.
Translated from Russian.
Includes bibliographical references and index.
ISBN 0-8493-9307-8
1. Vortex-motion. 2. Aerodynamics—Mathematical models.
I. Lifanov, I. K. (Ivan Kuz'mich) II. Title.
QA925.B44 1992
629.132'3—dc20 92-13460
CIP

Direct all inquiries to CRC Press, Inc., 2000 Corporate Blvd., N.W., Boca Raton, Florida 33431.

International Standard Book Number 0-8493-9307-8

Library of Congress Card Number 92-13460

Printed in the United States of America 1 2 3 4 5 6 7 8 9 0

Printed on acid-free paper

Authors' Preface to English Edition

The advent of high-speed supercomputers resulted in narrowing the gap between fundamental mathematical and applied problems and enhancing their interdependence. The emergence of a new powerful and versatile method of analysis—the numerical experiment—has brought together the physical essence of a problem, its mathematical formulation, and a numerical method of solution taking into account specific features of computers. Most efficient solutions proved to be natural descriptions of physical phenomena embracing the whole of the process of their development. The predominance of discrete and nonstationary approaches was established.

The ever-growing requirements of practice result in a constantly growing complexity of applied problems for which traditional numerical methods often prove to be inadequate. Quite often applied researchers win the competition in developing a rational approach to solving problems, because their reasoning is based on understanding the essence of a problem, whereas professional mathematicians treat it in a more formal way. A numerical experiment facilitates finding rationale in such an approach by creating favorable conditions for rigorous mathematical verification and generalization. The development of a method encompassing the three major aspects of the problem—physical, mathematical, and computational—is decisive for achieving success. It must be stressed that the role of mathematics is not restricted to verifying or generalizing a method: it and only it makes a method both strict and versatile as well as extendable beyond the limits of a family of problems.

This book is the result of 20 years of work by the authors and their pupils, who have traveled together along the aforementioned path. Even before that, the analysis of physical and mathematical peculiarities of aerodynamic problems resulted in elaboration of a rational approach to their solution; the correctness of the method was verified by general analysis and logical reasoning. Then the method was thoroughly verified and perfected by using systematic numerical experiments and tested by comparing exact solutions to special problems of aerodynamics with calculated data.

This stage of the development of the method of discrete vortices (MDV) and its application to solving both stationary and nonstationary problems of aerodynamics for a variety of lifting surfaces was summed up by S. M. Belotserkovsky in his doctoral thesis, which was approved in 1955. Here the MDV was treated as a method of solving singular integral equations, both one dimensional (airfoils, cascades, annular wings) and two dimensional (monoplane wings of arbitrary plan form).

Further development of the method called on profound mathematical verification and generalization that was implemented by I. K. Lifanov and permitted spreading the ideas of the method into neighboring areas. The first results of this stage in the development of the MDV were generalized in our monograph published in the Soviet Union in 1985. The present book incorporates all the materials of the monograph, which were revised, corrected, and further developed by the authors. We have also included additional material obtained since 1985.

The formation of the new trend, which was developed aggressively during the last two decades, was supported by regular fruitful discussions at the All-Union seminars directed by the authors.

While writing this book, we used theoretical and calculated data as well as the useful advice of V. A. Aparinov, N. G. Afendikova, V. A. Bushuev, Yu. V. Gandel, A. V. Dvorak, V. V. Demidov, V. A. Ziberov, A. F. Matveev, A. A. Mikhailov, N. M. Molyakov, M. I. Nisht, L. N. Poltavskii, A. P. Revyakin, E. B. Rodin, A. A. Saakyan, M. M. Soldatov, I. Ya. Timofeev, and S. D. Shapilov. To all of them we express our profound gratitude. We also thank V. A. Khokhryakov for translating the book.

Sergei M. Belotserkovsky
Ivan K. Lifanov

Contents

An Introduction to Singular Integral Equations in Aerodynamics[†]

The present book provides an effective direct method of the numerical solution of singular integral equations for both one and two (or more) dimensions and includes multiple integrals, especially as applied to separated and vortex flows in aerodynamics. The authors of the book are a professional mathematician (Ivan Lifanov) and a numerical experimentalist in aerodynamics (Sergei Belotserkovsky) who have collaborated for many years. The book[‡] represents a notable milestone in the brilliant 250-year history of the theory of functions of a complex variable and provides a strong impetus for future development of singular integral equations in aerodynamics.

With a general aim to popularize singular integral equations in aerodynamics, the purpose of this introduction is to elucidate these issues under the following headings:

1. A little bit of history
2. Lift force of a thin airfoil
3. Optimal airfoil problem

The latter may be of interest for those seeking future engineering applications of the method presented in the book.

[†] Prepared by Genady P. Cherepanov, Florida International University, Miami.

[‡] The content of this book is different from that of the recent book entitled *Two-Dimensional Separated Flows* (CRC Press, 1993) because the latter does not cover singular integral equations at all. However, they are connected by common ideas and form an excellent collection of desk references for every aerospace engineer.

I.1. A LITTLE BIT OF HISTORY

Both aerodynamics and the theory of functions of a complex variable originated in the work of the Russian academician Leonhard Euler more than two centuries ago. However, Euler seemed to kill his newly born hydrodynamics when he proved the theorem of zero drag of a body moving in a perfect fluid. Engineers lost interest in this subject, and it took more than a century to revive their interest after works by Helmholtz, Kirchhoff, Kelvin, Joukowski, Rayleigh, and other great scholars appeared. Their computations of drag and lift forces laid the foundation for future applications of aerodynamics in aviation and turbine construction.

The theory of functions of a complex variable was being intensively developed during this time, presumably as a theoretical branch of mathematics until 1900, but it became strongly applied in the twentieth century. I will mention here only some results of primary importance for singular integral equations.

A *Cauchy integral* is the complex function $\phi(z)$ of a complex variable z of the following shape:

$$\phi(z) = \frac{1}{2\pi i}\int_L \frac{\varphi(\tau)\,d\tau}{\tau - z}, \qquad \tau \in L, i = \sqrt{-1}. \tag{I.1}$$

Here L is a closed or unclosed contour in a z plane and $\varphi(\tau)$ is a continuous or discontinuous complex function.

The function of a complex variable is said to be analytic at a point if it is differentiable at the point any number of times and expandable into a convergent power series at the point. The Cauchy integral in Equation I.1 provides a function $\phi(z)$ that is analytic at any point z outside L and discontinuous on L, that is, $\phi(z)$ tends to different values $\phi^+(t)$ and $\phi^-(t)$ when z tends to the same point t of L from the left-hand or right-hand side of L, respectively, if one goes along L.

A *singular integral* is the complex function $\phi(t)$ of a complex variable t defined by the divergent improper integral

$$\frac{1}{2\pi i}\int_L \frac{\varphi(\tau)}{\tau - t}\,d\tau, \qquad t \in L, \tag{I.2}$$

understood in the sense of *Cauchy principal value* as

$$\begin{aligned}\phi(t) &= \frac{1}{2\pi i}\text{P.V.}\int_L \frac{\varphi(\tau)\,d\tau}{\tau - t} \\ &= \lim_{\epsilon \to 0} \frac{1}{2\pi i}\int_{L_\epsilon} \frac{\varphi(\tau)\,d\tau}{\tau - t}, \qquad t \in L, \tau \in L.\end{aligned} \tag{I.3}$$

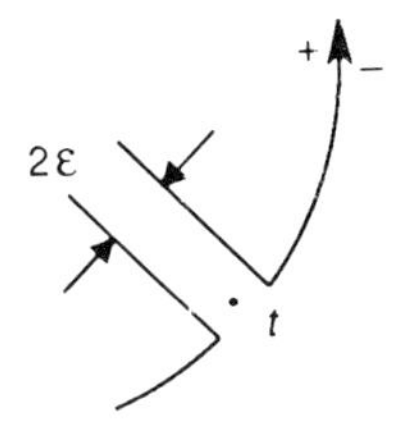

FIGURE I.1. The equilateral ϵ neighborhood of a point t excluded from the integration contour L participating in the definition of the Cauchy principal value. The arrow shows the direction of running L, and + and − show the location of discontinuity points on L relating to this direction.

Here L_ϵ is L without its vicinity of t cut off by a circle of a small radius ϵ centered at t (Figure I.1). For example, calculating the improper integral

$$\int_a^b \frac{dx}{x-c} = \lim_{\substack{\epsilon_1 \to 0 \\ \epsilon_2 \to 0}} \left[-\int_a^{c-\epsilon_1} \frac{dx}{c-x} + \int_{c+\epsilon_2}^b \frac{dx}{x-c} \right]$$

$$= \ln\frac{b-c}{c-a} + \lim_{\substack{\epsilon_1 \to 0 \\ \epsilon_2 \to 0}} \frac{\epsilon_1}{\epsilon_2}, \qquad a < c < b, \tag{I.4}$$

we get

$$\text{P.V.}\int_a^b \frac{dx}{x-c} = \ln\frac{b-c}{c-a}.$$

Hence, the improper integral does not exist (diverges) because the limit depends on the method of vanishing ϵ_1 and ϵ_2. However, a Cauchy principal value of the singular integral, when $\epsilon_1 = \epsilon_2 = \epsilon$, exists and equals the first term on the right-hand side of this equation, because the second term vanishes in this case. The same is valid for a Cauchy principal value of an arbitrary singular integral.

In 1873, the Russian mathematician Sokhotsky derived the basic equations

$$\phi^+(t) - \phi^-(t) = \varphi(t) \tag{I.5}$$

$$\phi^+(t) + \phi^-(t) = \text{P.V.}\frac{1}{\pi i}\int_L \frac{\varphi(\tau)\,d\tau}{\tau - t}. \tag{I.6}$$

Sokhotsky's work was unknown in the West, and in 1908 these equations were rederived by the German mathematician Plemeli. Sokhotsky equations allow mutual connection of the following famous boundary-value problems:

Riemann Problem:

$$\phi^+(t) = G(t)\phi^-(t) + g(t), \qquad t \in L. \tag{I.7}$$

Here $G(t)$ and $g(t)$ are some given functions, and $\phi^+(z)$ and $\phi^-(z)$ are analytic functions required to be found. In the case of unclosed L,

$$\phi^+(z) = \phi^-(z) = \phi(z).$$

Hilbert Problem:

$$a(s)u(s) + b(s)v(s) = c(s). \tag{I.8}$$

Here s is the length of arc on L, $a(s)$ and $c(s)$ are some given real functions, and $u(s)$ and $v(s)$ are the real and imaginary parts of an analytic function required to be found.

Singular Integral Equation Problems:
With a Cauchy kernel:

$$d(t)\varphi(t) + \frac{1}{\pi i}\text{P.V.}\int_L \frac{M(t,\tau)}{\tau - t}\varphi(\tau)\,d\tau = g(t), \qquad t \in L. \tag{I.9}$$

Here $d(t)$, $M(t,\tau)$, and $g(t)$ are some given complex functions and $\varphi(t)$ is a sought complex function.
With a Hilbert kernel:

$$a(s)u(s) - \frac{b(s)}{2\pi}\int_0^{2\pi} u(\sigma)\cot\frac{\sigma - s}{2}\,d\sigma = c(s). \tag{I.10}$$

All these problems can be reduced one to another; for example, Equation I.10 can be reduced to Equation I.8 in the same designations.

Poincaré, Hilbert, and Noether studied singular integral equations and established some general theorems by reducing them to Fredholm integral equations. Hilbert, Plemeli, and Carleman found explicit solutions to some important particular cases of these problems. The full explicit solution of the problems as formulated in Equations I.7–I.10 was given by Gakhov in 1936–1941 and Muskhelishvili in 1941–1946 (for full details, see Gakhov 1966 and Muskhelishvili 1946).

Similar problems for systems of singular integral equations reduced to the matrix Riemann problem of Equation I.7, where g, ϕ^+, and ϕ^- are n columns and G is an $m \times n$ matrix, are as yet not solved in explicit form.

In the general problem, explicit solutions exist only for some particular classes found by Gakhov (1952), Cherepanov (1962, 1965), and Khrapkov (1971).

As an illustration of the method, we consider Carleman's singular integral equation

$$\mu\varphi(x) + \frac{\lambda}{\pi i}\text{P.V.}\int_0^1 \frac{\varphi(\tau)\,d\tau}{\tau - x} = ig(x) \tag{I.11}$$

(λ and μ are positive constants; $0 < x < 1$; $g(x)$ and $\varphi(x)$ are real). The function

$$\Phi(z) = \frac{1}{2\pi i}\int_0^1 \frac{\varphi(\tau)\,d\tau}{\tau - z} \tag{I.12}$$

is introduced, which is analytic outside the unit cut (0, 1) along the real axis in the z plane.

Using Sokhotsky Equations I.5 and I.6, and the Cauchy integral in Equation I.12, Equation I.11 can be written as

$$\phi^+(x) + m\phi^-(x) = \frac{ig(x)}{\lambda + \mu}, \qquad 0 < x < 1,$$

$$m = \frac{\lambda - \mu}{\lambda + \mu}, \qquad -1 < m < 1. \tag{I.13}$$

This is a particular example of the Riemann problem. Let us find an auxiliary (canonical) solution $X(z)$ to the problem for the case where $g(x) = 0$:

$$X^+(z) + mX^-(z) = 0. \tag{I.14}$$

By knowing the properties of functions of type of z^{δ}, it is easy to obtain a solution to Equation I.14 in the form

$$x(z) = z^{\delta}(z-1)^{1-\delta}, \tag{I.15}$$

$$\delta = \frac{1}{2} - \frac{1}{2\pi i}\ln m, \qquad \lim_{z\to\infty}\frac{X(z)}{z} = 1.$$

Substituting $m = -X^+/X^-$ into Equation I.13, we get

$$\left(\frac{\phi}{X}\right)^+ - \left(\frac{\phi}{X}\right)^- = \frac{ig(x)}{(\lambda + \mu)X^+(x)}. \tag{I.16}$$

From physical considerations, it is important to specify the behavior of the solution near the ends $x = 0$ and $x = 1$. The function

$$F(z) = \frac{\phi(z)}{X(z)} - \frac{1}{2\pi(\lambda + \mu)} \int_0^1 \frac{g(\tau)\, d\tau}{X^+(\tau)(\tau - z)}, \tag{I.17}$$

in view of Equation I.16, will satisfy the condition equation

$$F^+(x) = F^-(x), \qquad 0 < x < 1, \tag{I.18}$$

and hence $F(z)$ is analytic everywhere in the z plane with the possible exception of points $z = 0$ and $z = 1$, where $F(z)$ can have poles, dependent on the specified solution behavior at these points. For example, if the only integrable singularities of the solution are physically permitted, these poles are simple; then, according to the Liouville theorem, we obtain

$$F(z) = \frac{C_1}{z} + \frac{C_2}{z - 1}, \tag{I.19}$$

where C_1 and C_2 are some constants.

Considering Equations I.17 and I.19, we obtain the solution in the form

$$\phi(z) = \frac{X(z)}{2\pi(\lambda + \mu)} \int_0^1 \frac{g(\tau)\, d\tau}{X^+(\tau)(\tau - z)} + X(z)\left(\frac{C_1}{z} + \frac{C_2}{z - 1}\right), \tag{I.20}$$

where

$$C_1 + C_2 = \frac{1}{(\lambda + \mu)2\pi i} \int_0^1 \frac{g(\tau)\, d\tau}{X^+(\tau)}.$$

The solution of the singular integral Equation I.11 is determined from Equation I.20 using Equation I.5. In this case, the solution depends on the arbitrary constant.

The opportunities of the explicit solution were intensively used in thousands of papers and books devoted to numerous applied problems of mathematical physics reduced to the problem in Equations I.7–I.10. For example, the bibliography of papers on hydrodynamics of jets and cavities in Gurevich (1965) includes nearly 700 main sources. The author also took part in these developments (see Cherepanov 1963, 1964).

Explicit methods were extremely popular in Russia from 1900 to 1960, and undoubtedly they furthered the success of large-scale engineering projects during that time, despite limited funding and human resources. As a matter of fact, the theory of functions of a complex variable gained more efficient engineering formulas than the rest of mathematics. Joukowski believed that "the task of mechanics was to find out explicit solutions." However, until recently, there were no direct numerical methods in singular integral equations, so that the latter were being quickly forgotten because the yield of explicit solutions was already collected.

Numerical method of discrete vortices, treated in detail in this book, enables us to use all the technical advantages of potentials and general solution. Its ease of use is greater than that of general numerical methods (for example, the method of finite elements), so that in cases when the latter require a supercomputer, a calculator sometimes suffices in the former.

I.2. LIFT FORCE OF A THIN AIRFOIL

Airfoils are main thrust and lift-generating structural elements of aerodynamical machines, including wings and blades.

Let us consider the irrotational steady polytropic flow of an inviscid compressible gas. Following Cherepanov (1987), the governing equations of gas dynamics can be written in the form of Γ integrals invariant with respect to closed integration surfaces:

$$\int_S \rho v_i n_i \, dS = 0, \qquad i, k = 1, 2, 3, \tag{I.21}$$

$$\int_S (\rho v_k v_i n_i + p n_k) \, dS = \Gamma_k, \tag{I.22}$$

$$\frac{p}{p_\infty} = \left(\frac{\rho}{\rho_\infty}\right)^{\varkappa}. \tag{I.23}$$

Here ρ and p are the density and pressure of a gas, respectively, v_i are the Cartesian components of gas velocity, p_∞ and v_∞ are the pressure and velocity of unperturbed gas flow, n_i are outer unit normal components to S, $\varkappa$ is the polytrope coefficient equal to the ratio of specific heat capacities in the case of adiabatic processes, S is an arbitrary closed surface in gas, and Γ_k are the Cartesian components of the equivalent outer force applied to bodies or singularities inside of S. The Γ_k also can be treated as the dissipation of gas energy inside of S, when the bodies have moved one unit length distance along the x_k axis. The Γ_ks equal zero if there are no bodies, field singularities, or shock waves inside of S. Repeated indices, as usual, mean summation.

Equation I.21 is the mass conservation law. Equation I.22 can be considered as both the energy or momentum conservation law. Equation I.23 is valid for a locally polytropic (e.g., adiabatic) process. When there are only gas particles inside of S, by applying the divergence theorem to Equations I.21 and I.22, one easily can derive the traditional differential equation system of gas dynamics. When there is a body inside of S, using Equation I.22 allows computation of the drag and lift forces acting upon the body. The S may be chosen in any way convenient for calculation.

For example, let a body of finite dimensions move with a subsonic speed in a gas that is at rest at infinity. If there are no vortices in the flow, we may move S into infinity and calculate drag and lift force using Equation I.22. The perturbed flow at infinity, according to the method of small perturbations, has an order

$$\Delta\rho = O(r^{-2}), \qquad \Delta p = O(r^{-2}), \qquad \Delta v_i = O(r^{-2}), \qquad \text{(I.24)}$$

where r is the distance from the body. From Equations I.22 it follows in this case that $\Gamma_k = 0$, which expresses the above-mentioned D'Alembert–Euler paradox. For infinite bodies (e.g., prisms), this paradox is not valid.

Now let us consider the two-dimensional problem of a thin airfoil moving in a gas. Engineering applications require the minimum possible perturbations of gas by a moving airfoil, or even maximum lift and minimum drag. The latter is provided by a thin shape profile close to a flat, infinitely thin plate moving with a small angle of attack (Figure I.2). In this case we designate

$$v_1 = v_\infty(1 + \bar{u}_1), \qquad v_2 = v_\infty \bar{u}_2, \qquad v_3 = 0,$$

$$p = p_\infty(1 + \bar{p}), \qquad \rho = \rho_\infty(1 + \bar{\rho}), \qquad \text{(I.25)}$$

where the dimensionless perturbed quantities are small:

$$\bar{u}_1 \ll 1, \qquad \bar{u}_2 \ll 1, \qquad \bar{p} \ll 1, \qquad \bar{\rho} \ll 1. \qquad \text{(I.26)}$$

Here ρ_∞, p_∞, and v_∞ are the density, pressure, and velocity of unperturbed gas flow at infinity with respect to the airfoil (the gas velocity direction at infinity coincides with the direction of the x_1 axis). Substituting Equations I.25 into Equations I.21–I.23 and omitting higher-order terms, in the two-

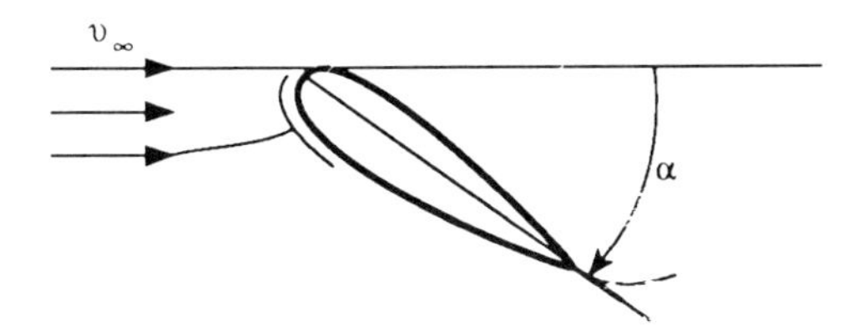

FIGURE I.2. Flow of a gas past a thin airfoil (α is the angle of attack or incidence, and v_∞ is the unperturbed speed of the gas).

dimensional case one may get

$$\int_L \bar{u}_i n_i \, dS = M_\infty^2 \int_L \bar{u}_1 n_1 \, dS, \tag{I.27}$$

$$\Gamma_1 = 0, \tag{I.28}$$

$$\Gamma_2 = \rho_\infty v_\infty^2 \int_L (\bar{u}_2 n_1 - \bar{u}_1 n_2) \, dS, \tag{I.29}$$

$$\bar{p} = \varkappa \bar{\rho} = -\varkappa M_\infty^2 \bar{u}_1, \tag{I.30}$$

$$\bar{\rho} = -M_\infty^2 \bar{u}_1, \qquad c_\infty^2 = \varkappa p_\infty / \rho_\infty, \qquad M_\infty = v_\infty / c_\infty.$$

Here M_∞ is the Mach number at infinity, c is the local sound speed, L is an arbitrary closed contour in gas, and Γ_1 and Γ_2 are the drag and lift forces acting upon a thin airfoil.

Equations I.27–I.30 form the gas dynamics equation system that holds for an arbitrary body at a sufficient distance from it and for very thin airfoils in the entire flow domain. Using the divergence theorem in the integrals of Equations I.27 and I.29 along a contour encompassing only fluid domains, we arrive at the differential equation system

$$(1 - M_\infty^2) \frac{\partial \bar{u}_1}{\partial x_1} + \frac{\partial \bar{u}_2}{\partial x_2} = 0,$$

$$\frac{\partial \bar{u}_1}{\partial x_2} - \frac{\partial \bar{u}_2}{\partial x_1} = 0. \tag{I.31}$$

Its solution can be written by means of the complex potential as

$$\bar{u}_1 = \operatorname{Re} W'(z), \tag{I.32}$$

$$\bar{u}_2 = -\sqrt{1 - M_\infty^2} \operatorname{Im} W'(z), \tag{I.33}$$

where

$$z = x_1 + ix_2\sqrt{1 - M_\infty^2}\,. \tag{I.34}$$

Here the function $W'(z)$ is analytic in the flow domain. It vanishes at infinity and hence its behavior for $z \to \infty$ has the form

$$W'(z) = \frac{\Gamma i}{2\pi z v_\infty}, \tag{I.35}$$

where Γ is a constant. It is easy to verify by direct calculation that Γ is real if there is no mass ejection or suction on the airfoil surface, and the value of Γ equals the circulation

$$\Gamma = \int_{L_\infty} v_1\, dx_1 + v_2\, dx_2 \tag{I.36}$$

consistent with a vortex of strength Γ at infinity. Using the fact that the integral in Equation I.29 is invariant with respect to L, we move L into infinity and utilize Equations I.32–I.35. By calculating the integral, we obtain the lift of the airfoil in the form

$$\Gamma_2 = -\frac{\rho_\infty v_\infty \Gamma}{1 - M_\infty^2}\,. \tag{I.37}$$

This formula for $M_\infty = 0$ was obtained by Joukowski in 1902.

Comparison of theoretical and experimental data shows that the linearized aerodynamic theory usually provides good results if the inequality condition

$$\frac{\Delta}{a\sqrt{1 - M_\infty^2}} \ll 1 \tag{I.38}$$

is met, where Δ and a are the thickness and chord of an airfoil, respectively.

In the coordinate system of Figure I.2, it is convenient to write the airfoil contour equation in the form

$$x_2 = -x_1 \tan \alpha + f_{\pm}(x_1), \qquad 0 < x_1 < a, \tag{I.39}$$

considering the thin airfoil to be close to the straight linear segment of length a, inclined to the x_1 axis with angle α, and issuing the coordinate origin. The functions $f_+(x)$ and $f_-(x)$ describe the upper and lower sides of the airfoil. The condition of nonpenetration on the airfoil contour means that the normal component of the fluid velocity equals zero or

$$v_1 \tan \varphi = v_2, \quad \text{where } \tan \varphi = x_2'(x_1) = -\tan \alpha + f_{\pm}'(x_1). \tag{I.40}$$

Using Equations I.25 and I.32–I.34, we arrive at

$$-\sqrt{1 - M_\infty^2}\, \operatorname{Im} W'(z) = [1 + \operatorname{Re} W'(z)][f_{\pm}'(x_1) - \tan \alpha]. \tag{I.41}$$

In view of $\alpha \ll 1$ and the condition of Equations I.26, it is convenient to simplify Equation I.41 and consider it to hold along cut (0, a) of the x_1 axis. Hence, we get

$$\operatorname{Im} W'(z) = \frac{\alpha - f_{\pm}'(x_1)}{\sqrt{1 - M_\infty^2}}, \qquad x_2 = 0, 0 < x_1 < a. \tag{I.42}$$

It is required to find $W'(z)$ that is analytic outside the cut (0, a) and vanishes at infinity, using the boundary value condition in Equation I.42.

Let us introduce the new functions

$$\Psi(z) = \tfrac{1}{2}[W'(z) + \overline{W}'(z)],$$

$$\Omega(z) = \tfrac{1}{2}[W'(z) - \overline{W}'(z)], \tag{I.43}$$

where

$$\overline{W}'(z) = \overline{W'(\bar{z})}.$$

It is evident that $\bar{z} \to x_1 - i0$ when $z \to x_1 + i0$. The functions $\Psi(z)$ and $\Omega(z)$ are analytic outside the same cut. In addition, on the x_1 axis outside the cut ($x_2 = 0$ and $x_1 < 0$ or $x_1 > a$),

$$\Psi(x_1) = \operatorname{Re} W'(x_1), \quad \text{that is,} \quad \operatorname{Im} \Psi(x_1) = 0, \tag{I.44}$$

$$\Omega(x_1) = i \operatorname{Im} W'(x_1), \quad \text{that is,} \quad \operatorname{Re} \Omega(x_1) = 0. \tag{I.45}$$

On the interval $0 < x_1 < a$ ($x_2 = 0$) of the x_1 axis, we obtain according to Equations I.42 and I.43,

$$\operatorname{Im} \Psi^{\pm}(x_1) = \mp \frac{f'_+(x_1) - f'_-(x_1)}{2\sqrt{1 - M_\infty^2}}, \qquad x_2 = 0, 0 < x_1 < a, \quad \text{(I.46)}$$

$$\operatorname{Im} \Omega^{\pm}(x_1) = \frac{2\alpha - [f'_+(x_1) + f'_-(x_1)]}{2\sqrt{1 - M_\infty^2}}, \qquad x_2 = 0, 0 < x_1 < a. \quad \text{(I.47)}$$

Let us represent the functions $\Psi(z)$ and $\Omega(z)$ using the following Cauchy integrals:

$$\Psi(z) = \frac{1}{2\pi} \int_0^a \frac{\psi(t)\, dt}{t - z}, \quad \text{(I.48)}$$

$$\Omega(z) = \frac{1}{2\pi i} \int_0^a \frac{\omega(t)\, dt}{t - z}. \quad \text{(I.49)}$$

Here $\psi(t)$ and $\omega(t)$ are unknown real functions sought in the interval $0 < t < a$. Equations I.48 and I.49 satisfy the boundary condition in Equations I.44 and I.45 at $z = x_1$ and $x_1 > a$ or $x_1 < 0$.

Using the Sokhotsky Equations I.5 and I.6, and the representation in Equations I.48 and I.49, the boundary condition in Equations I.46 and I.47 can be reduced to the

$$\psi(x_1) = -\frac{1}{\sqrt{1 - M_\infty^2}} [f'_+(x) - f'_-(x)], \quad \text{(I.50)}$$

$$\frac{1}{\pi} \int_0^a \frac{\omega(t)\, dt}{t - x_1} = -\frac{1}{\sqrt{1 - M_\infty^2}} [2\alpha - f'_+(x_1) - f'_-(x_1)],$$

$$0 < x_1 < a. \quad \text{(I.51)}$$

Substituting Equation I.50 into Equation I.48, we determine the function

$$\Psi(z) = -\frac{1}{2\pi\sqrt{1 - M_\infty^2}} \int_0^a \frac{f'_+(t) - f'_-(t)}{t - z}\, dt. \quad \text{(I.52)}$$

Expression I.51 is a singular integral equation coinciding with that of Equation I.11 at $\mu = 0$ and $\lambda = 1$. In this case, $m = 1$ and $\delta = \frac{1}{2}$. Let us assume that the thin airfoil has a sharp tail. Such airfoils usually applied in engineering are called Joukowski airfoils. They provide a vortex sheet separation locus just where the tail becomes the trailing edge, if the angle of attack is sufficiently small. In this case the fluid velocity and $W'(z)$ are finite at this edge, so that potential $\Omega(z)$ is finite at $z = a$. At the leading edge, $z = 0$, the velocity and $\Omega(z)$ are infinite, but integrable. Hence, the canonical solution to Equation I.14 will be provided in this case by the function

$$X(z) = \sqrt{\frac{z-a}{z}}, \qquad \lim_{z\to\infty} X(z) = 1. \tag{I.53}$$

Similarly to Equation I.20, the potential $\Omega(z)$ in this case can be found in the form

$$\Omega(z) = \frac{X(z)}{2\pi\sqrt{1-M_\infty^2}} \int_0^a \frac{2\alpha - f'_+(\tau) - f'_-(\tau)}{X^+(\tau)(\tau - z)}\, d\tau. \tag{I.54}$$

According to Equation I.43,

$$W'(z) = \Psi(z) + \Omega(z), \tag{I.55}$$

so that substituting here Equations I.52 and I.54, we get the solution in the following shape:

$$W'(z) = -\frac{1}{2\pi\sqrt{1-M_\infty^2}} \int_0^a \frac{f'_+(\tau) - f'_-(\tau)}{\tau - z}\, d\tau$$
$$+ \frac{X(z)}{2\pi\sqrt{1-M_\infty^2}} \int_0^a \frac{2\alpha - f'_+(\tau) - f'_-(\tau)}{X^+(\tau)(\tau - z)}\, d\tau. \tag{I.56}$$

From here, comparing with Equation I.35 at infinity we find the circulation

$$\Gamma = \frac{v_\infty}{\sqrt{1-M_\infty^2}} \int_0^a \frac{2\alpha - f'_+(\tau) - f'_-(\tau)}{|X(\tau)|}\, d\tau. \tag{I.57}$$

In the limiting particular case of a straight linear airfoil of infinitesimal thickness, when $f'_+(\tau) = f'_-(\tau) = 0$, we get from Equation I.57,

$$\Gamma = \frac{2\alpha a v_\infty}{\sqrt{1-M_\infty^2}} \int_0^1 \sqrt{\frac{x}{1-x}}\, dx = \frac{\pi \alpha a v_\infty}{\sqrt{1-M_\infty^2}}. \tag{I.58}$$

By Equations I.37 and I.58 the lift of this airfoil equals

$$\Gamma_2 = \frac{\pi \alpha a \rho_\infty v_\infty^2}{1 - M_\infty^2}. \tag{I.59}$$

In the case of relatively small velocity of flight when $M_\infty \ll 1$, the air can be considered incompressible, and the lift equals

$$\Gamma_2 = \pi \alpha a \rho_\infty v_\infty^2. \tag{I.60}$$

Compare this result with the exact solution for a very thin plate moving in an incompressible fluid with an arbitrary nonsmall angle of attack (see Gurevich 1965):

$$\Gamma_2 = \pi a \rho_\infty v_\infty^2 \sin \alpha. \tag{I.61}$$

Even at $\alpha = 30°$, the exact solution differs only by 5% from that of the linearized approach.

It should be noted that using the Bernoulli integral $p + \frac{1}{2}\rho v_i v_i = \text{const}$ in the flow domain of incompressible fluid, the invariant integral in Equation I.22 can be written in the simple form

$$\Gamma_k = \rho \int_S \left(-\frac{1}{2} v_i v_i n_k + v_i n_i v_k \right) dS, \tag{I.62}$$

$$v_i = \varphi_{,i}, \quad i,k = 1,2,3.$$

Here φ is the flow potential.

I.3. OPTIMAL AIRFOIL PROBLEM

According to Equation I.59 the lift of a thin airfoil increases when the flight speed grows until the local Mach number achieves unity at a critical point of the upper side of the airfoil. This occurs at a critical Mach $M_\infty = M_* < 1$. If $M_\infty > M_*$, a local supersonic zone and local shock waves develop near the critical point. At this regime, the drag is sharply increased.

If a moving body is "well-streamlined," like a thin plate moving with a very small angle of attack, so that there is no separation of a boundary layer and no formation of drag vortices in the fluid, the drag is defined solely by viscous friction in the boundary layer. For example, in the case of a thin plate under study, the drag equals

$$D = \frac{1.33 \rho_\infty v_\infty^2 a}{\sqrt{a v_\infty / \nu}}, \tag{I.63}$$

where ν is the kinematic viscosity. The dimensionless coefficients of lift and drag, c_L and c_D, are commonly used to describe the forces; as applied to a "well-streamlined" thin plate they are

$$c_L = \frac{\Gamma_2}{\frac{1}{2} a \rho_\infty v_\infty^2} = \frac{2\pi\alpha}{1 - M_\infty^2}, \qquad M_\infty = \frac{v_\infty}{c_\infty}$$

$$c_D = \frac{D}{\frac{1}{2} a \rho_\infty v_\infty^2} = 2.66 \sqrt{\frac{\nu}{a v_\infty}}. \tag{I.64}$$

Thus, when the flight speed grows, the coefficient of lift increases and the coefficient of drag decreases. Hence, for every thin airfoil there exists a certain optimal flight speed before the separation of drag vortices from the airfoil starts or local shock waves form.

It follows that the optimal airfoil shape should provide for a maximum flight speed without drag vortices separation and shock waves. This clear engineering requirement is not easy to formulate in a mathematically unique and strict way. The history of developing aircraft and turbines in this century is the factual record of many events devoted to solving this problem. Some mathematical efforts concerning the sonic barrier are described by Morawetz (1982).

Let us treat an approach to this problem formulated by Cherepanov (1973) and based on some engineering observations and numerical experiments. There are three main requirements to an optimal airfoil:

1. The tail of the airfoil should be a trailing edge, i.e., separation locus of a vortex sheet. To achieve this aim at different regimes of flight, the tail part should be an individually controlled hinged flap with a sharp cusped end.
2. The nose of the airfoil should coincide with the stagnation point and be sharply cusped, which helps eliminate some problems connected with large velocity gradients and separation of drag vortices. (There should be no other stagnation points different from the nose.) To achieve this aim at different regimes of flight, the nose part should be an individually controlled hinged flap with a sharp cusped spike.
3. The upper side of the middle part of the airfoil should provide for one and the same pressure at all the points at a virtual velocity of

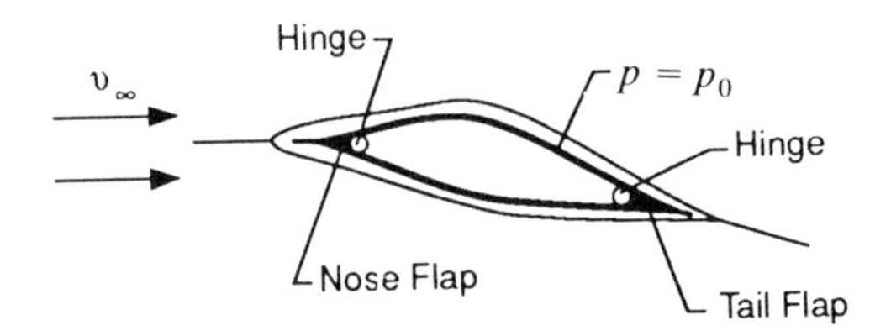

FIGURE I.3. Sketch of an optimal airfoil having a nose flap and a hinged tail flap. Pressure $p = p_0$ is uniform along the upper middle part of the airfoil. One streamline doubles at the nose and trails from the tail of the airfoil.

flight, which helps to maximize the critical Mach number at this flight regime.

A sketch of the optimal airfoil is shown in Figure I.3. One streamline doubles at the nose and emerges at the tail of the optimal airfoil.

Let us formulate the mathematical problem of an optimal airfoil in this approach for a thin airfoil. Boundary-value condition $p = p_0 = \text{const}$, according to Equations I.25 and I.30, can be written as

$$\bar{u}_1 = u_* \quad \text{where} \quad u_* = \frac{p_\infty - p_0}{\varkappa p_\infty M_\infty^2}. \tag{I.65}$$

Constant p_0 or u_* should be chosen from additional physical considerations. The boundary condition of nonpenetration holds along the entire surface of the airfoil. It means that on the arc of uniform pressure being sought, two boundary condition equations have to be satisfied.

Using Equation I.65 and a consideration analogous to that used for deriving Equation I.42, we find the corresponding boundary value problem of an optimal airfoil (see Figure I.4):

Upper side of nose flap:

$$z = x_1 + i0, \qquad 0 < x_1 < \epsilon_n a,$$

$$\operatorname{Im} W'(z) = \frac{\alpha_n - f'_{n_+}(x_1)}{\sqrt{1 - M_\infty^2}}. \tag{I.66}$$

Upper side of middle part of airfoil:

$$z = x_1 + i0, \qquad \epsilon_n a < x_1 < a(1 - \epsilon_t),$$

$$\operatorname{Re} W'(z) = u_*. \tag{I.67}$$

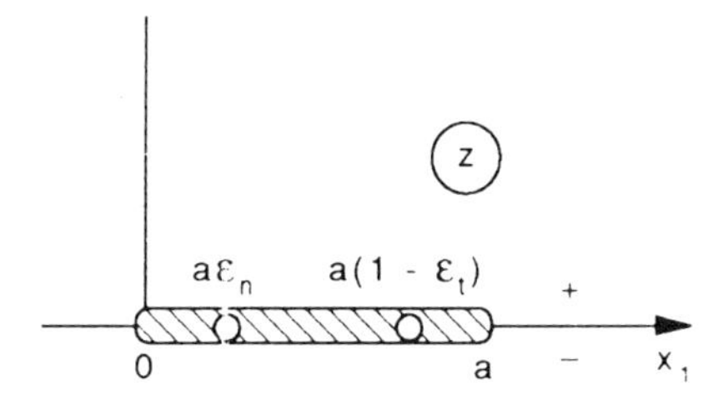

FIGURE I.4. The z plane outside the cut $(0, a)$ along the real axis having the points $x_1 = a\epsilon_n$ and $x_1 = a(1 - \epsilon_t)$ at which the mixed boundary value problem changes $(0 < \epsilon_n < 1,$ $0 < \epsilon_t < 1)$.

Upper side of tail flap:

$$z = x_1 + i0, \qquad a(1 - \epsilon_t) < x_1 < a,$$

$$\operatorname{Im} W'(z) = \frac{\alpha_t - f'_{t_+}(x_1)}{\sqrt{1 - M_\infty^2}}. \tag{I.68}$$

Lower side of nose flap:

$$z = x_1 - i0, \qquad 0 < x_1 < \epsilon_n a,$$

$$\operatorname{Im} W'(z) = \frac{\alpha_n - f'_{n_-}(x_1)}{\sqrt{1 - M_\infty^2}}. \tag{I.69}$$

Lower side of middle part of airfoil:

$$z = x_1 - i0, \qquad \epsilon_n a < x_1 < a(1 - \epsilon_t),$$

$$\operatorname{Im} W'(z) = \frac{\alpha_m - f'_m(x_1)}{\sqrt{1 - M_\infty^2}}. \tag{I.70}$$

Lower side of tail flap:

$$z = x_1 - i0, \qquad a(1 - \epsilon_t) < x_1 < a,$$

$$\operatorname{Im} W'(z) = \frac{\alpha_t - f'_{t_-}(x_1)}{\sqrt{1 - M_\infty^2}}. \tag{I.71}$$

[$f'_{n_+}(x_1)$, $f'_{n_-}(x_1)$, $f'_m(x_1)$, $f'_{t_+}(x_1)$, and $f'_{t_-}(x_1)$ are assumed to be given.] Here unknown constants ϵ_n, ϵ_t, α_n, α_m, and α_t determine the geometry of a broken chord line of an optimal airfoil consistent with the virtual flight velocity regime. The length and thickness of the nose and tail flaps as well as the form of the lower side of the middle part may be given depending on some free parameters. All these geometrical parameters and u_* should be chosen in such a way that all the additional condition equations are satisfied, including the four conditions of the finiteness of $W'(z)$ at points $z = 0, a, a\epsilon_n + i0$, and $a(1 - \epsilon_t) + i0$. Moreover, $W'(z)$ must vanish at infinity. Using the direct method advanced by Cherepanov (1964), the boundary problem of Equations I.66–I.71 can be solved in an explicit form. An explicit form solution also can be found using conformal mapping onto a canonical domain and the method treated previously in this chapter.

Although it is dependent on several free parameters, after $W'(z)$ is found, the unknown boundary of the upper middle part of the airfoil can be computed from the equation

$$\frac{dy}{dx_1} = \frac{v_2}{v_1} = \frac{\bar{u}_2}{1 + \bar{u}_1} = -\frac{\sqrt{1 - M_\infty^2}\,\mathrm{Im}\,W'(z)}{1 + \mathrm{Re}\,W'(z)}. \tag{I.72}$$

Here $x = y(x_1)$ is the equation sought for the middle part of the optimal airfoil. Equation I.72 expresses the coincidence of this boundary with a streamline. Because the right-hand side of Equation I.72 is a defined function of x_1, $y(x_1)$ may be found from this equation by simple integration over x_1. We provide the solution result to problem Equations I.66–I.71 only in one particular case when both the nose and tail flaps are not available, that is, $\epsilon_n = \epsilon_t = 0$:

$$W'(z) = u_* - \frac{u_*}{\sqrt{2}}\left[e^{\pi i/4}X_1(z) + e^{-\pi i/4}X_1^{-1}(z)\right] + \frac{1}{2\pi i\sqrt{1 - M_\infty^2}}\int_0^a \left[\frac{X_1(z)}{X_1^+(x)} - \frac{X_1^+(x)}{X_1(z)}\right]\frac{\alpha_m - f_m'(x)}{x - z}\,dx. \tag{I.73}$$

Here,

$$X_1(z) = \left(\frac{z - a}{z}\right)^{1/4}, \qquad \lim_{z\to\infty} X_1(z) = 1,$$

$X_1^+(x)$ is the value of $X_1(z)$ on the upper face of cut $(0, a)$ of the z plane having argument $\pi/4$ and α_m is the angle of attack.

The function $W'(z)$ is finite at $z = 0$ and $z = a$, if equations

$$\int_0^a \frac{\alpha_m - f'_m(x)}{x^{3/4}(a-x)^{1/4}}\,dx = \pi\sqrt{2}\,u_*\sqrt{1 - M_\infty^2}\,,$$

$$\int_0^a \frac{\alpha_m - f'_m(x)}{x^{1/4}(a-x)^{3/4}}\,dx = \pi\sqrt{2}\,u_*\sqrt{1 - M_\infty^2} \tag{I.74}$$

are met. The upper boundary of the optimal airfoil sought in this case is given in the form

$$y = \frac{1}{1+u_*}\int_0^{x_1}\left\{\frac{1}{2\pi}\int_0^a \frac{\sqrt{(a-t)x} - \sqrt{t(a-x)}}{[t(a-t)]^{1/4}[x(a-x)]^{1/4}} \cdot \frac{f'_m(t) - \alpha_m}{t-x}\,dt \right.$$
$$\left. + \frac{1}{\sqrt{2}}\sqrt{1 - M_\infty^2}\,u_*\left[\left(\frac{a-x}{x}\right)^{1/4} - \left(\frac{x}{a-x}\right)^{1/4}\right]\right\} dx. \tag{I.75}$$

Despite the explicit form of the solution, the problem of optimal airfoil existence, even in such an engineering formulation, has not, as yet, been solved because of difficulties in comprehensive analysis of complicated functions of several free parameters.

The methods developed in this book provide a good opportunity for solving the problem of an optimal airfoil using supercomputers. For example, let us show how the problem in Equations I.66–I.71 can be reduced to a singular integral equation. At first, we notice that the solution to boundary Equations I.66–I.71 may be written in the form of Equation I.56, where

$$f'_+(t) = \begin{cases} f'_{n_+}(t) & \text{when } 0 < t < a\epsilon_n, \\ y'(t) & \text{when } a\epsilon_n < t < a(1-\epsilon_t), \\ f'_{t_+}(t) & \text{when } a(1-\epsilon_t) < t < a, \end{cases}$$

$$f'_-(t) = \begin{cases} f'_{n_-}(t) & \text{when } 0 < t < a\epsilon_n, \\ f'_m(t) & \text{when } a\epsilon_n < t < a(1-\epsilon_t), \\ f'_{t_-}(t) & \text{when } a(1-\epsilon_t) < t < a. \end{cases}$$

The unknown function $y'(x_1)$ determines the upper middle part of the profile and all other functions are known. Then, by Sokhotsky formulas we find Re $W'(z)$ when $z \to x_1 + i0$ for $a\epsilon_n < x_1 < a(1-\epsilon_t)$ and equate it to u_* in view of Equation I.65. The resulting equation represents a singular integral equation with respect to $y'(x_1)$ and it belongs to the singular integral equations studied numerically in this book.

REFERENCES

Chaplygin, S. A. 1902. *On Cas Jets*. Moscow University Press.
Cherepanov, G. P. 1962. The solution to one linear Riemann problem, *Appl. Math. Mech.*, 26(5).
Cherepanov, G. P. 1963. The flow of an ideal fluid having a free surface in multiple connected domains, *Appl. Math. Mech.*, 27(4).
Cherepanov, G. P. 1964. Riemann–Hilbert problems of a plane with cuts, *Dokl. USSR Acad. Sci. (Math.)*, 156(2).
Cherepanov, G. P. 1965. On one case of the Riemann problem for several functions, *Dokl. USSR Acad. Sci. (Math.)*, 161(6).
Cherepanov, G. P. 1973. On optimal airfoil in ultimate subsonic regime.
Cherepanov, G. P. 1987. Invariant integrals, in *Fracture Mechanics of Rocks in Drilling*. Nedra Publishers, Moscow, pp. 8–20.
Gakhov, F. D. 1952. Riemann problem for systems having *n* pairs of functions, *Adv. Math. Sci.*, 7, No. 4(50).
Gakhov, F. D. 1966. *Boundary Value Problems*. Pergamon Press, London.
Gilbarg, D. and Shiffman, M. 1954. On bodies achieving extreme values of the critical Mach number, *J. Ration. Mech. Anal.*, 3(2).
Guderley, G. 1956. *Die Theorie der Naneschalgeschwindigkeit Strömungen*. Springer-Verlag, Berlin.
Gurevich, M. I. 1965. *The Theory of Jets in an Ideal Fluid*. Academic Press, New York.
Khrapkov, A. A. 1971. Problems of elastic equilibrium of infinite wedge with asymmetrical cut at the vertex, solved in explicit form, *Appl. Math. Mech.*, 35(6).
Morawetz, C. S. 1982. The mathematical approach to the sonic barrier, *Bull. Am. Math. Soc.*, 6(2).
Muskhelishvili, N. I. 1946. *Singular Integral Equations*. Noordhoff Publishers, Groningen.
Sedov, L. I. 1950. *Two-Dimensional Problems in Hydrodynamics and Aerodynamics*. Wiley, New York.

Introduction

The very essence of the method of discrete vortices considered in this book may be formulated as follows. A continuous vortex sheet, modeling both a lifting surface and a wake downstream of it, is replaced by a system of discrete vortices. Certain points are chosen at the lifting surface (called the reference points) where the nonpenetration condition, according to which the normal flow velocity component is equal to zero, is imposed. The problem of obtaining unknown circulations of discrete vortices reduces to a system of linear algebraic equations. The solution of the problem is not unique and may have singularities at the edges and at angular points of a lifting surface. The required class of solutions is determined by the physical nature of the problem and is singled out by an appropriate choice of the singularities. In the framework of the method of discrete vortices the choice is implemented in the so-called B condition of the method of discrete vortices. According to this condition, at the grid points nearest to where the solution must be unlimited, discrete vortices are placed and where the solution must be limited, the reference points are placed. In addition, the sums that replace singular integrals in the lifting surface theory must correspond to the Cauchy principal values of the integrals. To meet this condition the internal points must be placed midway between the vortices positioned at the surface. In this form, the method of discrete vortices was originally formulated in the doctoral thesis by S. M. Belotserkovsky in 1955 (Belotserkovsky 1955b).

Let us illustrate the ideas of the method by considering flow past a thin airfoil modeled by a vortex sheet whose strength is $\gamma(x)$ (see Figure 0.1). The problem reduces to the following characteristic singular integral equation of the first kind on segment $[-1, 1]$:

$$\frac{1}{2\pi}\int_{-1}^{1}\frac{\gamma(x)\,dx}{x-x_0}=f(x_0),\qquad x_0\in(-1,1). \tag{0.1}$$

Two grids are introduced: $x_k = -1 + kh$, $h = 2/(n+1)$, $k = 1,\dots,n$, and $x_{0k} = x_k + h/2$, $k = 0, 1, \dots, n$, for discrete vortices and reference points, respectively.

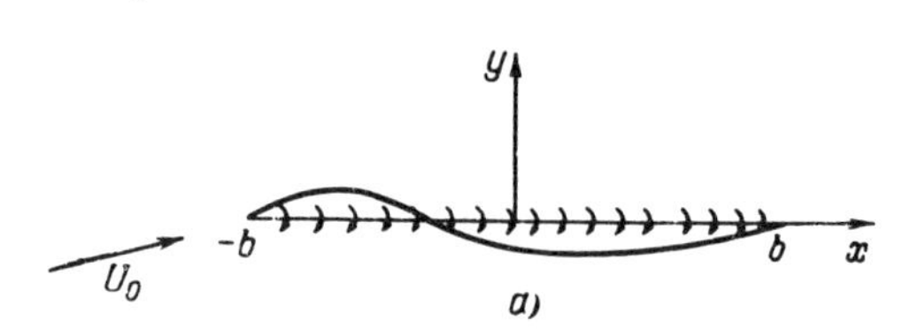

FIGURE 0.1. Vortex sheet simulating a thin slightly curved airfoil, positioned at the chord of the airfoil.

FIGURE 0.2. Mutual positions of discrete vortices and reference points in the case of a problem with nonzero total circulation.

Aerodynamics considers three modes of flow past a body: circulatory (i.e., with nonzero total circulation), noncirculatory (i.e., with zero total circulation), and limited everywhere. The first mode is based on the Chaplygin–Joukowski hypothesis about the finiteness of $\gamma(x)$ at the sharp-tail of the airfoil and is used for analyzing lifting properties of wings. Generally speaking, in this case the solution tends to infinity at the nose of a thin airfoil (see Figure 0.2), and Equation (0.1) is replaced by the following system of linear algebraic equations:

$$\sum_{k=1}^{n} \frac{\gamma_n(x_k)h}{x_{0j} - x_k} = f(x_{0j}), \qquad j = 1, \ldots, n. \tag{0.2}$$

The second mode is used when considering oscillations of an airfoil in a fluid at rest. In this case, both edges function under similar conditions, and the total circulation is equal to zero, whereas at the edges $\gamma(x)$ tends to infinity (Figure 0.3). The corresponding system of equations has the form:

$$\sum_{k=1}^{n} \frac{\gamma_n(x_k)h}{x_{0j} - x_k} = f(x_{0j}), \qquad j = 1, \ldots, n-1, \tag{0.3}$$

$$\sum_{k=1}^{n} \gamma_n(x_k)h = 0, \qquad j = n. \tag{0.4}$$

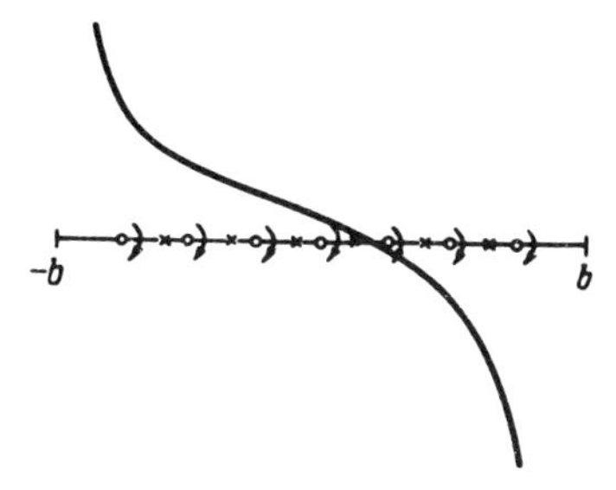

FIGURE 0.3. Mutual positions of discrete vortices and reference points in the case of a problem with zero total circulation.

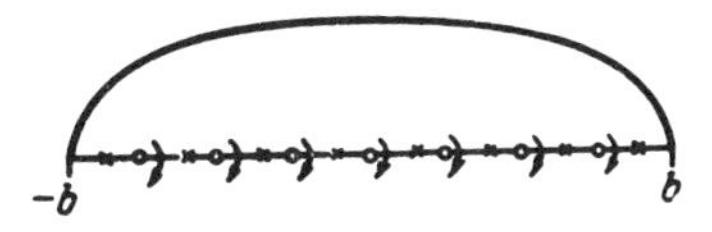

FIGURE 0.4. Mutual positions of discrete vortices and reference points in the case of a problem with a solution limited at both edges of a thin airfoil.

The latter equation spells out the condition of zero total circulation, and according to the B condition, vortices are placed at the grid points nearest to both edges.

The third mode of flow is used when one determines not only the lifting properties of a wing, but also can change the form of the wing in order to avoid flow separation at the nose, which occurs when $\gamma(x)$ tends to infinity there (Figure 0.4). In this case the integral equation is substituted by the following system of linear algebraic equations:

$$\gamma_{0n} + \sum_{k=1}^{n} \frac{\gamma_n(x_k)h}{x_{0j} - x_k} = f(x_{0j}), \qquad j = 0, 1, \ldots, n, \tag{0.5}$$

where γ_{0n} is an additional unknown related to the deformation of the wing for which the strength $\gamma(x)$ at the nose becomes zero: $\gamma(-1) = 0$. From the mathematical point of view, this unknown may be treated as a regularizing variable, because according to the B condition, the requirement of both $\gamma(-1)$ and $\gamma(1)$ being finite results in the appearance of an extraneous reference point.

The method of discrete vortices has the following important peculiarity: There is no need to make any assumptions about the way a solution tends to infinity at both the ends of a segment and at angular points. The required class of solution is singled out by choosing relative positions of two grids composed of discrete vortices and reference points.

This method is radically different when compared with other numerical methods of aerodynamics developed prior to it. The earliest work close to the direction of this book was the article published in 1932 by M. A. Lavrent'ev that presented a numerical solution to the singular integral equation for flow past a thin airfoil. The solution based on polynomial approximation was a forerunner of many methods in this area. In 1938, Multhopp's work was published, in which the solution to a one-dimensional singular integral equation was presented in the form $\gamma(x) = \omega(x)\varphi(x)$ where $\omega(x)$ accounted for the singularities of a solution and $\varphi(x)$ was a smooth function expanded into a series in the Chebyshev polynomials of the first kind. The right-hand side of the equation was expanded in the Chebyshev polynomials of the second kind. Then, the relationship between the Chebyshev polynomials of the first and second kinds was employed to compose a system of linear algebraic equations for the coefficients of the series. Later the idea was applied to a finite-span wing (Kolesnikov 1957, Falkner 1947). However, this approach was neither sufficiently profound nor applicable to unsteady problems as well as to wings of a complex plan form, to say nothing of an airplane as a whole.

On the other hand, the method of discrete vortices was criticized for applying rectangle-type quadrature formulas to singular integrals as well as for ignoring an explicit account of singularities at a wing's edges (Polyakhov, 1973). Because rigorous verification of the basic assumptions of the method was lacking, the authors and their followers concentrated their efforts on solving the problem and developing similar approaches for related areas (Lifanov and Polonskii 1975; Belotserkovsky et al. 1978; Lifanov 1978a, b, 1979a, b, 1980a, b, c, 1981a, b; Belotserkovsky and Lifanov 1981; Lifanov and Saakyan 1982; Lifanov and Matveev 1983; Matveev 1982a,b; Belotserkovsky et al. 1983; Gandel 1982, 1983).

This monograph sums up the first results of this work. It consists of four parts. The first part is dedicated to analyzing quadrature formulas for one-dimensional and multidimensional singular integrals. We start by considering the above-mentioned rectangle-rule formulas for one-dimensional integrals. In this case, the integration domain is a piecewise Lyapunov curve having angular points only. Also, interpolation-type quadratures for singular integrals on a segment–circle and with the Hilbert kernel, which allow construction of numerical methods for integral equations on the sets, are analyzed. We continue by considering quadrature formulas for Cauchy multiple singular integrals, certain two-dimensional integrals treated in the sense of the Hadamard finite value (Hadamard 1978) one often comes across in aerodynamics, and multidimensional singular integrals. Finally, the Poincaré–Bertrand formula for reversing the order of integration in multiple singular integrals is proved by employing quadrature relationships of the type used in the method of discrete vortices.

In the second part, direct numerical methods for solving singular integral equations with one- and multidimensional Cauchy integrals are constructed by using quadrature formulas derived in Part I. The situations encompassed are those for which the right-hand sides of integral equations may incorporate integrable power-law singularities at the end points and discontinuities of the same form or discontinuities of the first kind at inner points. The case when a solution has a singularity of the form $(q - t)^{-1}$ at an inner point was also studied. In fact, we succeeded in proving the existence theorems for solutions to characteristic singular integral equations of the first kind on a segment and circle (i.e., in developing an analog of the Fredholm theory).

It should be noted that the theory of one-dimensional singular integral equations has been thoroughly studied (Gakhov 1977, Muskhelishvili 1952). However, equations incorporating multiple Cauchy integrals were only considered for toruses (Gakhov 1977, Kakichev 1959). Therefore, we present analytic solutions to a wide class of characteristic singular integral equations of the second kind with multiple Cauchy integrals for the cases when integration domains are products of circles and segments. Then efficient numerical methods are developed to solve the equations.

In the third part, mathematical verification of the method of discrete vortices is presented for a wide class of linear steady problems of aerodynamics. The problem of stability of the developed numerical methods for solving singular integral equations is also considered. The actuality of the problem is related to the fact that the equations are unstable in uniform metrics often used in practice. The numerical methods are shown to be regularizing, with the number of reference points being the regularizing parameter.

Finally, the fourth part is dedicated to extending the developed numerical methods into alternative areas of interest: the theory of elasticity, electrodynamics, and mathematical physics, i.e., into any area in which the problems may be reduced, either naturally or by special measures, to singular integral equations. Prior to developing reliable stable methods for solving such equations, many attempts were made to reduce the problems to regular Fredholm integral equations of the second kind (in the case of plane problems of the theory of stability in Parton and Perlin 1982, 1984) or to integral equations of the first kind with logarithmic singularity (in the case of electrodynamic problems in Mitra 1977). However, the resulting regular Fredholm equations often have nonsingle-valued solutions, because the equations "set" on the spectrum. This results in significant difficulties while solving the problems numerically. The process of solution becomes unstable, and special methods must be employed for smoothing the resulting fluctuations (Mitra 1977). By reducing the problem to singular integral equations and using the approaches of this book, one may construct unified stable numerical methods for solving a wide class of various problems (Belotserkovsky 1967; Belotserkovsky and Nisht 1978; Belotserkovsky et al. 1983; Gandel 1982, 1983; Lifanov and Saakyan 1982). The fourth part of this book is dedicated to some novel applications of the theory. It also presents methods for deriving singular equations in the mentioned areas (Belotserkovsky et al. 1983; Gandel 1982, 1983; Lifanov and Saakyan 1982; Saakyan 1978).

Numerical methods representing a certain physical substance as a set of discrete physical objects (the method of discrete vortices presents a vortex layer as a set of discrete vortices) are presently called methods of discrete singularities.

The studies carried out by the authors after publishing the Russian version of this book required extending the method of discrete singularities to the class of singular integral operators with variable coefficients and Cauchy and Hilbert kernels. It was found that such index $\varkappa$ operators transform polynomials of degree n multiplied by the corresponding weight functions into polynomials of the degree $n - \varkappa$. Moreover, the same conclusion is also true for generalized polynomials. By using this property, one may construct numerical methods, similar to the method of discrete singularities, for singular integral equations (of algebraic degree of accuracy) with variable coefficients and the Cauchy kernel on a piecewise-continuous curve as well as with the Hilbert kernel. These mathematical findings provided the solution to aerodynamic problems for permeable airfoils as well as developed a new approach to calculating finite-thickness airfoils. It was found that if the boundary nonpenetration condition is transferred from the surface of an airfoil onto a certain middle line, then the problem of flow past an airfoil reduces to solving a system of two singular integral equations with variable coefficients.

Currently of great practical importance are the problems of flow past blunt (high-drag) bodies with sharp corners, such as buildings, cars, etc., in particular the problems of calculating virtual inertia of the bodies. To solve the problems it was necessary to consider anew the process of shedding of a sheet of free vortices from angular points. In the framework of the new scheme, discrete vortices are positioned at the angular points, also and are assumed to be free. When separated from angular points, the vortices trail along the local flow velocity vector. The preceding problems were solved for both two- and three-dimensional flows.

This book is also supplemented with a new chapter on the reduction of certain boundary value problems of mathematical physics to singular integral equations. The chapter follows.

Part I: Quadrature Formulas for Singular Integrals

1

Quadrature* Formulas of the Method of Discrete Vortices for One-Dimensional Singular Integrals

1.1. SOME DEFINITIONS OF THE THEORY OF ONE-DIMENSIONAL SINGULAR INTEGRALS

In the following text, only curves in a plane with the right-hand system of coordinates *OXY* are considered. Sometimes the points of the plane will be considered as complex numbers of the form $t = x + iy$, where $i = \sqrt{-1}$.

A curve will be called smooth if it is simple, i.e., does not intersect itself.

A line L will be called smooth and unclosed (an arc) if it can be presented in the following parametric form (Muskhelishvili 1952):

$$x = x(s), \qquad y = y(s), \qquad s_a \leq s \leq s_b, \tag{1.1.1}$$

where s_a and s_b are finite constants; $x(s)$, $y(s)$, $x'(s)$, and $y'(s)$ are continuous functions in the interval $[s_a, s_b]$; and the derivatives never vanish simultaneously. Different points of the curve L correspond to different values of the parameter s in $[s_a, s_b]$.

If points of the curve L are denoted by $t = x(s) + iy(s)$, then a one-to-one correspondence exists between t and s ($t \in L, s \epsilon [s_a, s_b]$) and $t'_s = x'(s) + iy'(s)$.

* The term "quadrature" is equivalent to the term "numerical integration" that is commonly used now (G.Ch.).

Sometimes the curve L will be denoted by ab, where $a = t(s_a)$ and $b = t(s_b)$.

A smooth curve will be called a smooth closed contour L if $x(s_b) = x(sa)$, $y(s_a) = y(s_b)$ and

$$x'(s_b - 0) = x'(s_a + 0).$$

$$y'(s_b - 0) = y'(s_a + 0).$$

Thus, in this case the functions $x(s)$, $y(s)$ and $x'(s)$, $y'(s)$ can be considered as periodic functions whose period is equal to $T = s_b - s_a$.

A smooth (simple) line is a set of a finite number of closed or unclosed smooth contours having no common points (the end points included).

A curve is referred to as piecewise smooth if it consists of a finite number of smooth unclosed curves having no common points (the only possible exception may be the end points). The curve will be said to possess angular nodes only, if at each of the nodes any two smooth curves meet at a nonzero angle, i.e., the nodes are not cusps.

In Appendix 1 to Muskhelishvili (1952) the following statement was proved: Let L be a simple piecewise smooth curve that contains angular points only (i.e., L is composed of a finite number of smooth unclosed curves $a_1a_2, a_2a_3, \ldots, a_{n-1}a_n$ so that the end point of each preceding smooth curve coincides with the initial point of the next one). Then the following inequality holds for any pair of points t_l, t_2 belonging to L:

$$K_0\sigma(t_1, t_2) \leq r(t_1, t_2) \leq \sigma(t_1, t_2), \tag{1.1.2}$$

where:

1. $\sigma(t_1, t_2)$ is the length of the part of curve L between the points t_1 and t_2. (If L is a closed curve, then $\sigma(t_1, t_2)$ corresponds to the shorter arc).
2. $r(t_1, t_2) = |t_1 - t_2|$, i.e., the distance between points t_1, t_2 in plane OXY.
3. The constant $K_0 \epsilon (0, 1)$ is independent of the positions of points t_1 and t_2 on the curve L.

Note that if a piecewise smooth curve L contains angular nodes only, then Inequality (1.1.2) is also satisfied for any points of L.

Definition 1.1.1. *A function $\varphi(t)$ (where t is generally a complex variable) meets the condition $H(\mu)$ (the Hölder condition of the degree μ) on a given set D of the values of the variable if for any two values belonging to the set, the* *following inequality holds*:

$$|\varphi(t_1) - \varphi(t_2)| \leq A|t_1 - t_2|^{\mu}, \tag{1.1.3}$$

where A and μ are positive numbers and $0 < \mu \leq 1$. The constant A is called the coefficient and μ the exponent of the condition $H(\mu)$. If we are not interested in the exponent μ, then the function $\varphi(t)$ is said to meet the condition H (or belong to the class H) on the set D and will be written in the form $\varphi(t)\epsilon H(\mu)$ or $\varphi(t)\epsilon H$.

Note that if $\varphi(t)\epsilon H(\mu)$, then $|\varphi(t)|\epsilon H(\mu)$.

The condition H may be generalized onto the case of a function of several variables: A function $\varphi(t_1, \ldots, t_n)$ meets the condition $H(\mu_1, \ldots, \mu_n)$ (or simply H) on the set D of the variables $t_1, t_2, \ldots, t_n$, if the following inequality is satisfied for any two points $(t_1', \ldots, t_n')$ and $(t_1'', \ldots, t_n'')$ in

$$|\varphi(t_1'', \ldots, t_n'') - \varphi(t_1', \ldots, t_n')| \leq A_1|t_1'' - t_1'|^{\mu_1} + \cdots + A_n|t_n'' - t_n'|^{\mu_n}, \tag{1.1.4}$$

where $A_i > 0, 0 < \mu_i \leq 1, i = 1, \ldots, n$.

From (1.1.4) it follows that if $\varphi(t_1, \ldots, t_n)\epsilon H$, then φ belongs to the class $H(\mu_k)$ with respect to a variable t_k, $k = 1, \ldots, n$, uniformly† with respect to all the other variables. The opposite statement is also true.

Stating in what follows that a function $\varphi(t_1, \ldots, t_n)$ meets the condition H with respect to each of the variables taken alone, we suppose that the condition is satisfied uniformly with respect to the other variables.

A smooth unclosed curve L is called a Lyapunov curve if the derivative $t'(s)$ meets the condition $H(\alpha)$ on $[s_a, s_b]$. Then, according to Muskhelishvili (1952), the function

$$f(s, s_0) = \frac{t - t_0}{s - s_0} = \frac{t(s) - t(s_0)}{s - s_0}$$

meets the condition $H(\alpha)$ in both variables s and s_0 and does not vanish on $[s_a, s_b]$.

A curve will be called a piecewise Lyapunov curve if all its smooth parts are Lyanpunov curves.

† In other words, $|\varphi(t_1, \ldots, t_k'', \ldots, t_n) - \varphi(t_1, \ldots, t_k', \ldots, t_n)| < A|t_k'' - t_k'|^{\mu_k}$, where the constant A is independent of $t_1, \ldots, t_n$.

Definition 1.1.2. *A function $\varphi(t)$ belongs to the class H^* on a piecewise smooth curve L if it has the form*

$$\varphi(t) = \varphi^*(t)/P_L^\nu(t), \qquad P_L^\nu(t) = \prod_{k=1}^{p} |t - c_k|^{\nu_k}. \tag{1.1.5}$$

where $\varphi^*(t) \epsilon H_0$ on L, i.e., φ belongs to the class H on each smooth part of the curve L, $0 \le \nu_k < 1$, and c_k, $k = 1, \ldots, p$, are nodes of the curve L.

Without loss of generality, it can be assumed that $\varphi^*(t) \epsilon H$ on L.

Let us recall the definition of a Cauchy singular integral along a piecewise smooth curve.

Definition 1.1.3. *Let point t_0 differ from any node of the curve L. In other words, let it be an internal point. Let us draw a circle centered at this point, with the radius of the circle $\epsilon > 0$ being so small that it intersects the curve at two points t' and t'' exactly. Let the arc l be denoted by $\overset{\frown}{t't''}$. Consider the integral*

$$\int_{L \setminus l} \frac{\varphi(t)\, dt}{t - t_0}.$$

If for $\epsilon \to 0$ the integral tends to a finite limit, then the limit is called the Cauchy principal value of the integral, that is,

$$I(t_0) = \lim_{\epsilon \to 0} \int_{L \setminus l} \frac{\varphi(t)\, dt}{t - t_0} = \int_L \frac{\varphi(t)\, dt}{t - t_o}. \tag{1.1.6}$$

Muskhelishvili (1952) proved that the class H^* of functions on a piecewise smooth curve L is invariant with respect to an integral in the sense of the Cauchy principal value (a singular integral). In other words, if $\varphi(t) \epsilon H^*$ on L, then $I(t_0) \epsilon H^*$ on L.

1.2. SINGULAR INTEGRAL OVER A CLOSED CONTOUR

Let us start by considering the singular integral

$$I(t_0) = \int_L \frac{\varphi(t)\, dt}{t - t_0} \tag{1.2.1}$$

around the circle L whose radius is equal to unity and which is centered at the origin of coordinates. Here $\varphi(t)$ is a class H function on L.

For simplicity, we start by considering the integral

$$I_0(t_0) = \int_L \frac{dt}{t - t_0},$$

for which it is known that (Muskhelishvili 1952)

$$I_0(t_0) = \pi i. \tag{1.2.2}$$

Let us choose on L two sets of nodes: $E = (t_k, k = 1, \ldots, n)$ and $E_0 = \{t_{0k}, k = 1, \ldots, n\}$, such that the points $t_k, k = 1, \ldots, n$, divide the circle into n equal parts, and point t_{0k} is the middle of the arc $\breve{t_k t_{k+1}}$, where $t_{n+1} = t_1$. In what follows the sets E and E_0 chosen in such a way will be called a canonic division of circle L.

Lemma 1.2.1. *For any point $t_{0j} \in E_0$ the following inequality is fulfilled*:

$$\left| \int_L \frac{dt}{t - t_{0j}} - \sum_{k=1}^{n} \frac{\Delta t_k}{t_k - t_{0j}} \right| \leq O\left(\frac{1}{n}\right). \tag{1.2.3}$$

where $\Delta t_k = t_{k+1} - t_k, k = 1, \ldots, n$. Let $O(1/n)$ or $O_\delta(1/n)$ be a quantity of the order of $1/n$. Hence, the right-hand side of the inequality equals B/n or B_δ/n where the constant B or B_δ is independent of n (and the constant B_δ depends on the parameter δ).

Because L is a unit circle centered at the origin of coordinates, one may write

$$t_k = e^{i\theta_k}, \qquad t_{0k} = e^{i\theta_{0k}},$$

where θ_k and θ_{0k} are polar angles of the points t_k and t_{0k}, respectively, and $k = 1, \ldots, n$.

Taking into account periodicity of the function $\exp(i\theta)$ and denoting η_m by $2\pi m/n - \pi/n, m = 1, \ldots, n$, we get

$$\sum_{k=1}^{n} \frac{\Delta t_k}{t_k - t_{0j}} = - \sum_{m=1}^{n} \frac{e^{i\eta_{m+1}} - e^{i\eta_m}}{1 - e^{i\eta_m}}$$

$$= + \sum_{m=1}^{n} \left[\cot\frac{\eta_m}{2} \cos\frac{\Delta\eta_m}{2} - \sin\frac{\Delta\eta_m}{2} \right.$$

$$\left. + i\left(\cos\frac{\Delta\eta_m}{2} + \cot\frac{\eta_m}{2} \sin\frac{\Delta\eta_m}{2} \right) \right] \sin\frac{\Delta\eta_m}{2}, \tag{1.2.4}$$

where $\Delta\eta_m = \eta_{m+1} - \eta_m = 2\pi/n, m = 1, \ldots, n$.

Note that the numbers $\eta_m/2, m = 1, \ldots, n$, are located symmetrically with respect to $\pi/2$, and hence,

$$\sum_{m=1}^{n} \cot\frac{\eta_m}{2} = 0. \tag{1.2.5}$$

From Equalities (1.2.4) and (1.2.5) it follows that

$$\sum_{m=1}^{n} \frac{\Delta t_k}{t_k - t_0 j} = n \sin^2\frac{\pi}{n} + i\frac{n}{2}\sin\frac{2\pi}{n} = i\pi + O\left(\frac{1}{n}\right). \tag{1.2.6}$$

Together with (1.2.2) this proves the validity of Inequality (1.2.3).

Note 1.2.1. The following estimate is true:

$$\sum_{k=1}^{n} \frac{|\Delta t_k|}{|t_{0j} - t_k|} \leq O(\ln n), \qquad j = 1, \ldots, n. \tag{1.2.7}$$

In fact, we note that

$$\sum_{k=1}^{n} \frac{\Delta t_k}{t_k - t_{0j}} = \sum_{m=1}^{n} \left[\cos\frac{\eta_m + 1}{2} = i\sin\frac{\eta_m + 1}{2}\right] \frac{\sin \Delta\eta_m/2}{\sin \eta_m/2}.$$

Hence,

$$\sum_{k=1}^{n} \frac{|\Delta t_k|}{|t_k - t_{0j}|} = \sum_{m=1}^{n} \frac{|\sin \Delta\eta_m/2|}{|\sin \eta_m/2|} \leq C \sum_{m=1}^{[n/2]+1} \frac{\Delta\eta_m}{2}\frac{2}{\eta_m} = O(\ln n). \tag{1.2.8}$$

where $[x]$ is the integer part of x.

Let us next analyze an analogous quadrature sum for the integral (1.2.1). Let the sets of points E and E_0 form a canonic division of the circle L. Then we designate

$$S_n(t_{0j}) = \sum_{k=1}^{n} \frac{\varphi(t_k)\,\Delta t_k}{t_k - t_{0j}}, \qquad j = 1, \ldots, n, \tag{1.2.9}$$

and formulate the following theorem.

Theorem 1.2.1. *Let $\varphi(t)$ satisfy the condition $H(\alpha)$ on L. Then the following inequality holds*:

$$|I(t_{0j}) - S_n(t_{0j})| \leq \theta(t_{0j}), \qquad j = 1, \ldots, n,$$

$$\theta(t_{0j}) = O\left(\frac{1}{n^\alpha}\ln n\right) + |\varphi(t_{0j})| O\left(\frac{1}{n}\right). \tag{1.2.10}$$

Proof. For the sake of convenience we put $t_{0j} = 1$. Then

$$|I(t_{0j}) - S_n(t_{0j})| \leq I_1 + I_2,$$

$$I_1 = \left| \int_L \frac{\varphi(t) - \varphi(1)}{t - 1} dt - \sum_{k=1}^{n} \frac{\varphi(t_k) - \varphi(1)}{t_k - 1} \Delta t_k \right|$$

$$I_2 = |\varphi(1)| \left| \int_L \frac{dt}{t - 1} - \sum_{k=1}^{n} \frac{\Delta t_k}{t_k - 1} \right|.$$

Inequality (1.2.3) gives an estimate for I_2. The expression for I_1 can be transformed in the following way:

$$I_1 \leq I_1' + I_1'' + I_1''',$$

$$I'_1 = \left| \sum_{k=1}^{n-1} \int_{t_k}^{t_{k+1}} \left[\frac{\varphi(t) - \varphi(1)}{t - 1} - \frac{\varphi(t_k) - \varphi(1)}{t_k - 1} \right] dt \right|,$$

$$I_1'' = \left| \int_{t_n}^{t_1} \frac{\varphi(t) - \varphi(1)}{t - 1} dt \right|, \qquad I_1''' = \left| \frac{\varphi(t_n) - \varphi(1)}{t_n - 1} \right| \frac{2\pi}{n}.$$

Because the function $\varphi(t)$ meets the condition $H(\alpha)$ on L,

$$I_1'' \leq \int_{t_n}^{t_1} \frac{|\varphi(t) - \varphi(1)|}{|t - 1|} |dt| \leq A \int_{t_n}^{t_1} |t - 1|^{1+\alpha} |dt|.$$

For a unit circle $|dt| = d\theta$ and

$$|t - 1| = |e^{i\theta} + 1| = |(\cos\theta - 1) + i\sin\theta| = 2|\sin\theta/2|.$$

Hence,

$$I_1'' \leq A2^{\alpha}\int_0^{\pi/n}\left(\sin\frac{\theta}{2}\right)^{-1+\alpha} d\theta \leq c_1\int_0^{\pi/n}\theta^{-1+\alpha}\,d\theta = O\left(\frac{1}{n^{\alpha}}\right).$$

For I_1''' one gets

$$I_1''' \leq A|t_n - 1|^{-1-\alpha}\frac{2\pi}{n} = 0\left(\frac{1}{n^{\alpha}}\right).$$

In order to estimate I_1' we use the transform

$$\frac{\varphi(t) - \varphi(t_{0j})}{t - t_{0j}} - \frac{\varphi(t_k) - \varphi(t_{0j})}{t_k - t_{0j}} = \frac{\varphi(t) - \varphi(t_k)}{t - t_{0j}} + \frac{[\varphi(t_k) - \varphi(t_{0j})](t_k - t)}{(t - t_{0j})(t_k - t_{0j})}, \tag{1.2.11}$$

which will be used often in what follows.

By (1.2.11) we have

$$I_1' \leq \sum_{k=1}^{n-1}\int_{t_k}^{t_{k+1}}\frac{|\varphi(t) - \varphi(t_k)|}{|t - 1|}|dt| + \sum_{k=1}^{n-1}\int_{t_k}^{t_{k+1}}\frac{|\varphi(t_k) - \varphi(1)|}{|t_i - 1|} \times \frac{|t_k - t|}{|t - 1|}|dt| = S_1 + S_2.$$

Because $\varphi(t)$ meets the condition $H(\alpha)$ on L and $t = \exp(i\theta)$,

$$S_1 \leq A\left(\frac{2\pi}{n}\right)^{\alpha}\sum_{k=1}^{n-1}\int_{t_k}^{t_{k-1}}\frac{|dt|}{|t - 1|} \leq C_2\frac{1}{n^{\alpha}}\int_{\pi/n}^{\pi/2}\frac{d\theta}{\theta} = O\left(\frac{\ln n}{n^{\alpha}}\right).$$

Finally, we get for the sum S_2,

$$S_2 \leq A\frac{2\pi}{n}\sum_{k=1}^{n-1}\frac{1}{|t_k - 1|^{1-\alpha}}\int_{t_k}^{t_{k+1}}\frac{|dt|}{|t - 1|} \leq C_3\frac{\pi}{n}\int_{\pi/n}^{\pi}\frac{d\theta}{\theta^{2-\alpha}} = O\left(\frac{1}{n^{\alpha}}\right).$$

By substituting the estimates for S_1 and S_2 into the inequality for I_1' and the estimates for I_1', I_1'', and I_1''' into the inequality for I_1, we see that

$$I_1 \leq O(n^{-\alpha} \ln n).$$

Thus, Theorem 1.2.1 is proved. ■

Definition 1.2.1. *A function $\varphi(t)$ belongs to the class Π of functions on L if it is of the form*

$$\varphi(t) = \frac{\psi(t)}{q - t},$$

where $\psi(t) \epsilon H$ on L and q is a certain fixed point on L. Note that one may write

$$\int_L \frac{\varphi(t)\, dt}{t - t_0} = \frac{1}{q - t_0} \left[\int_L \frac{\psi(t)\, dt}{t - t_0} - \int_L \frac{\psi(t)\, dt}{t - q} \right].$$

According to the latter formula, the following theorem is true.

Theorem 1.2.2. *Let $\varphi(t)$ belong to the class Π on the circle L and the sets E and E_0 form a canonic division of the circle L, with $q \epsilon E_0$ for $j = j_q$. Then the following inequality holds*:

$$|I(t_{0j}) - S_n(t_{0j})| \leq \theta(t_{0j}), \qquad j \neq j_q, j = 1, \ldots, n. \tag{1.2.12}$$

where the quantity $\theta(t_{0j})$ has the form

$$\theta(t_{0j}) = \frac{1}{|t_{0j} - q|} O\left(\frac{1}{n^{\alpha}} \ln n \right), \qquad j \neq j_q.$$

Obviously, the quantity $\theta(t_{0j})$ satisfies the inequalities

$$\theta(t_{0j}) \leq O_l\left(\frac{1}{n_1^{\lambda}} \right), \qquad \lambda_1 > 0, \tag{1.2.13}$$

for all $t_{0j} \epsilon L \setminus \mathrm{l}$, where l is, however, a small neighborhood of point q, and

$$\sum_{\substack{j=1 \\ j \neq j_q}}^{n} \theta(t_{0j}) |\Delta t_{0j}| \leq O\left(\frac{1}{n^{\lambda_2}} \right), \qquad \lambda_2 > O. \tag{1.2.14}$$

Evidently, the latter two inequalities may be made more accurate by replacing their right-hand sides by $O_1(n^{-\alpha} \ln n)$ and $O(n^{-\alpha} \ln n)$, respectively.

Note 1.2.2. If $\varphi(t, t_0) \epsilon H(\alpha)$ on L, then Inequality (1.2.10) also remains valid for the integral $\int(\varphi(t, t_0/(t - t_0))\, dt$. In other words,

$$\left| \int_L \frac{\varphi(t, t_{0j})\, dt}{t - t_{0j}} - \sum_{k=1}^{n} \frac{\varphi(t_k, t_{0j})}{t_k - t_{0j}} \right| \leq O\left(\frac{\ln n}{n^{\alpha}} \right). \tag{1.2.15}$$

Note 1.2.3. Let L_1 be a closed Lyapunov contour. Then a one-to-one correspondence $\tau = \tau(t)$ exists between points τ of the curve and points t of the standard (unit) circle L, such that the derivative $\tau'(t) = d\tau/dt$ belongs to the class $H(\beta)$ and does not vanish anywhere on L. If $\varphi(\tau)$ meets the condition $H(\alpha)$ on L_1, then by the formula of substitution of a variable in singular integrals (Muskhelishvili 1952), one gets

$$\int_{L_1} \frac{\varphi(\tau)\, d\tau}{\tau - \tau_0} = \int_L \frac{\psi(t, t_0)\, dt}{t - t_0}, \qquad \psi(t, t_0) = \frac{t - t_0}{\tau(t) - \tau(t_0)} \tau'(t) \varphi(\tau(t)).$$

Let us next analyze the canonic division of the circle L formed by sets E and E_0. The sets of the points $\tau_k = \tau(t)_k$, $t_k \epsilon E$, and $\tau_{0k} = \tau(t_{0k})$, $t_{0k} \epsilon E_0$, will be called the canonic division of the curve L_1. Let us take the sum

$$\sum_{k=1}^{n} \frac{\varphi(\tau_k)\, \Delta\tau_k}{\tau_k - \tau_{0j}} = \sum_{k=1}^{n} \frac{\varphi(\tau(t_k))(t_k - t_{0j})}{\tau(t_k) - \tau(t_{0j})} \frac{\Delta\tau_k}{\Delta t_k} \frac{\Delta t_k}{t_k - t_{0j}}.$$

Because

$$\frac{\Delta\tau_k}{\Delta t_k} = \frac{\tau(t_{k+1}) - \tau(t_k)}{t_{k+1} - t_k} = \tau'(\tilde{t}_k), \qquad \tilde{t}_k \in \breve{t_k t}_{k+1},$$

and $\tau'(t) \epsilon H(\beta)$ on L, hence $|\tau'(\tilde{t}_k) - \tau'(t_k)| \leq A|\tilde{t}_k - t_k|^{\beta}$, i.e., $\tau'(\tilde{t}_k) = \tau'(t_k) + O(|\tilde{t}_k - t_k|^{\beta})$.

Thus,

$$\sum_{k=1}^{n} \frac{\varphi(\tau_k)\, \Delta\tau_k}{\tau_k - \tau_{0j}} = \sum_{k=1}^{n} \frac{\varphi(\tau(t_k))(t_k - t_{0j})}{\tau(t_k) - \tau(t_{0j})} \frac{\tau'(t_k)\, \Delta t_k}{t_k - t_{0j}}$$

$$+ \sum_{k=1}^{n} \frac{\varphi(\tau(t_k))(t_k - t_{0j})}{\tau(t_k) - \tau(t_{0j})} O\left(|\tilde{t}_k - t_k|^{\beta}\right)$$

$$\times \frac{\Delta t_k}{t_k - t_{0j}} = S_j^1 + S_j^2.$$

By using Inequality (1.2.7), it can be readily shown that

$$|S_j^2| \leq O(n^{-b} \ln n).$$

Hence, in accordance with Inequality (1.2.15) and the preceding estimate for S_j^2,

$$\left| \int_{L_1} \frac{\varphi(\tau)\, d\tau}{\tau - \tau_{0j}} - \sum_{k=1}^{n} \frac{\varphi(\tau_k)\, \Delta\tau_k}{\tau_k - \tau_{0j}} \right| \leq O\left(\frac{1}{n^\lambda}\right), \qquad \lambda > 0. \quad (1.2.16)$$

Now let curve L be a set of p nonintersecting closed Lyapunov curves $L_1, \ldots, L_p$ and let the sets $E_m = \{\tau_k, k = n_{m-1} + 1, \ldots, n_m\}$ and $E_{0m} = \{\tau_{0k}, k = n_{m-1} + 1, \ldots, n_m\}$ form a canonic $N_m = n_m - n_{m-1}$ division of the curve L_m $m = 1, \ldots, p$; $n_0 = 0$. We denote

$$N = \min_{m=1,\ldots,p} N_m$$

and suppose that $N_m/N \leq R < +\infty$. We also denote

$$S_{n_p}(\tau_{0j}) = \sum_{k=1}^{n_p} \frac{\varphi(\tau_k)\, \Delta\tau_k}{\tau_k - \tau_{0j}}, \qquad j = 1, \ldots, n_p,$$

where $\Delta\tau_k = \tau_{k+1}$, $k = 1, \ldots, n_p$, $k \neq n_1, \ldots, n_p$, and $\Delta\tau_{nm} = \tau_{nm-1+1} - \tau_{nm}$, $m = 1, \ldots, p$.

The following theorem proves to be true.

Theorem 1.2.3. *Let $\varphi(\tau)$ meet condition H on curve L. Then for any point $\tau_{0j} \in \bigcup_{m=1}^{p} E_{0m}$, the following inequality holds:*

$$|I(\tau_{0j}) - S_{n_p}(\tau_{0j})| \leq O(1/N^\lambda), \qquad \lambda > 0. \qquad (1.2.17)$$

1.3. SINGULAR INTEGRAL OVER A SEGMENT*

Let us assume that in the singular integral (1.2.1), $L = [a, b]$ on the real axis, and the function $\varphi(t) \in H^*$ on L, i.e.,

$$\varphi(t) = \frac{\psi(t)}{(t-a)^\nu (b-t)^\mu},$$

* "Interval" is now commonly used for "segment" (G.Ch.)

where $\psi(t) \in H(\alpha)$ on $[a, b]$, $0 \leq \nu, \ \mu < 1$.

Also, let the points $t_0 = a, t_1, \ldots, t_n, t_{n+1} = b$ divide the segment $[a, b]$ into $n + 1$ equal parts $h = (b - a)/(n + 1)$ long, and the point t_{0j} be the middle of the segment $[t_j, t_{j+1}]$, $j = 0, 1, \ldots, n$. It will be said that the points of the sets $E = \{t_k, k = 1, \ldots, n\}$ and $E_0 = \{t_{0j}, j = 0, 1, \ldots, n\}$ form a canonic division of the segment $[a, b]$ with the subinterval equal to h.

The following lemma may be formulated.

Lemma 1.3.1. *Inequality*

$$\left| \int_a^b \frac{dt}{t - t_{0j}} - \sum_{k=1}^{n} \frac{h}{t_k - t_{0j}} \right| \leq \frac{hB}{(t_{0j} - a)(b - t_{0j})} \tag{1.3.1}$$

holds for any point $t_{0j} \in E_0$, where B is a certain constant.

Note that the inequality

$$\sum_{k=1}^{n} \frac{h}{|t_k - t_{0j}|} \leq O(|\ln h|) \tag{1.3.2}$$

is also true.

Lemma 1.3.2. *Let the function $\varphi(t) \in H(\alpha)$ on the segment $[a, b]$. Then for any point $t_{0j} \in E_0$, the following inequality holds:*

$$I_1 \equiv \left| \int_a^b \frac{\varphi(t) - \varphi(t_{0j})}{t - t_{0j}} \, dt - \sum_{k=1}^{n} \frac{\varphi(t_k) - \varphi(t_{0j})}{t_k - t_{0j}} h \right| \leq O(h^{\alpha} |\ln h|). \tag{1.3.3}$$

Proof. Let us implement the transformation

$$\begin{aligned} I_1 \leq & \left| \int_a^{t_1} \frac{\varphi(t) - \varphi(t_{0j})}{t - t_{0j}} \, dt \right| \\ & + \left| \sum_{\substack{k=1 \\ k \neq j}}^{n} \int_{t_k}^{t_{t+1}} \left(\frac{\varphi(t) - \varphi(t_{0j})}{t - t_{0j}} - \frac{\varphi(t_k) - \varphi(t_{0j})}{t_k - t_{0j}} \right) dt \right| \\ & + \left| \int_{t_j}^{t_{j+1}} \frac{\varphi(t) - \varphi(t_{0j})}{t - t_{0j}} \, dt \right| + \left| \frac{\varphi(t_j) - \varphi(t_{0j})}{t_j - t_{0j}} \right| h \\ = & \ I_1^1 + I_1^2 + I_1^3 + I_1^4. \end{aligned} \tag{1.3.4}$$

Because $\varphi(t)$ satisfies the condition $H(\alpha)$,

$$I_1^1 = O(h^\alpha), \qquad I_1^3 = O(h^\alpha), \qquad I_1^4 = O(h^\alpha).$$

The use of Formula (1.2.11) gives

$$\begin{aligned} I_1^2 \le & \sum_{\substack{k=1\\k\neq j}}^{n} \int_{t_k}^{t_{k+1}} \frac{|\varphi(t) - \varphi(t_k)|}{|t - t_{0j}|}\,dt \\ & + \sum_{k=1}^{n} \int_{t_k}^{t_{k+1}} \frac{|t_k - t|}{t - t_{0j}|} \frac{|\varphi(t_k) - \varphi(t_{0j})|}{|t_k - t_{0j}|}\,dt \\ = & \; S_1 + S_2. \end{aligned}$$

For S_1 one gets the estimate

$$S_1 \le Ah^\alpha \left(\int_{t_1}^{t_j} \frac{dt}{t_{0j} - t} + \int_{t_{j+1}}^{b} \frac{dt}{t - t_{0j}} \right) = O(h^\alpha |\ln h|).$$

For any k and j, the inequality

$$\frac{h}{|t_i - t_{0j}|} \le 2. \tag{1.3.5}$$

holds. Then

$$\begin{aligned} S_2 & \le Ah \sum_{\substack{k=1\\k\neq j}}^{n} \int_{t_k}^{t_{k+1}} \frac{dt}{|t - t_{0j}||t_k - t_{0j}|^{1-\alpha}} \\ & \le Ch^\alpha \sum_{\substack{k=1\\k\neq j}}^{n} \int_{t_k}^{t_{k+1}} \frac{dt}{|t - t_{0j}|} \\ & = O(h^\alpha |\ln h|). \end{aligned}$$

The validity of Lemma 1.3.2 is proved by substituting the estimates for S_1 and S_2 into the formula for I_1^2 and the estimates for $I_1^1, \ldots, I_1^4$ into the formula for I_1. ∎

Inequalities (1.3.1) and (1.3.3) allow formulation of the following theorem.

Theorem 1.3.1. *Let the function $\varphi(t) \epsilon H(\alpha)$ on segment $[a, b]$ and the sets E and E_0 form a canonic division of the segment. Then for any point $t_{0j} \epsilon E_0$, one has*

$$|I(t_{0j}) - S_n(t_{0j})| \leq O\left(\frac{h^\alpha |\ln h| + |\varphi(t_{0j})| h}{(t_{0j} - a)(b - t_{0j})} \right),$$

$$\text{where } S_n(t_{0j}) = \sum_{i=1}^{n} \frac{\varphi(t_i)}{t_i - t_{0j}}. \tag{1.3.6}$$

Let us finally prove the following theorem.

Theorem 1.3.2. *Let $\varphi(t) \epsilon H^*$ on the segment $[a, b]$. Let the sets E and E_0 form a canonic division of the segment. Then the inequality*

$$I \equiv \left| \int_a^b \frac{\varphi(t)\, dt}{t - t_{0j}} - \sum_{k=1}^{n} \frac{\varphi(t_k) h}{t_k - t_{0j}} \right| \leq \theta(t_{0j}), \qquad j = 0, 1, \ldots, n, \tag{1.3.7}$$

holds where the quantity $\theta(t_{0j})$ satisfies the inequalities:
(a) for all points $t_{0j} \epsilon [a + \delta, b - \delta]$,

$$\theta(t_{0j}) \leq O_\delta(h^{\lambda_1}), \qquad 0 < \lambda_1 \leq 1, \tag{1.3.8}$$

where $\delta > 0$ is, however, a small number;
(b) for all points $t_{0j} \epsilon [a, b]$,

$$\sum_{j=0}^{n} \theta(t_{0j}) |\Delta t_{0j}| \leq O(h^{\lambda_2}), \qquad 0 < \lambda_2 \leq 2, \tag{1.3.9}$$

where $|t_{0j}| = h$, $j = 0, 1, \ldots, n$.

Proof. We may write

$$I \leq \left| \int_a^b \frac{\varphi(t) - \varphi(t_{0j})}{t - t_{0j}} dt - \sum_{k=1}^{n} \frac{\varphi(t_k) - \varphi(t_{0j})}{t_k - t_{0j}} h \right|$$

$$+ |\varphi(t_{0j})| \left| \int_a^b \frac{dt}{t - t_{0j}} - \sum_{k=1}^{n} \frac{h}{t_k - t_{0j}} \right|$$

$$= I_1 + I_2.$$

Here I_2 can be estimated with the help of Inequality (1.3.1). To estimate I_1 we first observe that if $\varphi(t) \epsilon H^*$ on $[a, b]$, then

$$|\varphi(t_2) - \varphi(t_1)| \leq A\left[\frac{|t_2 - t_1|^\alpha}{(t_2 - a)^\nu (b - t_2)^\mu} + \frac{|\psi(t_1)||t_2 - t_1|^\nu}{(t_2 - a)^\nu (b - t_2)^\mu (t_1 - a)^\nu} + \frac{|\psi(t_1)||t_2 - t_1|^\mu}{(b - t_2)\mu(b - t_1)^\mu (t_1 - a)^\nu}\right]. \tag{1.3.10}$$

If I_1 is represented in the same way as in Inequality (1.3.4), then one gets for I_1^4,

$$I_1^4 \leq A_1\left[\frac{h^\alpha}{(t_{0j} - a)^\nu (b - t_{0j})^\mu} + \frac{|\psi(t_{0j})|h^\nu}{(t_{0j} - a)^{2\nu}(b - t_{0j})^\mu} + \frac{|\psi(t_{0j})|h^\mu}{(t_{0j} - a)^\nu (b - t_{0j})^{2\mu}}\right],$$

because for any $j = 1, \ldots, n$,

$$\frac{1}{2} \leq \frac{t_j - a}{t_{0j} - a}, \qquad \frac{t_j - b}{t_{0j} - b} \leq 2. \tag{1.3.11}$$

For I_1^3 one has

$$I_1^3 \leq A\left[\int_{t_j}^{t_{j+1}} \frac{dt}{(t - a)^\nu (b - t)^\mu |t - t_{0j}|^{1-\alpha}} + \frac{|\psi(t_{0j})|}{(t_{0j} - a)^\nu} \times \int_{t_j}^{t_{j+1}} \frac{dt}{(t - a)^\nu (b - t)^\mu |t - t_{0j}|^{1-\nu}} + \frac{|\psi(t_{0j})|}{(b - t_{0j})^\mu (t_{0j} - a)^\nu} \int_{t_j}^{t_{j+1}} \frac{dt}{(b - t)^\mu |t - t_{0j}|^{1-\mu}}\right].$$

Fikhtengoltz (1959) proved a theorem about the average value of an improper integral; it was formulated as follows. Let functions $f(x)$ and $g(x)$ be integrable on $[a, b]$, with $f(x)$ being limited, i.e., $-\infty < m \leq f(x) \leq M < +\infty$, and $g(x)$ changing its sign. Then the function $f(x)g(x)$ is integrable and

$$\int_a^b f(x)g(x)\,dx = \mu\int_a^b g(x)\,dx, \qquad m \leq \mu \leq M.$$

By applying the theorem to all $j = 1, \ldots, n-1$, one gets

$$I_1^3 \leq A\left[\frac{1}{\left(\xi_j^1 - a\right)^{\nu}\left(b - \xi_j^1\right)^{\mu}}\int_{t_j}^{t_{j+1}}\frac{dt}{|t - t_{0j}|^{1-\alpha}}\right.$$

$$+\frac{|\psi(t_{0j})|}{(t_{0j} - a)^{\nu}\left(\xi_j^2 - a\right)^{\nu}\left(b - \xi_j^2\right)^{\mu}}$$

$$\times\int_{t_j}^{t_{j+1}}\frac{dt}{|t - t_{0j}|^{1-\nu}} + \frac{|\psi(t_{0j})|}{(b - t_{0j})^{\mu}(t_{0j} - a)^{\nu}\left(b - \xi_j^3\right)^{\mu}}$$

$$\left.\times\int_{t_j}^{t_{j-1}}\frac{dt}{|t - t_{0j}|^{1-\mu}}\right] \leq A_2\left[\frac{h^{\alpha}}{(t_{0j} - a)^{\nu}(b - t_{0j})^{\mu}}\right.$$

$$\left.+\frac{|\psi(t_{0j})h^{\nu}}{(t_{0j} - a)^{2\nu}(b - t_{0j})^{\mu}} + \frac{|\psi(t_{0j})|h^{\mu}}{(t_{0j} - a)^{\nu}(b - t_{0j})^{2\mu}}\right],$$

because $\xi_j^k \epsilon [t_j, t_{j+1}], k + 1, 2, 3$, and hence, the ratios $(\xi_j^k - a)/(t_{0j} - a)$ and $(\xi_j^k - b)/(t_{0j} - b)$ are quantities limited for $j = 1, \ldots, n-1$. For $j = n$ in I_1^3 we again use the previously formulated theorem about the average value having preliminary divided the interval of integration into two segments: $[t_n, t_n + 3h/4]$ and $[t_n + 3h/4, b]$. Then we deduce that I_1^3 has the same estimate as $j = n$. Analogously, I_1^1 for $j = 0, 1, \ldots, n$ may be shown to have the same estimate.

Finally, we consider I_1^2. Similarly to I_1^2 in the proof of Inequality (1.3.3), we write

$$I_1^2 \leq S_1 + S_1.$$

By using Equation (1.3.10) and taking into account that $|\psi(t_k)|$ is limited on $[a, b]$, one gets for S_1,

$$S_1 \le A\left[h^\alpha \sum_{\substack{k=1\\k\neq j}}^{n} \int_{t_k}^{t_{k+1}} \frac{dt}{(t-a)^\nu (b-t)^\mu |t-t_{0j}|} + h^\nu \sum_{\substack{k\text{-}1\\k\neq j}}^{n} \frac{|\psi(t_k)|}{(t_k-a)^\nu}\right.$$

$$\times \int_{t_k}^{t_{k+1}} \frac{dt}{(t-a)^\nu (b-t)^\mu |t-t_{0j}|}$$

$$\left. + h^\mu \sum_{\substack{k=1\\k\neq j}}^{n} \frac{|\psi(t_k)|}{(t_k-a)^\nu (b-t_k)^\mu} \int_{t_k}^{t_{k+1}} \frac{dt}{(b-t)^\mu |t-t_{0j}|}\right]$$

$$\le A_3\left[h^\alpha \frac{1}{(b-t_j)^\mu} \int_{t_1}^{t_j} \frac{dt}{(t-a)^\nu |t-t_{0j}|}\right.$$

$$+ h^\alpha \frac{1}{(t_{j+1}-a)^\nu} \int_{t_{j+1}}^{b} \frac{dt}{(b-t)^\mu |t-t_{0j}|}$$

$$+ \frac{h^\nu}{(b-t_j)^\mu}\left(\frac{1}{(t_1-a)^\nu} \int_{t_1}^{t_2} \frac{dt}{(t-a)^\nu |t-t_{0j}|}\right.$$

$$\left. + \int_{t_1}^{t_j} \frac{dt}{(t-h-a)^{2\nu} |t-t_{0j}|}\right)$$

$$+ \frac{h^\nu}{(t_{j+1}-a)^{2\nu}} \int_{t_{j+1}}^{b} \frac{dt}{(b-t)^\mu |t-t_{0j}|}$$

$$+ \frac{h^\mu}{(b-t_j)^{2\mu}} \int_{t_1}^{t_j} \frac{dt}{(t-h-a)^\nu |t-t_{0j}|}$$

$$+ \frac{h^\mu}{(t_{j+1}-a)^\nu}\left(\int_{t_{j+1}}^{b-h} \frac{dt}{(b-t)^{2\mu} |t-t_{0j}|}\right.$$

$$\left.\left. + \frac{1}{(b-t_n)^\mu} \int_{t_n}^{n} \frac{dt}{(b-t)^\mu |t-t_{0j}|}\right)\right].$$

Next we note that

$$\int_{t_1}^{t_j} \frac{dt}{(t-a)^{\nu}|t-t_{0j}|} = \int_{t_1}^{(t_j+t_1)/2} \frac{dt}{(t-a)^{\nu}(t_{0j}-t)}$$

$$+ \int_{(t_j+t_1)/2}^{t_j} \frac{dt}{(t-a)^{\nu}(t_{0j}-t)}$$

$$\leq \frac{2}{(1-\nu)(t_{0j}-t_1+t_{0j}-t_j)}$$

$$\times\left[\left(\frac{t_j+t_1}{2}-a\right)^{1-\nu}-(t_1-a)^{1-\nu}\right]$$

$$+\frac{2^{\nu}}{(t_j-a+t_1-a)^{\nu}} \ln \frac{t_{0j}-t_j+t_{0j}-t_1}{2(t_{0j}-t_j)}$$

$$\leq \frac{A_4}{(t_{0j}-t_1)^{\nu}}+\frac{O(|\ln h|)}{(t_j-a)^{\nu}}$$

$$= \frac{O(|\ln h|)}{(t_{0j}-a)^{\nu}}, \qquad j=1,\ldots,n. \tag{1.3.12}$$

Analogously, by considering the integral

$$\int_{t_2}^{t_j} \frac{dt}{(t-h-a)^{2\nu}|t-t_{0j}|},$$

one concludes that (1) if $2\nu \leq 1$, then the estimate of the form (1.3.12) is valid and (2) if $2\nu > 1$, then by the inequality $1-2\nu<0$, Equation (1.3.12) becomes

$$\int_{t_2}^{t_j} \frac{dt}{(t-h-a)^{2\nu}|t-t_{0j}|} \leq \left[\frac{h^{1-2\nu}}{t_{0j}-a}+\frac{1}{(t_{0j}-a)^{2\nu}}\right] O(|\ln h|). \tag{1.3.13}$$

For all the other integrals entering the inequality for S_1 the estimate of the form (1.3.12) or (1.3.13) is valid if ν is substituted by μ and a by b.

Hence, finally we have

$$S_1 \leq \left[\frac{h^{\alpha}}{(t_{0j}-a)^{\nu}(b-t_{0j})^{\mu}} + \frac{h^{\nu}}{(t_{0j}-a)^{2\nu}(b-t_{0j})^{\mu}} + \frac{h^{\mu}}{(t_{0j}-a)^{\nu}(b-t_{0j})^{2\mu}} + \frac{h^{1-\nu}}{(t_{0j}-a)(b-t_{0j})^{\mu}} + \frac{h^{1-\mu}}{(t_{0j}-a)^{\nu}(b-t_{0j})}\right] O(|\ln h|).$$

Let us consider S_2. By applying Formula (1.3.10), one gets

$$S_2 \leq A_4 h\left[\frac{1}{(t_{0j}-a)^{\nu}(b-t_{0j})^{\mu}} \sum_{\substack{k=1\\k\neq j}}^{n} \int_{t_k}^{t_{k+1}} \frac{dt}{|t_k-t_{0j}|^{1-\alpha}|t-t_{0j}|} + \frac{1}{(t_{0j}-a)^{\nu}(b-t_{0j})^{\mu}} \sum_{\substack{k=1\\k\neq j}}^{n} \int_{t_k}^{t_{k+1}} \frac{dt}{(t_k-a)^{\nu}|t_k-t_0|^{1-\nu}|t-t_{0j}|} + \frac{1}{(b-t_{0j})^{\mu}} \sum_{\substack{k=1\\k\neq j}}^{n} \int_{t_k}^{t_{k+1}} \frac{dt}{(t_k-a)^{\nu}(b-t_k)^{\mu}|t_k-t_{0j}|^{1-\mu}|t-t_{0j}|}\right].$$

Again using the fact that $h/|t_i - t_{0j}| \leq 2$ for $j = 1, \ldots, n$ and any i as well as the estimates of the form (1.3.12), one gets

$$S_2 \leq \left[\frac{h^{\alpha}}{(t_{0j}-a)^{\nu}(b-t_{0j})^{\mu}} + \frac{h^{\nu}}{(t_{0j}-a)^{2\nu}(b-t_{0j})^{\mu}} + \frac{h^{\mu}}{(t_{0j}-a)^{\nu}(b-t_{0j})^{2\mu}} O(|\ln h|).\right.$$

By denoting $\eta(t_{0j}, \alpha, \beta) = (t_{0j} - a)^{-\alpha}(b - t_{0j})^{-\beta}$, the estimate for I may be written in the form

$$I \leq \theta(t_{0j}) = \theta_0(t_{0j}, \alpha, \nu, \mu) O(|\ln h|), \tag{1.3.14}$$

where

$$\theta_0(t_{0j}, \alpha, \nu, \mu) = \eta(t_{0j}, \nu, \mu)h^{\alpha}$$
$$+\eta(t_{0j}, 2\nu, \mu)h^{\nu} + \eta(t_{0j}, \nu, 2\mu)h^{\mu}$$
$$+\eta(t_{0j}, 1, \mu)h^{1-\nu} + \eta(t_{0j}, \nu, 1)h^{1-\mu}$$
$$+\eta(t_{0j}, 1+\nu, 1+\mu)h, \quad 0 < \nu, \mu < 1.$$

The quantity $\theta(t_{0j})$ entering the latter inequality readily can be shown to satisfy Conditions (1.3.8) and (1.3.9). ■

Note 1.3.1. In fact, Inequality (1.3.7) was proved subject to the condition $0 < \nu, \mu < 1$. However, the preceding considerations allow us to state (1) that if $\nu \neq 0, \mu = 0$, then the expression for $\theta_0(t_{0i}, \alpha, \nu, 0)$ has the form

$$\theta_0(t_{0j}, \alpha, \nu, 0) = \eta(t_{0j}, \nu, 0)h^{\alpha} + \eta(t_{0j}, 2\nu, 0)h^{\nu} + \eta(t_{0j}, 1, 0)h^{1-\nu}$$
$$+ \eta(t_{0j}, 1+\nu, 1)h. \tag{1.3.15}$$

and (2) that if $\nu = \mu = 0$, then the quantity $\theta(t_{0j})$ has the form of the right-hand side of Inequality (1.3.6).

Let us also make the following remark. If a function $\varphi(t)$ has the form

$$\varphi(t) = \frac{(b-t)^{\mu}}{(t-a)^{\nu}}\psi(t),$$

where $0 < \nu < 1$, $0 < \mu$, and $\psi(t)$ belongs to the class H on segment $[a, b]$, then Inequality (1.3.8) is fulfilled for all the points $t_{0j} \epsilon [a+\delta, b]$.

Note 1.3.2. In aerodynamics (see, for example Belotserkovskii 1967, Lifanov and Polonskii 1975) points t_k and t_{0j} are often chosen in the following way. Let us divide the segment $[a, b]$ into n equal parts each h long and denote the parts by $\Delta_k, k = 1, \ldots, n$. The points whose distances from the left end of the segment Δ_k are equal to $h/4$ and $3h/4$, respectively, will be denoted by t_k and $t_{0k}, k = 1, 2, \ldots, n$. According to Lifanov and Polonskii (1975), all the statements made for a singular integral on segment $[a, b]$ are also valid for the division. Calculations show that the latter scheme gives better results for model examples than the canonic division (see, for example, Belotserkovskii 1967). A more general statement is also valid. Inequalities (1.3.1), (1.3.3), (1.3.6), and (1.3.7) are also valid in the case when points belonging to the sets $E = \{t_k, k = 1,\ldots,n\}$ and $E_0 = \{t_{0j}, j = 0,1,\ldots,n\}$ do not form a canonic division but meet the condition

$$|t_{k+1} - t_k| = h, \qquad k = 1,\ldots,n-1,$$

$$|t_{0j+1} - t_{0j}| = h, \qquad j = 0,1,\ldots,n-1,$$

$$t_1 - t_{00} = \frac{h}{2}, \qquad t_{0k} = t_k + \frac{h}{2}, \qquad k = 1,\ldots,n,$$

$$t_{00} - a = hq_h^a, \quad b - t_{0n} = hq_h^b,$$

$$0 < p_1 \leq q_h^a, q_h^b \leq p_2 < +\infty,$$

where p_1 and P_2 are fixed numbers.

Such a situation takes place if a fixed point $g\epsilon(a, b)$ is desired to take a certain position with respect to division points, e.g., if we want to point q to belong to the set E or E_0 for an n (see Lifanov 1978a).

Let us show how a point q may be made to belong to E_0 for any n. To do this we divide segment $[a, b]$ into $n + 2$ parts by points t'_k, $k = 1,\ldots,n+1, t'_0 = a, t'_{n+2} = b$, and t'_{0k} (the middle segment $[t'_k, t'_{k+1}], k = 1,\ldots,n+1$). Let point q lie in the segment $[t'_{0j_q}, t'_{0j_q+1}]$. Let us displace the set of points $\{t'_k, k = 1,\ldots,n+1\} \cup \{t_{0j}, j = 0,1,\ldots,n+1\}$ as a rigid whole in such a way that the end of the segment $[t'_{0j_q}, t'_{0j_q} + 1]$ nearest to point q would coincide with the point. Let us next discard the end points belonging to the set $\{t'_k, k = 1,\ldots,n+1\}$ and $\{t'_{0j}, j = 0,1,\ldots,n+1\}$, from the end to which the displacement was performed. The remaining points of the displaced sets will be denoted by $t_k, k = 1,\ldots,n$, and $t_{0j}, j = 0,1,\ldots,n$. If point q coincided originally with one of the points t'_{0j}, then the original sets are denoted by E and E_0.

If q is required to belong to E, then points t_k and t_{0j} are constructed in a similar way.

In the case of flow past an airfoil with a flap, the sets E and E_0 are chosen in such a way that point q lies exactly halfway between the nearest points belonging to E and E_0. This can be done in the following way. Let $h = (b - a)/(n + 1)$. Let the point lying at the distance $h/4$ to the right of point q belong to the set E_0, whereas that lying at the same distance to the left belongs to the set E; the rest of the points of the sets E and E_0 are distributed with the step h starting from the chosen ones. The same result can be achieved by displacing the points of the canonic division previously described with the said step, the displacement being no more than $h/4$.

If points belonging to the sets E and E_0 take the places of each other, the preceding results for a singular integral remain valid. Hence, an integrand may be taken at the points of the set E_0, whereas the integral is evaluated at the points of the set E.

Note 1.3.3. All the results are also valid for the function

$$\varphi(t, \tau) = \frac{\psi(t, \tau)}{(t-a)^{\nu}(b-t)^{\mu}}, \tag{1.3.16}$$

if the function $\psi(t, \tau) \epsilon H$ on segment $[a, b]$ with respect to both variables or if the sum is constructed by using points of the sets E or E_0 only. Thus, if we denote

$$S_j' = \sum_{\substack{k=1 \\ k \neq j}}^{n} \frac{\varphi(t_k)h}{t_k - t_j}, \tag{1.3.17}$$

$$I_j' = \int_a^b \frac{\varphi(t)\, dt}{t - t_j}, \tag{1.3.18}$$

then an inequality of the form of (1.3.7) is valid for the absolute value of the difference $I_j' - S_j'$.

In Section 1.2 we introduced the class Π of functions for a circle. A similar class of functions on segment $[a, b]$ is defined as follows (Lifanov 1978a).

Definition 1.3.1. *A function $\varphi(t)$ belongs to the class Π on segment $[a, b]$ if it is of the form*

$$\varphi(t) = \frac{\psi(t)}{q - t}, \tag{1.3.19}$$

where function $\psi(t) \epsilon H^*$ on segment $[a, b]$, and $q \epsilon (a, b)$ is a fixed point.

The following theorem may be formulated.

Theorem 1.3.3. *Let $\varphi(t) \epsilon \Pi$ on segment $[a, b]$ and the sets E and E_0 be chosen in such a way that point q belongs to the set E_0 for any $n, q = t_{0j_q}$. Then,*

$$|I(t_{0j}) - S_n(t_{0j})| \leq \theta_q(t_{0j}), \qquad j \neq j_q, \tag{1.3.20}$$

where $\theta_q(t_{0j}) = |q - t_{0j}|^{-1}\theta(t_{0j})$, *and the quantity* $\theta(t_{0j})$ *possesses the same properties as in Inequality* (1.3.7). *Therefore, Inequality* (1.3.8) *holds for the quantity* $\theta_q(t_{0j})$ *for all points* t_{0j} *belonging to the set* $[a + \delta, q - \delta] \cup [q + \delta, b - \delta]$, *and Inequality* (1.3.9) *also holds in which the sum in the left-hand side is carried out for all* $j \neq j_q$.

1.4. SINGULAR INTEGRAL OVER A PIECEWISE SMOOTH CURVE

1. Let L be an unclosed Lyapunov curve. This means that a one-to-one correspondence $\tau = \tau(t)$ exists between points τ of L and points t of segment $[a, b]$, such that the derivative $\tau'(t) = d\tau/dt$ belongs to the class $H(\beta)$ on $[a, b]$ and does not vanish on the segment. Hence, in accordance with what was previously said, the function

$$\omega(t, t_0) = \frac{t - t_0}{\tau(t) - \tau(t_0)}$$

belongs to the class $H(\beta)$ with respect to both variables on $[a, b]$ and does not vanish on the segment.

Let us take on segment $[a, b]$ the sets $E = \{t_k, k = 1, \dots, n\}$ and $E_0 = \{t_{0k}, k = 1, \dots, n\}$ forming a canonic division of the segment with the step h. Then one may state that the sets $\tilde{E} = \{\tau_k = \tau(t_k), k = 1, \dots, n\}$ and $\tilde{E}_0 = \{\tau_{0k} = \tau(t_{0k}), k = 1, \dots, n\}$ form a canonic division of curve L with the step h. We denote $\bar{a} = \tau(a)$, $\bar{b} = \tau(b)$, and $S_n(\tau_{0j}) = \Sigma_{k=1}^{n} ((\varphi(\tau_k)\Delta\tau_k)/(\tau_k - \tau_{0j}))$, where $\Delta\tau_k = \tau_k(t_{k+1}) - \tau(t_k)$.
The following theorem may be proved.

Theorem 1.4.1. *Let* $\varphi(\tau)$ *belong to the class* H^* *on an open Lyapunov curve* L *and the sets* $\tilde{E} = \{\tau_k, k = 1, \dots, n\}$ *and* $\tilde{E}_0 = \{\tau_{0k}, k = 1, \dots, n\}$ *form a canonic division of the curve with the step* h. *Then the inequality*

$$|I(\tau_{0j}) - S_n(\tau_{0j}) \leq \theta(\tau_{0j}), \qquad j = 0, 1, \dots, n, \qquad (1.4.1)$$

holds where the quantity $\theta(\tau_{0j})$ *satisfies Inequalities* (1.3.8) *and* (1.3.9) *and is determined by Equation* (1.3.14) *if* $\eta(t_{0j}, \nu, \mu)$ *is substituted by the function* $\eta(\tau_{0j}, \nu, \mu) = |\tau_{0j} - \bar{a}|^{-\nu}|\bar{b} - \tau_{0j}|^{-\mu}$, *and* α *by the number* $\lambda = \min(\alpha, \nu\beta, \mu\beta)$.

The theorem can be proved by using the formula for substitution of the variable entering the singular integral and superposing functions of the class H (Muskhelishvili 1952).

If $\nu = 0$ or $\mu = 0$, then for the quantity $\theta(\tau_{0j})$ entering Inequality (1.4.1) one has to make the changes mentioned in Note 1.3.1 with respect to $\theta(t_{0j})$ entering Inequality (1.3.14).

If only the points τ_{0j} lying near, say, the end point $\bar{a}$ of the curve L are considered, then Inequality (1.4.1) may be corrected in the following way:

$$\left| \int_L \frac{\varphi(\tau)\, d\tau}{\tau - \tau_{0j}} - \sum_{k=1}^{n} \frac{\varphi(\tau_k)\, \Delta\tau_k}{\tau_k - \tau_{0j}} \right|$$

$$\leq \left[\frac{h^\lambda}{|\tau_{0j} - \bar{A}|^\nu} + \frac{h^\nu}{|\tau_{0j} - \bar{a}|^{2\nu}} + \frac{h^{1-\nu}}{|\tau_{0j} - \bar{a}} \right.$$

$$\left. + \frac{h}{|\tau_{0j} - \bar{a}|^{1+\nu}} \right] O(|\ln h|). \tag{1.4.2}$$

2. Now let curve L be a piecewise Lyapunov curve containing angular nodes only and consisting of l Lyapunov curves $L_1, \ldots, L_l$. On each segment $[a_m, b_m]$ mapped onto a curve $L_m, m = 1, \ldots, l$, we choose a canonic division with step h_m, formed by sets $E_m = \{t_k, k = n_{m-1} + 1, \ldots, n_m\}$ and $E_{0m} = \{t_{0j}, j = n_{m-1} + 0, n_{m-1} + 1, \ldots, n_m\}$, $n_0 = 0$.[†] We denote

$$h = \max h_m.$$

In what follows it is supposed that $h/h_m \leq R < +\infty$. Then the quantity $h_p/h_m, p = 1, \ldots, l$ also remains limited for $h \to 0$. Next we denote

$$S(\tau_{0j}) = \sum_{k=1}^{n_l} \frac{\varphi(\tau_k)\, \Delta\tau_k}{\tau_k - \tau_{0j}}, \qquad j = 0, 1, \ldots, n_1, n_1 + 0, \ldots,$$

$$n_2, \ldots, n_l,$$

$$\Delta\tau_k = \tau_{k+1} - \tau_k, \qquad k = 1, \ldots, n_1, k \neq n_1, \ldots, n_l,$$

$$\Delta\tau_{n_m} = \bar{b}_m - \tau_{n_m}, \qquad m = 1, \ldots, l, \tau_k = \tau_m(t_k),$$

$$k = n_{m-1} + 1, \ldots, n_m, \quad \tau_{0j} = \tau_m(t_{0j}),$$

$$j = n_{m-1} + 0, \ldots, n_m, \tag{1.4.3}$$

[†] This means that subscript j numbering points of the set E_{0m} on the curve L_m acquires the values $0, 1, \ldots, n_m - n_{m-1}$.

where $\tau_m(t)$ is segment $[a_m, b_m]$ mapped onto the curve $L_m = \bar{a}_m\bar{b}_m$. Then the following theorem may be formulated.

Theorem 1.4.2. *Let function $\varphi(\tau)\epsilon H^*$ on curve L. Then*

$$|I(\tau_{0j}) - S(\tau_{0j})| \leq \theta(\tau_{0j}), \qquad j = 0, 1, \ldots, n_1, n_1 + 0, \ldots, n_l, \quad (1.4.4)$$

where quantity $\theta(\tau_{0j})$ satisfies the inequalities:
(1) *for all points τ_{0j} belonging to the curve L' that is a portion of curve L devoid of nodes together with their close neighborhoods,*

$$\theta(\tau_{0j}) \leq O(h^{\lambda_1}), \qquad 0 < \lambda_1 \leq 1; \quad (1.4.5)$$

(2) *for all points τ_{0j} lying in the vicinity of the node $\bar{a}$,*

$$\theta(\tau_{0j}) \leq \left[\frac{h^{\lambda_2}}{|\tau_{0j} - \bar{a}|^{\nu}} + \frac{h^{\nu}}{|\tau_{0j} - \bar{a}|^{2\nu}} + \frac{h^{l-\nu}}{|\tau_{0j} - \bar{a}|}\right] O(|\ln h|), \quad (1.4.6)$$

if function $\varphi(\tau)$ has the form

$$\varphi(\tau) = \frac{\psi(\tau)}{|\tau - \bar{a}|^{\nu}}, \quad (1.4.7)$$

where $\psi(\tau)\epsilon H(\beta), 0 < \nu < 1$.

If $\nu = 0$, then the right-hand side of Equation (1.4.6) *becomes*

$$\left(h^{\lambda_2} + \frac{h}{|\tau_{0j} - a|}\right) O(|\ln h|).$$

Finally we note that Equations (1.4.5) *and* (1.4.6) *imply that*

$$\sum_{j=0}^{n_l} \theta(\tau_{0j}) |\Delta\tau_{0j}| \leq O(h^{\lambda_3}), \qquad 0 < \lambda_3 \leq 1. \quad (1.4.8)$$

Proof. One can write

$$|I(\tau_{0j}) - S(\tau_{0j})| \leq \sum_{m=1}^{l} |I_m(\tau_{0j}) - S_m(\tau_{0j})|,$$

where

$$I_m(\tau_{0j}) = \int_{L_m} \frac{\varphi(\tau)\, d\tau}{\tau - \tau_{0j}}, \qquad S_m(\tau_{0j}) = \sum_{k=n_{m-1}+1}^{n_m} \frac{\varphi(\tau_k)\, \Delta\tau_k}{\tau_k - \tau_{0j}}.$$

Let us take the point $\tau_{0j} \epsilon L_p \cap L'$. Then for $m = p$, Inequality (1.4.1) may be applied to $|I_m(\tau_{0j}) - S_m(\tau_{0j})|$, and for $m \neq p$ the conventional formula of rectangles may be employed. Hence, Inequality (1.4.5) is true. Next we prove the validity of Inequality (1.4.6). Let $\bar{a}$ be the common end point of smooth curves $L_1, \dots, L_{l1}, l_1 \leq l$. For points τ_{0j} in the neighborhood of node $\bar{a}$, one may write

$$|I(\tau_{0j}) - S(\tau_{0j})| \leq \sum_{m=1}^{l_1} |I_m(\tau_{0j}) - S_m(\tau_{0j})| + O(h^\lambda), \qquad 0 < \lambda \leq 1.$$

Because all the curves $L_1, \dots, L_{l1}$ are Lyapunov curves, Inequality (1.4.2) is true for $|I_p(\tau_{0j}) - S_p(\tau_{0j})|$ for any $\tau_{0j} \epsilon L_p, 1 \leq p \leq l_1$. Therefore, one has to consider $|I_m(\tau_{0j}) - S_m(\tau_{0j})|$ for $m = 1, \dots, l_1, m \neq p$. We have

$$I_m(\tau_{0j}) - S_m(\tau_{0j})| \leq \left| \int_{\bar{a}}^{\bar{b}} \frac{\varphi(\tau) - \varphi(\tau_{0j})}{\tau - \tau_{0j}} d\tau - \sum_{k=n_{m-1}+1}^{n_m} \frac{\varphi(\tau_k) - \varphi(\tau_{0j})}{\tau_k - \tau_{0j}} \Delta\tau_k \right| + |\varphi(\tau_{0j})| \left| \int_{\bar{a}}^{\bar{b}} \frac{d\tau}{\tau - \tau_{0j}} - \sum_{k=n_{m-1}+1}^{n_m} \frac{\Delta\tau_k}{\tau_k - \tau_{0j}} \right|,$$

where $\bar{b}$ is the other end of curve L_m and numeration of the points τ_k of the curve may be done in such a way that point $\bar{a}$ is the starting point on the curve L_m (in the direction of counting).

Muskhelishvili (1952) showed that if function $\psi(\tau)$ is uniquely defined on L and belongs to the class $H(\mu)$ on each of the smooth arcs converging at the end point $\bar{a}$, then the function belongs to the class $H(\mu)$ on L throughout the neighborhood of point $\bar{a}$ (if node $\bar{a}$ is a corner node of the curve L). Hence, the function $|\bar{a} - \tau|^\nu, 0 < \nu \leq 1$, belongs to the class $H(\nu)$ on the entire curve under consideration. Taking into account Inequality (1.1.12), one can reduce the estimate for $|I_m(\tau_{0j}) - S_m(\tau_{0j})|, m \neq p$, to estimates of integrals of the form

$$\int_a^b \frac{dt}{(t-a)^\nu (t - t_{0j})} \tag{1.4.9}$$

for $t_{0j} < a < b$. Having evaluated the latter integrals, one may return to the variable τ with the help of Inequality (1.1.2).

Gakhov (1977) showed that for the function

$$K(x,a) = \frac{(b-x)^{1-\mu}}{\pi} \int_a^b \frac{dt}{(b-t)^{1-\mu}(t-a)^{\mu}(t-x)}, \quad (1.4.10)$$

the following representation is valid:

$$K(x,a) = \frac{1}{(a-x)^{\mu} \sin \mu\pi}, \quad (1.4.11)$$

for $x < a$ and $0 < \mu < 1$. Hence,

$$\int_a^b \frac{dt}{(t-a)^{\nu}(t-t_{0j})} \leq O\left(\frac{1}{|t_{0j}-a|^{\nu}}\right) \quad (1.4.12)$$

for $t_{0j} < a, 0 < \nu < 1$, and

$$\int_{a+h}^b \frac{dt}{(t-a)^{2\nu}(t-t_{0j})} \leq \begin{cases} O\left(\dfrac{|\ln h|}{a-t_{0j}}\right), & 2\nu = 1, \\ O\left(\dfrac{h^{1-2\nu}}{a-t_{0j}}\right), & 1 < 2\nu < 2. \end{cases} \quad (1.4.13)$$

Inequalities (1.4.12) and (1.4.13) considered together with what was said when providing Inequality (1.3.7), demonstrate the validity of Inequality (1.4.4). ■

Note that for a function $\varphi(\tau)$ belong to the class H^* on L, i.e., for a function of the form

$$\varphi(\tau) = \psi(\tau)/P_L^{\nu}(\tau)$$

(see (1.1.5)), one can always assume that $\psi(\tau)$ vanishes at the nodes of a curve L and belongs to the class H in the vicinity of the nodes.

Also note that all the additional statements formulated upon proving Theorem 1.3.2 dealing with the choice of the grids E and E_0 and the function $\varphi(t,\tau)$ depending on a parameter (see (1.3.16)) can be generalized onto the case under consideration in the most natural way.

Let us now single out two special cases of Theorem 1.4.2 that will be of use for what follows.

Theorem 1.4.3. *Let curve L be a union of segments $[a, q]$ and $[q, b]$; let $\varphi(t)$ also belong to the class H^* on L ($a < q < b$). The sets $E = \{t_k, k = 1, \dots, n\}$ and $E_0 = \{t_{0k}, k = 1, \dots, n\}$ forming the division of segment $[a, b]$ with step h will be selected in such a way that point q lies halfway between the nearest points belonging to the sets E and E_0. Then*

$$\left| \int_a^b \frac{\varphi(t)\,dt}{t - t_{0j}} - \sum_{k=1}^{n} \frac{\varphi(t_k)h}{t_k - t_{0j}} \right| \le \theta(t_{0j}), \qquad j = 0, 1, \dots, n,$$

where the quantity $\theta(t_{0j})$ satisfies the inequalities:
(1) $\theta(t_{0j}) \le O_\delta(h^{\lambda_1}), 0 < \lambda_1 \le 1$, *for points* $t_{0j} \epsilon [a + \delta, q - \delta] \cup [q + \delta, b - \delta]$;
(2)

$$\sum_{j=0}^{n} \theta(t_{0j})h \le O(h^{\lambda_2}), 0 < \lambda_2 \le 1.$$

Theorem 1.4.4. *Let L be either a union of nonintersecting segments $[a_m, b_m]$, $m = 1, \dots, l$, or a union of segments $[a_m, q_m], [q_m, b_m], q_m \epsilon (a_m, b_m), m = 1, \dots, l$. In the former case, the sets $E_m = \{t_k, k = n_{m-1} + 1, \dots, n_m\}$ and $E_{0m} = \{t_{0k}, k = n_{m-1} + 1, \dots, n_m\}$ form a canonic division of the segment $[a_m, b_m]$ with step $h_m, m = 1, \dots, l, n_0 = 0$; in the later case, the sets are chosen in such a way that $q_m \epsilon E_{0m}$. Then*

$$\left| \int_L \frac{\varphi(t)\,dt}{t - t_{0j}} - \sum_{k=1}^{n_l} \frac{\varphi(t_k)\,\Delta t_k}{t_k - t_{0j}} \right| \le \theta(t_{0j}),$$

$$j = 0, 1, \dots, n_1, n_1 + 0, \dots, n_l,$$

where $\Delta t_k, h_m, k = n_{m-1} + 1, \dots, n_m, m = 1, \dots, l$, and the quantity $\theta(t_{0j})$ satisfies the inequalities:
(1) $\theta(t_{0j}) \le O(h^{\lambda_1}), 0 < \lambda_1 \le 1$ *for all points* $t_{0j} \epsilon \bigcup_{m=1}^{l} [a_m + \delta, b_m - \delta]$, *in the former case and* $t_{0j} \epsilon \bigcup_{m=1}^{l} ([a_m + \delta, q_m - \delta] \cup [q_m + \delta, b_m - \delta])$ *in the latter case;*
(2) $\sum_{j=0}^{n_l} \theta(t_{0j})|\Delta t_{0j}| < O(h^{\lambda_2}), 0 < \lambda_2 \le 1$, *where* $\Delta t_{0j} = h_m, j = n_{n-1} + 0, \dots, n_m, h = \max_{m=1,\dots,l} h_m$, *and it is supposed that* $h/h_m \le R < +\infty$.

1.5. SINGULAR INTEGRALS WITH HILBERT'S KERNEL

Consider an integral of the form

$$I(\theta_0) = \int_0^{2\pi} \cot \frac{\theta - \theta_0}{2} \varphi_1(\theta)\, d\theta, \tag{1.5.1}$$

where $\varphi_1(\theta)$ is a function with the period equal to 2π.

From Muskhelishvili (1952) it is known that

$$\int_L \frac{\varphi(t)\, dt}{t - t_0} = \frac{1}{2} \int_0^{2\pi} \cot \frac{\theta - \theta_0}{2} \varphi_1(\theta)\, d\theta + \frac{i}{2} \int_0^{2\pi} \varphi_1(\theta)\, d\theta, \tag{1.5.2}$$

where L is a unit-radius circle centered at the origin of coordinates, $t = \exp(i\theta)$, t_0-$\exp(i\theta_0)$, $\varphi_1(\theta) = \varphi[\exp(i\theta)]$.

Let the function $\varphi(t)$ belong to the class H on L and the sets $E = \{t_k = \exp(i\theta_k), k = 1, \dots, n\}$ and $E_0 = \{t_{0k} = \exp(i\theta_{0k}), k = 1, \dots, n\}$ form a canonic division of L. Then

$$\sum_{k=1}^{n} \frac{\varphi(t_k)\, \Delta t_k}{t_k - t_{0j}} \sum_{k=1}^{n} \frac{\varphi(e^{i\theta_k}) e^{i\theta_k}(e^{i\Delta\theta_k} - 1)}{e^{i\theta_k} - e^{i\theta_{0j}}}$$

$$= \sum_{k=1}^{n} \frac{\varphi_1(\theta_k) e^{i(\theta_k - \theta_{0j})/2}}{e^{i(\theta_k - \theta_{0j})/2} - e^{-i(\theta_k - \theta_{0j})/2}} \left(-2\sin^2 \frac{\Delta\theta_k}{2} + i \sin \Delta\theta_k \right)$$

$$= \sum_{k=1}^{n} \varphi_1(\theta_k) \left(\cot \frac{\theta_k - \theta_{0j}}{2} + i \right) \left(i \sin^2 \frac{\Delta\theta_k}{2} + \frac{1}{2} \sin \Delta\theta_k \right),$$

where $\Delta\theta_k = 2\pi/n$.

From Taylor's formula one gets

$$\sin \Delta\theta_k = \Delta\theta_k + O\big((\Delta\theta_k)^3\big).$$

Hence, we arrive at the equality

$$\sum_{k=1}^{n} \frac{\varphi(t_k)\, \Delta t_k}{t_k - t_{0j}} = \frac{1}{2} \sum_{k=1}^{n} \frac{\varphi_1(\theta_k)\, \Delta\theta_k}{\tan(\theta_k - \theta_{0j})/2}$$

$$+ \frac{i}{2} \sum_{k=1}^{n} \varphi_1(\theta_k)\, \Delta\theta_k + O\left(\frac{1}{n}\right) \tag{1.5.3}$$

Because the left-hand side of this equality well approximates the left-hand side of Equality (1.5.2), and $\sum_{k=1}^{n} \varphi_1(\theta_k)\,\Delta\theta_k$ is a good approximation for the integral $\int_0^{2\pi} \varphi_1(\theta)\,d\theta$, the following is true.

Theorem 1.5.1. *Let a function* $\varphi_1(\theta) \epsilon H$ *on segment* $[0, 2\pi]$ *and its period be equal to* 2π. *Let the points* $E = \{\theta_k, k = 1, \dots, n\}$ *and* $E_0 = \{\theta_{0k}, k = 1, \dots, n\}$ *be chosen on segment* $[0, 2\pi]$ *in the following way:* $\theta_{k+1} - \theta_k = 2\pi/n = h$, $k = 1, \dots, n-1$, $\theta_1 + 2\pi - \theta_n = h$, $\theta_{0k} = \theta_k + h/2$, $k = 1, \dots, n$, *i.e., points* $t_k = \exp(i\theta_k)$ *and* $t_{0k} = \exp(i\theta_{0k})$, $k = 1, \dots, n$, *form a canonic division of L. Then the inequality*

$$|I(\theta_{0j}) - S_n(\theta_{0j})| \leq O(n^{-\lambda} \ln n), \tag{1.5.4}$$

is true, where $S_n(\theta_{0j}) = \sum_{k=1}^{n} \cot((\theta_k - \theta_{0j})/2)\varphi_1(\theta_k)\,\Delta\theta_k$.

Note that functions $\varphi(t)$ and $\varphi_1(\theta)$ belong to a class $H(\lambda)$, each on its set.

1.6. UNIFICATION OF QUADRATURE AND DIFFERENCE FORMULAS

In aerodynamics (Bisplinghoff, Ashley, and Halfman 1955, Ashley and Landal 1967) one often has to deal with the Prandtl integrodifferential equation, which contains the integral

$$I(t_0) = \int_a^b \frac{\gamma'(t)\,dt}{t - t_0}, \tag{1.6.1}$$

where the function $\gamma'(t)$ belongs to the class H^* on $[a, b]$, $t_0 \in (a, b)$, i.e., $\gamma(t)$ belongs to the class H_1^* on $[a, b]$.

If $\gamma'(t) \epsilon H^*$ on $[a, b]$, then the integral

$$I_{(2)}(t_0) = \int_a^b \frac{\gamma(t)\,dt}{(t - t_0)^2} \tag{1.6.2}$$

may be reduced to integral (1.6.1). Here integral (1.6.2) is considered in the sense of Hadamard's finite value,

$$I_{(2)}(t_0) = \lim_{\epsilon \to 0} \left(\int_{L_*} \frac{\gamma(t)\,dt}{(t - t_0)^2} + \frac{2\gamma(t_0)}{\epsilon} \right), \tag{1.6.3}$$

where $L_* = L \setminus Q(\epsilon, t_0)$ and $Q(\epsilon, t_0) = (t_0 - \epsilon, t_0 + \epsilon)$.

From (1.6.3) we deduce that

$$I_{(2)}(t_0) = \frac{\gamma(a)}{a - t_0} - \frac{\gamma(b)}{b - t_0} + \int_a^b \frac{\gamma'(t)\,dt}{t - t_0}. \tag{1.6.4}$$

Let us next construct a quadrature-difference formula for approximate evaluation of integral (1.6.2). Let the sets $E = \{t_k, k = 1, \ldots, n\}$ and $E_0 = \{t_{0k}, k = 1, \ldots, n\}$ form a canonic division of segment $[a, b]$ with the step h. According to Section 1.3, it is natural to consider the quadrature formula

$$S(t_{0j}) = \sum_{k=1}^{n} \frac{\gamma'(t_k)h}{t_k - t_{0j}} - \frac{\gamma(b)}{b - t_{0j}} + \frac{\gamma(a)}{a - t_{0j}}, \qquad j = 0, 1, \ldots, n, \tag{1.6.5}$$

when considering integral (1.6.2).

Convergence of the latter quadrature formula to integral (1.6.1), and hence to integral (1.6.2), was analyzed in Section 1.3.

Let us continue by constructing the following quadrature sum for integral (1.6.2):

$$S_{(2)}(t_{0j}) = \sum_{k=0}^{n} \gamma(t_{0k}) \left[\frac{1}{t_k - t_{0j}} - \frac{1}{t_{k+1} - t_{0j}} \right], \qquad j = 0, 1, \ldots, n, \tag{1.6.6}$$

where $t_0 = a$, $t_{n+1} = b$. The formula may be derived as follows:

$$\int_a^b \frac{\gamma(t)\,dt}{(t - t_{0j})^2} = \sum_{k=0}^{n} \int_{t_k}^{t_{k+1}} \frac{\gamma(t)\,dt}{(t - t_{0j})^2} \approx \sum_{k=0}^{n} \gamma(t_{0k}) \int_{t_k}^{t_{k+1}} \frac{dt}{(t - t_{0j})^2}$$

$$= \sum_{k=0}^{n} \gamma(t_{0k}) \left[\frac{1}{t_k - t_{0j}} - \frac{1}{t_{k+1} - t_{0j}} \right].$$

The results obtained in Section 1.3 and Formula (1.6.4) allow us to formulate the following theorem.

Theorem 1.6.1. *Let a function $\varphi(t) \epsilon H_1^*$ on $[a, b]$ and points $t_0 = a, t_1, \ldots, t_n, t_{n+1} = b$ divide segment $[a, b]$ into $n + 1$ equal parts; let t_{0k} be the middle segment $[t_k, t_{k+1}]$, $k = 1, \ldots, n$. Then the inequality*

$$|I_{(2)}(t_{0j}) - S_{(2)}(t_{0j})| \leq \theta(t_{0j}) \tag{1.6.7}$$

holds for any $j = 0, 1, \ldots, n$, *where* $\theta(t_{0j})$ *has the same properties as in the Theorem* 1.3.2.

Note that quadrature formulas of the form $S_{(2)}(t_{0j})$ may also be constructed for the integral

$$I(t_0) = \int_a^b \frac{\gamma^{(m)}(t)\,dt}{t - t_0} \tag{1.6.8}$$

in the following way. Let

$$\int_a^b \frac{\gamma^{(m)}(t)\,dt}{t - t_0} \approx \sum_{k=1}^{n} \frac{\gamma^{(m)}(t_k)h}{t_k - t_{0j}} \approx \sum_{k=1}^{n} \frac{\gamma^{(m-1)}(t_{0k}) - \gamma^{(m-1)}(t_{0k-1})}{t_k - t_{0j}}.$$

In the latter sum we replace $\gamma^{(m-1)}(t_{0k})$ with the help of the difference scheme $(\gamma^{(m-2)}(t_{0k}) - \gamma^{(m-2)}(t_{0k-1})/h$. By continuing the procedure, one gets

$$\int_a^b \frac{\gamma^{(m)}(t)\,dt}{t - t_{0j}} \approx \sum_{k=1}^{n} \gamma(t_{0k})\,\omega_{kj}^{m}, \tag{1.6.9}$$

where $\gamma^{(m-1)}(t_{00}), \ldots, \gamma^{(0)}(t_{00}) \equiv \gamma(t_{00})$ enter ω_{kj}^{m}; in other words, are not substituted with the help of the difference scheme (see Matveev 1982).

2

Interpolation Quadrature Formulas for One-Dimensional Singular Integrals

2.1. SINGULAR INTEGRAL WITH HILBERT'S KERNEL

In order to construct quadrature formulas for the integral

$$I(\theta_0) = \int_{-\pi}^{\pi} \cot\frac{\theta - \theta_0}{2}\varphi(\theta)\,d\theta, \qquad \theta_0 \in [-\pi, \pi], \tag{2.1.1}$$

one has to recall the following facts.

In Luzin (1951) the inequalities

$$\int_0^{2\pi} \cot\frac{\theta - \theta_0}{2}\sin m\theta\,d\theta = 2\pi\cos m\theta_0, \qquad m = 1, 2, \ldots, \tag{2.1.2}$$

$$\int_0^{2\pi} \cot\frac{\theta - \theta_0}{2}\cos m\theta\,d\theta = -2\pi\sin m\theta_0, \qquad m = 0, 1, \ldots \tag{2.1.3}$$

were proved.

In addition, from trigonometry it is known that

$$\frac{\sin(2n+1)(\theta - \theta_k)/2}{2\sin(\theta - \theta_k)/2} = \frac{1}{2} + \cos(\theta - \theta_k) + \cdots + \cos n(\theta - \theta_k), \tag{2.1.4}$$

Hence, for any periodic function $\varphi(\theta)$, the trigonometric polynomial of the power n,

$$\varphi_n(\theta) = \sum_{k=0}^{2n} \frac{1}{2n+1}\varphi(\theta_k)\frac{\sin(2n+1)(\theta-\theta_k)/2}{\sin(\theta-\theta_k)/2}, \tag{2.1.5}$$

meets the condition

$$\varphi_n(\theta_k) = \varphi(\theta_k), \qquad k = 0,1,\ldots,2n, \tag{2.1.6}$$

if

$$\theta_k = \beta + \frac{2k}{2n+1}\pi, \qquad 0 \le \beta \le \frac{2\pi}{2n+1}, \qquad k = 0,1,\ldots,2n.$$

This can be proved by observing that

$$\sin\frac{2n+1}{2}\alpha = 0 \quad \text{for } \alpha = \frac{2k}{2n+1}\pi,\ k = 0,1,\ldots,2n,$$

$$\lim_{\alpha\to 0}\frac{\sin(2n+1)\alpha/2}{\sin\alpha/2} = 2n+1. \tag{2.1.7}$$

The quadrature formula for the integral (2.1.1) will be constructed in the following way:

$$\begin{aligned} S(\theta_0) &= \int_0^{2\pi}\cot\frac{\theta-\theta_0}{2}\varphi_n(\theta)\,d\theta \\ &= \int_0^{2\pi}\cot\frac{\theta-\theta_0}{2}\sum_{k=0}^{2n}\frac{1}{2n+1}\varphi(\theta_k)\frac{\sin(2n+1)(\theta-\theta_k)/2}{\sin(\theta-\theta_k)/2}\,d\theta \\ &= \frac{1}{2n+1}\sum_{k=0}^{2n}\varphi(\theta_k)\int_0^{2\pi}\cos\frac{\theta-\theta_0}{2}\,\frac{\sin(2n+1)(\theta-\theta_k)/2}{\sin(\theta-\theta_k)/2}\,d\theta. \end{aligned} \tag{2.1.8}$$

Hence, by employing (2.1.2)–(2.1.4), one gets

$$2\int_0^{2\pi}\cot\frac{\theta-\theta_0}{2}\,\frac{\sin(2n+1)(\theta-\theta_k)/2}{\sin(\theta-\theta_k)/2}\,d\theta$$

$$= 2\sum_{m=1}^{n} 2\pi \sin m(\theta_0-\theta_k)$$

$$= 2\pi\left[\cot\frac{\theta_k-\theta_0}{2} - \frac{\cos(2n+1)(\theta_k-\theta_0)/2}{\sin(\theta_k-\theta_0)/2}\right]. \qquad (2.1.9)$$

At the latter stage we have used the equality

$$\frac{1}{2} + \cos\theta + \cdots + \cos n\theta + i[\sin\theta + \cdots + \sin n\theta]$$

$$= \frac{1}{2} + e^{i\theta} + \cdots e^{in\theta} = \frac{\sin(2n+1)\theta/2}{2\sin\theta/2}$$

$$+\frac{i}{2}\left[\cos\frac{\theta}{2} - \frac{\cos(2n+1)\theta/2}{\sin\theta/2}\right]. \qquad (2.1.10)$$

By (2.1.9) the formula for $S(\theta_0)$ may be rewritten in the form

$$S(\theta_0) = \sum_{k=0}^{2n}\varphi(\theta_k)\left[\cos\frac{\theta_k-\theta_0}{2} - \frac{\cos(2n+1)(\theta_k-\theta_0)/2}{\sin(\theta_k-\theta_0)/2}\right]\frac{2\pi}{2n+1}. \qquad (2.1.11)$$

Note that the latter quadrature formula is accurate for any trigonometric polynomial of degree n, because in this case $\varphi_n(\theta) \equiv \varphi(\theta)$, and Formulas (2.18) and (2.1.11) provide an accurate value of the integral $I(\theta_0)$.

Let a function $\varphi(\theta)$ belong to the class $H_r(\alpha)$ on $[0, 2\pi]$, i.e., $\varphi^{(r)}(\theta) \in H(\alpha)$ and is a periodic function. Let us represent it in the form of Fourier series

$$\varphi(\theta) = \frac{a_0}{2} + \sum_{k=1}^{\infty}(a_k\cos k\theta + b_k\sin k\theta).$$

Then function $I(\theta_0)$ has the following expansion into Fourier series (see Luzin 1951):

$$I(\theta_0) = 2\pi \sum_{k=1}^{\infty} (b_k \cos k\theta_0 - a_k \sin k\theta_0). \tag{2.1.12}$$

Let us denote by $\Phi_n(\theta)$ the sum of the first n terms of Fourier series for the function $\varphi(\theta)$ and by $\Phi_n(I(\theta_0))$, the corresponding sum for $I(\theta_0)$. Because the singular integral over a circle preserves smoothness of its density (Gakhov 1977), the integral $I(\theta_0)$ possesses, due to (1.5.2), the same properties of smoothness as the function $\varphi(\theta)$ does. Hence, in the case under consideration $I(\theta_0) \in H_r(\alpha)$ on $[0, 2\pi]$. We have

$$\begin{aligned}
I(\theta_0) - S(\theta_0) &= \int_0^{2\pi} \cot\frac{\theta - \theta_0}{2}[\varphi(\theta) - \varphi_n(\theta)]\,d\theta \\
&= \int_0^{2\pi} \cot\frac{\theta - \theta_0}{2}[(\varphi(\theta) - \Phi_n(\theta)) + (\Phi_n(\theta) - \varphi_n(\theta))]\,d\theta \\
&= I(\theta_0) - \Phi_n(I(\theta_0)) + \sum_{k=0}^{2n} [\Phi_n(\theta_k) - \varphi(\theta_k)] \\
&\quad \times \left[\cot\frac{\theta_k - \theta_0}{2} - \frac{\cos(2n+1)(\theta_k - \theta_0)/2}{\sin(\theta_k - \theta_0)/2}\right]\frac{2\pi}{2n+1}.
\end{aligned} \tag{2.1.13}$$

In Berezin and Zhidkov (1962) it is shown that in the case under consideration,

$$|\varphi(\theta) - \Phi_n(\theta)| \le (3 + \ln n)E_n \tag{2.1.14}$$

for any θ from $[0, 2\pi]$, where E_n is the best possible approximation of the function $\varphi(\theta)$ by polynomials of degree n, i.e., $E_n = \inf_{P_n(\theta)} \max_{\theta \in [0, 2\pi]} |\varphi(\theta) - P_n(\theta)|$, where $P_n(\theta)$ is an arbitrary polynomial of degree n and

$$E_n \le \frac{12^{r+1} m}{n^{r+\alpha}}, \tag{2.1.15}$$

where M is a constant in the Hölder condition for the function $\varphi^{(r)}(\theta)$. Because $I(\theta_0) \in H_r(\alpha)$, an inequality of the form of (2.1.14) holds for $|I(\theta_0) - \Phi_n(I(\theta_0))|$.

Thus, from (2.1.10) and (2.1.13)–(2.1.15) one gets

$$|I(\theta_0) - S(\theta_0)| \leq O\left(\frac{3 + \ln n + n}{n^{r+\alpha}}\right) \tag{2.1.16}$$

for any $\theta_0 \epsilon [0, 2\pi]$.

Let us consider Formula (2.1.11) at the points

$$\theta_{0k} = \begin{cases} \theta_k + \dfrac{\pi}{2n+1}, & k = 1, \ldots, 2n, \text{ if } 0 \leq \beta \leq \dfrac{\pi}{2n+1} \\ \theta_k - \dfrac{\pi}{2n+1}, & k = 1, \ldots, 2n, \text{ if } \dfrac{\pi}{2n+1} \leq \beta \leq \dfrac{2\pi}{2n+1} \end{cases} \tag{2.1.17}$$

In this case points θ_{0m}, $m = 0, 1, \ldots, 2n$, are roots of the function $\cos((2n+1)/2)(\theta_k - \theta_0)$, because

$$\cos\frac{2n+1}{2}(\theta_k - \theta_{0m}) = \cos\left[\frac{2n+1}{2}\,\frac{2(k-m)+1}{2n+1}\pi\right] = 0.$$

Also note that

$$\sin\frac{1}{2}(\theta_k - \theta_{0m}) = \sin\frac{2(k-m) \pm 1}{2(2n+1)}\pi \neq 0 \tag{2.1.18}$$

for any $m, k = 1, \ldots, 2n$.

Thus,

$$S(\theta_{0m}) = \sum_{k=0}^{2n} \cot\frac{\theta_k - \theta_{0m}}{2}\varphi(\theta_k)\frac{2\pi}{2n+1} \tag{2.1.19}$$

at the points θ_{0m}.

In this case Formula (2.1.16) becomes

$$|I(\theta_{0m}) - S(\theta_{0m})| \leq O\left(\frac{\ln n}{n^{r+\alpha}}\right), \tag{2.1.20}$$

because it can be readily shown that

$$\sum_{k=0}^{2n} \left| \cot \frac{\theta_{0m} - \theta_k}{2} \right| \frac{2\pi}{2n+1} = O(\ln n) \tag{2.1.21}$$

for $m = 0, 1, \ldots, 2n$.

2.2. SINGULAR INTEGRAL ON A CIRCLE

Let function $\varphi(t)$ belong to the class $H_r(\alpha)$ on the unit-radius circle L centered at the origin of the system of coordinates. Let us consider the polynomial

$$\varphi_n(t) = \frac{1}{2n+1} \sum_{k=0}^{2n} \varphi(t_k) \frac{t^{2n+1} - t_k^{2n+1}}{(t - t_k) t^n t_k^n}, \tag{2.2.1}$$

where points t_k divide the circle L into $2n + 1$ equal parts. By dividing a polynomial by a polynomial it can be readily shown that

$$\varphi_n(t_k) = \varphi(t_k), \qquad k = 0, 1, \ldots, 2n, \tag{2.2.2}$$

because

$$\frac{1}{2n+1} \frac{t^{2n+1} - t_k^{2n+1}}{(t - t_k) t^n t_k^n} = \begin{cases} 1, & t = t_k,\ k = 0, 1, \ldots, 2n, \\ 0, & t = t_m,\ m \neq k,\ m = 0, 1, \ldots, 2n \end{cases} \tag{2.2.3}$$

The latter equality also follows from the relationship

$$\frac{\sin(2n+1)0/2}{\sin \theta/2} = 1 + 2(\cos\theta + \cdots + \cos n\theta)$$

$$= e^{-in\theta} + e^{-i(n-1)\theta} + \cdots + e^{-i\theta}$$

$$+ 1 + e^{i\theta} + \cdots e^{in\theta} = \frac{T^{-n} + t^{n+1}}{1 - t},$$

$$t = e^{i\theta}, \tag{2.2.4}$$

and Formula (2.1.17).

From (2.2.4) we deduce that

$$\frac{\sin(2n+1)(\theta - \theta_k)/2}{\sin(\theta - \theta_k)/2} = \frac{t^{2n+1} - t_k^{2n+1}}{(t - t_k) t^n t_k^n}, \qquad t_k = e^{i\theta_k}. \tag{2.2.4a}$$

Next we consider the singular integral

$$I(t_0) = \int_L \frac{\varphi(t)\,dt}{t - t_0},$$

where $\varphi(t)\epsilon H_r(\alpha)$ on the circle L. We denote

$$S(t_0) = \int_L \frac{\varphi_n(t)\,dt}{t - t_0} = \frac{1}{2n+1} \sum_{k=0}^{2n} \varphi(t_k) \int_L \frac{t^{2n+1} - t_k^{2n+1}}{(t - t_k)t^n t_k^n} \frac{dt}{t - t_0}. \quad (2.2.5)$$

By employing the formula (Gakhov 1977)

$$\frac{1}{\pi i} \int_L \frac{t^n\,dt}{t - t_0} = \begin{cases} t_0^n, & n \geq 0, \\ -t_0^n, & n < 0, \end{cases} \quad (2.2.6)$$

where $t = \exp(i\theta)$, $t_0 = \exp(i\theta_0)$, and the evident identity

$$\frac{1}{(t - t_k)(t - t_0)} = \frac{1}{t_0 - t_k}\left(\frac{1}{t - t_0} - \frac{1}{t - t_k}\right), \quad (2.2.7)$$

we get

$$\int_L \frac{t^{2n+1} - t_k^{2n+1}}{(t - t_k)t^n t_k^n} \frac{dt}{t - t_0} = \frac{1}{t_k - t_0}\left[\int_L \frac{t^{2n+1} - t_k^{2n+1}}{t^n t_k^n} \frac{dt}{t - t_k} - \int_L \frac{t^{2n+1} - t_k^{2n+1}}{t^n t_k^n} \frac{dt}{t - t_0}\right] = \frac{1}{t_k - t_0} \times \left[\pi i \frac{t_k^{n+1} + t_k^{2n+1} t_k^{-n}}{t_k^n} - \pi i \frac{t_0^{2n+1} + t_k^{2n+1}}{t_0^n t_k^n}\right]. \quad (2.2.8)$$

Thus, the quadrature sum for the integral $I(t_0)$ has the form

$$S(t_0) = \sum_{k=0}^{2n} \frac{\varphi(t_k)}{t_k - t_0} \frac{\pi i}{2_n + 1}\left[2t_k - \frac{t_0^{2n+1} + t_k^{2n+1}}{t_0^n t_k^n}\right]. \quad (2.2.9)$$

Note that if $\varphi(t) \epsilon H_r(\alpha)$ on L, then the function $\varphi_1(\theta) = \varphi[\exp(i\theta)]$ also belongs to the class $H_r(\alpha)$. Thus, one has

$$\int_0^{2\pi} \frac{\sin(2n+1)(\theta-\theta_k)/2}{\sin(\theta-\theta_k)/2}\, d\theta = \int_0^{2\pi} (1 + 2\cos(\theta-\theta_k) + \cdots + 2\cos n(\theta-\theta_k))\, d\theta = 2\pi \tag{2.2.10}$$

and, hence,

$$\left| \int_0^{2\pi} \varphi_1(\theta)\, d\theta - \sum_{k=0}^{2n} \varphi_1(\theta_k) \frac{2\pi}{2n+1} \right| \leq O\left(\frac{1}{n^{r+\alpha}}\right). \tag{2.2.11}$$

By employing Formulas (1.5.2), (2.1.16), and (2.2.11), the following inequality may be shown to be true:

$$|I(t_0) - S(t_0)| \leq O\left(\frac{\ln n + n}{n^{r+\alpha}}\right). \tag{2.2.12}$$

Let us now find the roots of the function $(t_0^{2n+1} + t_k^{2n+1})/(t_0^n t_k^n (t_0 - t_k))$. This can be done by observing that

$$\frac{t_0^{2n+1} + t_k^{2n+1}}{t_0^n t_k^n (t_0 - t_k)} = i \frac{\cos(2n+1)(\theta_0 - \theta_k)/2}{\sin(\theta_0 - \theta_k)/2}. \tag{2.2.13}$$

Hence, the function has the roots

$$t_{0m} = \exp\left\{\theta_m + \frac{\pi}{2n+1}\right\}, \qquad m = 0, 1, \ldots, 2n;$$

in other words, t_{0k} divides the arc $\breve{t_k t_{k+1}}$ into two equal parts. Hence, we have for the chosen points t_{0m},

$$S(t_{0m}) = \sum_{k=0}^{2n} \frac{\varphi(t_k)}{t_k - t_{0m}} \frac{2\pi i t_k}{2n+1}, \qquad m = 0, 1, \ldots, 2n. \tag{2.2.14}$$

Then, by using the identity

$$\frac{it_k}{t_k - t_{0m}} = \frac{1}{2}\cot\frac{\theta_k - \theta_{0m}}{2} + \frac{1}{2}i, \tag{2.2.15}$$

one may write

$$S(t_{0m}) = \frac{1}{2}\left[\sum_{k=0}^{2n}\cot\frac{\theta_k - \theta_{0m}}{2}\varphi_1(\theta_k) + i\sum_{k=0}^{2n}\varphi_1(\theta_k)\right]\frac{2\pi}{2n+1}. \tag{2.2.16}$$

Finally, by employing Formulas (1.5.2), (2.1.20), and (2.2.11), one gets

$$|I(t_{0m}) - S(t_{0m})| < O\left(\frac{\ln n}{n^{r+\alpha}}\right), \qquad m = 0, 1, \ldots, 2n. \tag{2.2.17}$$

2.3. SINGULAR INTEGRAL ON A SEGMENT

In the present section we shall consider integrals of the form

$$I(\alpha, \beta, t_0) = \int_{-1}^{1}\frac{\omega(t)\psi(t)\,dt}{t - t_0}, \tag{2.3.1}$$

where $\omega(t) = (1-x)^{\alpha}(1+x)^{\beta}$, $0 < |\alpha|, |\beta| < 1$, and the function $\psi(t)$ belongs to the class $H_r(\lambda)$ on segment $[-1, 1]$, i.e., $\psi^{(r)}(t) \epsilon H(\lambda)$ on $[-1, 1]$.

Because in aerodynamics and some other applications one often comes across the case $\alpha, \beta = \pm\frac{1}{2}$, we shall consider the integrals in greater detail. Let us start by constructing quadrature formulas for the general form of integral (2.3.1) (Korneichuk 1964, Lifanov and Saakyan 1982, Savruk 1981, Stark 1971).

Let $P_n(t)$ be a polynomial of the degree n belonging to the system $\{P_m(x),\ m = 0, 1, \ldots\}$ of polynomials orthogonal on $[-1, 1]$ with the weight $\omega(x)$, where $\omega(x)$ is a positive function integrable on $[-1, 1]$. In this case integral (2.3.1) will be denoted by $I(t_0)$. The roots of the polynomial $P_n(t)$ will be denoted by t_i, $i = 1, \ldots, n$, and the polynomial

$$\psi_n(t) = \sum_{k=1}^{n}\psi(t_k)\frac{P_n(t)}{(t - t_k)P_n'(t_k)} \tag{2.3.2}$$

by $\psi_n(t)$. It is evident that the latter polynomial meets the condition

$$\psi_n(t_k) = \psi(t_k), \qquad k = 1, \ldots, n, \tag{2.3.3}$$

because

$$\lim_{t\to t_k} \frac{P_n(t)}{(t-t_k)P_n'(t_k)} = 1.$$

By $S(t_0)$ we shall denote the quadrature sum obtained in the following way:

$$S(t_0) = \int_{-1}^{1} \frac{\omega(t)\psi_n(t)}{t-t_0}\,dt = \sum_{k=1}^{n} \psi(t_k)\int_{-1}^{1} \frac{\omega(t)P_n(t)\,dt}{(t-t_0)(t-t_k)P_n'(t_k)}$$

$$= \sum_{k=1}^{n} \frac{\psi(t_k)}{t_k-t_0}\left[\frac{Q_n(t_k)}{P_n'(t_k)} - \frac{Q_n(t_0)}{P_n'(t_k)}\right]. \tag{2.3.4}$$

where

$$Q_n(t_0) = \int_{-1}^{1} \frac{\omega(t)P_n(t)\,dt}{t-t_0}. \tag{2.3.5}$$

Let t_{0m}, $m = 1,\ldots,R$, be roots of the function $Q_n(t_0)$. Then, the formula for $S(t_0)$ at the points becomes

$$S(t_{0m}) = \sum_{k=1}^{n} \frac{\psi(t_k)a_k}{t_k-t_{0m}}, \qquad m = 1,\ldots,R, \tag{2.3.6}$$

where

$$a_k = \frac{Q_n(t_k)}{P_n'(t_k)}, \qquad k = 1,\ldots,n. \tag{2.3.7}$$

In Korneichuk (1964) inequality

$$|I(t_{0m}) - S(t_{0m})| \leq \frac{2}{\pi}E_{n-1}\alpha(1), \tag{2.3.8}$$

was proved, where $\alpha(t) = \int_{-1}^{1}\omega(\tau)\,d\tau$ and E_{n-1} is the best approximation of the function $\psi'(t)$ obtainable with the help of polynomials of degree $n-1$. Hence, if function $\psi(t)\epsilon H_r(\alpha)$ on segment $[-1,1]$, then $E_{n-1} \leq O(n^{-r-\alpha+1})$; accurate values of the constants are presented in Kantorovich (1952). In Stark (1971) it is shown that if $\psi(t)$ is a polynomial of degree less or equal to $2n$, then $S(t_{0m}) = I(t_{0m})$. Next we present the concrete form of Formula (2.3.6) for the cases when $\alpha, \beta = \pm \frac{1}{2}$.

1. Let $\alpha = \beta = -\frac{1}{2}$, i.e.,

$$\omega(t) = (1 - t^2)^{-1/2}. \tag{2.3.9}$$

The polynomials orthogonal with the weight on $[-1, 1]$ are Chebyshev polynomials of the first kind, i.e.,

$$P_n(t) = T_n(t) = \cos(n \arccos t). \tag{2.3.10}$$

Roots of polynomial $T_n(t)$ are nodes of the quadrature formula (2.3.6) and have the form

$$t_k = \cos\frac{2k - 1}{2n}\pi, \qquad k = 1, \dots, n, \tag{2.3.11}$$

and the functions $Q_n(t)$ are

$$Q_n(t) = \pi U_{n-1}(t),$$

$$U_{n-1}(t) = \frac{\sin(n \arccos t)}{\sin(\arccos t)}, \tag{2.3.12}$$

the latter expressions being Chebyshev polynomials of the second kind. In this case, roots of function $Q(t)$ have the form

$$t_{0m} = \cos m\pi/n, \qquad m = 1, \dots, n - 1. \tag{2.3.13}$$

Because

$$T_n'(t) = nU_{n-1}(t), \qquad n = 1, 2, \dots, \tag{2.3.14}$$

it follows from (2.3.7) that

$$a_k = \pi/n, \qquad k = 1, 2, \dots, n. \tag{2.3.15}$$

2. Let $\alpha = \beta = \frac{1}{2}$, i.e.,

$$\omega(t) = \sqrt{1 - t^2}. \tag{2.3.16}$$

Then $P_n(t)$ are Chebyshev polynomials of the second kind

$$P_n(t) = \frac{\sin[(n+1)\arccos t]}{\sin(\arccos t)} \tag{2.3.17}$$

whose roots are given by

$$t_k = \cos\frac{k}{n+1}\pi, \qquad k = 1, \ldots, n. \tag{2.3.18}$$

For the function $Q_n(t)$, one gets

$$Q_n(\cos\theta_0) = \int_0^{\pi} \frac{\sin\theta \sin(n+1)\theta \, d\theta}{\cos\theta - \cos\theta_0}$$

$$= \pi \frac{\sin n\theta_0 - \sin(n+2)\theta_0}{\sin\theta_0} = \pi\cos(n+1)\theta_0. \tag{2.3.19}$$

Thus,

$$Q_n(t_0) = \pi T_{n+1}(t_0) \tag{2.3.20}$$

and, hence,

$$t_{0m} = \cos\frac{2m-1}{2(n+1)}\pi, \qquad m = 1, \ldots, n+1. \tag{2.3.21}$$

Then by (2.3.7) we have

$$a_k = \frac{\pi}{n+1}\sin^2\frac{k}{n+1}\pi, \qquad k = 1, \ldots, n. \tag{2.3.22}$$

3. Finally, consider the case $\alpha = \frac{1}{2}$, $\beta = -\frac{1}{2}$, i.e.,

$$\omega(t) = \sqrt{\frac{1-x}{1+x}}. \tag{2.3.23}$$

In this case polynomials $P_n(t)$ are given by

$$P_n(t) = \frac{T_{n+1}(t) - T_n(t)}{1-t} \tag{2.3.24}$$

and the roots are given by

$$t_k = \cos\frac{2k}{2n+1}\pi, \qquad k = 1, 2, \ldots, n. \tag{2.3.25}$$

For the function $Q_n(t)$ one gets

$$Q_n(t) = \pi|U_n(t) - U_{n-1}(t)|. \tag{2.3.26}$$

The roots of the function correspond to the points

$$t_{0m} = \cos\frac{2m-1}{2n+1}\pi, \qquad m = 1, \ldots, n. \tag{2.3.27}$$

For the coefficients a_k one has

$$a_k = \frac{4\pi}{2n+1}\sin^2\frac{k}{2n+1}\pi, \qquad k = 1, \ldots, n. \tag{2.3.28}$$

Note that inequality

$$|I(t_0) - S(t_0)| \le \frac{C_r \ln n}{\left(\sqrt{1-\eta^2} + |t_0|/n\right)n^{r+\alpha}},$$

was proved in Sanikidze (1970) for case 1, if function $\psi(t) \epsilon H_r(\alpha)$. Here $\eta = (1 + |t_0|)/2$, and the constant C_r is independent of n, t, and ψ. A more convenient estimate,

$$|I(t_0) - S(t_0)| \le O\left(\frac{\ln n}{n^{r+\alpha}}\right)\frac{1}{\sqrt{1-t_0^2}},$$

was obtained for the case in Sheshko (1976).

3

Quadrature Formulas for Multiple and Multidimensional Singular Integrals

3.1. MULTIPLE CAUCHY INTEGRALS

Let $L_1, \ldots, L_n$ be piecewise smooth plane curves. Following Gakhov (1977), their topological product will be referred to as $L = L_1 \times L_2 \times \cdots \times L_n$. Let us consider a function $\varphi(t) = \varphi(t^1, \ldots, t^n)$ defined on the frame L. A point $t = (t^1, \ldots, t^n)$ will be called an internal point of the frame L if point t^k is not a *nodal point* of curve L_k, $k = 1, \ldots, n$. Let point t_0 be an internal point of the frame L. In the plane of curve L_k we draw a circle of radius ϵ_k centered at point t_0^k and denote by l_k the part of curve L_k lying inside the circle (ϵ_k is supposed to be so small that l_k is a smooth unclosed arc).

The limit

$$\Phi(t_0) = \lim_{\epsilon_1, \ldots, \epsilon_n \to 0} \int_{L_*} \frac{\varphi(t^1, \ldots, t^n)\, dt^1 \cdots dt^n}{(t^1 - t_0^1) \cdots (t^n - t_0^n)},$$

$$L_* = (L_1 \setminus l_1) \times \cdots \times (L_n \setminus l_n), \tag{3.1.1}$$

where $\epsilon_1, \ldots, \epsilon_n$ tend to zero irrespective of each other, will be called a *Cauchy integral* (a multiple singular integral) of the function $\varphi(t^1, \ldots, t^n)$ over frame L at point $t_0 = (t_0^1, \ldots, t_0^n)$.

The limit will be denoted by

$$\Phi(t_0) = \int_{L_1\times\cdots\times L_n} \frac{\varphi(t^1,\ldots,t^n)\,dt^1\cdots dt^n}{(t^1-t_0^1)\cdots(t^n-t_0^n)} = \int_L \frac{\varphi(t)\,dt}{((t-t_0))},$$

where

$$((t-t_0)) = (t^1-t_0^1)\cdot\cdots\cdot(t^n-t_0^n), \qquad dt = dt^1\cdots dt^n. \quad (3.1.1a)$$

Similarly to a one-dimensional singular integral, it can be shown that $\Phi(t_0)$ exists at point t_0 if function $\varphi(t) = \varphi(t^1,\ldots,t^n)$ meets the condition H within a certain neighborhood of point $t_0 = (t_0^1,\ldots,t_0^n)$ belonging to the frame L. On the other hand, Equation (3.1.1) allows us to consider the multiple Cauchy integral as an iterated integral, i.e., one may write

$$\int_{L_1\times\cdots\times L_n} \frac{\varphi(t^1,\ldots,t^n)\,dt^1\cdots dt^n}{(t^1-t_0^1)\cdots(t^n-t_0^n)}$$

$$= \int_{L_1} \frac{dt^1}{t^1-t_0^1}\left(\cdots\left(\int_{L_n} \frac{\varphi(t^1,\ldots,t^n)\,dt^n}{t^n-t_0^n}\right)\cdots\right). \quad (3.1.2)$$

Let us demonstrate this in the example of two variables. Let function $\varphi(1^t, t^2)$ meet the condition $H(\mu_1,\mu_2)$ on $L_1\times L_2$. We denote (Gakhov 1977)

$$\varphi_1 = \varphi(t^1,t_0^2) - \varphi(t_0^1,t_0^2), \qquad \varphi_2 = \psi(t_0^1,t^2) - \varphi(t_0^1,t_0^2),$$

$$\varphi_{1\,2} = \varphi(t^1,t^2) - \varphi(t^1,t_0^2) - \varphi(t_0^1,t^2) + \varphi(t_0^1,t_0^2).$$

Then the following formulas are valid:

$$|\varphi_1| \le A_1|t^1-t_0^1|^{\mu_1}, \qquad |\varphi_1| \le A_2|t^2-t_0^2|^{\mu_2},$$

$$|\varphi_{1\,2}| \le 2A_1|t^1-t_0^1|^{\mu_1}, \quad |\varphi_{1\,2}| \le 2A_1|t^2-t_0^2|^{\mu_2}, \quad (3.1.3)$$

and hence,

$$|\varphi_{1\,2}| < A_{1\,2}|t^1-t_0^1|^{\alpha\mu_1}|t^2-t_0^2|^{(1-\alpha)\mu_1}, \qquad 0\le\alpha\le 1. \quad (3.1.4)$$

Note also that the

$$\varphi(t^1,t^2) - \varphi(t_0^1,t_0^2) = \varphi_1 + \varphi_2 + \varphi_{1\,2}$$

holds. Therefore

$$\iint_{L_1 \times L_2} \frac{\varphi(t^1, t^2)\, dt^1\, dt^2}{(t^1 - t_0^1)(t^2 - t_0^2)} = \int_{L_2} \frac{dt^2}{t^2 - t_0^2} \int_{L_1} \frac{\varphi_1\, dt^1}{t^1 - t_0^1}$$

$$+ \int_{L_1} \frac{dt^1}{t^1 - t_0^1} \int_{L} \frac{\varphi_2\, dt^2}{t^2 - t_0^2}$$

$$+ \iint_{L_1 \times L_2} \frac{\varphi_{1\,2}\, dt^1\, dt^2}{(t^1 - t_0^1)(t^2 - t_0^2)}$$

$$+ \varphi(t_0^1, t_0^2) \int_{L_1} \frac{dt^1}{t^1 - t_0^1} \int_{L_2} \frac{dt^2}{t^2 - t_0^2}. \qquad (3.1.5)$$

Here all the right-hand-side integrals exist either as conventional improper integrals or as one-dimensional Cauchy integrals.

Formula (3.1.2) allows us to transfer many properties of one-dimensional Cauchy integrals onto multiple Cauchy integrals. We shall start by considering one-dimensional singular integrals whose density depends on parameter τ,

$$\Phi(t_0, \tau) = \int_L \frac{\varphi(t, \tau)\, dt}{t - t_0}, \qquad (3.1.6)$$

where $\varphi(t, \tau)$ meets the condition $H(\mu)$ with respect to t on L and the condition $H(\nu)$ with respect to τ on a limited set of T.

The following theorem is valid.

Theorem 3.1.1. *Function $\Phi(t_0, \tau)$ meets condition $H(\mu, \nu - \epsilon)$ on the set $L' \times T$ where L' is a smooth part of the line L having no common ends with the latter, and $\epsilon > 0$ is a however small number.*

Proof. Muskhelishvili (1952) has shown that $\Phi(t_0, \tau)$ meets the condition $H(\mu)$ on L' for $\tau \in T$. Thus, we have to show only that the function meets the condition $H(\nu - \epsilon)$ with respect to τ for any $t_0 \in L$.

Consider the difference

$$\Phi(t_0, \tau + h) - \Phi(t_0, \tau) \equiv \int_L \frac{\varphi(t, \tau + h) - \varphi(t, \tau)}{t - t_0}\, dt.$$

Let us introduce the representation (Gakhov 1977)

$$\varphi(t,\tau+h)-\varphi(t,\tau)=\varphi_{12}+\varphi_2,$$

$$\varphi_{12}=[\varphi(t,\tau+h)-\varphi(t_0,\tau+h)]$$

$$-[\varphi(t,\tau)-\varphi(t_0,\tau)],$$

$$\varphi_2=\varphi(t_0,\tau+h)-\varphi(t_0,\tau).$$

Then

$$\Phi(t_0,\tau+h)-\Phi(t_0,\tau)=\int_L\frac{\varphi_{12}}{t-t_0}dt+\varphi_2\int_L\frac{dt}{t-t_0}=\Phi_{12}+\Phi_2.$$

Because $\varphi(t,\tau)$ meets condition $H(\mu,\nu)$ on $L\times T$, and $\int_L dt/(t-t_0)$ is a limited quantity on L',

$$|\Phi_2|\le O(|h|^\nu). \tag{3.1.7}$$

On the other hand, according to Gakhov (1977),

$$|\varphi_{12}|\le 2A|t-t_0|^\mu, \qquad |\varphi_{12}|\le 2B|h|^\nu.$$

From the latter inequalities it follows that

$$|\varphi_{12}|\le A_{12}|t-t_0|^{\alpha\mu}|h|^{(1-\alpha)\nu}, \qquad \alpha\in[0,1].$$

If $\alpha=\epsilon/\nu$, then

$$|\varphi_{12}|\le A_{12}|t-t_0|^{\epsilon\mu/\nu}|h|^{\nu-\epsilon},$$

where $\epsilon>0$ is a however small number.

The latter inequality for Φ_{12} implies that

$$|\Phi_{12}|\le\frac{A_{12}}{2\pi}|h|^{\nu-\epsilon}\int_L\frac{|dt|}{|t-t_0|^{1-\epsilon\mu/\nu}}\le C|h|^{\nu-\epsilon}\frac{1}{\epsilon}.$$

Because α lies between 0 and 1, $0<\epsilon\le\nu$. Now let $|h|$ be such that $\epsilon=-1/\ln|h|\le\nu$, that is $|h|\le\exp(-\nu^{-1})$. Then

$$|\Phi_{12}|\le C_1|h|^\nu|\ln|h||. \tag{3.1.8}$$

Thus, the validity of our statement concerning function $\Phi(t_0, \tau)$ follows from estimates (3.1.7) and (3.1.8). ■

From Theorem 3.1.1 we get the following theorem.

Theorem 3.1.2. *Let L be a smooth closed curve and function $\varphi(t, \tau_1, \ldots, \tau_m)$ meet the condition $H(\mu, \nu_1, \ldots, \nu_m)$ on the set $L_* = L \times T_1 \times \cdots \times T_m$, where T_k is a limited set of values of the variable τ_k, $k = 1, \ldots, m$. Then the function*

$$\Phi(t_0, \tau_1, \ldots, \tau_m) = \int_L \frac{\varphi(t, \tau 1, \ldots, \tau_m)\, dt}{t - t_0}$$

meets the condition $H(\mu, \nu_1 - \epsilon_1, \ldots, \nu_m - \epsilon_m)$ on L_*, where ϵ_k are however small positive numbers $k = 1, \ldots, m$.

If L is a segment, then the following theorem is true.

Theorem 3.1.3. *Let function $\varphi(t, \tau)$ be of the form*

$$\varphi(t, \tau) = \frac{\varphi^*(t, \tau)}{(t - a)^{\nu}}, \tag{3.1.9}$$

where $0 < \nu < 1$ and $\varphi^*(t, \tau) \in H(\alpha, \beta)$ on the set $[a, b] \times T$, where T is a limited set of values of τ. Then the function

$$\Omega(t_0, \tau) = (t_0 - a)^{\nu} \int_a^b \frac{\varphi^*(t, \tau)\, dt}{(t - a)^{\nu}(t_0 - t)} \tag{3.1.10}$$

belongs to the class $H(\lambda, \beta - \epsilon)$ for all points $(t_0, \tau) \in [a, c] \times T$, where $a < c < b$, $0 < \lambda < 1$, and $\epsilon > 0$ is a however small number.

Proof. Similarly to Theorem 3.1.1, it suffices to consider the difference $\Omega(t_0, \tau + h) - \Omega(t_0, \tau)$ presented, in analogy to function $\Phi(t_0, \tau)$, in the form $\Omega_{12} + \Omega_2$. Because $\varphi^*(t, \tau) \in H(\alpha, \beta)$ on $[a, b] \times T$, and the function

$$\Omega_3 = (t_0 - a)\nu \int_a^b \frac{dt}{(t - a)^{\nu}(t_0 - t)}$$

belongs to the class H on $[a, c]$ (see Muskhelishvili 1952), $|\Omega_2| \leq B|h|^{\beta}$, $t_0 \in [a, c]$. Similarly, we deduce for Ω_{12} that $|\Omega_{12}| \leq B_1 |h|^{\beta - \epsilon} \epsilon^{-1}$ for $0 < \epsilon \leq \beta$.

Consideration of the latter two inequalities together with the fact that $\Omega(t_0, \tau) \in H$ with respect to t_0 and uniformly with respect to $\tau \in T$ (see Muskhelishvili 1952) completes the proof of the theorem. ■

Note 3.1.1. Taking into account the results obtained in Muskhelishvili (1952), Theorem 3.1.3 may be generalized to the case of an arbitrary piecewise smooth curve L for which a is a node and $\nu = \nu_1 + \nu_{2^i}$, $0 < \nu_1 < 1$.

Let us now return to the multiple Cauchy integrals. The following theorem follows from Theorem 3.1.2.

Theorem 3.1.4. *Let function $\varphi(t^1, \ldots, t^n)$ meet the condition $H(\mu_1, \ldots, \mu_n)$ on an n-dimensional torus $L = L_1 \times \cdots \times L_n$, where L_k, $k = 1, \ldots, n$, are closed smooth curves. Then, singular integral $\Phi(t_0^1, t_0^2, \ldots, t_0^n)$ belongs to the class $H(\mu_1 - \epsilon, \ldots, \mu_n - \epsilon)$ on L where $\epsilon > 0$ is a however small number.*

We shall prove the theorem for the two-dimensional case. By using (3.1.2), $\Phi(t_0^1, t_0^2)$ may be presented in the form

$$\Phi(t_0^1, t_0^2) = \int_{L_1} \frac{\varphi(t^1, t_0^2)\, dt^1}{t^1 - t_0^1}, \qquad \psi(t^1, t_0^2) + \int_{L_2} \frac{\varphi(t^1, t^2)\, dt}{t^2 - t_0^2}.$$

Using Theorem 3.1.1 about singular integrals depending on a parameter, we deduce that $\psi(t^1, t_0^2) \in H(\mu_1 - \epsilon, \mu_2)$ and $\Phi(t_0^1, t_0^2 \in H(\mu_1 - \epsilon, \mu_2 - \epsilon)$.

If the function $\varphi(t) = \varphi(t^1, \ldots, t^n)$ has the form

$$\varphi(t) = \frac{\varphi^*(t)}{p_{L_1}^{\nu^1}(t^1) \cdot \cdots \cdot p_{L_n}^{\nu^n}(t^n)}, \tag{3.1.11}$$

where $\varphi^* \in H$ on L, $P_{L_k}^{\nu_k}(t^k) = \prod_{i_k=1}^{m_k} |t^k - c_{i_k}^k|$, $c_{i_k}^k$, $i_k = 1, \ldots, m_k$, are all the nodes of the curve L_k, and $\nu^k = (\nu_1^k, \ldots, \nu_{m_k}^k)$, $0 \le \nu_{i_k}^k < 1$, $i_k = 1, \ldots, m_k$, $k = 1, \ldots, n$, then it will be called a *function of the class* H^* on the frame $L = L_1 \times \cdots \times L_n$ (where L_k, $k = 1, \ldots, n$, are plane, piecewise smooth curves).

Now Theorem 3.1.3 and Note 3.1.1 allow us to formulate the following statement.

Theorem 3.1.5. *The class H^* of functions on the frame $L = L_1 \times \cdots \times L_n$ (where L_k, $k = 1, \ldots, n$, are piecewise smooth curves) is invariant with respect to the operation of taking a multiple Cauchy integral; in other words, if* *a function $\varphi(t)$ belongs to the class H^* on L, then $\Phi(t_0)$ also belongs to the class on L.*

Let us consider the question of changing the order of integration in multiple Cauchy integrals. While doing so we will use (3.1.2), which presents a multiple Cauchy integral as an iterated integral.

Let

$$I(\tau^1, \tau^2) = \iint_{L_1 \times L_2} \frac{dt_0^1\, dt_0^2}{(t_0^1 - \tau^1)(t_0^2 - \tau^2)} \iint_{L_1 \times L_2} \frac{\varphi(t_0^1, t_0^2, t^1, t^2)\, dt^1\, dt^2}{(t^1 - t_0^1)(t^2 - t_0^2)},$$

where L_1 and L_2 are piecewise smooth curves and $\varphi(t_0^1, t_0^2, t^1, t^2)$ has the form

$$\varphi(t_0, t) = \frac{\varphi^*(t_0, t)}{\prod_{k_1=1}^{m_1} |t^1 - c_{k_1}^1|^{\nu_{k_1}^1} |t_0^1 - c_{k_1}^1|^{\lambda_{k_1}^1} \times \prod_{k_2=1}^{m_2} |t^2 - c_{k_2}^2|^{\nu_{k_2}^2} |t_0^2 - c_{k_2}^2|^{\lambda_{k_2}^2}}, \tag{3.1.12}$$

where $\varphi^*(t_0, t)$ meets the condition H on L with respect to t_0 and t, $t_0 = (t_0^1, t_0^2)$, $t = (t^1, t^2)$ are points lying on L, $\nu_{k_1}^1, \lambda_{k_1}^1, \nu_{k_2}^2, \lambda_{k_2}^2 \geq 0$ and $\nu_{k_i}^i + \lambda_{k_i}^i < 1$, $c_{k_i}^i$, $k_i = 1, \ldots, m_i$ denote all the nodes of curve L_i, $i = 1, 2$.

Let us present $I(\tau^1, \tau^2)$ in the form

$$I(\tau^1, \tau^2) = \int_{L_2} \frac{dt_0^2}{t_0^2 - \tau^2} \left(\int_{L_1} \frac{dt_0^1}{t_0^1 - \tau^1} \int_{L_1} \frac{\psi(t_0^1, t_0^2, t^1, t_0^2) dt^1}{t^1 - t_0^1} \right),$$

$$\psi(t_0^1, t_0^2, t^1, t_0^2) = \int_{L_2} \frac{\varphi(t_0^1, t_0^1, t^1, t^2)\, dt^2}{t^2 - t_0^2}. \tag{3.1.13}$$

By the Poincaré–Bertrand formula (Muskhelishvili 1952) one has

$$\int_{L_1} \frac{dt_0^1}{t_0^1 - \tau^1} \int_{L_1} \frac{\psi(t_0^1, t_0^2, t^1, t_0^2)\, dt^1}{t^1 - t_0^1}$$

$$= -\pi^2 \psi(\tau^1, t_0^2, \tau^1, t_0^2) + \int_{L_1} dt^1 \int_{L_1} \frac{\phi(t_0^1, t_0^2, t^1, t_0^2)\, dt_0^1}{(t_0^1 - \tau^1)(t^1 - t_0^1)}. \tag{3.1.14}$$

Let us substitute the latter expression into Formula (3.1.13) and use the expression for ψ together with the fact that the order of integration over L_1 and L_2 may be changed:

$$I(\tau^1, \tau^2) = \int_{L_2} \frac{dt_0^2}{t_0^2 - \tau^2} \int_{L_2} \frac{-\pi^2 \varphi(\tau^1, t_0^2, \tau^1, t^2)\, dt^2}{t^2 - t_0^2}$$

$$+ \int_{L_2} \frac{dt_0^2}{t_0^2 - \tau^2} \int_{L_2} \frac{\psi_1(\tau^1, t_0^2, \tau^1, t^2) dt^2}{t^2 - t_0^2},$$

$$\psi_1(\tau^1, t_0^2, \tau^1, t^2) = \int_{L_1} dt^1 \int_{L_1} \frac{\varphi(t_0^1, t_0^1, t^1, t^2)\, dt_0^1}{(t_0^1 - \tau^1)(t^1 - t_0^1)}. \tag{3.1.15}$$

Finally, after using again the Poincaré–Bertrand formula for one-dimensional Cauchy integrals, one gets

$$\iint_{L_1 \times L_2} \frac{dt_0^1\, dt_0^2}{(t_0^1 - \tau^1)(t_0^2 - \tau^2)} \iint_{L_1 \times L_2} \frac{\varphi(t_0^1, t_0^2, t^1, t^2)\, dt^1\, dt^2}{(t^1 - t_0^1)(t^2 - t_0^2)}$$

$$= \iint_{L_1 \times L_2} dt^1\, dt^2 \iint_{L_1 \times L_2} \frac{\varphi(t_0^1, t_0^2, t^1, t^2)\, dt_0^1\, dt_0^2}{(t_0^1 - \tau^1)(t_0^2 - \tau^2)(t^1 - t_0^1)(t^2 - t_0^2)}$$

$$- \pi^2 \int_{L_2} dt^2 \int_{L_2} \frac{\varphi(\tau^1, t_0^2, \tau^1, t^2)\, dt_0^2}{(t_0^2 - \tau^2)(t^2 - t_0^2)}$$

$$- \pi^2 \int_{L_1} dt^1 \int_{L_1} \frac{\varphi(t_0^1, \tau^2, t^1, \tau^2)\, dt_0^1}{(t^1 - t_0^1)(t_0^1 - \tau^1)}$$

$$+ \pi^4 \varphi(\tau^1, \tau^2, \tau^1, \tau^2). \tag{3.1.16}$$

Now by applying the method of mathematical induction one can easily derive the formula for changing the order of integration in multiple Cauchy integrals of an arbitrary dimension.

Theorem 3.1.6. *Let function $\varphi(t)$ be a function of the class H^* on frame L. Then the following Poincaré–Bertrand formula for changing the order of* *integration in multiple Cauchy integrals is valid:*

$$\int_L \frac{dt_0}{((t_0 - \tau))} \int_L \frac{\varphi(t_0, t)\, dt}{((t - t_0))}$$

$$= \sum_{p=0}^{n} (-\pi^2)^p \sum_{k_1 \neq \cdots \neq k_p = 1} \int_{L_{k_1,\ldots,k_p}} dt_{k_1,\ldots,k_p}$$

$$\times \int_{L_{k_1,\ldots,k_p}} \frac{\varphi\left({}_\tau t_{0k_1,\ldots,k_p}, {}_\tau t_{k_1,\ldots,k_p}\right) dt_{0k_1,\ldots,k_p}}{((t_0 - \tau))_{k_1,\ldots,k_p}((t - t_0))_{k_1,\ldots,k_p}}, \qquad (3.1.17)$$

where $dt_{k_1,\ldots,k_p}$ is the product of dt with discarded multipliers corresponding to $k_1, \ldots, k_p$. A similar convention is valid for $L_{k_1}, \ldots, {}_{k_p}$, and if $p = 0$, then no terms are discarded, whereas ${}_\tau t_{k_1}, \ldots, {}_{k_p}$ means that the coordinates $t_{k_1}, \ldots, t_{k_p}$ must be substituted by coordinates $\tau_{k_1}, \ldots, \tau_{k_p}$ (a similar convention is assumed for ${}_\tau t_{0k_1}, \ldots, {}_{k_p}$).

Because the identify

$$\int_L \frac{dt_0}{(t - t_0)(t_0 - \tau)} \equiv 0,$$

holds for a smooth unclosed curve, the identity

$$\int_L \frac{dt_0}{((t_0 - \tau))} \int_L \frac{\varphi(t)\, dt}{((t - t_0))} \equiv (-\pi^2)^n \varphi(\tau). \qquad (3.1.18)$$

holds for a function $\varphi(t^1, \ldots, t^n) \in H$ on the frame L ($L_1, \ldots, L_n$ are smooth closed curves).

3.2. ABOUT SOME SINGULAR INTEGRALS FREQUENTLY USED IN AERODYNAMICS

In the steady problem of subsonic flow of perfect incompressible fluid past a finite-span wing, one has to consider the integral (Belotserkovskii and Lifanov 1981, Bisplinghoff, Ashley, and Halfman 1955)

$$A(x_0, z_0) = \iint_\sigma \frac{\gamma(x, z)}{(z_0 - z)^2} \left(1 + \frac{x_0 - x}{\sqrt{(x_0 - x)^2 + (z_0 - z)^T}}\right) dx\, dz, \qquad (3.2.1)$$

where σ is a part of the plane OXZ occupied by a plane wing.

Before defining integral (3.2.1), it should be noted that integral (1.6.2) may be defined as (see Equation (1.6.3))

$$I_{(2)}(x_0) = \lim_{\epsilon \to 0} \left(\int_{L \setminus Q(\epsilon, t_0)} \frac{\gamma(x)\, dx}{(x_0 - x)^2} - \frac{2\gamma(x_0)}{\epsilon} \right). \qquad (3.2.2)$$

Let us show that integral $I_{(2)}(x_0)$ exists for any function $\gamma(x) \in H_1(\alpha)$. In fact, in this case we have

$$|\gamma(x) - \gamma(x_0) - \gamma'(x_0)(x - x_0)| \leq B|x - x_0|^{1+\alpha}, \qquad (3.2.3)$$

and hence, the following equality holds:

$$\int_a^b \frac{\gamma(x)\, dx}{(x_0 - x)^2} = \int_a^b \frac{\gamma(x) - \gamma(x_0) - \gamma'(x_0)(x - x_0)}{(x_0 - x)^2}\, dx$$
$$+ \gamma(x_0) \int_a^b \frac{dx}{(x_0 - x)^2} - \gamma'(x_0) \int_a^b \frac{dx}{x_0 - x}.$$

Here the first integral on the right-hand side exists as an absolutely integrable improper integral, whereas the third one exists in the sense of the principal Cauchy value. For the second integral we have

$$\int_a^b \frac{dx}{(x_0 - x)^2} = \lim_{\epsilon \to 0} \left(\int_{L \setminus l} \frac{dx}{(x_0 - x)^2} - \frac{2}{\epsilon} \right)$$
$$= \lim_{\epsilon \to 0} \left(\frac{1}{x_0 - x} \bigg|_a^{x_0 - \epsilon} + \frac{1}{x_0 - x} \bigg|_{x_0 + \epsilon}^{b} - \frac{2}{\epsilon} \right)$$
$$= \lim_{\epsilon \to 0} \left(-\frac{1}{x_0 - a} + \frac{1}{x_0 - b} + \frac{2}{\epsilon} - \frac{2}{\epsilon} \right)$$
$$= \frac{1}{x_0 - b} - \frac{1}{x_0 - a}. \qquad (3.2.4)$$

Thus, integral $I_{(2)}(x_0)$ is also defined for any function $\gamma(x) \in H_1^*$ because in this case $\gamma(x) \in H_1$ in the neighborhood of any point $x_0 \in (a, b)$. If $\gamma(b) = \gamma(a) = 0$, then we have the equality

$$\int_a^b \frac{\gamma(x)\, dx}{(x_0 - x)^2} = -\int_a^b \frac{\gamma'(x)\, dx}{x_0 - x}.$$

Upon noting that

$$\int_{L\setminus l} \frac{(x-x_0)^k}{(x_0-x)^m}\,dx = \frac{(-1)^k}{m-k-1}\left[\frac{1-(-1)^{m-k-1}}{\epsilon^{m-k-1}} + \frac{1}{(x_0-b)^{m-k-1}} - \frac{1}{(x_0-a)^{m-k-1}}\right],$$

one can define the integral

$$I_{(m)}(x_0) = \int_a^b \frac{\gamma(x)\,dx}{(x_0-x)^m}$$

in the following way:

$$I_{(m)}(x_0) = \lim_{\epsilon\to 0}\left[\int_{L\setminus l} \frac{\gamma(x)\,dx}{(x_0-x)^m} - \sum_{k=0}^{m-2} \frac{(-1)^k}{k!}\gamma^{(k)}(x_0)\frac{1}{m-k-1}\frac{1-(-1)^{m-k-1}}{\epsilon^{m-k-1}}\right],$$

where $\gamma^{(0)}(x_0) \equiv \gamma(x_0)$.

Integral $I_{(m)}(x_0)$ exists for any function $\gamma(x) \in H^*_{m-1}$, because in this case $\gamma(x) \in H_{m-1}$ in a certain neighborhood of point x_0 and the inequality

$$\left|\gamma(x) - \gamma(x_0) - \frac{1}{1!}\gamma'(x_0)(x-x_0) - \cdots - \frac{1}{(m-1)!}\gamma^{(m-1)}(x_0)(x-x_0)^{m-1}\right| \le B|x-x_0|^{m-1+\alpha}.$$

holds. The following formula of integration by parts is valid for integral $I_{(m)}(x_0)$:

$$\int_a^b \frac{\gamma(x)\,dx}{(x_0-x)^m} = \frac{1}{m-1}\left[\frac{\gamma(b)}{(x_0-b)^{m-1}} - \frac{\gamma(a)}{(x_0-a)^{m-1}} - \int_a^b \frac{\gamma'(x)\,dx}{(x_0-x)^{m-1}}\right].$$

Let us return to considering integral $A(x_0, z_0)$ starting from the simplest case when region σ is a rectangle: $\sigma = [-b, b] \times [-l, l]$. Integral $A(x_0, z_0)$ will be defined by generalizing Formula (3.2.2) in a natural way:

$$A(x_0, z_0) = \lim_{\epsilon \to 0} \left[\iint_{\sigma \setminus Q(\epsilon, z_0)} \frac{\psi(x, x_0, z, z_0)}{(z_0 - z)^2} \, dx \, dz \right.$$

$$\left. - \frac{2}{\epsilon} \int_{I_{z_0}} \psi(x, x_0, z_0, z_0) \, dx \right], \tag{3.2.5}$$

where $Q(\epsilon, z_0) = \{(x, z): |z - z_0| < \epsilon\}$, $I_{z_0} = \sigma \cap L_{z_0}$, and L_{z_0} is a straight line given by the equation $z = z_0$, and

$$\psi(x, x_0, z, z_0) = \gamma(x, z) \left(1 + \frac{x_0 - x}{\sqrt{(x_0 - x)^2 + (z_0 - z)^2}} \right).$$

Theorem 3.2.1. *In the case under consideration integral $A(x_0, z_0)$ exists for any function $\gamma(x, z)$ and is such that $\gamma_z'(x, z) \equiv \partial \gamma(x, z) / \partial z$ belongs to the class H^* on σ.*

Proof. Suppose that $\gamma_z'(x, z) \in H(\alpha)$ on σ. Then by Formula (3.2.3) one has

$$|\gamma(x, z) - \gamma(x, z_0) - \gamma_2'(x, z_0)(z - z_0)| \leq B|z - z_0|^{1 + \alpha}. \tag{3.2.6}$$

Let us rewrite formula (3.2.5) in the form

$$A(x_0, z_0) = \lim_{\epsilon \to 0} \int_{-b}^{b} dx \left[\int_D \frac{\gamma(x, z) - \gamma(x, z_0) - \gamma_z'(z - z_0)}{(z_0 - z)^2} \right.$$

$$\times K_1(x, x_0, z, z_0) \, dz + \gamma(x, z_0) \int_D \frac{K_1(x, x_0, z, z_0)}{(z_0 - z)^2} \, dz$$

$$- \gamma_z'(x, z_0) \int_D \frac{K_1(x, x_0, z, z_0)}{z_0 - z} \, dz$$

$$\left. - \frac{2}{\epsilon} \gamma(x, z_0) K_1(x, x_0, z_0, z_0) \right],$$

where $D = D(\epsilon, z_0) \equiv [-l, z_0 - \epsilon] \cup [z_0 + \epsilon, l]$,

$$K_1(x, x_0, z, z_0) \equiv 1 + \frac{x_0 - x}{\sqrt{(x_0 - x)^2 + (z_0 - z)^2}}. \tag{3.2.7}$$

As far as $|K_1(x, x_0, z, z_0)| \leq 2$, we deduce by Inequality (3.2.6) that the first double integral under the limit sign is absolutely convergent. Then

$$\int_D \frac{K_1(x, x_0, z, z_0)}{z_0 - z} dz = F_1\Big|_{z=-1}^{z=z_0-\epsilon} + F_1\Big|_{z=z_0+\epsilon}^{z=l} = 2\ln\left|\frac{z_0 - l}{z_0 + l}\right| - \ln\left|\frac{K_2(x, x_0, l, z_0)}{K_2(x, x_0, -l, z_0)}\right|, \tag{3.2.8}$$

where

$$F_1 = F_1(x, x_0, z, z_0) = 2\ln|z_0 - z| - \ln|K_2(x, x_0, z, z_0)|,$$

$$K_2 = x_0 - x + \sqrt{(x_0 - x)^2 + (x_0 - z)^2}.$$

In a similar way,

$$\gamma(x, z_0)\int_D \frac{K_1(x, x_0, z, z_0)}{(z_0 - z)^2} dz = F_2\Big|_{z=-l}^{z=z_0-\epsilon} + F_2\Big|_{z=z_0+\epsilon}^{z=l}$$

$$= 2\frac{\gamma(x, z_0)}{\epsilon}\left(1 + \frac{\sqrt{(x_0 - x)^2 + \epsilon^2}}{x_0 - x}\right) + \frac{K_2(x, x_0, l, z_0)}{(z_0 - l)(x_0 - x)} - \frac{K_2(x, x_0, -l, z_0)}{(z_0 + l)(x_0 - x)}, \tag{3.2.9}$$

where

$$F_2 = F_2(x, x_0, z, z_0) = \frac{K_2(x, x_0, z, z_0)}{(z_0 - z)(x_0 - x)}.$$

Next we note that

$$\lim_{\epsilon \to 0} \frac{\sqrt{(x_0 - x)^2 + \epsilon^2} - |x_0 - x|}{\epsilon(x_0 - x)} = 0, \qquad x_0 \neq x. \tag{3.2.10}$$

Formulas (3.2.8)–(3.2.10) demonstrate validity of the theorem for the functions $\gamma(x, z)$, such that $\gamma_z'(x, z) \in H$ on σ. However, from the same considerations it follows that the theorem is also applicable to functions $\gamma(x, z)$ when $\gamma_z'(x, z) \in H^*$ on σ.

Note that $A(x_0, z_0)$ may be written in the form of an iterated integral,

$$A(x_0, z_0) = \int_{-b}^{b} dx \int_{-l}^{l} \frac{\gamma(x, z)}{(z_0 - z)^2} K_1(x, x_0, z, z_0)\, dz, \tag{3.2.11}$$

where the inner integral may be taken by parts. Thus, we get

$$A(x_0, z_0) = \int_{-b}^{b} \frac{\psi_1(x, x_0, l, z_0)}{(z_0 - l)(x_0 - x)}\, dz - \int_{-b}^{b} \frac{\psi_1(x, x_0, -l, z_0)}{(z_0 + l)(x_0 - x)}$$

$$- \int_{-b}^{b} \int_{-l}^{l} \frac{\partial \gamma(x, z)}{\partial z}$$

$$\times \frac{x_0 - x + \sqrt{(x_0 - x)^2 + (z_0 - z)^2}}{(x_0 - x)(z_0 - z)}\, dx\, dz, \tag{3.2.12}$$

where

$$\psi_1(x, x_0, z, z_0) = \gamma(x, z) K_2(x, x_0, z, z_0).$$

In the condition $\gamma(x, l) \equiv \gamma(x, -l) \equiv 0$ is met (this is the case one has to deal with in aerodynamics), then integral $A(x_0, z_0)$ may be presented in the form

$$A(x_0, z_0) = \int_{-b}^{b} \int_{-l}^{l} \frac{\partial \gamma(x, z)}{\partial z}$$

$$\times \frac{x_0 - x + \sqrt{(x_0 - x)^2 + (z_0 - z)^2}}{(x_0 - x)(x_0 - z)}\, dx\, dz. \tag{3.2.13}$$

Let region σ be limited by the straight lines $z = -l$ and $z = l$ as well as by the lines $x_2 = x_2(z) \geq x_1 = x_1(z)$ belonging to the class H_1 on $[-l, l]$. Then integral $A(x_0, z_0)$ will again be defined by Formula (3.2.5); however, now it has the form

$$A(x_0, z_0) = \lim_{\epsilon \to 0} \left[\int_D \frac{\psi_2(x_0, z, z_0)}{(z_0 - z)^2} \, dz - \frac{2}{\epsilon} \psi_2(x_0, z_0, z_0) \right], \quad (3.2.14)$$

where

$$\psi_2(x_0, z, z_0) = \int_{x_1(z)}^{x_2(z)} \gamma(x, z) K_1(x, x_0, z, z_0) \, dx.$$

We see that $A(x_0, z_0)$ exists if function $\partial \psi_2(x_0, z, z_0)/\partial z \in H^*$ on $[-l, l]$ uniformly with respect to x_0 and z_0. However, the criterion is rather inconvenient for checking. ■

Consider a special case which we shall have to deal with in aerodynamics.

Definition 3.2.1. *Region σ will be called a canonic trapezoid if it is limited by the straight lines $z = -l$, $z = l$ and the straight lines*

$$x^-(z) = a^0 + zb^0, \qquad x^+(z) = a^1 + zb^1, \qquad x^+(z) > x^-(z). \quad (3.2.15)$$

In what follows, $x^-(z)$ will be called the nose and $x^+(z)$ will be called the tail, as is customary in aerodynamics (Belotserkovsky and Lifanov 1981).

Let us reduce the case of the canonic trapezoid to that of a rectangle. This can be done by considering mapping F of the rectangle $D = [0, 1] \times [-l, l]$ belonging to the plane OX^1z, onto region σ, the mapping being defined by the formulas

$$x = x(x^1, z) = x^1[x^+(z) - x^-(z)] + x^-(z),$$

$$z = z. \quad (3.2.16)$$

Point (x_0^1, z_0) transforms into point $(x_0, z_0) = (x(x_0^1, z_0), z_0)$, and the Jacobian J of the mapping is given by the formula

$$J(z) = x^+(z) - x^-(z). \quad (3.2.17)$$

For a canonic trapezoid σ, integral $A(x_0, z_0)$ exists for any function $\gamma(x, z)$, such that function $\gamma_z'(x(x^1, z), z) \in H^*$ on rectangle D.

To prove this statement we change the variable appearing in (3.2.14) in the integral for ψ^2, with the help of the first of Equations (3.2.16). Then we get

$$A\big(x(x_0^1, z_0), z_0\big) = \lim_{\epsilon \to 0} \int_0^1 dx^1 \Bigg[\int_D \frac{\psi\big(x(x^1, z), x(x_0^1, z_0), z, z_0\big)}{(z_0 - z)^2} \times J(z)\, dz - \frac{2}{\epsilon} \psi\big(x(x^1, z_0), x(x_0^1, z_0), z_0, z_0\big) J(z_0) \Bigg]. \tag{3.2.18}$$

We assume that the function $\varphi(x^1, z) = \gamma(x(x^1, z), z)J(z)$ belongs to the class H_1 on D. Let us transform (3.2.18) in a manner similar to Formula (3.2.5). To do this, in Equation (3.2.7) we substitute x' for x, $\varphi(x', z)$ for $\gamma(x, z)$, and $K_1(x(x', z), x(x_0', z_0), z, z_0)$ for $K_1(x, x_0, z, z_0)$. Similarly to Formula (3.2.7) one has to calculate the integrals obtained from Integrals (3.2.8) and (3.2.9) by replacing function $|K_1(x, x_0, z, z_0)| \leq 2$ by $K_1(x(x^1, z), x(x_0^1, z_0), z, z_0)$. To do this we first observe that

$$x(x^1, z) = a(x^1) + zb(x^1),$$

$$a(x^1) = a^0 + x^1(a^1 - a^0),\ b(x^1) = b^0 + x^1(b^1 - b^0). \tag{3.2.19}$$

From (3.2.19) one gets

$$x(x_0^1, z_0) - x(x^1, z) = \lambda(x^1, x_0^1, z_0) - (z_0 - z)b(x^1),$$

$$\lambda \equiv x(x_0^1, z_0) - x(x^1, z_0) = J(z_0)(x_0^1 - x^1). \tag{3.2.20}$$

Then

$$\int \frac{x(x_0^1, z_0) - x(x^1, z)}{(z_0 - z)^2 \sqrt{\left[x(x_0^1, z_0) - x(x^1, z)\right]^2 + (z_0 - z)^2}}\, dz = \frac{\sqrt{\left[x(x_0^1, z_0) - x(x^1, z)\right]^2 + (z_0 - z)^2}}{(z_0 - z)\lambda(x^1, x_0^1, z_0)} + C. \tag{3.2.21}$$

The latter formula may be checked easily by differentiating the right-hand side with respect to z.

From (3.2.20) and (3.2.21) we deduce

$$\int_D \frac{K_1\big(x(x^1, z), x(x_0^1, z_0), z, z_0\big)}{(z_0 - z)^2}\, dz$$

$$= F_3(x^1, x_0^1, l, z_0) - F_3(x^1, x_0^1, l, z_0)$$

$$+ \frac{\sqrt{[\lambda(x^1, x_0^1, z_0) + \epsilon b(x^1)]^2}}{\epsilon\lambda}$$

$$+ \frac{\sqrt{[\lambda(x^1, x_0^1, z_0) - \epsilon b(x^1)]^2}}{\epsilon\lambda} + \frac{2}{\epsilon}, \qquad (3.2.22)$$

where

$$F_3(x^1, x_0^1, z, z_0) = \frac{1}{z_0 - z}$$

$$\times\left(1 + \frac{\sqrt{[x(x_0^1, z_0) - x(x^1, z)]^2 + (z_0 - z)^2}}{\lambda}\right).$$

From the latter formula it follows that the limit

$$\lim_{\epsilon\to 0}\left[\int_D \frac{K_1\big(x(x^1, z), x(x_0^1, z_0), z, z_0\big)\, dz}{(z_0 - z)^2} - \frac{2}{\epsilon}(1 + \operatorname{sign}\lambda)\right]$$

$$= F_4(x^1, x_0^1, z_0) \qquad (3.2.23)$$

exists and is a function integrable in the sense of the Cauchy principal value on segment [0, 1] of axis OX^1.

In a similar way it can be shown that the limit

$$\lim_{\epsilon\to 0}\int_D \frac{K_1\big(x(x^1, z), x(x_0^1, z_0), z, z_0\big)\, dz}{z_0 - z}$$

exists and is an absolutely integrable function on segment [0, 1] of axis OX^1. Thus, the preceding statement for a canonic trapezoid is proved.

Note that if σ is a canonic trapezoid, then the following equations hold:

$$A(x_0, z_0) = A\big(x(x_0^1, z_0), z_0\big)$$

$$= \int_0^1 dx^1 \int_{-l}^{l} \frac{\varphi(x^1, z)}{(z_0 - z)^2} K_1\big(x(x^1, z), x(x_0^1, z_0), z, z_0\big)\, dz$$

$$= \int_0^1 \left[\frac{\varphi(x^1, l) K_3(x^1, x_0, +l, z_0)}{(z_0 - l) J(z_0)(x_0^1 - x^1)} \right.$$

$$\left. - \frac{\varphi(x^1, -l) K_3(x^1, x_0^1, -l, z_0)}{(z_0 + l) J(z_0)(x_0^1 - x^1)} \right] dx^1$$

$$- \int_0^1 \int_{-l}^{l} \frac{\partial \varphi(x^1, z)}{\partial z} \frac{K_3(x^1, x_0^1, z, z_0)}{J(z_0)(x_0^1 - x^1)(z_0 - z)}\, dx^1\, dz, \tag{3.2.24}$$

where

$$K_3(x^1, x_0^1, z, z_0) = J(z_0)(x_0^1 - x^1)$$
$$+ \sqrt{\big[x(x_0^1, z_0) - x(x^1, z)\big]^2 + (z_0 - z)^2}\,.$$

3.3. QUADRATURE FORMULAS FOR MULTIPLE SINGULAR INTEGRALS

Let us consider the multiple singular Cauchy integral (see Equation (3.1.1a)):

$$I(t_0) = \int_L \frac{\varphi(t)\, dt}{((t - t_0))}\,.$$

1. Similarly to Chapter 1, we start considering quadrature formulas for integral $I(t_0)$ beginning with the case when the frame L is a torus, i.e., a product of closed Lyapunov curves. The following theorem is true.

Theorem 3.3.1. *Let function $\varphi(t) \in H$ on an m-dimensional torus L, and the sets $E^k = \{t_{ik}^k,\ i_k = 1, \dots, n_k\}$ and $E^{0k} = \{t_{0ik}^k,\ i_k = 1, \dots, n_k\}$ form a* *canonic division of closed Lyapunov curve L_k, $k = 1, \dots, m$. Then inequality*

$$\left| \int_L \frac{\varphi(t)\, dt}{((t - t_{0j}))} - \sum_{i=1}^{n} \frac{\varphi(t_i)\, \Delta t_i}{((t_i - t_{0j}))} \right| \le O\left(\frac{1}{N^\beta} \right) \tag{3.3.1}$$

is valid for any point $t_{0j} = (t_{0j_1}^1, \dots, t_{0j_m}^m)$ belonging to torus L. In the latter formula $\beta > 0$, $n = \max(n_1, \dots, n_k)$, and $N/n_k \le R < +\infty$ for

$$N \to \infty, \qquad k = 1, \dots, m, \qquad \sum_{i=1}^{n} = \sum_{i_1=n}^{n_1} \cdots \sum_{i_m+1}^{n_m},$$

$$\Delta t_i = \Delta t_{i_1}^1 \cdot \dots \cdot \Delta t_{i_m}^m.$$

To prove the theorem we consider the case when $m = 2$, and L_1 and L_2 are circles centered at the origin of the coordinates. One may write

$$\begin{aligned} &|I(t_{0j_1}^1, t_{0j_2}^2) - S(t_{0j_1}^1, t_{0j_2}^2)| \\ &\quad \le \left| \int_{L_1} \frac{\psi_1(t^1, t_{0j_2}^2)\, dt^1}{t^1 - t_{0j_1}^1} - \sum_{i_1=1}^{n_1} \frac{\psi_1(t_{i_1}^1, t_{0j_2}^2)}{t_{i_1}^1 - t_{0j_1}^1} - \Delta t_{i1}^1 \right| \\ &\qquad + \sum_{i_1=1}^{n_1} \frac{|\Delta t_{i_1}^1|}{|t_{i_1}^1 - t_{0j_1}^1|} \left| \int_{L_2} \frac{\varphi(t_{i_1}^1, t_{i_2}^2)\, dt^2}{t^2 - t_{0j_2}^2} - \sum_{i_2=1}^{n_2} \frac{\varphi(t_{i_1}^1, t_{i_2}^2)\, \Delta t_{i_2}^2}{t_{i_2}^2 - t_{0j_2}^2} \right|, \end{aligned}$$

where

$$\psi_1(t^1, t_0^2) = \int_{L_2} \frac{\varphi(t^1, t^2)\, dt^2}{t^2 - t_0^2}.$$

By using the results of Section 3.1, we get for singular integrals depending on a parameter,

$$|I(t_{0j_1}^1, t_{0j_2}^2) - S(t_{0j_1}^1, t_{0j_2}^2)| \le O\left(\frac{\ln n_1}{n_1^{\alpha_1 - \epsilon}} \right) + O(\ln n_1) O\left(\frac{\ln n_2}{n_2^{\alpha_2}} \right). \tag{3.3.2}$$

Because $N/n_k \le R < +\infty$, $k = 1, 2$, for N tending to $+\infty$, the latter formula terminates the proof of Theorem 3.3.1 for the special case under consideration. The note made in Section 1.4 concerning integrals over Lyapunov closed curves and Inequality (1.2.16) prove the validity of inequalities of the form (3.3.2), and hence, that of Theorem 3.3.1 for the general case. However, α_1, α_2 entering Inequality (3.3.2) must be replaced taking into account smoothness of the curves L_1 and L_2.

Note that according to Inequality (1.2.17), Theorem 3.3.1 remains valid in the case when L_k, $k = 1, \ldots, m$, is a set of nonintersecting Lyapunov curves.

2. Let now the frame L in multiple integral $I(t_0)$ be an m-dimensional parallelepiped, i.e., $L = L_1 \times \cdots \times L_m$, where L_k, $k = 1, \ldots, m$, is an open Lyapunov curve. Then the following theorem is true.

Theorem 3.3.2. *Let function $\varphi(t) \in H^*$ on m-dimensional parallelepiped L, and the sets E^k and E^{0k} form a canonic division of curve L_k with the step h_k. Then the inequality*

$$\left| \int_L \frac{\varphi(t)\, dt}{((t - t_{0j}))} - \sum_{i=1}^{n} \frac{\varphi(t_i)\, \Delta t_i}{((t_i - t_{0j}))} \right| \leq \theta(t_{0j}) \tag{3.3.3}$$

holds for any point $t_{0j} = (t^1_{0j_1}, \ldots, t^m_{0j_m} \in) E_0 = E^{0\,1} \times \cdots \times E^{0m}$, where the quantity $\theta(t_{0j})$ possesses the properties:

1. *For all points $t_{0j} \in L'_1 \times \cdots \times L'_m$, where L'_k is the part of curve L_k containing no ends of the curve together with their certain neighborhoods,*

$$\theta(t_{0j}) \leq O(h^{\lambda_1}), \qquad \lambda_1 > 0. \tag{3.3.4}$$

2. *For all points*

$$\sum_{j=0}^{n} \theta(t_{0j}) h_1 \cdot \cdots \cdot h_m \leq O(h^{\lambda_2}), \qquad \lambda_2 > 0, \tag{3.3.5}$$

where $h = \max(h_1, \ldots, h_m)$ and $h/h_k \leq R < +\infty$ for $h \to 0$, $k = 1, \ldots, m$.

Proof. Let us consider the case $m = 2$. Let $L_k = [a_k, b_k]$, $k = 1, 2$, be segments of the real axis. Because $\varphi(t) \in H^*$ on L,

$$\varphi(t^1, t^2) = \frac{\psi(t^1, t^2)}{p_{L_1}^{\nu^1}(t^1) p_{L_2}^{\nu^2}(t^2)},$$

$$p_{L_k}^{\nu^k}(t^k) = (t^k - a_k)^{\nu_1^k} (b_k - t^k)^{\nu_2^k}, \qquad k = 1, 2, 0 < \nu_1^k, \nu_2^k < 1.$$

By Theorem 3.1.3 concerning integrals depending on a parameter, function

$$\psi(t^1, t_0^2) = \int_{a_2}^{b_2} \frac{\varphi(t^1, t^2)\, dt^2}{(t^2 - a_2)^{\nu_1^2}(b_2 - t)^{\nu_2^2}(t^2 - t_0^2)}$$

has the same form as the density of the corresponding singular integral, i.e.,

$$\psi_1(t^1, t_0^2) = \frac{\varphi_1(t^1, t_0^2)}{p_{L_2}^{\nu^2}(t_0^2)}, \quad \text{where } \varphi_1(t^1, t_0^2) \in H(\alpha_1 - \epsilon_1, \lambda_2),\ \lambda_2 > 0.$$

Hence, one may write

$$\begin{aligned}
&|I(t_{0j_1}^1, t_{0j_2}^2) - S(t_{0j_1}^1, t_{0j_2}^2)| \\
&\quad \leq \left| \int_{a_1}^{b_1} \frac{\psi_1(t^1, t_{0j_2}^2)\, dt^1}{p_{L_i}^{\nu^1}(t^1)(t^1 - t_{0j_1}^1)} - \sum_{i_1=1}^{n_1} \frac{\psi_1(t_{i_1}^1, t_{0j_2}^2) h_1}{p_{L_1}^{\nu^1}(t_{i_1}^1)(t_{i_1}^1 - t_{0j_1}^1)} \right| \\
&\quad + \sum_{i_1=1}^{n_1} \frac{h_1}{p_{L_1}^{\nu^1}(t_{i_1}^1)|t_{i_1}^1 - t_{0j_1}^1|} \\
&\quad \times \left| \int_{a_2}^{b_2} \frac{\psi(t_{i_1}^1, t^2)\, dt^2}{p_{L_2}^{\nu^2}(t^2)(t^2 - t_{0j_2}^2)} - \sum_{i_2=1}^{n_2} \frac{\psi(t_{i_1}^1, t_{i_2}^2) h_2}{p_{L_2}^{\nu^2}(t_{i_2}^2)(t_{i_2}^2 - t_{0j_2}^2)} \right|.
\end{aligned}$$

Taking into account the comments made in the closing part of Section 1.3 concerning Function (1.3.16), as well as Inequalities (1.3.7) and (1.3.14), one gets

$$\begin{aligned}
&|I(t_{0j_1}^1, t_{0j_2}^2) - S(t_{0j_1}^1, t_{0j_2}^2)| \\
&\quad \leq \frac{\theta_1(t_{0j_1}^1)}{p_{L_2}^{\nu^2}(t_{0j_2}^2)} + \frac{O(|\ln h_1|)}{p_{L_1}^{\nu^1}(t_{0j_1}^1)} \theta_2(t_{0j_2}^2).
\end{aligned} \tag{3.3.6}$$

By using Formula (1.3.14) for $\theta_1(t_{0j_1}^1)$ and $\theta_2(t_{0j_2}^2)$ together with the relationship $h/h_k \leq R < +\infty$ for $h \to 0$, $k = 1, 2$, for we prove the validity of Inequality (3.3.3).

3. Now let the frame L be a product of piecewise Lyapunov curves containing corner nodes only. With the help of Inequality (1.4.4), Inequality (3.3.3) can be generalized in this case in a natural way.

3.4. QUADRATURE FORMULAS FOR THE SINGULAR INTEGRAL FOR A FINITE-SPAN WING

Let us start considering quadrature formulas for the integral $A(x_0, z_0)$ for a finite-span wing beginning with a rectangular wing. In this case

$$A(x_0, z_0) = \int_{-b}^{b} dx \int_{-l}^{l} \frac{\gamma(x, z)}{(z_0 - z)^2}$$

$$\times \left(1 + \frac{x_0 - x}{\sqrt{(x_0 - x)^2 + (z_0 - z)^2}}\right) dx\, dz.$$

Let points $x_1, \ldots, x_n$ and $x_{0\,0}, x_{0\,1}, \ldots, x_{0n}$ form a canonic division of segment $[-b, b]$ with the step h_1, and points $z_1, \ldots, z_N$ and $z_{0\,0}, z_{0\,1}, \ldots, z_{0N}$ form a canonic division of segment $[-l, l]$ with the step h_2. We suppose that $h/h_k \le R < +\infty$, $k = 1, 2$; $h = \max(h_1, h_2)$. By applying the principle of constructing quadrature formulas for a singular integral along the x coordinate and the principle of constructing the sum $S_{(2)}(t_{0j})$ entering Formula (1.6.6) along the z axis, one can consider the following quadrature sum for $A(x_0, z_0)$:

$$S(x_{0j}, z_{0m}) = \sum_{i=1}^{n} \sum_{k=0}^{N} \frac{\gamma(x_i, z_{0k})}{x_{0j} - x_i} \left[\frac{K_2(x_i, x_{0j}, z_{k+1}, z_{0m})}{z_{0m} - z_{k+1}} \right.$$

$$\left. - \frac{K_2(x_i, x_{0j}, z_k, z_{0m})}{z_{0m} - z_k} \right] h_1, \tag{3.4.1}$$

where

$$K_2(x, x_0, z, z_0) = x_0 - x + \sqrt{(x_0 - x)^2 + (z_0 - z)^2},$$

$$z_0 = -l, \; z_{N+1} = l.$$

Note that Formula (3.4.1) may be presented in the form

$$S(x_{0j}, z_{0m}) = \sum_{i=1}^{n} \sum_{k=0}^{N} \gamma(x_i, z_{0k}) \int_{z_k}^{z_{k+1}} \frac{K_1(x_i, x_{0j}, z, z_{0m})}{(z_{0m} - z)^2} dz\, h_1, \tag{3.4.2}$$

where

$$K_1(x, x_0, z, z_0) = 1 + \frac{x_0 - x}{\sqrt{(x_0 - x)^2 + (z_0 - z)^2}}.$$

Now we can prove the following theorem.

Theorem 3.4.1. *Let region σ in (3.2.1) be a rectangle and the function $\gamma(x, z)$ be such that $\gamma_z'(x, z)$ belongs to the class H^* on σ. Then inequality*

$$|A(x_{0j}, z_{0m}) - S(x_{0j}, z_{0m})| \leq \theta(x_{0j}, z_{0m}) \tag{3.4.3}$$

holds for any point (x_{0j}, z_{0m}), $j = 0, 1, \ldots, n$, $m = 0, 1, \ldots, N$, where $\theta(x_{0j}, z_{0m})$ is characterized by the following relationships:

1. *For all points $(x_{0j}, z_{0m}) \in [-b + \delta, b - \delta] \times [-l + \delta, l - \delta]$, where $\delta > 0$ is a however small number,*

$$\theta(x_{0j}, z_{0m}) \leq O(h^{\lambda_1}), \qquad 0 < \lambda_1 \leq 1. \tag{3.4.4}$$

2. *For all points (x_{0j}, z_{0j}),*

$$\sum_{j=0}^{n} \sum_{m=0}^{N} \theta(x_{0j}, z_{0m}) h_1 h_2 \leq O(h^{\lambda_2}), \qquad 0 < \lambda_2 \leq 1. \tag{3.4.5}$$

Proof. Let us rewrite Formula (3.4.1) as

$$\begin{aligned} S(x_{0j}, z_{0m}) = & \sum_{i=1}^{n} \frac{\gamma(x_i, z_{0N}) K_2(x_i, x_{0j}, l, z_{0m})}{(z_{0m} - l)(x_{0j} - x_i)} h_1 \\ & - \sum_{i=1}^{n} \frac{\gamma(x_i, z_{0\,0}) K_2(x_i, x_{0j}, -l, z_{0m})}{(z_{0m} + l)(x_{0j} - x_i)} h_1 \\ & - \sum_{i=1}^{n} \sum_{k=1}^{n} \frac{\gamma(x_i, z_{0k}) - \gamma(x_i, z_{0k-1})}{h_2} \\ & \times \frac{K_2(x_i, x_{0j}, z_k, z_{0m})}{(z_{0m} - z_k)(x_{0j} - x_i)} h_1 h_2. \end{aligned} \tag{3.4.6}$$

Comparing the latter formula with (3.2.12) for integral $A(x_{0j}, z_{0m})$ and using the results obtained in Sections 1.3 and 3.3, we deduce that Inequality (3.4.3) is true. ■

Now let the domain of integration of integral $A(x_0, z_0)$ (see (3.2.1)) be a canonic trapezoid σ. By using the presentation of the integral given by Formula (3.2.24) for the case of a canonic trapezoid, one may construct the required quadrature sum in the following way.

Let us consider points $x_1^1, \ldots, x_n^1$ and points $x_{0\,0}^1, x_{0\,1}^1, \ldots, x_{0n}^1$ forming a canonic division of segment $[0,1]$ with the step h_1 and points $z_1, \ldots, z_N$ and $z_{0\,0}, z_{0\,1}, \ldots, z_{0N}$ forming a canonic division of segment $[-l, l]$ with the step h_2. We suppose that $h/h_k \leq R < +\infty$, $k = 1, 2$; $h = \max(h_1, h_2)$, $h \to 0$.

Denote by $B_{ik}(x_{ik}, z_k)$, $B_{i0k}(x_{i0k}, z_{0k})$, and $B_{0i0k}(x_{0i0k}, z_{0k})$ the points of canonic trapezoid σ that are corresponding images of points $B_{ik}^1(x_i^1, z_k)$, $B_{i0ik}^1(x_i^1, z_{0k})$, and $B_{0i0k}^1(x_{0i}^1, z_{0k})$ of rectangle D for the mapping F given by Formula (3.2.16). In analogy to sum (3.4.1), let us consider the sum

$$\begin{aligned} S(x_{0j}, z_{0m}) &= S\left(x\left(x_{0j}^1, z_{0m}\right), z_{0m}\right) \\ &= \sum_{i=1}^{n} \sum_{k=0}^{N} \frac{\varphi(x_i^1, z_{0k})}{J(z_{0m})\left(x_{0j}^1 - x_i^1\right)} \\ &\quad \times \left[\frac{K_3\left(x_i^1, x_{0j}^1, z_{k+1}, z_{0m}\right)}{z_{0m} - z_{k+1}} - \frac{K_3\left(x_i^1, x_{0j}^1, z_k, z_{0m}\right)}{z_{0m} - z_k} \right] h_1, \end{aligned} \tag{3.4.7}$$

which is a digital analog of integral $A(x_0, z_0)$ given by (3.2.24).

When the sum $S(x_{0j0m}, z_{0m})$ is written in a form analogous to that of the sum $S(x_{0j}, z_{0m})$ in Formula (3.4.6) and using the presentation of integral $A(x_0, z_0)$ in Formula (3.2.24), we come to the conclusion that the following theorem is true.

Theorem 3.4.2. *Let region σ be a canonic trapezoid and function $\gamma(x, z)$ be such that*

$$\frac{\partial}{\partial z}\left(\gamma(x(x^1, z), z) J(z)\right) \in H^*$$

on rectangle $D = [0,1] \times [-l, l]$. Then inequality

$$|A(x_{0j0m}, z_{0m}) - S(x_{0j0m}, z_{0m})| \leq \theta(x_{0j0m}, z_{0m}) = \theta_1\left(x_{0j}^1, z_{0m}\right). \tag{3.4.3$'$}$$

holds for any point (x_{0j0m}, z_{0m}), $j = 0, 1, \ldots, n$, $m = 0, 1, \ldots, N$, *where* $\theta_1(x^1_{0j}, z_{0m})$ *satisfies Inequalities* (3.4.4) *and* (3.4.5).

Note that Formula (3.4.7) also may be written in the form

$$S(x_{0j0m}, z_{0m}) = \sum_{i=1}^{n} \sum_{k=0}^{N} \varphi(x^1_i, z_{0k}) h_1$$

$$\times \int_{z_k}^{z_{k+1}} \frac{K_1\big(x(x^1, z), x\big(x^1_{0j}, z_{0m}\big), z, z_{0m}\big)}{(z_{0m} - z)^2}\, dz. \quad (3.4.8)$$

The following note is also of interest. Because the equation $x = x(x^1, z)$ (see (3.2.16)) provides a straight line for any fixed value of x^1, the number $x_0 - x(x^1, z_0)$ is proportional to the distance between point (x_0, z_0) and the straight line. Consider integral $A(x_0, z_0)$ over canonic trapezoid σ at point (x_0, z_0) lying outside σ but with $z_0 \in (-l, l)$. Then Formula (3.2.21) becomes

$$\int \frac{\big[x_0 - x(x^1, z)\big]\, dz}{(z_0 - z)^2 \sqrt{\big[x_0 - x(x^1, z)\big]^2 + (z_0 - z)^2}}$$

$$= \frac{\sqrt{\big[x_0 - x(x^1, z)\big]^2 + (z_0 - z)^2}}{\big[x_0 - x(x^1, z_0)\big](z_0 - z)} + c, \quad (3.4.9)$$

where $|x_0 - x(x^1, z_0)| \geq \delta > 0$. Hence, in the case under consideration, $A(x_0, z_0)$ reduces to the set of a conventional one-dimensional integral with respect to x^1 and a two-dimensional integral with a singularity of the form $(z_0 - z)^{-1}$.

Note also that if function $\gamma(x, z)$ entering the integral

$$A(x_0, z_0) = \int_{-b}^{b} dx \int_{-l}^{l} \frac{\gamma(x, z)}{(z_0 - z)^2}$$

$$\times \left(1 + \frac{x_0 - x}{\sqrt{(x_0 - x)^2 + (z_0 - z)^2}}\right) dx\, dz$$

possesses the property described by

$$\int_{-b}^{b} \gamma(x, z)\, dx \equiv 0 \tag{3.4.10}$$

(realized, for instance, in the case of zero-circulation flow past a rectangular wing), then $A(x_0, z_0)$ may be written in the form

$$A(x_0, z_0) = \int_{-b}^{b}\int_{-l}^{l} \frac{\gamma(x, z)}{(z_0 - z)^2} \frac{x_0 - x}{\sqrt{(x_0 - x)^2 + (z_0 - z)^2}}\, dx\, dz. \tag{3.4.11}$$

Next we rewrite the latter integral as a repeated one and take the inner integral by parts with respect to x. Then, taking into account Identity (3.4.10) one gets

$$A(x_0, z_0) = \int_{-l}^{l}\int_{-b}^{b} \frac{Q(x, z)}{\left[(x_0 - x)^2 + (z_0 - z)^2\right]^{3/2}}\, dx\, dt. \tag{3.4.12}$$

Here we neither validated the possibility of integration by parts nor specified the exact sense of the integral entering the latter formula. Let us give the following definition.

Definition 3.4.1. *Let functions $f(x)$ and $g(x)$ be defined on segment $[a, b]$ and possess the following properties: $f'(x)$ is continuous on $[a, b]$, whereas function $g(x)$ is unintegrable on $[a, b]$, continuous on $(a, b]$, and has the only singularity at point a. However, there exists function $G(x)$ absolutely integrable on $[a, b]$, such that $G'(x) = g(x)$. Then we assume that by definition,*

$$I_{G(x)}(f(x)) = \int_a^b f(x) g(x)\, dx$$

$$= \lim_{\epsilon \to 0}\left[\int_{a+\epsilon}^{b} f(x) g(x)\, dx + f(a + \epsilon) G(a + \epsilon)\right]. \tag{3.4.13}$$

However, if $f(x) \in H_1(\alpha)$ on $[a, b]$, the function $g(x)$ has a singularity at point x_0 of the interval (a, b), and the function $G(x), G'(x) \equiv g(x)$, is either absolutely integrable or may be presented in the form $\varphi(x, x_0)/(x$

$- x_0)$ where $\varphi(x, x_0) \in H$ on $[a, b]$, then

$$\int_a^b f(x)g(x)\,dx = -\int_a^b f'(x)G(x)\,dx + f(b)G(b) - f(a)G(a), \tag{3.4.14}$$

where the right-hand-side integral is assumed to be either an integral of an absolutely integrable function or an integral in the sense of the Cauchy principal value. It can be readily shown that Formula (3.4.14) defines the integral irrespective of the choice of the function's origin. Also, the defined integral operator is linear.

Hence, if the point under consideration lies inside the domain, one can define two-dimensional integrals in the manner similar to Formula (3.4.14). Then one gets

$$A(x_0, z_0) = \int_{-b}^{b}\int_{-l}^{l} \frac{Q(x,z)}{\left[(x_0 - x)^2 + (z_0 - z)^2\right]^{3/2}}\,dx\,dz$$

$$= \int_{-l}^{l} dx \left(-\frac{Q(b,z)(x_0 - b)}{(z_0 - z)^2\sqrt{(x_0 - b)^2 + (z_0 - z)^2}} \right.$$

$$\left. + \frac{Q(-b,z)(x_0 + b)}{(z_0 - z)^2\sqrt{(x_0 + b)^2 + (z_0 - z)^2}} \right)$$

$$+ \int_{-b}^{b} dx \int_{-l}^{l} \frac{Q'_x(x,z)(x_0 - x)}{(z_0 - z)^2\sqrt{(x_0 - x)^2 + (z_0 - z)^2}}\,dz$$

$$= -Q(b,l)\frac{\sqrt{(x_0 - b)^2(z_0 - l)^2}}{(x_0 - b)(z_0 - l)}$$

$$+ Q(b,-l)\frac{\sqrt{(x_0 - b)^2 + (z_0 + l)^2}}{(x_0 - b)(z_0 + l)}$$

$$+ Q(-b,l)\frac{\sqrt{(x_0 + b)^2 + (z_0 - l)^2}}{(x_0 + b)(z_0 - l)}$$

$$-Q(-b,-l)\frac{\sqrt{(x_0+b)^2+(z_0+l)^2}}{(x_0+b)(z_0+l)}$$

$$+\int_{-l}^{l}\left[\frac{Q_z'(b,z)\sqrt{(x_0-b)^2+(z_0-z)^2}}{(x_0-b)(z_0-z)}\right.$$

$$\left.-\frac{Q_z'(-b,z)\sqrt{(x_0+b)^2+(z_0-z)^2}}{(x_0+b)(z_0-z)}\right]dz$$

$$+\int_{-b}^{b}\left[\frac{Q_x'(x,l)\sqrt{(x_0-x)^2+(z_0-l)^2}}{(x_0-x)(z_0-l)}\right.$$

$$\left.-\frac{Q_x'(x,-l)\sqrt{(x_0-x)^2+(z_0+l)^2}}{(x_0-x)(z_0+l)}\right]dx$$

$$-\int_{-b}^{b}\int_{-l}^{l}\frac{Q''_{xz}\sqrt{(x_0-x)^2+(z_0-z)^2}}{(x_0-x)(z_0-z)}\,dx\,dz \tag{3.4.15}$$

If $G''_{xz}(x,z)\in H$ on the rectangle $[-b,b]\times[-l,l]$, then all the integrals entering this formula are taken in the sense of one-dimensional–two-dimensional Cauchy integrals. In the special case of $G(x,z)\equiv C=$ const, one has

$$A(x_0,z_0)=C\left[-\frac{\sqrt{(x_0-b)^2+(z_0-l)^2}}{(x_0-b)(z_0-l)}\right.$$

$$+\frac{\sqrt{(x_0-b)^2+(z_0+l)^2}}{(x_0-b)(z_0+l)}$$

$$+\frac{\sqrt{(x_0+b)^2+(z_0-l)^2}}{(x_0+b)(z_0-l)}$$

$$\left.+\frac{\sqrt{(x_0+b)^2+(z_0+l)^2}}{(x_0+b)(z_0+l)}\right]. \tag{3.4.16}$$

Therefore the quadrature sum for $A(x_0, z_0)$ will be constructed as follows. Let points $x_0 = -b, x_1, \ldots, x_n, x_{n+1} = b$ and $x_{00}, x_{01}, \ldots, x_{0n}$ form a canonic division of the segment $[-b, b]$ with the step h_1, whereas points $z_0 = -l, z_1, \ldots, z_N, z_{N+1} = l$ and $z_{00}, z_{01}, \ldots, z_{0N}$ form a canonic division of the segment $[-l, l]$ with the step h_2. Then we shall designate

$$
\begin{aligned}
S(x_{0j}, z_{0m})
&= \sum_{i=0}^{n} \sum_{k=0}^{N} Q(x_{0j}, z_{0k}) \int_{x_i}^{x_{i+1}} \int_{z_k}^{z_{k+1}} \frac{dx\,dz}{\left[(x_{0j} - x)^2 + (z_{0m} - z)^2\right]^{3/2}} \\
&= \sum_{i=0}^{n} \sum_{k=0}^{N} Q(x_{0j}, z_{0k}) \left[-\frac{\sqrt{(x_{0j} - x_{i+1})^2 + (z_{0m} - z_{k+1})^2}}{(x_{0j} - x_{i+1})(z_{0m} - z_{k+1})} \right. \\
&\quad + \frac{\sqrt{(x_{0j} - x_i)^2 + (z_{0m} - z_{k+1})^2}}{(x_{0j} - x_i)(z_{0m} - z_{k+1})} \\
&\quad + \frac{\sqrt{(x_{0j} - x_{i+1})^2 + (z_{0m} - z_k)^2}}{(x_{0j} - x_{i+1})(z_{0m} - z_k)} \\
&\quad \left. - \frac{\sqrt{(x_{0j} - x_i)^2 (z_{0m} - z_k)^2}}{(x_{0j} - x_i)(z_{0m} - z_k)} \right].
\end{aligned}
\tag{3.4.17}
$$

It can be shown readily that by rearranging the terms, the latter sum may be presented as a quadrature sum for the integral $A(x_{0j}, z_{0m})$ written in the form given in (3.4.15). This can be done by paying attention to the signs of the summands appearing in Formula (3.4.17) and corresponding to the point (x_i, z_k) [see Figure 3.1 where the summands are shown by points and the crosses correspond to points (x_{0j}, z_{0m})].

3.5. QUADRATURE FORMULAS FOR MULTIDIMENSIONAL SINGULAR INTEGRALS

Consider the singular integral (Mikhlin 1948)

$$
v(x_0) = \int_D \frac{f(x_0, \theta)}{r^2} u(x)\, dx, \tag{3.5.1}
$$

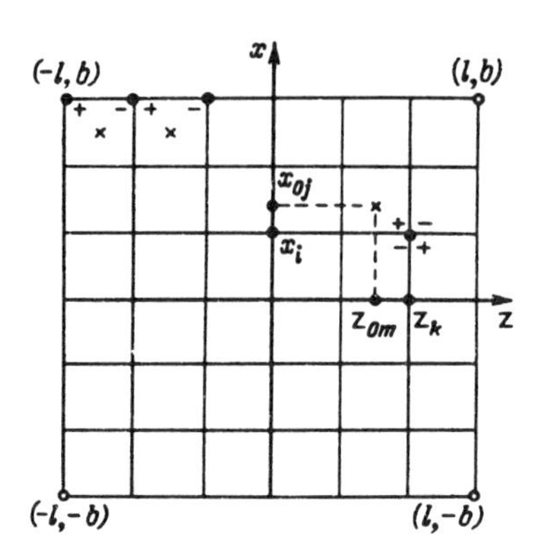

FIGURE 3.1. Distribution of signs before the summands of the quadrature formula (3.4.17), corresponding to point (x_i, z_k).

where:

1. D is a closed limited region of the plane E_2 for whose boundary Jordan's two-dimensional measure (Fikhtengoltz 1959) is equal to zero.
2. $x = (x^1, x^2)$ and $x_0 = (x_0^1, x_0^2)$ are points of region D, $r = |x - y|$, and $\theta = (x - y)/r$ is a point of the unit circle σ in E_2.
3. The density of singular integral $u(x)$ is absolutely integrable on D and meets condition $H(\alpha)$ on any closed set F lying inside D.
4. Characteristic $f(x_0, \theta)$ is limited and continuous with respect to θ for a fixed x_0.

If x_0 lies inside D, then (3.5.1) is an integral in the sense

$$v(x_0) = \lim_{\epsilon \to 0} \int_{D \setminus O(x_0, \epsilon)} \frac{f(x_0, \theta)}{r^2} u(x)\, dx, \tag{3.5.2}$$

where $O(x_0, \epsilon)$ is an ϵ neighborhood of point x_0. In Mikhlin (1948) it was shown that to ensure the existence of integral (3.5.1) in the sense of the principal value (3.5.2), it is necessary and sufficient to meet the condition

$$\int_\sigma f(x_0, \theta)\, d\sigma = 0. \tag{3.5.2'}$$

We start considering quadrature formulas for integral (3.5.1) beginning with the case when D is represented by rectangle $[a^1, b^1] \times [a^2, b^2] = I^2$. Let points $x_0^k = a^k, x_1^k, \ldots, x_{nk}^k, x_{n_{k+1}}^k = b^k$ and points $x_{0\,0}^k, x_{0\,1}^k, \ldots, x_{0nk}^k$ form a canonic division of segment $[a^k, b^k]$ when $h/h_k \le r < +\infty$ for $h \to 0$, when $h = \max\{h_1, h_2]$, $h_k = (b^k - a^k)/(n_k + 1)$, $k = 1, 2$.

Let us consider the following quadrature sum for the integral (3.5.1):

$$S(x_{0k}^1, x_{0m}^2) = \sum_{i=0}^{n_1} \sum_{j=0}^{n_2} \frac{f_{if}^{km}}{\left(r_{if}^{km}\right)^2} u\left(x_i^1, x_j^2\right) h_1 h_2, \tag{3.5.3}$$

where

$$f_{ij}^{km} = f\left(x_{0k}^1, x_{0m}^2, \theta_{ij}^{km},\right), \qquad \theta_{ij}^{km} = \frac{(x_i^1 - x_{0k}^1)\bar{l}_1 + (x_i^2 - x_{0m}^2)\bar{l}_2}{r_{ij}^{km}},$$

$$r_{ij}^{km} = \sqrt{(x_i^1 - x_{0k}^1)^2 + \left(x_j^2 - x_{0m}^2\right)^2},$$

and $\bar{l}_1$ and $\bar{l}_2$ are unit vectors of the axes OX^1 and OX^2, respectively.

Definition 3.5.1. *A characteristic $f(x_0, \theta)$ possesses property l if it is odd with respect to axis OX_0^1 or OX_0^2, where OX_0^k is given by the equation $x^k = x_0^k$, $k = 1, 2$. In other words, if points $(x_1^1, x_1^2$ and (x_2^1, x_2^2) are symmetric with respect to one of the axes, then*

$$f(x_0, \theta_1) = -f(x_0, \theta_2), \tag{3.5.4}$$

where $\theta_i = (x_0 - x_i)/|x_0 - x_i|$, $i = 1, 2$, or $f(x_0, \theta)$ is odd with respect to point $x_0 = (x_0^1, x_0^2)$.

The following theorem is true.

Theorem 3.5.1. *Let the characteristic $f(x_0, \theta)$ of integral* (3.5.1) *possess property l and be continuously integrable with respect to Cartesian coordinates of points x and $\theta \in \sigma$. Let there also be $u(x) \in H(\alpha)$ on I^2. Then the inequality*

$$|v(x_{0k}^1, x_{0m}^2) - S(x_{0k}^1, x_{0m}^2)| \le \theta(x_{0k}^1, x_{0m}^2), \tag{3.5.5}$$

holds where the quantity $\theta(x_{0k}^1, x_{0m}^2)$ *satisfies the relationship*

$$\theta(x_{0k}^1, x_{0m}^2) = O_F(h^\alpha |\ln h|) \tag{3.5.6}$$

for all points belonging to the closed set F lying inside rectangle I^2 *at a nonzero distance from its boundary.*

Consider the absolute value of the difference

$$\begin{aligned}
P_{km} &= |v(x_{0k}^1, x_{0m}^2) - S(x_{0k}^1, x_{0m}^2)| \\
&= \left| \iint_{I^2} \frac{f^{km}}{(r^{km})^2} u(x^1, x^2)\, dx^1\, dx^2 \right. \\
&\qquad \left. - \sum_{i=0}^{n_1} \sum_{j=0}^{n_2} \frac{f_{ij}^{km}}{\left(r_{ij}^{km}\right)^2} u\left(x_i^1, x_j^2\right) h_1 h_2 \right| \\
&\le \left| \iint_{I^2} \frac{f^{km}}{(r^{km})^2} (u - u_{km})\, dx^1\, dx^2 \right. \\
&\qquad \left. - \sum_{i=0}^{n_1} \sum_{j=0}^{n_2} \frac{f_{ij}^{km}}{\left(r_{ij}^{km}\right)^2} (u_{ij} - u_{km}) h_1 h_2 \right| \\
&\qquad + |u_{km}| \left| \iint_{I^2} \frac{f^{km}}{(r^{km})^2}\, dx^1\, dx^2 - \sum_{i=0}^{n_1} \sum_{j=0}^{n_2} \frac{f_{ij}^{km}}{\left(r_{ij}^{km}\right)^2} h_1 h_2 \right| \\
&= {}_1P + {}_2P,
\end{aligned} \tag{3.5.7}$$

where

$$u = u(x^1, x^2), \qquad u_{km} = u(x_{0k}^1, x_{0m}^2), \qquad u_{ij} = u\left(x_i^1, x_j^2\right),$$

$$r^{km} = \sqrt{(x^1 - x_{0k}^1)^2 + (x^2 - x_{0m}^2)^2}, \qquad f^{km} = f(x_{0k}^1, x_{0m}^2, \theta^{km}).$$

The expression for ${}_1P$ may be reduced to

$$
\begin{aligned}
{}_1P &\le \sum_{i=0}^{n_1}\sum_{j=0}^{n_2}\iint_{\Pi_{ij}} \frac{|f_{ij}^{km}|\cdot|u-u_{ij}|}{\left(r_{ij}^{km}\right)^2}\,dx^1\,dx^2 \\
&\quad + \sum_{i=0}^{n_1}\sum_{j=0}^{n_2}\iint_{\Pi_{ij}} \frac{|u-u_{km}|\cdot|f^{km}|\cdot\left(r_{ij}^{km}\right)^2-(r^{km})^2|}{(r^{km})^2\left(r_{ij}^{km}\right)^2}\,dx^1dx^2 \\
&\quad + \sum_{i=0}^{n_1}\sum_{j=0}^{n_2}\iint_{\Pi_{ij}} \frac{|u-u_{km}||f^{km}-f_{ij}^{km}|}{\left(r_{ij}^{km}\right)^2}\,dx^1\,dx^2 \\
&= {}_{11}P + {}_{12}P + {}_{13}P,
\end{aligned}
\tag{3.5.8}
$$

where Π_{ij} is a partial rectangle from the division I^2 whose lower left angular point coincides with point $M = (x_i^1, x_j^2)$.

Let us consider each sum from ${}_1\lambda P$, $\lambda = 1, 2, 3$:

$$
\begin{aligned}
{}_{11}P &\le A\sqrt{2}\,h^\alpha \sum_{i=0}^{n_1}\sum_{j=0}^{n_2}\iint_{\Pi_{ij}} \frac{dx^1\,dx^2}{\left(r_{ij}^{km}\right)^2} \\
&= A\sqrt{2}\,h^\alpha\left[8\frac{h_1h_2}{\left(\sqrt{h_1^2+h_2^2}\right)^2} + \sum_{\substack{i=0\\ i\ne k,\\ j\ne m,}}^{n_1}\sum_{\substack{j=0\\ k+1\\ m+1}}^{n_2}\iint_{\Pi_{ij}}\frac{dx^1\,dx^2}{\left(r_{ij}^{km}\right)^2}\right] \\
&\le A\sqrt{2}\,h^\alpha\left[64 + 4\int_0^{2\pi}d\varphi\int_{h/\sqrt{2}}^{\Delta}\frac{dr}{r}\right] \\
&= A\sqrt{2}\,h^\alpha\left[64 + 8\pi\left(\ln\Delta - \ln\frac{h}{\sqrt{2}}\right)\right] = O(h^\alpha \ln h),
\end{aligned}
\tag{3.5.9}
$$

where $|u - u_{ij}| \le A\sqrt{2}\,h^\alpha$ and Δ is the diameter of the integration domain.

Hence, we have

$$
\begin{aligned}
{}_{12}P &\le AB\sum_{i=0}^{n_1}\sum_{j=0}^{n_2} \frac{(r^{km})^\alpha\left(r_{ij}^{km}-r^{km}\right)\left(r_{ij}^{km}+r^{km}\right)}{(r^{km})^2\left(r_{ij}^{km}\right)^2}\,dx^1\,dx^2 \\
&\le ABC\cdot\sqrt{2}\,h\sum_{i=0}^{n_1}\sum_{j=0}^{n_2}\iint_{\Pi_{ij}}\frac{dx^1\,dx^2}{(r^{km})^{2-\alpha}\,r_{ij}^{km}}
\end{aligned}
$$

$$\leq A_1 h \cdot 4\pi \left(\frac{1}{\sqrt{h_1^2 + h_2^2}} \int_0^{3h} \frac{dr}{r^{1-\alpha}} + 2\int_{h/\sqrt{2}}^{\Delta} \frac{dr}{(r)^{2-\alpha}} \right)$$

$$= A_1 \cdot 2\pi h \left[\frac{2}{\alpha} \frac{3^\alpha h^\alpha}{\sqrt{h_1^2 + h_2^2}} + \frac{4}{1-\alpha} \left(\frac{-1}{\Delta^{1-\alpha}} + \frac{2^{1-\alpha} h^\alpha}{h} \right) \right] = O(h^\alpha), \tag{3.5.10}$$

where

$$|r_{ij}^{km} - r^{km}| < Cr_{ij} = C\sqrt{(x^1 - x_i^1)^2 + \left(x^2 - x_j^2\right)^2} \leq C\sqrt{2}\,h, r^{km}/r_{ij}^{km}$$

$$\leq \left(r_{ij}^{km} + h\sqrt{2}\,\right)/r_{ij}^{km} \leq 3$$

for any k, m, i, j.

Finally, for ${}_{13}P$ one gets

$${}_{13}P \leq A_2 \sum_{i=0}^{n_1} \sum_{j=0}^{n_2} \iint_{\Pi_{ij}} \frac{h\, dx^1\, dx^2}{\left(r_{ij}^{km}\right)^{2-\alpha}} \leq O(h). \tag{3.5.11}$$

In order to analyze the behavior of difference ${}_2P$ we note that the position of $M(x_i^1, x_j^2)$ with respect to each of the points $M_0 = (x_{0k}^1, x_{0m}^2)$ has the following property. If point M_0 is assumed to be the origin of coordinates and one draws through it axes which are parallel to the original coordinate axes, then for any point M_{ij} from the neighborhood of point M_{km} there exists points $M_{i_1 j_1}$ symmetric to point M_{ij} with respect to both point M_{km} and each of the axes. Because function $f(x_0, \theta)$ possesses the property l, its values at points M_1 and M_2 positioned symmetrically with respect to either axes OX_0^1, OX_0^2 or point M_0 are equal in magnitude but opposite in sign. Thus, we have

$${}_2P = |u_{km}| \left| \iint_{I^2 \setminus O(\delta, M_{km})} \frac{f^{km}}{\left(r^{km}\right)^2} dx^1\, dx^2 \right.$$

$$\left. - \sum_{\substack{i=0 \\ \Pi_{ij} \not\subset F_{km}^{\delta}}}^{n_1} \sum_{j=0}^{n_2} \frac{f_{ij}^{km}}{\left(r_{ij}^{km}\right)^2} h_1 h_2, \right. \tag{3.5.12}$$

where F_{km}^{δ} is a set composed of all the rectangles lying within the δ neighborhood $O(\delta, M_{km})$ of point M_{km}, whose left lower angles form a set symmetric with respect to both the point itself and the axes OX_{0k}^{1} or OX_{0m}^{2}. Thus,

$$
{}_2P \leq L\left|\iint_{\phi_{km}^{\delta}} \frac{f^{km}}{\left(r^{km}\right)^2}\,dx^1\,dx^2 - \sum_{\substack{i=0\\ \Pi_{ij}\not\subset \Pi\setminus O(\delta, M_{km})}}^{n_1}\sum_{j=0}^{n_2} \frac{f_{ij}^{km}}{\left(r_{ij}^{km}\right)^2}h_1h_2 \right.
$$

$$
\left. + \sum_{i=0}^{n_1}\sum_{j=0}^{n_2}\iint_{\Pi_{ij}\subset\Pi\setminus[O(\delta, M_{km})\cup\phi_{km}^{\delta}]}\left[\frac{f^{km}}{\left(r^{km}\right)^2} - \frac{f_{ij}^{km}}{\left(r_{ij}^{km}\right)^2}\right]dx^1\,dx^2\right|, \tag{3.5.13}
$$

where $\phi_{km}^{\delta} = \Pi \setminus \cup_{s,t}\Pi_{s,t}$, $\Pi_{s,t} \subset \Pi \setminus O(\delta, M_{km})$, and $L = \max_{(x^1,x^2)\in I^2}|u(x^1, x^2)|$.

In view of the properties of function $f(x_0, \theta)$ we have

$$
{}_2P \leq LC_1\left(\frac{1}{\delta}\pi 4\sqrt{2}\,h \frac{\delta}{\left(\delta - h\sqrt{2}\right)^2}\pi 4\sqrt{2}\,h + O(h)\right) = O(h). \tag{3.5.14}
$$

because an analysis of the latter sum entering Formula (3.5.13), similar to that done when calculating ${}_{12}P$ and ${}_{13}P$, allows one to conclude that the sum is of the order of h. The preceding estimates of ${}_{1i}P$, $i = 1, 2, 3$, and ${}_2P$ demonstrate the validity of Inequality (3.5.5).

Note that Inequality (3.5.5) remains valid if the rectangle I^2 is substituted by an arbitrary closed limited plane region D whose boundary has zero two-dimensional Jordan measure. If such is the case, then the quadrature sum must be composed as follows. Consider rectangle Π incorporating region D and divided by straight lines parallel to the coordinate axes and separated from each other by the distance h_i. The rectangle fully belonging to region D will be numbered and denoted by Π_i, $i = 1, \ldots, n$. By x_{0i} we denote the center of a rectangle Π_i, and by x_i we denote the left lower angle of the rectangle. The points thus selected are used for constructing the quadrature.

However, if characteristic $f(x_0, \theta)$ does not possess the property l, then Inequality (3.5.5) may be invalid for an arbitrary relationship between h_1 and h_2. For example, function $f(x_0, \theta) = \cos 2\varphi$ (where φ is the polar angle of point x with respect to point x_0) does not possess the property l, and for it Inequality (3.5.5) is valid for $h_1 = h_2$ only. In fact, $\cos 2(\varphi +$

$\pi/2) = \cos(2\varphi + \pi) = -\cos 2\varphi$. On the other hand, all the points M_{ij} lying within a circle centered at point M_{km} may be divided into pairs in which a point is obtained from another point by means of rotation through angle $\pi/2$ about point M_{km}.

Inequality (3.5.5) is also likely to be invalid for the function $f(x_0, \theta) = \cos 4\varphi$ for $h_1 = h_2 = h$. In fact,

$$\int_K \frac{\cos 4\varphi}{r^2}\, dx^1\, dx^2 = 0, \tag{3.5.15}$$

where K is a circle of radius R centered at the origin of coordinates. Consider the square I^2 circumscribing the circle K, whose sides are parallel to the coordinate axes. Next consider a division of I^2 into small squares whose sides are equal to h and whose centers coincide with respective origins of coordinates. We assume that

$$V_h = \sum_{i=1}^{n} \frac{\cos 4\varphi_i}{r_i^2} h^2.$$

Calculations carried out for $h_1 = R/5$ and $h_2 = R/10$ have shown that

$$V_{h1} = -8 + 0.552, \qquad V_{h2} = -8 + 0.486.$$

3.6. EXAMPLES OF CALCULATING SINGULAR INTEGRALS

1. The following equality is valid:

$$I(x_0) = \int_{-1}^{1} \frac{\sqrt{1-x^2}\, dx}{x_0 - x} = \pi x_0, \qquad x_0 \in (-1, 1). \tag{3.6.1}$$

The latter integral was calculated by using the quadrature formula

$$I_n(x_{0j}) = \sum_{i=1}^{n} \frac{\sqrt{1-x_i^2}\, h}{x_{0j} - x_i}, \qquad j = 1, \ldots, n+1, \tag{3.6.2}$$

for equally spaced grid points with the step $h = 2/(n+1)$. The calculated results are presented in Figure 3.2, where the solid line

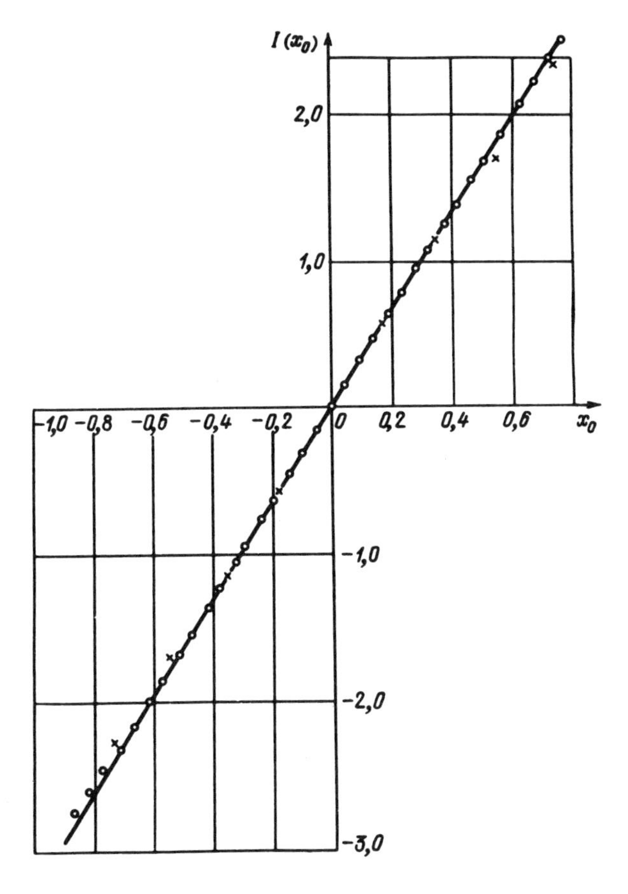

FIGURE 3.2. Calculation of singular integral in (3.6.1) with the help of quadrature formula (3.6.2). The solid line corresponds to the exact value, ×× to $h = 2/11$ and ∞ to $h = 2/17$.

corresponds to the accurate value, and ×× and ∞ correspond to $h = 2/11$ and $2/41$, respectively.

2. For an integral of the form

$$I(x_0, z_0) = \frac{1}{\pi^2} \int_{-1}^{1} \int_{-1}^{1} \sqrt{\frac{(1-x)(1-z)}{(1+x)(1+z)}} \frac{dx\,dz}{(x_0 - x)(z_0 - z)} = 1, \tag{3.6.3}$$

calculations were carried out with the help of equally spaced data. The results are shown in Figure 3.3, where the solid curve corresponds to the accurate value ×× and ∞ correspond to $h = 2/9$ and

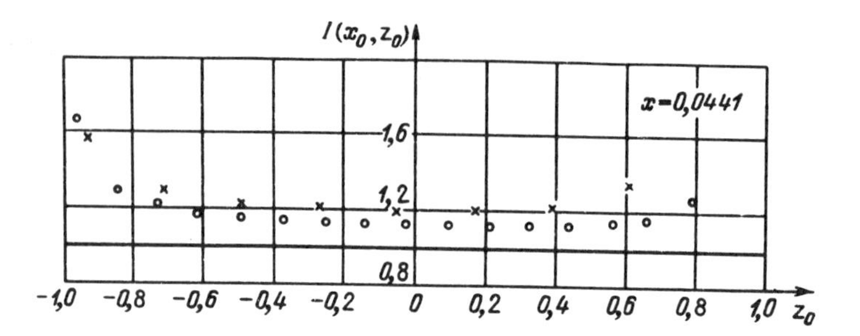

FIGURE 3.3. Calculation of singular integral in (3.6.3) by using equally spaced grid points for $x_0 = 0.0441$. The solid line corresponds to the exact value, $\times\times$ to $h = 2/11$, and $\circ\circ$ to $h = 2/17$.

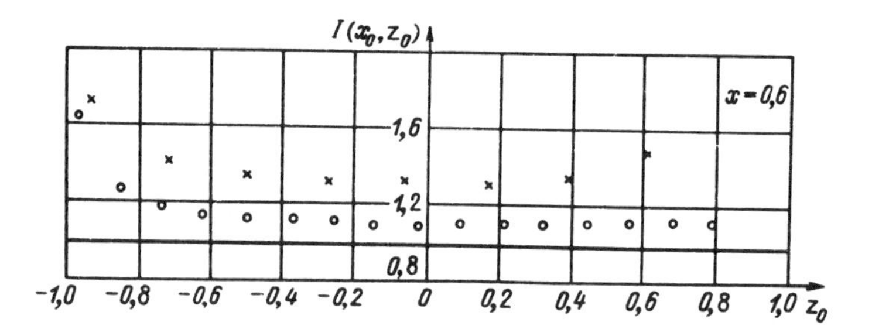

FIGURE 3.4. Calculation of singular integral in (3.6.3) by using equally spaced grid points for $x_0 = 0.6$. The solid line corresponds to the exact value, $\times\times$ to 8 points and $\circ\circ$ to 16 points.

2/17, respectively (for $x_0 = 0.0641$), and in Figure 3.4, where the solid curve corresponds to the accurate value and $\times\times$ and $\circ\circ$ correspond to 8 and 16 points, respectively. For calculating we used the formula

$$I_{n_1, n_2}(x_{0j}, z_{0m}) = \frac{1}{\pi^2} \sum_{i=1}^{n_1} \sum_{k=1}^{n_2} \sqrt{\frac{(1 - x_i)(1 - z_k)}{(1 + x_i)(1 + z_k)}} \times \frac{h_1 h_2}{(x_{0j} - x_i)(z_{0m} - z_k)}, \tag{3.6.4}$$

$j = 1, \ldots, n_1$, $m = 1, \ldots, n_2$. It was supposed that $h_1 = h_2 = h$.

3. Calculations of integral in (3.6.3) carried out with the help of equally spaced grid points corresponding to the weight function under consideration have produced more accurate results as compared with the

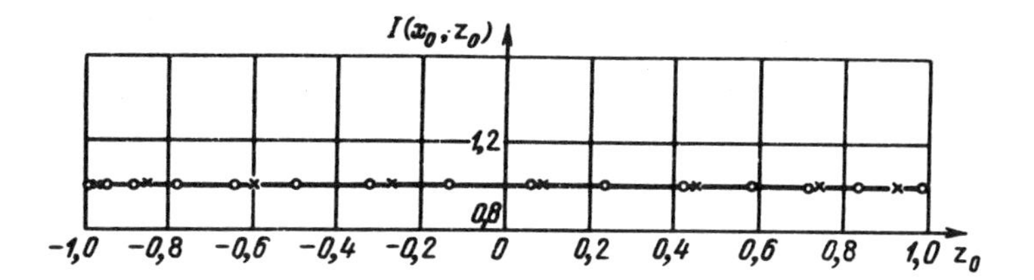

FIGURE 3.5. Calculation of singular integral in (3.6.3) by using unequally spaced grid points for an arbitrary x_0. The solid line corresponds to the exact value, × × to 8 points and ∘∘ to 16 points.

equally spaced data. We used the quadrature formula

$$I_{n_1,n_2}(x_{0j}, z_{0m}) = \sum_{i=1}^{n_1} \sum_{k=1}^{n_2} \frac{a_i b_k}{(x_{0j} - x_i)(z_{0m} - z_k)},$$

$$x_i = z_i = \cos\frac{2\pi i}{2n+1}, \qquad x_{0i} = z_{0i} = \cos\frac{2i-1}{2n+1}\pi,$$

$$a_i = b_i = \frac{4\pi}{2n+1}\sin^2\frac{\pi i}{2n+1} \quad \text{for } n_1 = n_2 = n. \tag{3.6.5}$$

The results presented in Figure 3.5, where the solid curve corresponds to the accurate solution, and × × and ∘∘ correspond to 8 and 16 points, respectively, have shown that

$$|I(x_{0j}, z_{0m}) - I_{n_1,n_2}(x_{0j}, z_{0m})| \le 10^{-4} \tag{3.6.6}$$

for any j and m.

However, if the weight function is unknown or is of a complex form, unequally spaced grid points are preferable for obtaining preliminary data.

4

Poincaré–Bertrand Formula

4.1. ONE-DIMENSIONAL SINGULAR INTEGRALS

In Muskhelishvili (1952) the Poincaré–Bertrand formula was derived. It regulates the sequence of integration in one-dimensional singular integrals:

$$\int_L \frac{dt}{t - t_0} \int_L \frac{\varphi(t, \tau)\, d\tau}{\tau - t} = \int_L d\tau \int_L \frac{\varphi(t, \tau)\, dt}{(t - t_0)(\tau - t)} - \pi^2 \varphi(t_0, t_0), \quad (4.1.1)$$

where L is a piecewise smooth curve, and function $\varphi(t, \tau)$ has the form

$$\varphi(t, \tau) = \frac{\varphi^*(t, \tau)}{\Pi(t, \tau)}, \quad (4.1.2)$$

where $\varphi^*(t, \tau)$ is a function belonging to the class H_0 on L (i.e., a function meeting the condition H on each of the smooth arcs forming L) and

$$\Pi(t, \tau) = \prod_{k=1}^{n} |t - c_k|^{\alpha_k} |\tau - c_k|^{\beta_k},$$

where c_k, $k = 1, \ldots, n$, are all the nodes of the curve L, $\alpha_k, \beta_k \geq 0$, and $\alpha_k + \beta_k < 1$.

In this section we prove Formula (4.1.1) for a piecewise Lyapunov curve L containing only angular nodes by means of singular integral quadrature formulas.

Let L be a segment $[a, b]$. Then (4.1.2) becomes

$$\varphi(t, \tau) = \frac{\varphi^*(t, \tau)}{(t - a)^{\nu_1} (\tau - a)^{\nu_2} (b - t)^{\mu_1} (b - \tau)^{\mu_2}}, \quad (4.1.3)$$

where $\varphi^*(t, \tau) \in H(\alpha)$ on $[a, b] \times [a, b]$, $\nu_k, \mu_k \geq 0$, $k = 1, 2$, $\nu_1 + \nu_2 < 1$, and $\mu_1 + \mu_2 < 1$.

Consider the integral

$$A(t_0) = \int_a^b \frac{dt}{t - t_0} \int_a^b \frac{\varphi(t, \tau)\, dt}{\tau - t}, \qquad t_0 \in (a, b). \tag{4.1.4}$$

Let us choose two sets of points, $E = \{\tau_k,\ k = 1, \ldots, n\}$ and $E_0 = \{t_j,\ j = 1, \ldots, n\}$, on the segment $[a, b]$, forming such a canonic division of the segment $[a, b]$ with the step h that point t_0 belongs to the set E for $k = k_0$.

Consider the sums

$$\mathfrak{B}_n(t_0) = \sum_{j=1}^{n} \frac{\Delta t_j}{t_j - t_0} \sum_{k=1}^{n} \frac{\varphi(t_j, \tau_k)\, \Delta\tau_k}{\tau_k - t_j}, \tag{4.1.5}$$

where $\Delta t_j = \Delta\tau_k = h$, $k, j = 1, \ldots, n$, and

$$B_n(t_0) = \sum_{j=1}^{n} \frac{\Delta t_j}{t_j - t_0} \int_a^b \frac{\varphi(t_j, \tau)\, dr}{\tau - t_j}. \tag{4.1.6}$$

Then taking into account (1.3.14) and (4.1.3), one gets

$$\begin{aligned} |B_n(t_0) - \mathfrak{B}_n(t_0)| \leq \sum_{j=1}^{n} \frac{h}{|t_j - t_0|} \Big[&\eta(t_j, \nu_1 + \nu_2, \mu_1 + \mu_2) h^{\alpha} \\ &+ \eta(t_j, 2\nu_2 + \nu_1, \mu_1 + \mu_2) h^{\nu_2} \\ &+ \eta(t_j, \nu_1 + \nu_2, \mu_1 + 2\mu_2) h^{\mu_2} \\ &+ \eta(t_j, 1 + \nu_1, \mu_1 + \mu_2) h^{1-\nu_2} \\ &+ \eta(t_j, \nu_1 + \nu_2, 1 + \mu_1) h^{1-\mu_2} \\ &+ \eta(t_j, 1 + \nu_1 + \nu_2, 1 + \mu_1 + \mu_2) h \Big] O(|\ln h|). \end{aligned} \tag{4.1.7}$$

After removing the brackets and taking into account that point t_0 is fixed within (a, b), we see that all the resulting sums are of the order of h^{λ}

(where $\lambda > 0$ is a certain fixed number), i.e.,

$$|B_n(t_0) + \mathfrak{B}_n(t_0)| \leq O_{t_0}(h^{\lambda}), \qquad \lambda > 0. \tag{4.1.8}$$

In Muskhelishvili (1952) it was shown that

$$\varphi_1(t) \equiv \int_a^b \frac{\varphi(t, \tau)\, d\tau}{\tau - t} \in H^*$$

on $[a, b]$. Therefore, taking into account that point t_0 is fixed and is one of the points τ_k, $k = 1, \ldots, n$, we get

$$|A(t_0) - B_n(t_0)| \leq O_{t_0}(h^{\lambda_1}), \qquad \lambda_1 > 0. \tag{4.1.9}$$

where λ_1 is a fixed number.

From (4.1.8) and (4.1.9) it follows that

$$|A(t_0) - \mathfrak{B}_n(t_0)| \leq O_{t_0}(h^{\lambda_2}), \qquad \lambda_2 > 0, \tag{4.1.10}$$

i.e.,

$$\lim_{n \to \infty} \mathfrak{B}_n(t_0) = A(t_0). \tag{4.1.11}$$

Let us now transform the sum $\mathfrak{B}_n(t_0)$ in the following way:

$$\begin{aligned} \mathfrak{B}_n(t_0) &= \sum_{k=1}^{n} \Delta\tau_k \sum_{j=1}^{n} \frac{\varphi(t_j, \tau_k)\, \Delta t_j}{(t_j - t_0)(\tau_k - t_j)} = \sum_{\substack{k=1 \\ k \neq k_0}}^{n} \frac{\Delta\tau_k}{\tau_k - t_0} \\ &\times \left[\sum_{j=1}^{n} \frac{\varphi(t_j, \tau_k)\, \Delta t_j}{t_j - t_0} - \sum_{j=1}^{n} \frac{\varphi(t_j, \tau_k)\, \Delta t_j}{t_j - \tau_k} \right] \\ &+ \Delta\tau_{k_0} \sum_{j=1}^{n} \frac{\varphi(t_j, t_0)\, \Delta t_j}{(t_j - t_0)(t_0 - t_j)} \\ &= \mathfrak{B}_n^1(t_0) + \mathfrak{B}_n^2(t_0). \end{aligned} \tag{4.1.12}$$

By removing the square brackets in $\mathfrak{B}_n^1(t_0)$ and taking into account the remarks made with respect to the sum S_j' and integral I_j' in Formulas

(1.3.17) and (1.3.18), respectively, one gets

$$\lim_{n\to\infty} \mathfrak{B}_n^1(t_0) = \int_a^b \frac{d\tau}{\tau - t_0}\int_a^b \frac{\varphi(t,\tau)\,dt}{t - t_0} - \int_a^b \frac{d\tau}{\tau - t_0}\int_a^b \frac{\varphi(t,\tau)\,dt}{t - \tau}$$

$$= \int_a^b d\tau \int_a^b \frac{\varphi(t,\tau)\,dt}{(t - t_0)(\tau - t)}. \tag{4.1.13}$$

Next we consider $\mathfrak{B}_n^2(t_0)$. We have

$$\mathfrak{B}_n^2(t_0) = \sum_{j=1}^{n} \frac{\varphi(t_j, t_0)\,\Delta t_j\,\Delta\tau_{k_0}}{(t_j - t_0)^2} = -\varphi(t_0,t_0)\mathfrak{B}_n^3 - \mathfrak{B}_n^4,$$

$$\mathfrak{B}_n^3 = \sum_{j=1}^{n} \frac{\Delta t_j\,\Delta\tau_{k_0}}{(t_j - t_0)^2}, \qquad \mathfrak{B}_n^4 = \sum_{j=1}^{n} \frac{\left[\varphi(t_j,t_0) - \varphi(t_0,t_0)\right]\Delta t_j\,\Delta\tau_{k_0}}{(t_j - t_0)^2}. \tag{4.1.14}$$

Let us divide the sum $\mathfrak{B}_n^4(t_0)$ into two summands, one of which summation is carried out over points $t_j \in [(a + t_0/2, (b + t_0)/2] = \tilde{I}$, and the other over all the rest of t_j. In other words

$$\mathfrak{B}_n^4(t_0) = \sum_{t_j\in[a,b]\setminus\tilde{I}} \psi(t_j,t_0)\,\Delta t_j\,\Delta\tau_{k_0} + \sum_{t_j\in\tilde{I}} \psi(t_j,t_0)\,\Delta t_j\,\Delta\tau_{k_0} = D_1 + D_2,$$

$$\psi(t_j,t_0) = \left[\varphi(t_j,t_0) + \varphi(t_0,t_0)\right]/(t_j - t_0)^2. \tag{4.1.15}$$

By (4.1.3) we have

$$|D_1| \le C_1 h\left[\int_a^{(a+t_0)/2} \frac{dt}{(t-a)^{\nu_1}} + \int_{(b+t_0)/2}^{b} \frac{dt}{(b-t)^{\mu_1}}\right] = O(h), \tag{4.1.16}$$

because $0 \le \nu_1, \mu_1 < 1$.

As previously noted, function $\varphi(t, t_0)$ belongs to the class $H(\alpha)$ with respect to t on $\tilde{I}$. Hence,

$$|D_2| \le C_2 \sum_{t_j\in\tilde{I}} \frac{\Delta t_j\,\Delta\tau_{k_0}}{|t_j - t_0|^{2-\alpha}} \le C_2 h^\alpha \sum_{j=1}^{n} \frac{1}{|j - k_0 - \frac{1}{2}|^{2-\alpha}}$$

$$\le C_2 h^\alpha 2 \sum_{k=0}^{\infty} \frac{1}{(k + \frac{1}{2})^{2-\alpha}} = O(h^\alpha).$$

Thus, for $\mathfrak{B}_n^4(t_0)$ we have

$$|\mathfrak{B}_n^4(t_0)| \leq O(h^\alpha). \tag{4.1.17}$$

Note that if $\alpha = 1$, then $\mathfrak{B}_n^4(t_0) = O(h|\ln h|)$.

Hence, for $\mathfrak{B}_n^3(t_0)$ we have

$$\mathfrak{B}^3(t_0) = \sum_{j=1}^{n} \frac{1}{(j - k_0 - \frac{1}{2})^2} = 2\sum_{k=0}^{m} \frac{1}{(k + \frac{1}{2})^2} + \sum_{k=m+1}^{M} \frac{1}{(k + \frac{1}{2})^2}, \tag{4.1.18}$$

where $m = \min(k_0, n - k_0 - 1)$ and $M = \max(k_0, n - k_0 - 1)$.

Because t_0 is a fixed point, $n \to \infty$ and $m \to \infty$ imply that $M \to m$. The series $\Sigma_{k=0}^{\infty}(k + \frac{1}{2})^{-2}$ converges because the second summand in the latter formula tends to zero for $n \to \infty$. Also, it is shown in Hardy (1949) that

$$\sum_{k=0}^{\infty} \frac{1}{(2k + 1)^2} = \frac{\pi^2}{8}. \tag{4.1.19}$$

Hence,

$$\lim_{n\to\infty} \mathfrak{B}_n^3(t_0) = \pi^2. \tag{4.1.20}$$

Then from (4.1.14), (4.1.17), and (4.1.20), it follows that

$$\lim_{n\to\infty} \mathfrak{B}_n^2(t_0) = -\pi^2\varphi(t_0, t_0). \tag{4.1.21}$$

Thus, Equations (4.1.11)–(4.1.14), (4.1.17), and (4.1.21) demonstrate the validity of the Poincaré–Bertrand formula for a singular integral over a segment.

It should be noted that the proof relies mainly on the grid points lying in the neighborhood of point t_0 or, more so, on the relative positions of points τ_k and t_j within the said neighborhood. Hence, we deduce that a similar proof is valid in the case when L is a piecewise Lyapunov curve containing angular nodes only, in particular for a circle. In fact, if we choose two arbitrary points a_1 and a_2 lying on a circle ($a_1 \neq t_0$, $a_2 \neq t_0$) and assume that they are nodes, then we get a curve for which the Poincaré–Bertrand formula has been proved already. For a circle the formula may be proved in a straightforward manner by using the quadrature formula presented in Section 5.1.

In what follows we will come across the following situation when we have to represent integral $A(t_0)$ over $L = [a, b]$, as well as both sides of the Poincaré–Bertrand formula, by quadrature sums.

Let the sets $E = \{\tau_k, k = 1, \dots, n\}$ and $E_0 = \{t_j, j = 0, 1, \dots, n\}$ form a canonic division of segment $[a, b]$ with the step h. Let integral $A(t_0)$ be considered at points $t_{0m} = \tau_m$, $m = 1, \dots, n$, i.e.,

$$A(\tau_m) = \int_a^b \frac{dt}{t - \tau_m} \int_a^b \frac{\varphi(t, \tau)\, dt}{\tau - t}. \tag{4.1.22}$$

In a similar way we form the sum

$$\mathfrak{B}_n(\tau_m) = \sum_{j=0}^{n} \frac{\Delta t_j}{t_j - \tau_m} \sum_{k=1}^{n} \frac{\varphi(t_j, \tau_k)\, \Delta\tau_k}{\tau_k - t_j}. \tag{4.1.23}$$

Then the following theorem is true.

Theorem 4.1.1. *Let function $\varphi(t, \tau)$ be of the form* (4.1.3) *on $[a, b]$. Then*

$$|A(\tau_m) - \mathfrak{B}_n(\tau_m)| \le \theta(\tau_m), \qquad m = 1, \dots, n, \tag{4.1.24}$$

where quantity $\theta(\tau_m)$ satisfies Inequalities (1.3.8) *and* (1.3.9).

Finally, we observe that the Poincaré–Bertrand formula remains valid if function $\varphi(t, \tau)$ is replaced by function $\varphi(t, \tau, \xi)$ of the form

$$\varphi(t, \tau, \xi) = \frac{\varphi^*(t, \tau, \xi)}{\prod_{k=1}^{p} |t - c_k|^{\beta_k} |\tau - c_k|^{\lambda_k}}, \tag{4.1.25}$$

where $\varphi^*(t, \tau, \xi) \in H$ as a function of three variables within the region under consideration, and the set of values of ξ is limited. Note that point ξ may be a point of an m-dimensional space. In the latter case the Poincaré–Bertrand formula becomes

$$\int_L \frac{dt}{t - t_0} \int_L \frac{\varphi(t, \tau, \xi)}{\tau - t}\, d\tau = \int_L d\tau \int_L \frac{\varphi(t, \tau, \xi)\, dt}{(t - t_0)(\tau - t)} - \pi^2 \varphi(t_0, t_0, \xi). \tag{4.1.26}$$

4.2. MULTIPLE SINGULAR CAUCHY INTEGRALS

Let us start considering the Poincaré–Bertrand formula for multiple singular integrals with the two-dimensional case.

Let function $\varphi(t_0, t)$, where $t_0 = (t_0^1, t_0^2)$ and $t = (t^1, t^2)$, be defined on $L = L_1 \times L_2$ and have the form (3.1.12). The curves L_1 and L_2 are piecewise Lyapunov and contain angular nodes only. Let us show that in this case Formula (4.1.1) is valid.

1. Let $L_1 = [a_1, b_1]$ and $L_2 = [a_2, b_2]$. Consider the integral

$$A(t_0^1, t_0^2) = \iint_{L_1 \times L_2} \frac{dt^1\, dt^2}{(t^1 - t_0^1)(t^2 - t_0^2)}$$

$$\times \iint_{L_1 \times L_2} \frac{\varphi(t^1, t^2, \tau^1, \tau^2)\, d\tau^1\, d\tau^2}{(\tau^1 - t^2)(\tau^2 - t^2)}.$$

On segment $[a_p, b_p]$ we take sets $E^p = \{\tau_{k_p}^p, k_p = 1, \ldots, n_p\}$ and $E_0^p = \{t_{j_p}^p, j_p = 0, 1, \ldots, n_p\}$, which form a division of segment $[a_p, b_p]$ and are such that for $k_p = k_p^0$ point t_0^p is a point of the set E^p, $p = 1, 2$. After introducing again the sum

$$\mathfrak{B}_{n_1 n_1}(t_0^1, t_0^2) = \sum_{j_1=1}^{n_1} \sum_{j_2=1}^{n_2} \frac{\Delta t_{j_1}^1\, \Delta_{j_2}^2}{\left(t_{j_1}^1 - t_0^1\right)\left(t_{j_2}^2 - t_0^2\right)}$$

$$\times \sum_{k_1=1}^{n_1} \sum_{k_2=1}^{n_2} \frac{\varphi\left(t_{j_1}^1, t_{j_2}^2, \tau_{k_1}^1, \tau_{k_2}^2\right) \Delta\tau_{k_1}^1\, \Delta\tau_{k_2}^2}{\left(\tau_{k_1}^1 - t_{j_1}^1\right)\left(\tau_{k_2}^2 - t_{j_2}^2\right)}$$

and using Formulas (3.3.3) and (3.3.6) we deduce that

$$\lim_{n_1 n_2 \to \infty} \mathfrak{B}_{n_1 n_2}(t_0^1, t_0^2) = A(t_0^1, t_0^2). \tag{4.2.1}$$

Let us transform the sum $\mathfrak{B}_{n_1 n_2}(t_0^1, t_0^2)$ as

$$\mathfrak{B}_{n_1 n_2}(t_0^1, t_0^2) = \sum_{k_1=1}^{n_1} \sum_{k_2=1}^{n_2} \Delta\tau_{k_1}^1\, \Delta\tau_{k_2}^2 \sum_{j_1=1}^{n_1} \sum_{j_2=1}^{n_2} \frac{\varphi\left(t_{j_1}^1, t_{j_2}^2, \tau_{k_1}^1, \tau_{k_2}^2\right)}{\left(t_{j_1}^1 - t_0^1\right)\left(t_{j_2}^2 - t_0^2\right)}$$

$$\times \frac{\Delta t_{j_1}^1\, \Delta_{j_2}^2}{\left(\tau_{k_1}^1 - t_{j_1}^1\right)\left(\tau_{k_2}^2 - t_{j_2}^2\right)}$$

$$= \sum_{\substack{k_1=1 \\ k_1 \neq k_1^0,}}^{n_1} \sum_{\substack{k_2=1 \\ k_2 \neq k_2^0}}^{n_2} \frac{\Delta\tau_{k_1}^1\, \Delta\tau_{k_2}^2}{\left(\tau_{k_1}^1 - t_0^1\right)\left(\tau_{k_2}^2 - t_0^2\right)} S_{n_1 n_2}$$

$$-\Delta\tau^1_{k^0_1}\sum_{\substack{k_1=1\\k_2\neq k^0_2}}^{n_2}\Delta\tau^2_{k_2}\sum_{j_1=1}^{n_1}\sum_{j_2=1}^{n_2}$$

$$\times\frac{\varphi\left(t^1_{j_1},t^2_{j_2},t^1_0,\tau^2_{k_2}\right)\Delta t^1_{j_1}\,\Delta t^2_{j_2}}{\left(t^1_{j_1}-t^1_0\right)^2\left(t^2_{j_2}-t^2_0\right)\left(\tau^2_{k_2}-t^2_{j_2}\right)}$$

$$+\Delta\tau^2_{k^0_2}\sum_{\substack{k_1=1\\k_1\neq k^0_1}}^{n_1}\Delta\tau^1_{k_1}\sum_{j_1=1}^{n_1}\sum_{j_1=1}^{n_2}$$

$$\times\frac{\varphi\left(t^1_{j_1},t^2_{j_2},\tau^1_{k_1},t^2_0\right)\Delta^1_{j_1}\,\Delta t^2_{j_2}}{-\left(t^2_{j_2}-t^2_0\right)^2\left(t^1_{j_1}-t^1_0\right)\left(\tau^1_{k_1}-t^1_{j_1}\right)}$$

$$+\Delta\tau^1_{k^0_1}\,\Delta\tau^2_{k^0_2}\sum_{j_1=1}^{n_1}\sum_{j_2=1}^{n_2}\frac{\varphi\left(t^1_{j_1},t^2_{j_2},t^1_0,t^2_0\right)\Delta t^1_{j_1}\,\Delta t^2_{j_2}}{\left(t^1_{j_1}-t^1_0\right)^2\left(t^2_{j_1}-t^2_0\right)^2}$$

$$=S^1_{n_1n_2}+S_{n_2}+S_{n_1}+S, \tag{4.2.2}$$

where

$$S_{n_1n_2}=\sum_{j_1=1}^{n_1}\sum_{j_2=1}^{n_2}\left[\frac{1}{\left(t^1_{j_1}-t^1_0\right)\left(t^2_{j_2}-t^2_0\right)}-\frac{1}{\left(t^1_{j_1}-t^1_0\right)\left(t^2_{j_2}-\tau^2_{k_2}\right)}\right.$$

$$\left.-\frac{1}{\left(t^1_{j_1}-\tau^1_{k_1}\right)\left(t^2_{j_2}-t^2_0\right)}+\frac{1}{\left(t^1_{j_1}-\tau^1_{k_1}\right)\left(t^2_{j_2}-\tau^2_{k_2}\right)}\right]$$

$$\times\varphi\left(t^1_{j_1},t^2_{j_2},\tau^1_{k_1},\tau^2_{k_2}\right)\Delta t^1_{j_1}\,\Delta t^2_{j_2}. \tag{4.2.3}$$

By removing the square brackets in $S_{n_1n_2}$ and using Inequality (3.3.3) as was done for $\mathfrak{B}_{n_1n_2}(t^1_0,t^2_0)$, we get for $S^1_{n_1n_2}$

$$\lim_{n_1,n_2\to\infty}S^1_{n_1n_2}=\iint_{L_1\times L_2}d\tau^1\,d\tau^2\iint_{L_1\times L_2}$$

$$\times\frac{\varphi(t^1,t^2,\tau^1,\tau^2)\,dt^1\,dt^2}{(t^1-t^1_0)(t^2-t^2_0)(\tau^1-t^1)(\tau^2-t^2)}. \tag{4.2.4}$$

For S_{n_2} we have

$$S_{n_2} = \sum_{\substack{k_2=1 \\ k_2 \neq k_2^0}}^{n_2} \Delta\tau_{k_2}^2 \sum_{j_2=1}^{n_2} \frac{\varphi\left(t_0^1, t_{j_2}^2, t_0^1, \tau_{k_2}^2\right) \Delta t_{j_2}^2}{\left(t_{j_2}^2 - t_0^2\right)\left(\tau_{k_2}^2 - t_{j_2}^2\right)} \sum_{j_1=1}^{n_1} \frac{\Delta\tau_{k_1^0}^1 \Delta t_{j_1}^1}{-\left(t_{j_1}^1 - t_0^1\right)^2}$$

$$+ \sum_{\substack{k_2=1 \\ k_2 \neq k_2^0}}^{n_2} \Delta\tau_{k_2}^2 \sum_{j_2=1}^{n_2} \frac{\Delta t_{j_2}^2}{\left(t_{j_2}^2 - t_0^2\right)\left(\tau_{k_2}^2 - t_{j_2}^2\right)}$$

$$\times \left\{ \sum_{j_1=1}^{n_1} \frac{\Delta\tau_{k_1^0}^1 \Delta t_{j_1}^2}{-\left(t_{j_1}^1 - t_0^1\right)^2} \left[\varphi\left(t_{j_1}^1, t_{j_2}^2, t_0^1, \tau_{k_2}^2\right) - \varphi\left(t_0^1, t_{j_2}^2, t_0^1, \tau_{k_2}^2\right)\right] \right\}$$

$$= S_{n_2}^1 + S_{n_2}^2. \tag{4.2.5}$$

From Section 4.1 it follows that

$$\lim_{n_1, n_2 \to \infty} S_{n_2}^1 = -\pi^2 \int_{L_2} d\tau^2 \int_{L_2} \frac{\varphi(t_0^1, t^2, t_0^1, \tau^2)\, dt^2}{(t^2 - t_0^2)(\tau^2 - t^2)}. \tag{4.2.6}$$

The sum $S_{n_2}^2$ will be estimated similarly to the sum $\mathfrak{B}_n^4(t_0)$ (see the preceding section): namely, we split the sum P_{n_2} between the braces in the formula for $S_{n_2}^2$ into the sum over the points $t_{j_1}^1$ belonging to neighborhood $O(t_0^1)$ whose closure does not contain points a_1 and b_1 and into the sum over the points $t_{j_1}^1$ lying outside the neighborhood.

Remember that

$$\varphi\left(t_{j_1}^1, t_{j_2}^2, t_0^1, \tau_{k_2}^2\right) - \varphi\left(t_0^1, t_{j_2}^2, t_0^1, \tau_{k_2}^2\right)$$

$$= \frac{\eta\left(t_{j_2}^2, \beta_1^2, \beta_2^2\right)\eta\left(\tau_{k_2}^2, \lambda_1^2, \lambda_2^2\right)}{(t_0^1 - a_1)^{\lambda_1^1}(b_1 - t_0^1)^{\lambda_2^1}}$$

$$\times \left[\frac{\varphi^*\left(t_{j_1}^1, t_{j_2}^2, t_0^1, \tau_{k_2}^2\right)}{\left(t_{j_1}^1 - a_1\right)^{\beta_1^1}\left(b_1 - t_{j_1}^1\right)^{\beta_2^1}} - \frac{\varphi^*\left(t_0^1, t_{j_2}^2, t_0^1, \tau_{k_2}^2\right.}{(t_0^1 - a_1)^{\beta_1^1}(b_1 - t_0^1)^{\beta_2^1}} \right]. \tag{4.2.7}$$

Therefore,

$$|P_{n_2}| \le O(h^\beta)\eta(t_2^2, \beta_1^1, \beta_2^2)\eta\left(\tau_{k_2}^2, \lambda_1^1, \lambda_2^2\right), \qquad \beta > 0. \tag{4.2.8}$$

Now with the use of the estimates presented in Section 1.3 for a singular integral over a segment, one gets

$$\lim_{n_1, n_2 \to \infty} |S_{n_2}^2| \le \lim_{n_1, n_2 \to \infty} O(h^\lambda \ln^2 h) = 0, \qquad \lambda > 0. \tag{4.2.9}$$

From (4.25), (4.2.6), and (4.2.9) we deduce that

$$\lim_{n_1, n_2 \to \infty} S_{n_2} = -\pi^2 \int_{L_2} d\tau^2 \int_{L_2} \frac{\varphi(t_0^1, t^2, t_0^1, \tau^2)\, dt^2}{(t^2 - t_0^2)(\tau^2 - t^2)}. \tag{4.2.10}$$

In a similar way it may be proved that

$$\lim_{n_1, n_2} S_{n_1} = -\pi^2 \int_{L_1} d\tau_1 \int_{L_1} \frac{\varphi(t^1, t_0^2, \tau^2, t_0^2)\, dt^1}{(t^1 - t_0^1)(\tau^1 - t^1)}, \tag{4.2.11}$$

$$\lim_{n_1, n_2 \to \infty} S = \pi^4 \varphi(t_0^1, t_0^2, t_0^1, t_0^2). \tag{4.2.12}$$

The validity of Formula (3.1.16) for the case under consideration follows from (4.2.2)–(4.2.4) and (4.2.10)–(4.2.12).

If L_1 and L_2 are assumed to be arbitrary piecewise Lyapunov curves containing angular points only, then by generalizing the analysis presented in Section 4.1, Formula (3.1.16) may be shown to be valid in this case too. Now it is clear that Formula (3.1.17) may also be shown to be valid although the calculations will, naturally, be more complicated.

Note that if function $\varphi(t, \tau)$ has the form (3.1.12) and points $\tau_{k_p}^p$, $k_p = 1, \ldots, n_p$ and $t_{j_p}^p$, $j_p = 0, 1, \ldots, n_p$, form a canonic division of unclosed Lyapunov curve L_p, $p = 1, 2$ with the step h_p, then

$$|A(\tau_{k_1}^1, \tau_{k_2}^2) + \mathfrak{B}_{n_1 n_2}(\tau_{k_1}^1, \tau_{k_2}^2)| \le \theta(\tau_{k_1}^1, \tau_{k_2}^2), \tag{4.2.13}$$

$k_p = 1, \ldots, n_p$, $p = 1, 2$, where the quantity $\theta(\tau_{k_1}^1, \tau_{k_2}^2)$ satisfies inequalities of the form (3.3.4) and (3.3.5). This statement may be proved similarly to Inequality (4.1.24).

Part II: Numerical Solution of Singular Integral Equations

5

Equation of the First Kind on a Segment and / or a System of Nonintersecting Segments

5.1. CHARACTERISTIC SINGULAR EQUATION ON A SEGMENT: UNIFORM DIVISION*

Consider the characteristic singular equation

$$\int_a^b \frac{\varphi(t)\,dt}{t_0 - t} = f(t_0). \tag{5.1.1}$$

According to Gakhov (1977) and Muskhelishvili (1952), the index $\kappa = 0, 1, -1$ solution to the latter equation is given by the formula

$$\varphi(t) = -\frac{1}{\pi^2} R_\kappa(t) \left[\int_a^b R_\kappa^{-1}(t_0) \frac{f(t_0)\,dt_0}{t - t_0} - \nu_\kappa \pi C \right], \tag{5.1.2}$$

where $\nu_1 = 1$, $\nu_0 = \nu_{-1} = 0$, and

$$R_0(t) = \sqrt{\frac{b - t}{t - a}}, \qquad R_1(t) = \frac{1}{\sqrt{(b - t)(t - a)}},$$

$$R_{-1}(t) = R_1^{-1}(t).$$

*This means equally spaced data or equally spaced grid points (G. Ch.).

In what follows we do not consider especially solutions with a singularity of the form $\sqrt{(t-a)/(b-t)}$ because the alterations to be done in this case will be seen clearly.

Let the sets $E = \{t_k, k = 1, \ldots, n\}$ and $E_0 = \{t_{0j}, j = 0, 1, \ldots, n\}$ form a canonic division of the segment $[a, b]$ with the step h. The following theorem is true.

Theorem 5.1.1. *Let function $f(t)$ belong to the class $H(\lambda)$ on $[a, b]$. Then, between solutions to the systems of linear algebraic equations,*

$$\sum_{k=1}^{n} \frac{\varphi_n(t_k)h}{t_{0j} - t_k} = f(t_{0j}), \qquad j = 1, \ldots, n; \tag{5.1.3}$$

$$\sum_{k=1}^{n} \frac{\varphi_n(t_k)h}{t_{0j} - t_k} = f(t_{0j}), \qquad j = 1, \ldots, n-1,$$

$$\sum_{k=1}^{n} \varphi_n(t_k)h = C; \tag{5.1.4}$$

$$\gamma_{0n} + \sum_{k=1}^{n} \frac{\varphi_n(t_k)h}{t_{0j} - t_k} = f(t_{0j}), \qquad j = 0, 1, \ldots, n, \tag{5.1.5}$$

and the index $\kappa = 0, 1, -1$ solution $\varphi(t)$ to Equation (5.1.2), the following inequality holds:

$$|\varphi(t_k) - \varphi_n(t_k)| \leq \theta_n(t_k), \qquad k = 1, \ldots, n. \tag{5.1.6}$$

In (5.1.6), the quantity $\theta_n(t_k)$ satisfies:

1. *The inequalities*

$$\theta_n(t_k) \leq O_\delta(H^{\lambda_1}), \qquad \lambda_1 > 0, \tag{5.1.7}$$

   for all points $t_k \in [a + \delta, b - \delta]$, where $\delta > 0$ is however small;
2. *The inequalities*

$$\sum_{k=1}^{n} \theta_n(t_k)h \leq O(h^{\lambda_2}), \qquad \lambda_2 < 0, \tag{5.1.8}$$

   for all points $t_k \in [a, b]$.

Proof. Note that due to results presented in Section 1.3 for quadrature formulas for a singular integral on a segment, Systems (5.1.3)–(5.1.5) approximate Equation (5.1.1). For the determinants of Systems (5.1.3)–(5.1.5) one gets

$$D_\kappa^{(n)} = h^n \Delta_\kappa^{(n)}, \qquad \kappa = 0, 1, -1,$$

$$\Delta_0^{(n)} = \begin{vmatrix} \dfrac{1}{t_{01} - t_1} & \cdots & \dfrac{1}{t_{01} - t_n} \\ \cdots & \cdots & \cdots \\ \dfrac{1}{t_{0n} - t_1} & \cdots & \dfrac{1}{t_{0n} - t_n} \end{vmatrix}$$

$$\Delta_1^{(n)} = \begin{vmatrix} \dfrac{1}{t_{01} - t_1} & \cdots & \dfrac{1}{t_{01} - t_n} \\ \cdots & \cdots & \cdots \\ \dfrac{1}{t_{0n-1} - t_1} & \cdots & \dfrac{1}{t_{0n-1} - t_n} \\ 1 & \cdots & 1 \end{vmatrix},$$

$$\Delta_{-1}^{(n)} = \begin{vmatrix} 1 & \dfrac{1}{t_{00} - t_1} & \cdots & \dfrac{1}{t_{00} - t_n} \\ \cdots & \cdots & \cdots & \cdots \\ 1 & \dfrac{1}{t_{0n} - t_1} & \cdots & \dfrac{1}{t_{0n} - t_n} \end{vmatrix}. \tag{5.1.9}$$

It can be readily shown (Proskuryakov 1967) that

$$\Delta_\kappa^{(n)} = \frac{\prod\prod_{1 \le m < p \le n}(t_m - t_p)\prod\prod_{1-\zeta(-\kappa) \le m < p \le n-\zeta(\kappa)}(t_{0p} - t_{0m})}{\prod_{m=1-\zeta(-\kappa)}^{n-\zeta(\kappa)}\prod_{p=1}^{n}(t_{0m} - t_p)}, \tag{5.1.10}$$

where $\zeta(x) = 1$ for $x > 0$ and $\zeta(x) = 0$ for $x \le 0$. Let us prove this statement for $\Delta_0^{(n)}$, because for the remaining two determinants it can be proved analogously (with some evident alterations).

Subtract the bottom row of determinant $\Delta_0^{(n)}$ from all the preceding rows and take out factor $1/(t_{0n} - t_k)$, $k = 1, \ldots, n$, from all the columns and factor $t_{0n} - t_{0m}$, $m = 1, \ldots, n-1$, from all the rows. Then we have

$$\Delta_0^{(n)} = \prod_{k=1}^{n} \frac{1}{t_{0n} - t_k} \prod_{m=1}^{n-1} (t_{0n} - t_{0m}) \begin{vmatrix} \frac{1}{t_{01} - t_1} & \ldots & \frac{1}{t_{01} - t_n} \\ \cdots & \cdots & \cdots \\ \frac{1}{t_{0n-1} - t_1} & \ldots & \frac{1}{t_{0n-1} - t_n} \\ 1 & \ldots & 1 \end{vmatrix}$$

Now we subtract the last column of the last determinant from all the preceding columns, take out common factors from each row and column, and expand the resulting determinant in the last row. Then we get

$$\Delta_0^{(n)} \frac{\prod_{m=1}^{n-1}(t_{0n} - t_{0m})(t_m - t_n)}{\prod_{m,p=1}^{n}(t_{0n} - t_p)(t_{0m} - t_n)} \Delta_0^{(n-1)}.$$

The proof of the formula for $\Delta_0^{(n)}$ is terminated by employing the method of mathematical induction.

From (5.1.10) it is seen that $D_{k \neq 0}^{(n)}$ for any n. By employing the Cramér rule for solving a system of linear algebraic equations, one gets

$$\varphi_n(t_k) = \frac{D_{\kappa,k}^{(n)}}{D_\kappa^{(n)}} = \frac{1}{h} \sum_{j=1-\zeta(-\kappa)}^{n-\zeta(\kappa)} \frac{(-1)^{j+k} \Delta_{\kappa(j,k)}^{(n)}}{\Delta_\kappa^{(n)}} f(t_{0j})$$

$$+ \nu_\kappa \frac{(-1)^{n+k}}{h} \frac{\Delta_{1(n,k)}^{(n)}}{\Delta_1^{(n)}} C, \tag{5.1.11}$$

where ν_κ is given by (5.1.2) and $\zeta(x)$ is given by (5.1.10); $D_{\kappa,k}^{(n)}$ and $\Delta_{\kappa,k}^{(n)}$ are obtained from $D_\kappa^{(n)}$ and $\Delta_\kappa^{(n)}$, respectively, by replacing the kth column by the column of free terms from the right-hand side of the system, and $\Delta_{\kappa(j,k)}^{(n)}$ is obtained from $\Delta_\kappa^{(n)}$ by deleting the jth row and the kth column. Analogously, one gets for $\Delta_{\kappa(j,k)}^{(n)}$,

$$\Delta_{\kappa(j,k)}^{(n)} = \frac{\prod\prod_{\substack{1 \le m < p \le n \\ m.p \neq k}} (t_m - t_p) \prod\prod_{\substack{1-\zeta(-\kappa) \le m < p \le n-\zeta(\kappa) \\ m.p \neq j}} (t_{0p} - t_{0m})}{\prod_{\substack{m=1-\zeta(-\kappa) \\ m \neq j}}^{n-\zeta(\kappa)} \prod_{\substack{p=1 \\ p \neq k}}^{n} (t_{0m} - t_p)}. \tag{5.1.12}$$

Hence, Equation (5.1.11) becomes

$$\varphi_n(t_k) = -\frac{1}{h} I^{(n)}_{\kappa,k} \sum_{j=1-\zeta(-\kappa)}^{n-\zeta(\kappa)} \frac{1}{h} I^{(n)}_{\kappa,0j} \frac{f(t_{0j})h}{t_k - t_{0j}} + \nu_\kappa \frac{1}{h} I^{(n)}_{1,k} C,$$

$$k = 1,\dots,n, \quad (5.1.13)$$

where

$$I^{(n)}_{0,k} = \prod_{m=1}^{n} (t_{0m} - t_k) \Bigg/ \prod_{\substack{m=1 \\ m \neq k}}^{n} (t_m - t_k),$$

$$I^{(n)}_{0,0j} = \prod_{m=1}^{n} (t_{0j} - t_m) \Bigg/ \prod_{\substack{m=1 \\ m \neq j}}^{n} (t_{0j} - t_{0m}),$$

$$I^{(n)}_{1,k} = I^{(n)}_{0,k} \frac{1}{t_{0n} - t_k}, \qquad I^{(n)}_{1,0j} = I^{(n)}_{0,0j}(t_{0n} - t_{0j}),$$

$$I^{(n)}_{-1,k} = I^{(n)}_{0,k}(t_k - t_{00}), \qquad I^{(n)}_{-1,0j} = I^{(n)}_{0,0j} \frac{1}{t_{0j} - t_{00}}.$$

Next we get

$$\frac{1}{h} I^{(n)}_{0,k} = \frac{1}{2} P_{k-1} P_{n-k},$$

$$P_{k-1} = \prod_{m=1}^{k-1} \left(1 + \frac{t_{0m} - t_m}{t_m - t_k}\right), \qquad P_{n-k} = \prod_{m=k+1}^{n} \left(1 + \frac{t_{0m} - t_m}{t_m - t_k}\right),$$

$$P_0 = 1, \quad (5.1.14)$$

$$\frac{1}{h} I^{(n)}_{0,0j} = \frac{1}{2} P_{0j-1} P_{0,n-j},$$

$$P_{0,j-1} = \prod_{m=1}^{j-1} \left(1 + \frac{t_{0m} - t_m}{t_{0j} - t_{0m}}\right),$$

$$P_{0,n-1} = \prod_{m=j+1}^{n} \left(1 + \frac{t_{0m} - t_m}{t_{0j} - t_{0m}}\right), \qquad P_{0,0} = 1. \qquad (5.1.15)$$

Remember that in the case under consideration,

$$t_k = a + kh, \qquad t_{0k} - t_k = \frac{h}{2}, \qquad h = \frac{b-a}{n+1},$$

$$t_m - t_k = t_{0m} - t_{0k} = h(m-k). \tag{5.1.16}$$

Hence, in accordance with (5.1.14)–(5.1.16),

$$P_{k-1} = \prod_{m=1}^{k-1}\left(1 - \frac{1/2}{m}\right), \qquad P_{n-k} = \prod_{m=1}^{n-k}\left(1 + \frac{1/2}{m}\right). \tag{5.1.17}$$

The work of Hardy (1949) gives the following formula pertaining to the theory of gamma functions:

$$\frac{(1+\beta)(2+\beta)\cdots(n+\beta)}{1\cdot 2\cdot\cdots\cdot n} = \frac{n^{\beta}}{\Gamma(1+\beta)} + O(n^{\beta-1}). \tag{5.1.18}$$

The latter formula may be presented in the form

$$\prod_{m=1}^{n}\left(1 + \frac{\beta}{m}\right) = \frac{n^{\beta}}{\Gamma(1+\beta)} + O(n^{\beta-1}).$$

Because we need quantities on the order of n in Formula (5.1.18), we can write

$$\prod_{m=1}^{n}\left(1 + \frac{\beta}{m}\right) = \frac{(n+1)^{\beta}}{\Gamma(1+\beta)} + O\big((n+1)^{\beta-1}\big). \tag{5.1.18$'$}$$

By putting $\beta = \pm\frac{1}{2}$, we get

$$P_{k-1} = \frac{K^{-1/2}}{\Gamma(1/2)} + O(k^{-3/2}),$$

$$P_{n-k} = \frac{(n-k+1)^{1/2}}{\Gamma(3/2)} + O\big((n-k+1)^{-1/2}\big) \tag{5.1.19}$$

The work of Fikhtengoltz (1959) gives the formulas

$$\Gamma(1/2) = \sqrt{\pi}, \qquad \Gamma(a+1) = \Gamma(a)a, \qquad \Gamma(3/2) = \sqrt{\pi}/2.$$

Thus,

$$P_{k-1}P_{n-k} = \frac{2}{\pi}\sqrt{\frac{n-k+1}{k}} + O\left(k^{-1/2}(n-k+1)^{-1/2}\right)$$

$$+ O\left(k^{-3/2}(n-k+1)^{1/2}\right). \qquad (5.1.20)$$

In a similar way,

$$P_{0j-1} = \prod_{m=1}^{j-1}\left(1 + \frac{1/2}{m}\right), \qquad P_{0n-j} = \prod_{m=1}^{n-j}\left(1 - \frac{1/2}{m}\right),$$

$$P_{0j-1}P_{0n-j} = \frac{2}{\pi}\sqrt{\frac{j+1/2}{n+j+1/2}} + O\left(\left(j+\frac{1}{2}\right)^{-1/2}\left(n-j+\frac{1}{2}\right)^{-1/2}\right)$$

$$+ O\left(\left(j+\frac{1}{2}\right)^{1/2}\left(n-j+\frac{1}{2}\right)^{-3/2}\right). \qquad (5.1.21)$$

Note that using (5.1.16) we have

$$\sqrt{b-t_k} = \sqrt{hn-k+1}, \qquad \sqrt{t_k-a} = \sqrt{hk},$$

$$\sqrt{b-t_{0j}} = \sqrt{hn-j+1/2}, \qquad \sqrt{t_{0j}-a} = \sqrt{hj+1/2}. \quad (5.1.22)$$

In accordance with the latter formulas, Equations (5.1.20) and (5.1.21) may be written in the form

$$P_{k-1}P_{n-k} = \frac{2}{\pi}\sqrt{\frac{b-t_k}{t_k-a}} + A_{0(n,k)}, \qquad (5.1.23)$$

$$A_{0(n,k)} = O\left(\frac{h}{\sqrt{(b-t_k)(t_k-a)}}\right)$$

$$+O\left(\frac{h(b-t_k)^{1/2}}{(t_k-a)^{3/2}}\right),$$

$$P_{0j-1}P_{0n-j} = \frac{2}{\pi}\sqrt{\frac{t_{0j}-a}{b-t_{0j}}} + B_{0(n,j)},$$

$$B_{0(n,j)} = O\left(\frac{h}{\sqrt{(t_{0j}-a)(b-t_{0j})}}\right) + O\left(\frac{h(t_{0j}-a)^{1/2}}{(b-t_{0j})^{3/2}}\right). \tag{5.1.24}$$

Similarly, for $\kappa = 1, -1$ we have

$$\frac{1}{h} I^{(n)}_{\kappa,k} = \frac{1}{\pi}[(t_k - a)(b - t_k)]^{-\kappa/2} + A_{\kappa(n,k)},$$

$$A_{\kappa(n,k)} = O\left(\frac{h}{[(t_k - a)(b - t_k)]^{3/2-\zeta(-\kappa)}}\right), \tag{5.1.25}$$

$$\frac{1}{h} I^{(n)}_{\kappa,0j} = \frac{1}{\pi}\left[(t_{0j} - a)(b - t_{0j})\right]^{\kappa/2} + B_{\kappa(n,0j)},$$

$$B_{\kappa(n,0j)} = O\left(\frac{h}{\left[(t_{0j} - a)(b - t_{0j})\right]^{3/2-\zeta(\kappa)}}\right). \tag{5.1.26}$$

By substituting (5.1.23)–(5.1.26) into (5.1.13) and acting in a way analogous to the method used when proving Theorem 1.3.2, one gets

$$\varphi_n(t_k) = -\frac{1}{\pi^2} R_\kappa(t_k) \int_a^b R_\kappa^{-1}(t_0) \frac{f(t_0)\, dt_0}{t_k - t_0} + \nu_\kappa R_\kappa(t_k) C + \theta_\kappa(t_k),$$

$$k = 1, \dots, n, \tag{5.1.27}$$

where the quantity $|\theta_\kappa(t_k)|$ satisfies Inequalities (5.1.7) and (5.1.8). Thus, Theorem 5.1.1 is proved. ■

Note 5.1.1. Let us consider in greater detail the unknown γ_{0n} appearing in System (5.1.5). If the system is devoid of γ_{0n}, then the number of equations in it exceeds the number of unknowns and the system, as a rule, has no solution. On the other hand, the system must approximate Equation (5.1.1), which has a unique solution with index $\kappa = -1$ if the condition

$$\int_a^b \frac{f(t_0)\, dt_0}{\sqrt{(t_0 - a)(b - t_0)}} = 0 \tag{5.1.28}$$

is met (Gakhov 1977, Muskhelishvili 1952). Therefore, the level of disagreement in System (5.1.5) with zero γ_{0n} must decrease. Hence, if the system with γ_{0n} has a solution, then $\lim_{n\to\infty}\gamma_{0n} = 0$. However, the statement has to be proved. Thus, the variable γ_{0n} makes System (5.1.5) well-posed and, therefore, will be called a regularizing factor. To find γ_{0n} we use the Cramér rule again:

$$\gamma_{0n} = \sum_{j=0}^{n} \frac{1}{h} I_{-1,0j}^{(n)} f(t_{0j})h = \frac{1}{\pi} \sum_{j=0}^{n} \frac{f(t_{0j})h}{\sqrt{(t_{0j}-a)(b-t_{0j})}} + O(x^{1/2}). \tag{5.1.29}$$

Thus $\lim_{n\to\infty}\gamma_{0n} = 0$ if and only if the index $\kappa = -1$ solution for Equation (5.1.1) does exist. Hence, the behavior of γ_{0n} can be used as an indicator of the existence of the index $\kappa = -1$ solution.

Note 5.1.2. In applications (such as aerodynamics, elasticity, etc.), it is often required to calculate not the function $\varphi(t)$ itself, but the integral $\int_a^b \psi(t)\varphi(t)\,dt$, where $\psi(t) \in H$ on $[a, b]$. From Inequalities (5.1.6) and (5.1.8), it follows that the preceding integral may be calculated by using the rectangle rule with respect to points t_k, $k = 1, \ldots, n$, taking at the points not the function $\varphi(t)$, but the values $\varphi_n(t_k)$ for which the inequality

$$\left| \int_a^b \psi(t)\varphi(t)\,dt - \sum_{k=1}^{n} \psi(t_k)\varphi_n(t_k)h \right| \le O(h^{\lambda_3}), \qquad \lambda_3 > 0, \tag{5.1.30}$$

holds. In fact, one has

$$\begin{aligned}
&\left| \int_a^b \psi(t)\varphi(t)\,dt - \sum_{k=1}^{n} \psi(t_k)\varphi_n(t_k)h \right| \\
&\quad \le \left| \int_a^{a+h} \psi(t)\varphi(t)\,dt \right| \\
&\quad + \left| \int_{a+h}^{b} \psi(t)\varphi(t)\,dt - \sum_{k=1}^{n} \psi(t_k)\varphi(t_k)h \right| \\
&\quad + \left| \sum_{k=1}^{n} \psi(t_k)(\varphi(t_k) - \varphi_n(t_k))h \right| \\
&\quad \le O(h^{\lambda_4}) + M \sum_{k=1}^{n} \theta_n(t_k)h \le O(h^{\lambda_3}),
\end{aligned}$$

where $M = \max_{t\in[a,b]} |\psi(t)|$.

Note 5.1.3. Theorem 5.1.1 remains valid in the case where function $f(t)$ belongs not to the class H on segment $[a, b]$, but to the class H^*, i.e., has the form

$$f(t) = \frac{\psi(t)}{(t-a)^{\nu}(b-t)^{\mu}}, \tag{5.1.31}$$

where $\psi(t) \in H$ on $[a, b]$. In this case Theorem 5.1.1 remains valid for System (5.1.3) for $0 \leq \nu < 1$, $0 \leq \mu < \frac{1}{2}$, for System (5.1.4) for $0 \leq \nu, \mu < 1$ and for System (5.1.5) for $0 \leq \nu, \mu < \frac{1}{2}$. The fact that under these restrictions Functions (5.1.2) are solutions to Equation (5.1.1) is a consequence of the Poincaré–Bertrand formula.

Note 5.1.4. When solving the problem of steady flow past an airfoil with a flap, one has to consider the situation when function $f(t) \in H$ on segments $[a, q]$ and $[q, b]$ and suffers discontinuity of the first kind at the point q. If the sets $E = \{t_k, k = 1, \ldots, n\}$ and $E_0 = \{t_{0j}, j = 0, 1, \ldots, n\}$ are chosen in such a way that the point q lies in the middle between the nearest points belonging to sets E and E_0 (see Note 1.3.2), then Theorem 5.1.1 remains true, but Inequality (5.1.7) holds only for all points $t_k \in [a + \delta, q + \delta] \cup [q - \delta, \mathrm{b} - \delta]$. In fact, for Theorem 5.1.1, the relative positions of points t_k, $k = 1, \ldots, n$, and t_{0j}, $j = 0, 1, \ldots, n$ are of importance. Also, for these points the quadrature formulas for a singular integral on a segment analyzed in Section 1.3 must be valid.

The results of Section 1.3 and Notes 5.1.3 and 5.1.4 allow us to formulate the following theorem.

Theorem 5.1.2. *Let function $f(t)$ be of the form*

$$f(t) = \frac{\psi(t)}{(t-a)^{\nu}|q-t|^{\beta}(b-t)^{\mu}}, \tag{5.1.32}$$

where $\psi(t) \in H$ on $[a, b]$, $a < q < b$. Also let sets E and E_0 be chosen on segment $[a, b]$ with the step $h = (b - a)/(n + i)$ as indicated in Note 5.1.4. *Then Relationship* (5.1.6) *is valid in the following cases*:

1. *For $0 \leq \nu, \beta < 1$ and $0 \leq \mu < \frac{1}{2}$ between a solution to the system of linear algebraic equations* (5.1.3) *and the index $\kappa = 0$ solution of Equation* (5.1.1) *in Equation* (5.1.2);
2. *For $0 \leq \nu, \beta, \mu < 1$ between a solution to the system of linear algebraic equations* (5.1.4) *and the index $\kappa = 1$ solution of Equation* (5.1.1);
3. *For $0 \leq \nu, \mu < \frac{1}{2}$ and $0 \leq \beta < 1$ between a solution to the system of linear algebraic equations* (5.1.5) *and the index $\kappa = -1$ solution of Equation* (5.1.1) *in Equation* (5.1.2). *In Equation* (5.1.6) *the quantity*

$\theta(t_k)$ for all points $t_k \in [a+\delta, q-\delta] \cup [q+\delta, b-\delta]$ (where positive δ is however small) satisfies inequality

$$\theta(t_k) \le O(h^{\lambda_1}), \qquad 0 < \lambda_1 \le 1, \tag{5.1.33}$$

and Inequality (5.1.8) *for all points $t_k \in [a, b]$.*

Note that if $f(t)$ suffers a discontinuity of the first kind at the point q, then it may be presented in the form (5.1.32), where positive β is however small. In this case Solutions (5.1.2) to Equation (5.1.1) have a logarithmic singularity at point q. However, calculations show that good results may be obtained if point q coincides with one of the points t_{0j} for $j = j_q$, and the right-hand side in Systems (5.1.3)–(5.1.5) is chosen in the following way: $f(t_{0j})$ for $j \ne j_q$ and $[f(t_{0j_q} - 0) + f(t_{0j_q} + 0)]/2$ for $j = j_q$, where $f(t_{0j_q} - 0)$ and $f(t_{0j_q} + 0)$ are one-sided limits of the function $f(t_0)$ at point q.

From (5.1.2) the index $\kappa = 1$ solution $\varphi(t)$ to Equation (5.1.1) is seen to depend on an arbitrary constant C, which is an integral of the solution over segment $[a, b]$. However, according to Formula (5.1.2) the constant and hence, solution $\varphi(t)$ are uniquely defined if $\varphi(q)$ is fixed at a point $q \in (a, b)$. In order to find a numerical value of the solution from System (5.1.4), one has to find C_q and substitute it into the last equation of the system. Finding the corresponding constant C_q is a tiresome business, the more so if the equation incorporates a regular part. In the following text, we will show how one can do without the constant C_q, by using function $\varphi(q)$ only. The following theorem is true.

Theorem 5.1.3. *Let $f(t)$ belong to the class H on $[a, b]$ and the sets E and E_0 form a canonic division of the segment $[a, b]$ with the step h. Let us denote by t_{k_q} the point of the set E that lies at the shortest distance to the left from point $q \in (a, b)$. The Relationship* (5.1.6) *holds between a solution to the system of linear algebraic equations*

$$\sum_{\substack{k=1 \\ k \ne k_q}}^{n} \frac{\varphi_n(t_k)h}{t_{0j} - t_k} = f(t_{0j}) + \frac{\varphi(q)h}{t_{k_q} - t_{0j}}, \qquad j = 1, \ldots, n-1, \tag{5.1.34}$$

and the index $\kappa = 1$ solution $\varphi(t)$ to Equation (5.1.1):

$$\varphi(t) = -\frac{q-t}{\pi^2\sqrt{(t-a)(b-t)}} \int_a^b \frac{\sqrt{(t_0-a)(b-t_0)}\, f(t_0)\, dt_0}{(t-t_0)(q-t_0)} + \varphi(q)\sqrt{\frac{(q-a)(b-q)}{(t-a)(b-t)}}. \tag{5.1.35}$$

Proof. Similarly to Theorem 5.1.1 one may show that the determinant of System (5.1.34) is nonzero and apply the Cramér rule

$$\varphi_n(t_k) = -\left(t_k - t_{k_q}\right)\frac{1}{h}I_{1,k}^{(n)}\sum_{j=1}^{n}\frac{1}{h}I_{1,0j}^{(n)}\left[f(t_{0j}) + \frac{\varphi(q)h}{t_{k_q} - t_{0j}}\right]$$

$$\times\frac{h}{(t_k - t_{0j})\left(t_{0j} - t_{k_q}\right)},$$

$$k = 1,\ldots,n,\ k \neq k_q. \quad (5.1.36)$$

By denoting the first sum in Formula (5.1.13) for $\kappa = 1$ by $S_1(t_k)$, one may write

$$\varphi_n(t_k) = S_1(t_k) - S_1\left(t_{k_q}\right) - \frac{1}{h}I_{1,k}^{(n)}\sum_{j=1}^{n}\frac{1}{n}I_{1,0j}^{(n)}$$

$$\times\frac{\varphi(q)hh}{\left(t_{k_q} - t_{0j}\right)(t_k - t_{0j})}$$

$$+\frac{1}{h}I_{1,k}^{(n)}\sum_{j=1}^{n}\frac{1}{h}I_{1,0j}^{(n)}\frac{\varphi(q)hh}{\left(t_{k_q} - t_{0j}\right)^2}$$

$$= S_1(t_k) - S_1\left(t_{k_q}\right) - M_1 + M_2.$$

By employing the procedure used when proving the Poincaré–Bertrand formula, one gets

$$M_2 = \varphi(q)\sqrt{\frac{(q-a)(b-q)}{(t_k - a)(b - t_k)}} + O\left(\frac{h^{1/2}}{\sqrt{(t_k - a)(b - t_k)}}\right). \quad (5.1.37)$$

Then taking into account that

$$-\int_a^b\frac{\sqrt{(t_0 - a)(b - t_0)}\,dt_0}{t_{k_q} - t_0} + \int_a^b\frac{\sqrt{(t_0 - a)(b - t_0)}\,dt_0}{t_k - t_0}$$

$$= \pi\left(t_k - t_{k_q}\right),$$

we get

$$|M_1| \le \left[|\varphi(q)|h + \frac{h|\varphi(q)|}{t_k - t_{kq}} \frac{h^{1/2}}{\sqrt{(t_k - a)(b - t_k)}} \right] O(|\ln h|). \quad (5.1.38)$$

The validity of the proof of the theorem follows from Formulas (5.1.37) and (5.1.38) and the procedure used for proving Theorem 5.1.1. ■

In the case of the problem of flow past an airfoil with ejection of or sucking air (considered in greater detail later), it is necessary to solve numerically Equation (5.1.1) in the class of functions having the form

$$\varphi(t) = \frac{\psi(t)}{q - t},$$

where $\psi(t) \in H$ in the neighborhood of point $q \in (a, b)$.

In the subsequent Theorems 5.1.4 and 5.1.5, the sets E and E_0 will be taken in such a way that the point q would be located within the set E_0 for $j = j_q$.

Theorem 5.1.4. *Let function $f(t) \in H$ on $[a, b]$ and the value of function $\psi(t)$ be known at point q. Then, between a solution to the system of linear algebraic equations*

$$\sum_{\substack{k=1 \\ k \ne j_q}}^{n} \frac{\varphi_n(t_k)h}{t_{0j} - t_k} = f(t_{0j}) - \frac{h\psi(q)}{(t_{0j} - t_{jq})(Q - t_{j_q})},$$

$$j = 1, \ldots, n, \; j \ne j_q, \quad (5.1.39)$$

and the solution $\varphi(t)$ to Equation (5.1.1),

$$\varphi(t) = -\frac{1}{\pi^2} \sqrt{\frac{b - t}{t - a}} \left[\int_a^b \sqrt{\frac{t_0 - a}{b - t_0}} \frac{f(t_0)\, dt_0}{t - t_0} - \frac{\pi^2}{q - t} \sqrt{\frac{q - a}{b - q}} \psi(q) \right],$$

(5.1.40)

Relationship (5.1.6) *exists in which the quantity $\theta(t_k)$ satisfies Inequality*

(5.1.7) *for all* $t_k \in [a+\delta, q-\delta] \cup [q+\delta, b-\delta]$, *where positive* δ *is however small.*

Proof. Similarly to the preceding theorems, we will show that the determinant of System (5.1.39) is nonzero for any n. Therefore, by applying the Cramér rule to the system, one gets

$$\varphi_n(t_k) = -\frac{1}{h} I_{0,k}^{(n)} \frac{1}{q - t_k} \Bigg[\sum_{j=1}^{n} \frac{1}{h} I_{0,0j}^{(n)} \frac{(q - t_{0j}) f(t_{0j}) h}{t_k - t_{0j}}$$

$$- \sum_{j=1}^{n} \frac{1}{h} I_{0,0j}^{(n)} \frac{q - t_{0j}}{t_{j_q} - t_{0j}} f(t_{0j}) h - \psi(q) \sum_{j=1}^{n} \frac{1}{h} I_{0,0j}^{(n)} \frac{h^2}{\left(t_{j_q} - t_{0j}\right)^2}$$

$$+ 2\psi(q) \frac{t_k - q}{t_k - t_{j_q}} \sum_{j=1}^{n} \frac{1}{h} I_{0,0j}^{(n)} \left(\frac{1}{t_k - t_{0j}} - \frac{1}{t_{j_q} - t_{0j}} \right) h \Bigg]. \quad (5.1.41)$$

Now by using the proofs of Theorem 5.1.1 and the Poincaré–Bertrand formula we get

$$\varphi_n(t_k) = \varphi(t_k) + \tilde{\theta}(t_k),$$

where the quantity $|\tilde{\theta}(t_k)|$ satisfies the necessary relationships.

In a similar way the following theorem may be proved.

Theorem 5.1.5. *Let function* $f(t) \in H$ *on* $[a, b]$. *Then Relationship* (5.1.6) *holds just as in the case of Theorem* 5.1.4, *between solutions to the system of linear algebraic equations*

$$\sum_{k=1}^{n} \frac{\varphi_n(t_k) h}{t_{0j} - t_k} = f(t_{0j}), \qquad j = 0, 1, \ldots, n,\ j \neq j_q, \qquad (5.1.42)$$

$$\sum_{k=1}^{n} \frac{\varphi_n(t_k) h}{t_{0j} - t_k} = f(t_{0j}), \qquad j = 1, \ldots, n,\ j \neq j_q,$$

$$\sum_{k=1}^{n} \varphi_n(t_k) h = C, \qquad j = n + 1, \qquad (5.1.43)$$

and solutions to Equation (5.1.1), *respectively,*

$$\varphi(t) = -\frac{1}{\pi^2}\frac{\sqrt{(t-a)(b-t)}}{q-t}\int_a^b \frac{(q-t_0)f(t_0)\,dt_0}{\sqrt{(t_0-a)(b-t_0)}\,(t-t_0)}, \quad (5.1.44)$$

$$\varphi(t) = -\frac{1}{\pi^2}R_0(t)\frac{1}{q-t}\left[\int_a^b R_0^{-1}(t_0)\frac{(q-t_0)f(t_0)\,dt_0}{t-t_0} - \pi C\right]. \quad (5.1.45)$$

where for the latter case

$$\int_a^b \varphi(t)\,dt = C. \quad (5.1.46)$$

Note that when considering Theorem 5.1.4, one can assume that the required solution is known not at the point t_{j_q}, but at point t_{j_q+1}. Calculated results presented in Figure 10.5 confirm this conclusion.

Because function $\psi(t)$ belongs to the class H in the neighborhood of point q, the systems mentioned in Theorems 5.1.4 and 5.1.5 may be constructed as follows. First we choose sets E and E_0 on segment $[a, b]$ that form the latter's canonic division. Point t_{0j_q} is assumed to coincide with a point from the set E_0 located at the shortest distance from point q. Then the systems have the same form. Estimates between the corresponding accurate solutions to Equation (5.1.1) and solutions to the systems preserve their character.

As far as a wing may incorporate a number of air intakes, the previously formulated results may be naturally extended onto the case of finding numerical solutions to Equation (5.1.1) of the form

$$\varphi(t) = \frac{\psi(t)}{(q_1-t)\cdots(q_m-t)}, \quad (5.1.46')$$

where $q_1, \ldots, q_m$ are fixed points in the interval (a, b) and $\psi(t)$ is a function belonging to the class H in certain neighborhoods of the points. Then the following theorem is true.

Theorem 5.1.6. *Let function* $f(t) \in H$ *on* $[a, b]$ *and the values of the function* $\psi(t)$ *be known at the points* $q_1, \ldots, q_m$. *Then, Relationship* (5.1.6) *holds between a solution to the system of linear algebraic equations*

$$\sum_{\substack{k=1 \\ k \neq j_{q_1}, \ldots, j_{q_m}}} \frac{\varphi_n(t_k)h}{t_{0j} - t_k} = f(t_{0j}) - \frac{\varphi(q_1)h}{\left(t_{0j} - t_{j_{q1}}\right)\left(t_{0j_{q_1}} - t_{j_{q1}}\right)} - \cdots$$

$$- \frac{\psi(q_m)h}{\left(t_{0j} - t_{j_{q_m}}\right)\left(t_{0j_{q_m}} - t_{j_{q_m}}\right)},$$

$$j = 1, \ldots, n,\ j \neq j_q, \ldots, j_{q_m}, \quad (5.1.47)$$

and solution to Equation (5.1.1),

$$\varphi(t) = -\frac{1}{\pi^2}\sqrt{\frac{b-t}{t-a}} \int_a^b \sqrt{\frac{t_0 - a}{b - t_0}} \frac{f(t_0)\,dt_0}{t - t_0}$$

$$+ \sqrt{\frac{b-t}{t-a}} \left[\sqrt{\frac{q_1 - a}{b - q_1}} \frac{\psi(q_1)}{q_1 - t} + \cdots + \sqrt{\frac{q_m - a}{b - q_m}} \frac{\psi(q_m)}{q_m - t} \right],$$

$$(5.1.48)$$

where the quantity $\theta(t_k)$ *satisfies Inequality* (5.1.17) *for all points* $t_k \in [a + \delta, q_1 - \delta] \cup [q_1 + \delta, q_2 - \delta] \cup \cdots \cup [q_m + \delta, b - \delta]$, *where positive* δ *is a sufficiently small number, and Inequality* (5.1.8), *where* $q_1 < q_2 < \cdots < q_m$.

However, it should be noted that points t_k, $k = 1, \ldots, n$, and t_{0j}, $j = 0, 1, \ldots, n$, are chosen as in the case of Theorem 5.1.4 if points $q_1, \ldots, q_m$ are equally spaced from each other or are as in the case of the note to Theorem 5.1.5. The fact that the Functions (5.1.40), (5.1.44), (5.1.45), and (5.1.48) are solutions to Equation (5.1.1) may be checked straightforwardly.

Let us next consider Equation (5.1.1) on segment $[-b, b]$, which is symmetric with respect to the origin of coordinates. In other words, we consider the equation

$$\int_{-b}^{b} \frac{\varphi(t)\,dt}{t_0 - t} = f(t_0). \quad (5.1.1')$$

According to Gakhov (1977) and Muskhelishvili (1952), all the solutions to Equation 5.1.1′) are given by the formula

$$\varphi(t) = -\frac{1}{\pi^2\sqrt{b^2 - t^2}}\left[\int_{-b}^{b} \frac{\sqrt{b^2 - t_0^2}\, f(t_0)\, dt_0}{t - t_0} - \pi C\right]. \quad (5.1.49)$$

It is quite evident that if $f(t_0)$ is an odd function, then a solution to Equation (5.1.1′) is an even function, and if $f(t_0)$ is an even function, then a solution to Equation (5.1.1′) is an odd function for $C = 0$ only.

This circumstance can be employed for decreasing the order of a system of linear algebraic equations used for solving Equation (5.1.1′) numerically, by a factor of 2.

Theorem 5.1.7. *Let $f(t) \in H$ on $[-b, b]$ and be an even function on the segment. Then, Relationship* (5.1.6) *holds between a solution to the system of linear algebraic equations,*

$$\sum_{k=1}^{m} \varphi_n(t_k)\left[\frac{1}{t_{0j} - t_k} - \frac{1}{t_{0j} + t_k}\right]h = f(t_{0j}), \qquad j = 1, \ldots, m, \quad (5.1.50)$$

and solution $\varphi(t)$ of Equation (5.1.1′),

$$\varphi(t) = -\frac{1}{\pi^2\sqrt{b^2 - t^2}}\int_{-b}^{b} \frac{\sqrt{b^2 - t_0^2}\, f(t_0)\, dt_0}{t - t_0}. \quad (5.1.51)$$

Here the sets $E = \{t_k, k = 1, \ldots, 2m + 1\}$ and $E_0 = \{t_{0j}, j = 0, 1, \ldots, 2m + 1\}$ form a canonic division of segment $[-b, b]$ with the step h.

Analogously, if $f(t_0)$ is an odd function and $n = 2m$, then the system

$$\sum_{k=1}^{m} \varphi_n(t_k)\left(\frac{1}{t_{0j} + t_k} + \frac{1}{t_{0j} - t_k}\right)h = f(t_{0j}), \qquad j = 1, \ldots, m - 1,$$

$$\sum_{k=1}^{m} \varphi_n(t_k)h = \frac{1}{2}C \quad (5.1.52)$$

must be taken. If one desires to obtain the index $\kappa = -1$ solution, then the system

$$\sum_{k=1}^{m} \varphi_n(t_k)\left[\frac{1}{t_{0j}+t_k} + \frac{1}{t_{0j}-t_k}\right]h = f(t_{0j}), \qquad j = 0, 1, \ldots, m-1, \tag{5.1.53}$$

must be considered.

As an application of Theorem 5.1.1, we consider the issue of solving numerically the equation

$$\int_{-1}^{1} \frac{\varphi'(t)\,dt}{t_0 - t} = f(t_0), \tag{5.1.54}$$

whose solutions are all given by the formula

$$\varphi(t) = -\frac{1}{\pi^2}\int_{-1}^{1}\frac{d\tau}{\sqrt{1-\tau^2}}\int_{-1}^{1}\frac{\sqrt{1-t_0^2}\,f(t_0)\,dt_0}{\tau - t_0}$$
$$+\frac{C_1}{\pi}\left[\arcsin t + \frac{\pi}{2}\right] + C_2,$$

$$\varphi(-1) = C_2,\ \varphi(1) = C_1 + C_2. \tag{5.1.55}$$

Let us present the construction procedure for a numerical method for obtaining a solution to the equation which is zero at the ends of a segment —a situation most often occurring in applications. Let points $t_1 = -1$, $t_2, \ldots, t_n\ t_{n+1} = 1$ divide segment $[-1, 1]$ into equal parts h long and point t_{0j} be the middle of segment $[t_j, t_{j+1}]$, $j = 1, \ldots, n$.

Theorem 5.1.8. *Let function $f(t_0)$ belong to the class H on $[-1, 1]$. Then between a solution to the system of linear algebraic equations*

$$\sum_{k=1}^{n} \varphi_n(t_{0k})\left(\frac{1}{t_{0j}-t_k} - \frac{1}{t_{0j}-t_{k+1}}\right) = f(t_{0j}), \qquad j = 1, \ldots, n, \tag{5.1.55'}$$

and the solution $\varphi(t)$ *of Equation* (5.1.54) *vanishing at the ends of segment* $[-1,1]$, *the relationship*

$$|\varphi_n(t_{0k}) - \varphi(t_{0k})| \le O\left(\frac{1}{n^\lambda}\right), \qquad k = 1,\ldots,n, \tag{5.1.56}$$

holds where $\lambda > 0$.

Proof. System (5.1.55′) is equivalent to the system

$$\sum_{k=1}^{n+1} \frac{\varphi_n(t_{0k}) - \varphi_n(t_{0k-1})}{h} \frac{h}{t_{0j} - t_k} = f(t_{0j}), \qquad j = 1,\ldots,n,$$

$$\sum_{k=1}^{n+1} \frac{\varphi_n(t_{0k}) - \varphi_n(t_{0k-1})}{h} h = 0, \qquad j = n+1, \tag{5.1.57}$$

where $\varphi_n(t_{00}) = \varphi_n(t_{0n+1} = 0$.

The latter system of equations is seen to coincide with System (5.1.4) for $C = 0$. Therefore,

$$\frac{\varphi_n(t_{0k}) - \varphi_n(t_{0k-1})}{h} = -\frac{1}{h} I_{1,k}^{(n+1)} \sum_{j=1}^{n} \frac{1}{h} I_{1,0j}^{(n+1)} \frac{f(t_{0j})h}{t_k - t_{0j}},$$

$$k = 1,\ldots,n+1,$$

or

$$\varphi_n(t_{0k}) = -\sum_{i=1}^{k} \frac{1}{h} I_{1,k}^{(n+1)} h \sum_{j=1}^{n} \frac{1}{h} I_{1,0j}^{(n+1)} \frac{f(t_{0j})h}{t_k - t_{0j}}, \qquad k = 1,\ldots,n, \tag{5.1.58}$$

because $\varphi_n(t_{00}) = \varphi_n(t_{0n+1}) = 0$.

Hence, by Formulas (5.1.25) and (5.1.26) we have

$$\varphi_n(t_{0k}) = -\frac{1}{\pi^2} \sum_{i=1}^{k} \frac{h}{\sqrt{1-t_i^2}} \sum_{j=1}^{n} \frac{\sqrt{1-t_{0j}^2}\, f(t_{0j})h}{t_i - t_{0j}} + O\left(\frac{\ln n}{\sqrt{n}}\right),$$

$$k = 1,\ldots,n. \tag{5.1.59}$$

This formula proves the theorem because function

$$\varphi(t) = -\frac{1}{\pi^2}\int_{-1}^{t}\frac{d\tau}{\sqrt{1-\tau^2}}\int_{-1}^{1}\frac{\sqrt{1-t_0^2}f(t_0)\,dt_0}{\tau - t_0} \tag{5.1.60}$$

gives the required solution to Equation (5.1.54). ■

Note that the equality

$$\sum_{i=1}^{n+1}\frac{1}{h}I_{1,i}^{(n+1)}h\sum_{j=1}^{n}\frac{1}{h}I_{1,0j}^{(n+1)}\frac{f(t_{0j})h}{t_i - t_{0j}} = 0 \tag{5.1.61}$$

holds. Also note that from the proof of Theorem 5.1.8 it follows that (5.1.55′) is not an ill-conditioned system, and the matrix inverse with respect to the system's matrix is given by Formula (5.1.58). This may be demonstrated straightforwardly by using the Hadamard criterion according to which the absolute value of a diagonal term in each row is larger than the sum of the absolute values of all the rest terms of the row. In fact, we have

$$a_{jk} = \frac{1}{t_{0j} - t_k} - \frac{1}{t_{0j} - t_{k+1}} = -\int_{t_k}^{t_{k+1}}\frac{dt}{(t_{0j} - t)^2}, \qquad k, j = 1, \ldots, n. \tag{5.1.62}$$

Therefore if $t_{0j} \not\subset [t_k, t_{k+1}]$, then $a_{jk} > 0$, $a_{jj} < 0$, and $-\int_{-1}^{1}(dt/(t_{0j} - t)^2) > 0$. Hence,

$$\int_{-1}^{1}\frac{dt}{(t_{0j} - t)^2} = \sum_{\substack{k=1,\\ k\neq j}}^{n} a_{jk} + a_{jj} < 0. \tag{5.1.63}$$

Finally, from the latter equality we deduce that

$$-a_{jj} = |a_{jj}| > \sum_{\substack{k=1\\ k\neq j}}^{n} a_{jk} = \sum_{\substack{k=1\\ k\neq j}}^{n} |a_{jk}|. \tag{5.1.64}$$

In a similar way the following theorem may be proved.

Theorem 5.1.9. *Let function $f(t) \in H$ on $[-1,1]$. Then between a solution to the system of linear algebraic equations*

$$-\frac{C_2}{t_{0j}-t_1} + \sum_{k=1}^{n} \varphi_n(t_{0k})\left(\frac{1}{t_{0j}-t_k} - \frac{1}{t_{0j}-t_{k+1}}\right) + \frac{C_1+C_2}{t_{0j}-t_{n+1}} = f(t_{0j}),$$

$$j = 1,\ldots,n, \quad (5.1.65)$$

and solution $\varphi(t)$ of Equation (5.1.54) *given by Formula* (5.1.55), *Relationship* (5.1.56) *holds.*

To prove the theorem it suffices to note that System (5.1.65) is equivalent to the system

$$\sum_{k=1}^{n+1} \frac{\varphi_n(t_{0k}) - \varphi_n(t_{0k-1})}{h} \frac{h}{t_{0j}-t_k} = f(t_{0j}), \qquad j = 1,\ldots,n,$$

$$\sum_{k=1}^{n+1} \frac{\varphi_n(t_{0k}) - \varphi_n(t_{0k-1})}{h} h = C_1, \qquad j = n+1, (5.1.66)$$

where $\varphi_n(t_{00}) = C_2$ and $\varphi_n(t_{0n+1}) = C_1 + C_2$.

By using Theorem 5.1.1 one gets

$$\varphi_n(t_{0k}) = -\sum_{i=1}^{k} \frac{1}{h} I_{1,k}^{(n+1)} h \sum_{j=1}^{n} \frac{1}{h} I_{1,0j}^{(n+1)} \frac{f(t_{0j})h}{t_i - t_{0j}}$$

$$+C_1 \sum_{i=1}^{k} \frac{1}{h} I_{1,k}^{(n+1)} h + C_2. \quad (5.1.67)$$

Note that by (5.1.61) and condition $\varphi_n(t_{0n+1}) = C_1 + C_2$ one gets from (5.1.67) for $k = n+1$,

$$\sum_{i=1}^{n+1} I_{1,i}^{(n+1)} = 1. \quad (5.1.68)$$

In Matveev (1982) the latter result was generalized onto the equation

$$\int_{-1}^{1} \frac{\varphi^{(m)}(t)\, dt}{t_0 - t} = f(t_0). \quad (5.1.69)$$

5.2. CHARACTERISTIC SINGULAR EQUATION ON A SEGMENT: NONUNIFORM DIVISION*

In this section we consider questions of solving numerically Equations (5.1.1′) for $b = 1$,

$$\int_{-1}^{1} \frac{\varphi(t)\,dt}{t_0 - t} = f(t_0)$$

by using the representation

$$\varphi(t) = \omega(t)\psi(t), \tag{5.2.1}$$

where $\omega(t) = (1-t)^{\pm 1/2}(1+t)^{\pm 1/2}$ and $\psi(t)$ is supposed to belong to the class H on $[-1, 1]$. Therefore, for constructing a system of linear algebraic equations providing a numerical solution to Equation (5.1.1′) in a class of functions, we will use the interpolation-type quadrature formulas presented in Section 2.3.

Let us start by considering in greater detail the case $\kappa = 1$ when $\omega(t) = (1-t^2)^{-1/2}$.

Theorem 5.2.1. *Let function $f(t) \in H$ on $[-1, 1]$. Then between a solution to the system of linear algebraic equations*

$$\sum_{k=1}^{n} \frac{\psi_n(t_k)a_k}{t_{0j} - t_k} = f(t_{0j}), \qquad j = 1, \ldots, n-1,$$

$$\sum_{k=1}^{n} \psi_n(t_k)a_k = C, \qquad j = n \tag{5.2.2}$$

[*where* $a_k = \pi/n$; $t_k = \cos(2k-1)\pi/(2n)$, $k = 1, \ldots, n$; $t_{0j} = \cos j\pi/n$, $j = 0, 1, \ldots, n-1$] *and the function*

$$\psi(t) = -\frac{1}{\pi^2}\int_{-1}^{1} \frac{\sqrt{1-t_0^2}\,f(t_0)\,dt_0}{t - t_0} + \frac{C}{\pi}, \tag{5.2.3}$$

the relationship

$$|\psi(t_k) - \psi_n(t_k)| \le R_n(t_k) \tag{5.2.4}$$

*This means unequally spaced data or unequally spaced grid points (G. Ch.).

holds, where $R_n(t_k)$ is the error of approximation of the singular integral appearing in Equation (5.2.3) *at point t_k by using the interpolation-type quadrature formula constructed for points t_{0j}, $j = 0, 1, \dots, n-1$, in Section* 2.3 (*see Equation* (2.3.6)).

Proof. The analysis carried out when proving Theorem 5.1.1 shows that a solution to System (5.2.2) is of the form

$$\psi_n(t_k) = -\frac{1}{a_k} I_{1,k}^{(n)} \left[\sum_{j=1}^{n-1} I_{1,0j}^{(n)} \frac{f(t_{0j})}{t_i - t_{0j}} - C \right], \qquad k = 1, 2, \dots, n. \quad (5.2.5)$$

The expressions for $I_{1,k}^{(n)}$ and $I_{1,0j}^{(n)}$ may be conveniently presented in the form

$$I_{1,k}^{(n)} = \prod_{m=1}^{n-1} (t_k - t_{0m}) \bigg/ \prod_{\substack{m=1 \\ m \neq k}}^{n} (t_k - t_m),$$

$$I_{1,0j}^{(n)} = -\prod_{m=1}^{n} (t_{0j} - t_m) \bigg/ \prod_{\substack{m=1 \\ m \neq j}}^{n-1} (t_{0j} - t_{0m}). \quad (5.2.6)$$

By representing the polynomials as a product of linear factors, one gets (see Section 2.3)

$$P_n(t) = T_n(t) = B_n \prod_{m=1}^{n} (t - t_m),$$

$$Q_n(t) = -\pi U_{n-1}(t) = -\pi B_n \prod_{m=1}^{n-1} (t - t_{0m}),$$

$$P_n'(t_k) = B_n \sum_{\substack{m=1 \\ m \neq k}}^{n} (t_k - t_m),$$

$$Q_n(t_k) = -\pi B_n \prod_{m=1}^{n-1} (t_k - t_{0m}), \quad (5.2.7)$$

where $P_n(t)$ is a polynomial whose roots are used to construct a quadrature formula for the singular integral entering the equation $Q_n(t)$ is a polynomial whose roots are collocation points, and B_n is a coefficient before the senior power of the variable in $P_n(t)$. Thus, from Equations (2.3.7), (5.2.6), and (5.2.7), one has

$$I_{1,k}^{(n)} = a_k/\pi, \qquad k = 1, \dots, n. \tag{5.2.8}$$

Note that the equality of the coefficients before the senior powers of the variable in polynomials $T_n(t)$ and $U_{n-1}(t)$ follows from the formulas

$$T_n(t) = \frac{n}{2} \sum_{m=0}^{|n/2|} \frac{(-1)^m (n-m-1)!}{m!(n-2m)!} (2t)^{n-2m}, \qquad n = 1, 2, \dots,$$

$$U_n(t) = \sum_{m=0}^{|n/2|} \frac{(-1)^m (n-m)!}{m!(n-2m)!} (2t)^{n-2m}, \qquad n = 1, 2, \dots. \tag{5.2.9}$$

Consider integral

$$\Phi_1(t) = \int_{-1}^{1} \frac{\sqrt{1-t_0^2}\, f(t_0)\, dt_0}{t - t_0}. \tag{5.2.10}$$

In this case (see Equations (2.3.17) and (2.3.20) $P_{n-1}(t) = U_{n-1}(t)$ and $Q_{n-1}(t) = \pi T_n(t)$. Therefore, the quadrature for $\Phi_1(t)$ at points t_k corresponding to the roots of function $Q_{n-1}(t)$ will have the form

$$\hat{\Phi}_1(t_k) = \sum_{j=1}^{n-1} \frac{f(t_{0j}) b_j}{t_k - t_{0j}}, \qquad k = 1, \dots, n,$$

$$b_j = -\frac{Q_{n-1}(t_{0j})}{P'_{n-1}(t_{0j})} = -\pi \frac{T_n(t_{0j})}{U'_{n-1}(t_{0j})}, \qquad j = 1, \dots, n-1. \tag{5.2.11}$$

From Equations (5.2.6), (5.2.7), and (5.2.12), it follows that

$$I_{1,0j}^{(n)} = b_j/\pi, \qquad j = 1, \dots, n-1. \tag{5.2.12}$$

By comparing the equality

$$\psi_n(t_k) = -\frac{1}{\pi^2} \left[\sum_{j=1}^{n-1} \frac{f(t_{0j}) b_j}{t_k - t_{0j}} - \pi C \right], \qquad k = 1, \dots, n, \tag{5.2.13}$$

which follows from (5.2.5), (5.2.8), and (5.2.12) with (5.2.3), we conclude the proof of the theorem. ■

From Stark (1971) it follows that the formula

$$\varphi_n(t_k) = -\frac{1}{\pi^2\sqrt{1-t_k^2}}\left[\sum_{j=1}^{n-1}\frac{f(t_{0j})b_j}{t_k - t_{0j}} - \pi C\right], \qquad k = 1,\ldots,n, \tag{5.2.14}$$

gives an accurate value of the solution $\varphi(t)$ at points t_k if $f(t)$ is a polynomial of power less than or equal to $2(n-1)$. In the general case we have

$$|\varphi_n(t_k) - \varphi(t_k)| \le \frac{1}{\sqrt{1-t_k^2}}R_n(t_k).$$

Next we consider the solution to Equation (5.1.1′): $b = 1$ for $\kappa = 1$ and 0, i.e., for the cases when $\omega(t) = \sqrt{1-t^2}$ or $\omega(t) = \sqrt{(1-t)/(1+t)}$, respectively. Thus, the following theorem is true.

Theorem 5.2.2. *Let function $f(t) \in H$ on $[-1,1]$. Then between solutions of the systems of linear algebraic equations*

$$\lambda_{0n} + \sum_{k=1}^{n}\frac{\psi_n(t_k)a_k}{t_{0j} - t_k} = f(t_{0j}), \qquad j = 1,\ldots,n+1, \tag{5.2.15}$$

$$a_k = \frac{\pi}{n+1}\sin^2\frac{k\pi}{n+1}, \qquad t_k = \cos\frac{k\pi}{n+1}, \qquad k = 1,\ldots,n+1,$$

$$t_{0j} = \cos\frac{(2-j-1)\pi}{2(n+1)}, \qquad j = 1,\ldots,n+1;$$

$$\sum_{k=1}^{n}\frac{\psi_n(t_k)a_k}{t_{0j} - t_k} = f(t_{0j}), \qquad j = 1,\ldots,n, \tag{5.2.16}$$

$$a_k = \frac{4\pi}{2n+1}\sin^2\frac{k\pi}{2n+1}, \qquad t_k = \cos\frac{2k\pi}{2n+1}, \qquad k = 1,\ldots,n,$$

$$t_{0j} = \cos\frac{2j-1}{2n+1}\pi, \qquad j = 1,\ldots,n,$$

and the functions $\psi(t)$,

$$\psi(t) = -\frac{1}{\pi^2}\int_{-1}^{1}\frac{f(t_0)\,dt_0}{\sqrt{1-t_0^2}\,(t-t_0)}, \tag{5.2.17}$$

$$\psi(t) = -\frac{1}{\pi^2}\int_{-1}^{1}\sqrt{\frac{1+t_0}{1-t_0}}\,\frac{f(t_0)\,dt_0}{t-t_0}, \tag{5.2.18}$$

Relationship (5.2.4) *holds.*

Note that if one has to take an approximate solution as a function of t, then it is necessary to employ Formula (2.3.2), which may be presented in the form

$$\psi_n(t) = \sum_{k=1}^{n}\psi_n(t_k)\frac{P_n(t)}{(t-t_k)P_n'(t_k)},$$

where for Theorem 5.2.1,

$$P_n(t) = T_n(t) = \cos(n\arccos t),$$

and for Theorem 5.2.2,

$$P_n(t) = U_n(t) = \frac{\sin[(n+1)\arccos t]}{\sin(\arccos t)}$$

for System (5.2.15) and

$$P_n(t) = \frac{T_{n+1}(t) - T_n(t)}{1-t}$$

for System (5.2.16).

In the preceding section it was indicated that in some applications (Belotserkovsky 1967) one has to consider a solution to the characteristic equation when its value is known at a point $q \in (-1, 1)$. In this case the following analog of Theorem 5.1.3 may be formulated: A solution to the system of linear algebraic equations

$$\sum_{\substack{k=1\\k\neq k_q}}^{n}\frac{\psi_n(t_k)a_k}{t_{0j}-t_k} = f(t_{0j}) + \frac{\psi(q)a_{k_q}}{t_{k_q}-t_{0j}}, \qquad j = 1,\ldots,n-1 \tag{5.2.19}$$

(where a_k, t_k, $k = 1, \ldots, n$, and t_{0j}, $j = 0, 1, \ldots, n - 1$, are defined similarly to Theorem 5.2.1, and k_q is the number of point t_k located at the shortest distance from point q) converges uniformly to the function

$$\psi(t) = -\frac{q - t}{\pi^2} \int_{-1}^{1} \frac{\sqrt{1 - t_0^2}\, f(t_0)\, dt_0}{(t - t_0)(q - t_0)} + \psi(q). \qquad (5.2.20)$$

In fact, a solution to System (5.2.19) may be presented in the form

$$\psi_n(t_k) = -\left(t_k - t_{k_q}\right) \frac{1}{a_k} I_{1,k}^{(n)} \sum_{j=1}^{n} \frac{1}{b_j} I_{1,0j}^{(n)} \left[f(t_{0j}) + \frac{\psi(q) a_{k_q}}{t_{k_q} - t_{0j}} \right]$$

$$\times \frac{bj}{(t_k - t_{0j})\left(t_{0j} - t_{k_q}\right)},$$

$$k = 1, \ldots, n, k \neq k_q. \qquad (5.2.21)$$

Now we introduce the representation of $\psi_n(t_k)$ in the form

$$\psi_n(t_k) = S_1(t_k) - S_1\left(t_{k_q}\right) - M_1 + M_2,$$

where quantity M_2 may be represented as $M_2 = \psi(q) + O(n^{-1})$ and $|M_1| \leq O(n^{-1})$. Thus, if function $f(t)$ has a limited second derivative, then $|\psi_n(t_k) - \psi(t_k)|$ is also a quantity on the order of n^{-1}.

If, finally, a solution is sought in the class of functions $\varphi(t) = \psi(t)/(q - t)$, $\psi(t) = \sqrt{(1 - t)/(1 + t)}\, u(t)$, and the value of $u(q)$ is known at a point $q \in (-1, 1)$, then the following analog of Theorem 5.1.4 may be formulated.

Between a solution to the system of linear algebraic equations

$$\sum_{\substack{k=1 \\ k \neq j_q}}^{n} \frac{V_n(t_k) a_k}{t_{0j} - t_k} = f(t_{0j}) - \frac{u(q) a_{k_q}}{\left(t_{0j} - t_{j_q}\right)\left(t_{0j_q} - t_{j_q}\right)},$$

$$j = 1, \ldots, n, j \neq j_q \qquad (5.2.22)$$

[where a_k, t_k, and t_{0j} are defined similarly to System (5.2.6) and j_q is the number of point t_{0j} located at the shortest distance from point q, $v(t) =$

$u(t)/(q - t)$] and the function

$$V(t) = -\frac{1}{\pi^2}\int_{-1}^{1}\sqrt{\frac{1+t_0}{1-t_0}}\frac{f(t_0)\,dt_0}{t-t_0} + \frac{u(q)}{q-t}, \tag{5.2.23}$$

the relationship

$$|V_n(t_k) - v(t_k)| \leq \frac{1}{q-t_k}O\left(\frac{1}{n}\right). \tag{5.2.24}$$

holds [if the function $f(t_0)$ is smooth enough]. This statement can be proved by using Formula (5.1.41) and the procedure for proving the Poincaré–Bertrand formula.

One can also formulate analogs of Theorems 5.1.5 and 5.1.6.

Similarly to Note 5.1.2, we remark that by employing solutions to the systems of linear algebraic equations (5.2.19) and (5.2.22), one may properly calculate the integrals $\int_{-1}^{1}\eta(t)\varphi(t)\,dt$, where $\eta(t)$ is a function belonging to the class H on $[-1,1]$.

5.3. FULL EQUATION ON A SEGMENT

In this section we consider numerical solution of the full equation of the first kind on a segment

$$\int_{-1}^{1}\frac{\varphi(t)\,dt}{t_0-t} + \int_{-1}^{1}k(t_0,t)\varphi(t)\,dt = f(t_0). \tag{5.3.1}$$

At the start we assume that functions $f(t)$ and $k(t_0,t)$ belong to the class H on the corresponding domains of definition.

By solving Equation (5.3.1) with respect to its characteristic part, we deduce that it is equivalent to the Fredholm equation of the second kind in the sense of obtaining the index κ solutions (Gakhov 1977, Muskhelishvili 1952):

$$\varphi(t) + \int_{-1}^{1}N_\kappa(t,\tau)\varphi(\tau)\,d\tau = f_{1,\kappa}(t), \tag{5.3.2}$$

where

$$N_\kappa(t,\tau) = -\frac{1}{\pi^2}R_\kappa(t)\int_{-1}^{1}R_\kappa^{-1}(t_0)\frac{k(t_0,\tau)}{t-t_0}\,dt_0,$$

$$f_{1,\kappa}(t) = -\frac{1}{\pi^2}R_\kappa(t)\left[\int_{-1}^{1}R_\kappa^{-1}(t_0)\frac{f(t_0)\,dt_0}{t-t_0} - T_k C\right],$$

and $T_1 = \pi$, $T_0 = T_{-1} = 0$. In addition, the condition

$$\int_{-1}^{1} R_{-1}^{-1}(t) f(t)\, dt = \int_{-1}^{1} \left(\int_{-1}^{1} R_{-1}^{-1}(t) k(t,\tau)\, dt \right) \varphi(\tau) d\tau \quad (5.3.3)$$

must be met for $\kappa = -1$.

Note that the kernel $N_\kappa(t, \tau)$ has the form

$$N_\kappa(t,\tau) = \frac{\Phi(t,\tau)}{(1-t)^\alpha (1+t)^\beta}, \quad (5.3.4)$$

where α and β are equal either to 0 or to $\frac{1}{2}$, and function $\Phi(t, \tau)$ is continuous on the set $[-1,1] \times [-1,1]$. Note that Fredholm's theory of constructing approximate solutions (Goursat 1934, Privalov 1935) is fully applicable to integral equations of the second kind with kernels of the form (5.3.4). However, by means of an appropriate substitution of the variable, Equation (5.3.2) may be directly reduced to the Fredholm equation of the second kind with a continuous kernel. This allows transformation of the system of linear algebraic equations for Equation (5.3.1) (derived by employing the numerical method under consideration) into a system of linear algebraic equations for a Fredholm equation of the second kind equivalent to Equation (5.3.1) in a fixed class of solutions.

The following theorem is true.

Theorem 5.3.1. *Let functions $f(t)$ and $k(t_0, t)$ entering Equation* (5.3.1) *belong to the class H on the sets* $[-1,1]$ *and* $[-1,1] \times [-1,1]$, *respectively. Also let the equation have a unique solution in the class of functions corresponding to a fixed index κ (for $\kappa = 1$ the integral of a solution is assumed to be known). Then, between solutions to the systems of linear algebraic equations*

$$\sum_{k=1}^{n} \frac{\varphi_n(t_k)h}{t_{0j} - t_k} + \sum_{k=1}^{n} k(t_{0j}, t_k)\varphi_n(t_k)h = f(t_{0j}), \qquad j = 1, \ldots, n, \quad (5.3.5)$$

$$\sum_{k=1}^{n} \frac{\varphi_n(t_k)h}{t_{0j} - t_k} + \sum_{k=1}^{n} k(t_{0j}, t_k)\varphi_n(t_k)h = f(t_{0j}),$$

$$j = 1, \ldots, n-1,$$

$$\sum_{k=1}^{n} \varphi_n(t_k)h = C, \quad (5.3.6)$$

$$\gamma_{0n} + \sum_{k=1}^{n} \frac{\varphi_n(t_k)h}{t_{0j} - t_k} + \sum_{k=1}^{n} k(t_{0j}, t_k)\varphi_n(t_k)h = f(t_{0j}),$$

$$j = 0, 1, \ldots, n, \quad (5.3.7)$$

and the corresponding solutions to Equation (5.3.1), *relationship Equation* (5.1.6) *holds where the quantity* $\theta(t_k)$ *satisfies inequalities Equations* (5.1.7) *and* (5.1.8). *Here the sets* $E = \{t_k, k = 1, \ldots, n\}$ *and* $E_0 = \{t_0, j = 0, 1, \ldots, n\}$ *form a canonic division of the segment* $[-1, 1]$.

Proof. On the left-hand side of Systems (5.3.5)–(5.3.7), we leave the summands corresponding to a characteristic singular integral equation; all the other terms are transferred to the right-hand side. By employing the results of Theorem 5.1.1, we deduce that the systems under consideration are equivalent to the following systems ($\kappa = 0, 1, -1$):

$$\varphi_n(t_k) + \sum_{m=1}^{n} \tilde{N}_\kappa(t_k, t_m)\varphi_n(t_m)h = \tilde{f}_{1,\kappa}(t_k), \qquad k = 1, \ldots, n, \quad (5.3.8)$$

where

$$\tilde{N}_\kappa(t_k, t_m) = -\frac{1}{n} I_{\kappa,k}^{(n)} \sum_{j=1-\zeta(-\kappa)}^{n-\zeta(\kappa)} I_{\kappa,0j}^{(n)} \frac{k(t_{0j}, t_m)}{t_k - t_{0j}} h,$$

$$\tilde{f}_{1,\kappa}(t_k) = -\frac{1}{h} I_{\kappa,k}^{(n)} \left[\sum_{j=1-\zeta(-\kappa)}^{n-\zeta(\kappa)} I_{\kappa,0j}^{(n)} \frac{t(t_{0j})h}{t_k - t_{0j}} - T_\kappa C \right]$$

[for the definition of $\zeta(x)$, see (5.1.10)].

We proceed with a more detailed proof for $\kappa = 0$, because the other two cases are quite similar. From (5.1.27) it follows that by multiplying both sides of System (5.3.8) by the factor $(1 - t_k)^{1/4}(1 + t_k)^{3/4}$, denoting the product of the factor and $\varphi_n(t_k)$ by $\tilde{\psi}_n(t_k)$, and considering the resulting system of linear algebraic equations, one deduces that the latter system approximates the integral Fredholm equation of the second kind with a limited kernel:

$$\tilde{\varphi}(t_1) + \int_0^{\tilde{a}} \tilde{\tilde{N}}_\kappa(t_1, \tau_1)\tilde{\varphi}(\tau_1)\,d\tau_1 = \tilde{\tilde{f}}_{1,\kappa}(t_1), \qquad (5.3.9)$$

where

$$\tilde{\tilde{N}}_\kappa(t_1, \tau_1) = N_\kappa(t(t_1), \tau(\tau_1))(1 - t(t_1))^{1/4}(1 + t(t_1))^{3/4},$$

$$\tilde{\tilde{f}}_{1,\kappa}(t_1) = f_{1,\kappa}(t(t_1))(1 - t(t_1))^{1/4}(t + t(t_1))^{3/4},$$

$$\tilde{\varphi}(\tau_1) = \varphi(\tau(\tau_1))(1 - \tau(\tau_1))^{1/4}(1 + \tau(\tau_1))^{3/4},$$

$$d\tau_1 = \frac{d\tau}{(1 - \tau)^{1/4}(1 + \tau)^{3/4}},$$

$$\tau_1 = \int_{-1}^{\tau} \frac{d\tilde{\tau}}{(1 - \tilde{\tau})^{1/4}(1 + \tilde{\tau})^{3/4}} = \tau_1(\tau),$$

$$\tilde{a} = \int_{-1}^{1} \frac{d\tilde{\tau}}{(1 - \tilde{\tau})^{1/4}(1 + \tilde{\tau})^{3/4}}.$$

According to (5.1.27) the order of approximation has the form

$$\tilde{\theta}(t_k) \le \left[h^\lambda(t_k + 1)^{1/4} + \frac{h^{1/2}(1 + t_k)^{1/4}}{(1 - t_k)^{1/4}} + \frac{h}{(1 - t_k)^{3/4}(1 + t_k)^{3/4}}\right] O(|\ln h|). \tag{5.3.10}$$

From the theory of numerical methods for integral Fredholm equations of the second kind with a continuous kernel (Kantorovich and Krylov 1952) it follows that the order of approximation $\tilde{\varphi}_n(t_k)$ of function $\tilde{\varphi}(t)$ is the same. By returning to functions $\varphi(t)$ and $\varphi_n(t)$, we terminate the proof of the theorem for the case $\kappa = 0$. For $\kappa = 1$ and -1 the theorem is proved in a similar way. ■

Note 5.3.1. From the proof of Theorem 5.3.1 it is seen that the theorem stays true if the kernel $k(t_0, t)$ is of the form $k_1(t_0, t)/|t_0 - t|^\alpha$, where $0 \le \alpha < 1$, and $k_1(t_0, \mathrm{t})$ belong to the class H on a rectangle. In this case kernel $N_\kappa(t, \tau)$ has the same form (this can be shown by using the Poincaré–Bertrand formula for changing the order of integration and properties of singular integrals). Therefore, by passing to iterated kernels, one can derive a Fredholm equation with a continuous kernel. Obviously, all the operations may also be implemented in the discrete form. In accordance with the preceding analysis, Theorem 5.3.1 may be reformulated for the equation

$$\int_{-1}^{1} \frac{k(t_0, t)\varphi(t)\,dt}{t_0 - t} = f(t_0), \qquad (5.3.11)$$

where both $k(t_0, t)$ and $f(t)$ belong to the class H and $k(t_0, t_0) \neq 0$, $t_0 \in [-1, 1]$.

In this case, the construction of systems of linear algebraic equations does not require passing to a singular integral of the form (5.3.1). Thus, to find a numerical index $\kappa = 1, 0$, or -1 solution to Equation (5.3.11) one has to consider, respectively, the systems of linear algebraic equations

$$\sum_{k=1}^{n} \frac{K(t_{0j}, t_k)\varphi_n(t_k)h}{t_{0j} - t_k} = f(t_{0j}), \qquad j = 1, \ldots, n, \qquad (5.3.5')$$

$$\sum_{k=1}^{n} \frac{k(t_{0j}, t_k)\varphi_n(t_k)h}{t_{0j} - t_k} = f(t_{0j}), \qquad j = 1, \ldots, n-1,$$

$$\sum_{k=1}^{n} \varphi_n(t_k)h = C, \qquad j = n, \qquad (5.3.6')$$

$$\lambda_{0n} + \sum_{k=1}^{n} \frac{k(t_{0j}, t_k)\varphi_n(t_k)h}{t_{0j} - t_k} = f(t_{0j}), \qquad j = 0, 1, \ldots, n. \quad (5.3.7')$$

Finally, we note that the preceding method of analysis allows us to apply all the notes and theorems made and/or proved for the characteristic singular equation presented in Section 5.1 to Equations (5.3.1) and (5.3.11).

The analysis may be simplified if one applies the methods of numerical solution developed for the characteristic singular equation in Section 5.2 and based on using unequally spaced grid points. In this case the system of linear algebraic equations of the form (5.3.8) converts from the very beginning into a system of linear algebraic equations for an integral Fredholm equation of the second kind with a continuous kernel. Thus, one may formulate the following theorem.

Theorem 5.3.2. *Let functions $f(t)$ and $K(t_0, t)$ belong to the class H on* $[-1, 1]$ *and* $[-1, 1) \times [-1, 1]$, *respectively. Then, Relationship* (5.2.4) *holds* *between solutions to the systems of linear algebraic equations*

$$\sum_{k=1}^{n} \frac{\psi_n(t_k)a_k}{t_{0j} - t_k} + \sum_{k=1}^{n} K(t_{0j}, t_k)\psi_n(t_k)a_k = f(t_{0j}), \qquad j = 1, \ldots, n, \tag{5.3.12}$$

where

$$a_k = \frac{4\pi}{2n+1}\sin^2\frac{k}{2n+1}\pi, \qquad t_k = \cos\frac{2k}{2n+1}\pi, \qquad k = 1, \ldots, n,$$

$$t_{0j} = \cos\frac{2j-1}{2n+1}\pi, \qquad j = 1, \ldots, n,$$

$$\sum_{k=1}^{n} \frac{\psi_n(t_k)a_k}{t_{0j} - t_k} + \sum_{k=1}^{n} K(t_{0j}, t_k)\psi_n(t_k)a_k = f(t_{0j}), \qquad j = 1, \ldots, n-1,$$

$$\sum_{k=1}^{n} \psi_n(t_k)a_k = C, \qquad j = n, \tag{5.3.13}$$

where

$$a_k = \frac{\pi}{n}, \qquad t_k = \cos\frac{2k-1}{2n}\pi, \qquad k = 1, \ldots, n,$$

$$t_{0j} = \cos\frac{j}{n}\pi, \qquad j = 1, \ldots, n-1,$$

$$\lambda_{0n} + \sum_{k=1}^{n} \frac{\psi_n(t_k)a_k}{t_{0j} - t_k} + \sum_{k=1}^{n} K(t_{0j}, t_k)\psi_n(t_k)a_k = f(t_{0j}),$$

$$j = 1, \ldots, n+1, \tag{5.3.14}$$

where

$$a_k = \frac{\pi}{n+1}\sin^2\frac{k}{n+1}\pi, \qquad t_k = \cos\frac{k}{n+1}\pi, \qquad k = 1, \ldots, n,$$

$$t_{0j} = \cos\frac{2j-1}{2(n+1)}\pi, \qquad j = 1, \ldots, n+1,$$

and the corresponding functions $\psi(t)$, *determining in Equation* (5.2.1) *the index* κ *solution to Equation* (5.3.1). *In Relationship* (5.2.4) $R_n(t_k)$ *is an approximation error for the corresponding integrals in the Fredholm equation of the second kind.* (*The approximation formulas are applied for points* t_0, $j = 0, 1, \ldots, n - \kappa$.)

In a similar way one may write systems of linear algebraic equations for singular integral Equation (5.3.11).

Note 5.3.2. Fredholm's results (Goursat 1934, Privalov 1935) are also valid for a system of integral equations of the second kind if it has a unique solution and the kernels are regular (continuous). Therefore, the previously formulated results are valid for a system of singular integral equations if it has a unique solution of a fixed index and is subjected to corresponding additional conditions; also, a system of linear algebraic equations for the characteristic part may be transposed as was done in the case of a single singular equation.

Note 5.3.3. While calculating steady flow past an infinite-span wing with a flap, one has to consider a characteristic singular integral equation on the segment $[0, 1]$ under the following conditions. If a flap is deflected, then $f(t)$ suffers a discontinuity of the first kind at point q (the point of the flap's deflection). However, if the flap stays undeflected, then $f(t) \in H$ on the segment of integration. For concrete calculations it is desirable to have a method for considering both situations from a single point of view. On the other hand, hinge moments at the point of a flap's deflection are calculated by using nodes lying on the flap only. Therefore, if a flap is small (i.e., the segment $[q, 1]$ is short), then it incorporates a small number of nodes, and it is difficult to ensure satisfactory accuracy for calculating the moments. At the same time, it is desirable to construct such a computational procedure for which segments $[0, q]$ and $[q, 1]$ contain the same number of nodes. Here we have succeeded in realizing the following idea.

First we choose a mapping of segment $[0, 1]$ onto segment $[-1, 1]$ that is infinitely integrable (or has at least derivatives up to the order $r \geq 2$ where the rth derivative is limited). The first derivative does not vanish, and point q is mapped onto point 0 on $[-1, 1]$. Next we choose on segment $[-1, 1]$ either uniform or nonuniform grids subject to the condition that point 0 is a point of the set E_0. The inverse mapping distributes the nodes in the desired manner. One may choose the mapping

$$t(\tau) = \frac{q(1+\tau)}{(1+\tau)(2q-1) + 2(1-q)}, \tag{5.3.15}$$

where $q \in (0,1)$, $t \in [0,1]$, and $\tau \in [-1,1]$.

We see that $t(-1) = 0$, $t(1) = 1$, $t(0) = q$, and $t'(\tau) > 0$ for any $\tau \in [-1,1]$. Let us substitute the variable in Equation (5.1.1), using Formula (5.3.15) ($a = 0, b = 1$). Then we get

$$\int_{-1}^{1} \frac{\varphi(t(\tau))t'(\tau)\, d\tau}{t(\tau) - t(\tau_0)} = f(t(\tau_0)), \tag{5.3.16}$$

as seen in Equation (5.3.11). For consideration of the solution limited at the point 1 and unlimited at the point -1, we must examine the linear algebraic system

$$\sum_{i=1}^{2n-1} \frac{\Gamma_i}{t_i - t_{0j}} = f_j, \qquad j = 1, \ldots, 2n-1, \tag{5.1.17}$$

where

$$t_i = t(\tau_i), \qquad t_{0j} = t(\tau_{0j}), \qquad \tau_{0j} = -1 + jh, \qquad h = 1/n,$$

$$\tau_j = \tau_{0j} - h/2, \qquad j = 1, \ldots, 2n-1,$$

$$\Gamma_i = \varphi(t_i)t'(\tau_i)h, \qquad f_j = f\big(t(\tau_{0j})\big), \quad j = 1, \ldots, 2n-1, \qquad j \neq n,$$

$$f_n = [f(q-0) + f(q+0)]/2.$$

In applications it is often necessary to construct a direct numerical method, allowing us to have different numbers of grid points on different parts of a segment of integration. To our mind this can be done by finding suitable sets of standard mappings of a segment onto another segment, possessing the following properties:

1. The first derivative belonging to the class H does not vanish, or there exist higher-order derivatives.
2. Portions of a segment are mapped onto equal or approximately equal segments.

Note 5.3.4. By using results of numerical solution of Equation (5.1.54) and the results of this section, we deduce that for the Prandtl equation

$$\int_{-1}^{1} \frac{d\Gamma(t)}{dt} \frac{dt}{t_0 - t} - a(t_0)\Gamma(t_0) = f(t_0), \qquad t_0 \in (-1, 1), \quad (5.3.18)$$

one has to consider the following system of linear algebraic equations (see (5.1.56)):

$$-\frac{C_2}{t_{0j} - t_1} + \sum_{k=1}^{n} \Gamma_n(t_{0k}) \left(\frac{1}{t_{0j} - t_k} - \frac{1}{t_{0j} - t_{k+1}} \right)$$

$$+ \frac{C_1 + C_2}{t_{0j} - t_{n+1}} - a(t_{0j})\Gamma_n(t_{0j}) = f(t_{0j}), \qquad j = 1, \dots, n. \quad (5.3.19)$$

Relationship (5.1.56) is valid for a solution of the latter system and a solution to the Prandtl equation meeting the condition $\Gamma(-1) = C_2$ and $\Gamma(1) = C_1 + C_2$.

5.4. EQUATION ON A SYSTEM OF NONINTERSECTING SEGMENTS

Let us next consider a singular integral equation of the form

$$\int_L \frac{\varphi(t)\, dt}{t_0 - t} = f(t_0), \quad (5.4.1)$$

where L is a set of l nonintersecting segments $[A_1, B_1], \dots, [A_l, B_l]$.

According to Gakhov (1977) and Muskhelishvili (1952) this equation may have index $\kappa = -l, -l+1, \dots, -1, 0, 1, \dots, l$ determined by the number of segment ends where the sought after solution is limited. Following the procedure proposed by Muskhelishvili (1952), let us renumerate the ends of segments $[A_1, B_1], \dots, [A_l, B_l]$ in an arbitrary manner and denote them by $c_1, c_2, \dots, c_{2l}$. By $\eta(c_1, \dots, c_q)$ we denote the index $\kappa = l - q$ class of solutions to Equation (5.4.1), which, for $f(t) \in H$ on L, are limited at the ends $c_1, \dots, cq$ and unlimited at the ends $c_{q+1}, \dots, c_{2l}$.

On segment $[A_m, B_m]$ we take the sets $E_m = \{t_k, k = n_{m-1} + 1, \dots, n_m\}$ and $E_{0m} = \{t_{0j}, j = n_{m-1} + 0, n_{m-1} + 1, \dots, n_m\}$ forming a canonic division of the segment with the step h_m ($m = 1, \dots, l$; $n_0 = 0$). Let us denote $E = \bigcup_{m=1}^{l} E_m$ and $E_0 = \bigcup_{m=1}^{l} E_{0m}$ and assume that the sets E and E_0

form a canonic division of the curve L if the relationship

$$\frac{h}{h_{ni}} \leq R < +\infty, \qquad m = 1, \ldots, l, \tag{5.4.2}$$

holds, where $h = \max(h_1, \ldots, h_l)$, $h \to 0$. Note that according to relationship (5.4.2) the numbers n_l and $N_m = n_m - n_{m-1}$, $m = 1, \ldots, l$, for $n \to \infty$ are such that a ratio of any two of them is a limited quantity. In what follows we assume that Relationship (5.4.2) is always valid.

By $P(q)$ we denote the set of points t_{0j} obtained by excluding from the set E_0 points $c_{q+1}, \ldots, c_{2l}$ nearest to the ends.

Similarly to Section 5.1, the following theorems may be proved.

Theorem 5.4.1. *Let function $f(t) \in H$ on L. Then, between a solution to the system of linear algebraic equations*

$$\sum_{k=1}^{n_l} \frac{\varphi_{n_l}(t_k)\, \Delta t_k}{t_{0j} - t_k} = f(t_{0j}), \qquad t_{0j} \in P(q),$$

$$\sum_{k=1}^{n_l} t_k^{\epsilon} \varphi_{n_l}(t_k)\, \Delta t_k = C_\epsilon, \qquad \epsilon = 0, 1, \ldots, \kappa - 1, 1 \leq \kappa \leq l, \tag{5.4.3}$$

and the index $\kappa \geq 0$ solution $\varphi(t)$ belonging to the class $\eta(c_1, \ldots, c_q)$ and meeting the condition

$$\int_L t^\epsilon \varphi(t)\, dt = C_\epsilon, \tag{5.4.4}$$

the relationship

$$|\varphi(t_k) - \varphi_{n_l}(t_k)| \leq \theta(t_k), \qquad k = 1, \ldots, n_l, \tag{5.4.5}$$

holds where $\theta(t_k)$ satisfies the inequalities:

1. *For all points $t_k \in \bigcup_{m=1}^{1} [A_m + \delta, B_m - \delta]$, where positive δ is however small,*

$$\theta(t_k) \leq O(h^{\lambda_1}), \qquad 0 < \lambda_1 \leq 1. \tag{5.4.6}$$

2. *For all points $t_k \in L$,*

$$\sum_{k=1}^{n_l} \theta(t_k)\, \Delta t_k \leq O(h^{\lambda_2}), \qquad 0 < \lambda_2 \leq 1. \tag{5.4.7}$$

Theorem 5.4.2. *Let function $f(t) \in H$ on L. Then Relationship* (5.4.5) *holds between a solution to the system of linear algebraic equations*

$$\sum_{\epsilon=0}^{-\kappa-1} t_{0j}^{\epsilon}\gamma_{\epsilon} + \sum_{k=1}^{n_l} \frac{\varphi_{n_l}(t_k)\,\Delta t_k}{t_{0j}-t_k} = f(t_{0j}), \qquad t_{0j} \in P(q), \quad (5.4.8)$$

and the index $\kappa < 0$ solution $\varphi(t)$ belonging to the class $\eta(c_1, \ldots, c_q)$, $l - q = \kappa < 0$, and meeting the conditions

$$\int_L t^{\epsilon} R_{\kappa}^{-1}(t) f(t)\, dt = 0, \qquad \epsilon = 0, 1, \ldots, -\kappa - 1, \qquad (5.4.9)$$

where $R_{\kappa}(t)$ is a characteristic function belonging to the class $\eta(c_1, \ldots, c_q)$. Here the quantities $\gamma_{\epsilon}(\epsilon = 0, 1, \ldots, -\kappa - 1)$ are regularizing factors.

Detailed proofs of the latter two theorems are presented by Lifanov (1981).

Note 5.4.1. Theorems 5.4.1 and 5.4.2 remain true if function $f(t) \in H$ on L and can tend to infinity of the order of $\alpha \in [0, 1]$ at the ends $c_{q+1}, \ldots, c_{2l}$ and to infinity of the order of β, $\beta \in [0, \frac{1}{2}$ at the ends $c_1, \ldots, c_q$. Here the class $\eta(c_1, \ldots, c_q)$ corresponds to the class of solutions to Equations (5.4.1) that have a singularity of order less than $\frac{1}{2}$ at the ends $c_1, \ldots, c_q$, and either equal or more than $\frac{1}{2}$ at the ends $c_{q+1}, \ldots, c_{2l}$.

If, in addition, function $f(t)$ suffers a discontinuity of the form $|c - t|^{-\nu}$, $\nu \in [0, 1)$, on a segment $[A_m, B_m] \ni c$, then the sets E_{0m} and E_m must be chosen on the segment as was done when considering Theorem 5.1.2. However, if function $f(t)$ suffers a discontinuity of the first kind at point c, then both sets E_{0m} and E_m on segment $[A_m, B_m]$ and the system of linear algebraic equations for the points belonging to the segment should be composed as indicated in the addition to Theorem 5.1.2.

In problems of electrodynamics (Gendel 1982, 1983) it is more convenient to require that Equation (5.4.1) meet the conditions

$$\int_{L_k} \varphi(t)\, dt = C_k, \qquad k = 1, \ldots, m, m \leq l, \qquad (5.4.10)$$

on all segments L_k from L for which a solution is unlimited at both ends.

The system of conditions (5.4.10) may be readily shown to single out a unique solution to Equation (5.4.1). For the sake of convenience, we rewrite Equality (5.2.1) for the index κ solution on segment $[-1, 1]$ in the form: $\varphi_{\kappa}(t) = \omega_{\kappa}(t)\psi_{\kappa}(t)$, $\omega_{\pm 1}(t) = (1 - t^2)^{\mp 1/2}$, $\omega_0(t) = \sqrt{(1-t)/(1+t)}$, and $\psi_{\kappa}(t) \in H$ on $[-1, 1]$. Then the system of linear algebraic equations (5.1.3)–(5.1.5) may be written in the form ($\kappa = 0, 1, -1$):

$$\zeta(-\kappa)\gamma_{0n} + \sum_{k=1}^{n} \frac{\varphi_{\kappa,n}(t_k)h}{t_{0j} - t_k} = f(t_{0j}), \qquad j = 1, \ldots, n - \kappa,$$

$$\sum_{k=1}^{n} \zeta(\kappa)\varphi_{\kappa,n}(t_k)h = \zeta(\kappa)C, \tag{5.4.11}$$

where $\zeta(x) = 1$ for $x > 0$ and $\zeta(x) = 0$ for $\chi \leq 0$, and points t_{0j} in System (5.1.6) are renumerated from 1 to $n + 1$. For $\zeta(\kappa) = 0$ (i.e, for $\kappa = 0$ or -1) the latter equation is an identity and may be ignored.

Hence, the systems of linear algebraic equations (5.2.2), (5.2.15), and (5.2.16) may be rewritten in the form

$$\zeta(-\kappa)\gamma_{0n} + \sum_{k=1}^{n} \frac{\psi_{\kappa,n}(t_k)a_k}{t_{0j} - t_k} = f(t_{0j}), \qquad j = 1, \ldots, n - \kappa,$$

$$\sum_{k=1}^{n} \zeta(\kappa)\psi_{\kappa,n}(t_k)a_k = \zeta(\kappa)C, \tag{5.4.12}$$

where $a_k, t_k, k = 1, \ldots, n$, and $t_{0j}, j = 0, 1, \ldots, n - \kappa$, are chosen depending on the index κ as indicated in the preceding systems, namely, points t_k and t_{0j} are the roots of polynomials $P_{\kappa,n}(t)$ and $Q_{\kappa,n}(t)$, respectively.

In what follows we represent Equation (5.4.1) as a system of singular integral equations on segment $[-1, 1]$. Therefore, in accordance with terminology used in the theory of such systems (Muskhelishvili 1952), we say that a solution $\varphi(t)$ to Equation (5.4.1) has index $\kappa = (\kappa_1, \ldots, \kappa_l)$, $\kappa_m = 1, 0, -1$, and $m = 1, \ldots, l$, if it is (1) unlimited at both ends, (2) unlimited at an end, or (3) limited at both ends of segment $[A_m, B_m]$. The solution will be denoted by $\varphi_\kappa(t)$. Let us consider the mapping $g_m(\tau)$ of segment $[-1, 1]$ onto segment $[A_m, B_m]$, where

$$g_m(\tau) = \frac{B_m - A_m}{2}\tau + \frac{B_m + A_m}{2}, \qquad m = 1, \ldots, l. \tag{5.4.13}$$

Denote

$$\varphi_{\kappa_m}(t) = \varphi_\kappa(t)|_{t \in [A_m, B_m]} = \omega_{\kappa_m}(t)\psi_{\kappa,m}(t),$$

$$f_m(t) = f(t)|_{t \in [A_m, B_m]}, \qquad m = 1, \ldots, l. \tag{5.4.14}$$

We use a uniform division on each of the segments $[A_m, B_m]$, $m = 1, \ldots, l$. By employing the sets $E_m = \{t_{m,k}\ k = 1, \ldots, n_m\}$ and $E_{m,0} = \{t_{m\,0j}, j =$

$0, 1, \ldots, n_m\}$, $m = 1, \ldots, l$, we choose a canonic division of segment $[A_m, B_m]$ with the step h_m. Then the following theorem is true.

Theorem 5.4.3. *Let function $f(t) \in H$ on L. Then, between a solution to the system of linear algebraic equations*

$$\zeta(-\kappa_m)\gamma_{0n_m} + \sum_{k=1}^{n_m} \frac{\varphi_{\kappa_m, n_m}(t_{m,k})h_m}{t_{m,0j} - t_{m,k}} + \sum_{\substack{p=1 \\ p \neq m}}^{l} \sum_{k=1}^{n_p} \frac{\varphi_{\kappa_p, n_p}(t_{p,k})h_p}{t_{m\,0j} - t_{p,k}}$$

$$= f_m(t_{m,0j}), \qquad j = 1, \ldots, n_m - \kappa_m, \qquad m = 1, \ldots, l,$$

$$\sum_{k=1}^{n_m} \zeta(\kappa_m)\varphi_{\kappa_m, n_m}(t_{m,k})h_m = \zeta(\kappa_m)C_m, \qquad m = 1, \ldots, l, \quad (5.4.15)$$

and a solution $\varphi_\kappa(t)$ to Equation (5.4.1) *for which the values of the integrals are known for the segments composing L, on which it has the index* 1, *Relationship* (5.4.5) *holds.*

Let us next consider unequally spaced grid points on segments $[A_m, B_m]$, composed of points $t_{m,k}$, $k = 1, \ldots, n_m$, $t_{m,k} = g_m(\tau_k)$, where τ_k are the roots of polynomial $\mathrm{F}_{\kappa_m,n_m}(\tau)$ from the system of polynomials orthogonal on $[-1, 1]$ with weight $\omega_{\kappa_m}(\tau)$ and points $t_{m,0j}$, $j = 1, \ldots, n_m - k_m$ and $t_{m,0j} = g_m(\tau_{0j})$, where τ_{0j} are the roots of polynomial $Q_{\kappa_m, n_m}(\tau)$ defined by Equality (2.3.5) through $\mathrm{P}_{\kappa_m,n_m}(\tau)$. Then the following theorem is true.

Theorem 5.4.4. *Let function $f(t) \in H$ on L. Then, between a solution to the system of linear algebraic equations*

$$\zeta(-\kappa_m)\gamma_{0n_m} + \sum_{k=1}^{n_m} \frac{\psi_{\kappa_m, n_m}(t_{m,k})a_{mk}}{t_{m,0j} - t_{m,k}} + \sum_{\substack{p=1 \\ p \neq m}}^{l} \sum_{k=1}^{n_p} \frac{\psi_{\kappa_p, n_p}(t_{p,k})a_{pk}}{t_{m,0j} - t_{p,k}}$$

$$= f_m(t_{m,0j}), \qquad j = 1, \ldots, n_m - \kappa_m, \; m = 1, \ldots, l,$$

$$\sum_{k=1}^{n_m} \zeta(\kappa_m)\psi_{\kappa_m, n_m}(t_{m,k})a_{mk} = \zeta(\kappa_m)C_m, \qquad m = 1, \ldots, l, \quad (5.4.16)$$

where $a_{mk} = (B_m - A_m)a_k/2$, $a_k = Q_{k_m, n_m}(t_{m,k}/P'_{\kappa_m, n_m}(t_{m,k})$, and function $\psi_\kappa(t)$ determining the solution $\varphi_\kappa(t)$, Relationship (5.2.4) *holds on each segment $[A_m, B_m]$, where $R_{m,n_m}(t_{m,k})$ is an error of approximating a singular integral on L.*

Theorems 5.4.3 and 5.4.4 are proved in a similar way. Mappings (5.4.13) allow us to consider Equation (5.4.1) as a system of l singular integral equations on $[-1, 1]$ which has a unique solution, subject to the corresponding additional conditions (the value of an integral of the solution is known on those segments of L on both ends of which the solution is unlimited, i.e., on which the index is equal to 1). Hence, the system (Muskhelishvili 1952) is equivalent to a system of integral Fredholm equations of the second kind, which also has a unique solution. Therefore, by repeating the procedure of passing to a system of integral Fredholm equations of the second kind in discrete form, we conclude that the systems of linear algebraic equations (5.4.15) and (5.4.16) are equivalent to the systems of linear algebraic equations for this system of integral Fredholm equations of the second kind. The passage is possible due to the fact that Systems (5.4.11) and (5.4.12) are solvable for any $\kappa = 1, 0, -1$.

Consider the following full singular integral equation of the first kind on a system of nonintersecting segments:

$$\int_L \frac{\varphi(t)\, dt}{t_0 - t} + \int_L K(t_0, t)\varphi(t)\, dt = f(t_0). \tag{5.4.17}$$

For this equation analogs of all theorems proved in this section for Equation (5.4.1) are valid. However, the following sums must be added to the linear algebraic equations:

$$\sum_{k=1}^{n_l} K(t_{0j}, t_k)\varphi_{n_l}(t_k)\,\Delta t_k$$

to Systems (5.4.3) and (5.4.8);

$$\sum_{p=1}^{l} \sum_{k=1}^{n_p} K(t_{m,0j}, t_{p,k})\varphi_{\kappa_p, n_p}(t_{p,k})h_p$$

to system (5.4.15);

$$\sum_{p=1}^{l} \sum_{k=1}^{n_p} K(t_{m,0j}, t_{p,k})\psi_{\kappa_p, n_p} a_{p,k}$$

to System (5.4.16). If functions $f(t)$ and $K(t_0, t)$ belong to the class H on the corresponding sets or $K(t_0, t)$ has the form $K_1(t_0, t)/|t_0 - t|^\alpha$, $0 \le \alpha < 1$, where $K_1(t_0, t) \in H$ on $L \times L$, then the systems of linear algebraic equations are composed on equally spaced uniform grid points by using standard canonic divisions of segments $[A_m, B_m]$ forming L. However, if

$f(t)$ incorporates singularities as indicated in Note 5.4.1, then these grids on the corresponding segment must be chosen in accordance with the note.

5.5. EXAMPLES OF NUMERICAL SOLUTION OF THE EQUATION ON A SEGMENT

Figure 5.1 shows a numerical solution ($\triangle\,\triangle$, $h = 2/11$; $\times\times$, $h = 2/21$; $\circ\circ$, $h = 2/41$) of the equation

$$\int_{-1}^{1} \frac{\gamma(x)\,dx}{x_0 - x} = \pi, \tag{5.5.1}$$

whose accurate solution (see the solid line) for a uniform division is given by

$$\gamma(x) = \sqrt{\frac{1-x}{1+x}}\,. \tag{5.5.2}$$

We see that an increase in the number of points results in the numerical solution converging to the accurate one. The numerical solution is found from the system

$$\sum_{i=1}^{n} \frac{\gamma_n(x_i)h}{x_{0j} - x_i} = \pi, \qquad j = 1,\ldots,n, \tag{5.5.3}$$

where $x_i = -1 + h_i$, $x_{0i} = x_i + h/2$, $i = 1,\ldots,n$, and $h = 2/(n+1)$.

Figure 5.2 demonstrates numerical solutions of the equation

$$\int_{-1}^{1} \frac{\gamma(x)\,dx}{x_0 - x} = -\pi, \tag{5.5.4}$$

whose accurate solution (for the same division) is given by

$$\gamma(x) = \frac{x}{\sqrt{1-x^2}}\,. \tag{5.5.5}$$

The numerical solution was found from the system

$$\sum_{i=1}^{n} \frac{\gamma_n(x_i)h}{x_{0j} - x_i} = -\pi, \qquad j = 1,\ldots,n-1,$$

$$\sum_{i=1}^{n} \gamma_n(x_i)h = 0, \qquad j = n. \tag{5.5.6}$$

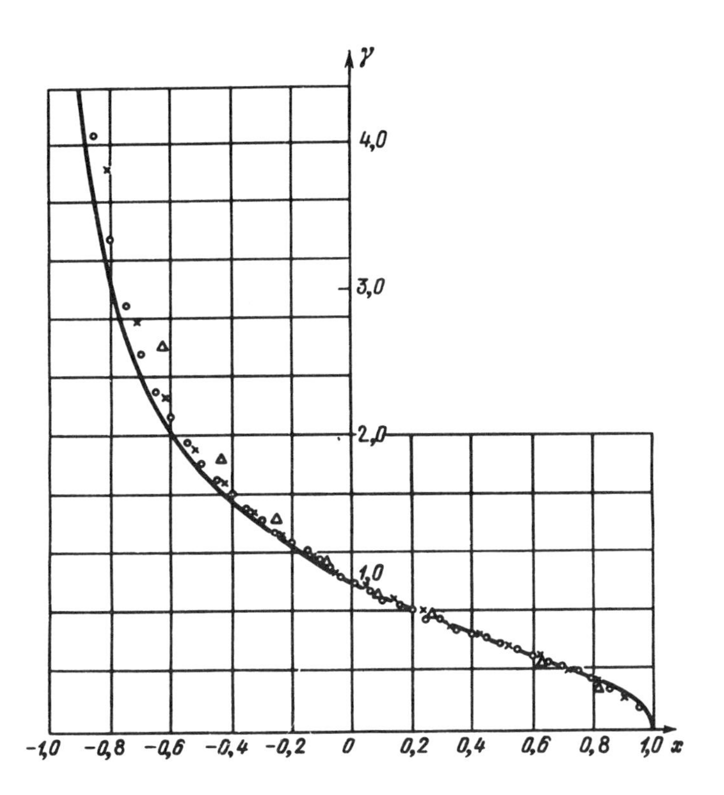

FIGURE 5.1. Index $\kappa = 0$ numerical solution to Equation (5.5.1) for a uniform division. The solid line corresponds to the exact solution, $\triangle\,\triangle$ to $h = 2/11$, $\times\times$ to $h = 2.21$, and $^{\circ\circ}$ to $h = 2/41$.

Finally, Figure 5.3 shows numerical solutions of the equation

$$\int_{-1}^{1} \frac{\gamma(x)\,dx}{x_0 - x} = \pi x_0, \tag{5.5.7}$$

whose accurate solution is given by

$$\gamma(x) = \sqrt{1 - x^2} \tag{5.5.8}$$

(for the same division). The numerical solution was found from the system

$$\gamma_{0n} + \sum_{i=1}^{n} \frac{\gamma_n(x_i)h}{x_{0j} - x_i} = \pi x_0 j, \qquad j = 0, 1, \ldots, n.$$

Figures 5.4–5.6, where the solid lines correspond to accurate solutions $\times\times$ to $n = 10$, and $^{\circ\circ}$ to $n = 20$, show numerical solutions of the same

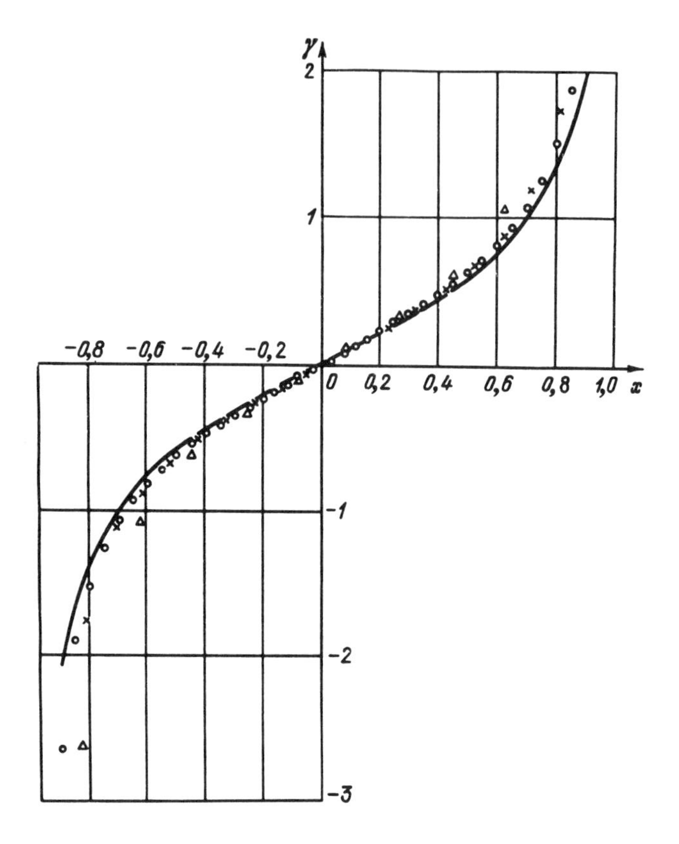

FIGURE 5.2. Index $\kappa = 1$ numerical solution to Equation (5.5.4) for a uniform division. The solid line corresponds to the exact solution, △ △ to $h = 2/11$, × × to $h = 2.21$, and ○○ to $h = 2/41$.

equations at unequally spaced grid points. In this case solution $\gamma(x)$ is represented in the form $\omega(x)u(x)$, and the systems of linear equations are constructed with respect to the values of function $(u(x)$ at the roots of the corresponding polynomials. Thus, numerical solution (5.5.2) is found by considering the system (see Figure 5.4).

$$\sum_{i=1}^{n} \frac{u_n(x_i)a_i}{x_{0j} - x_i} = \pi, \qquad j = 1, \dots, n,$$

$$x_i = \cos\frac{2i}{2n+1}\pi, \qquad a_i = \frac{4\pi}{2n+1}\sin^2\frac{i}{2n+1}\pi,$$

$$x_{0j} = \cos\frac{2j-1}{2n+1}\pi,$$

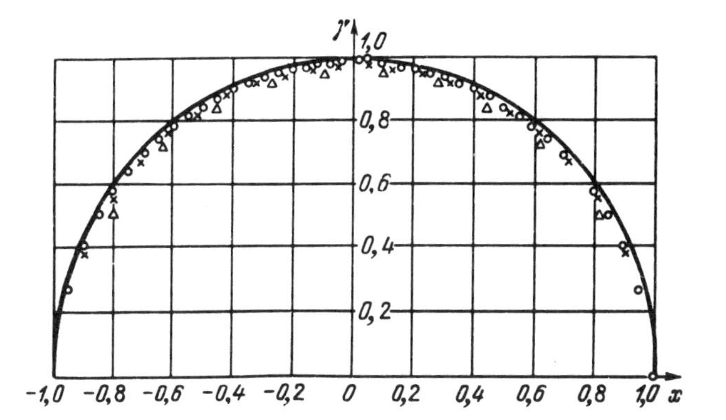

FIGURE 5.3. Index $\kappa = -1$ numerical solution to Equation (5.5.7) for a uniform division. The solid line corresponds to the exact solution, △ △ to $h = 2/11$, × × to $h = 2/21$, and ∘∘ to $h = 2/41$.

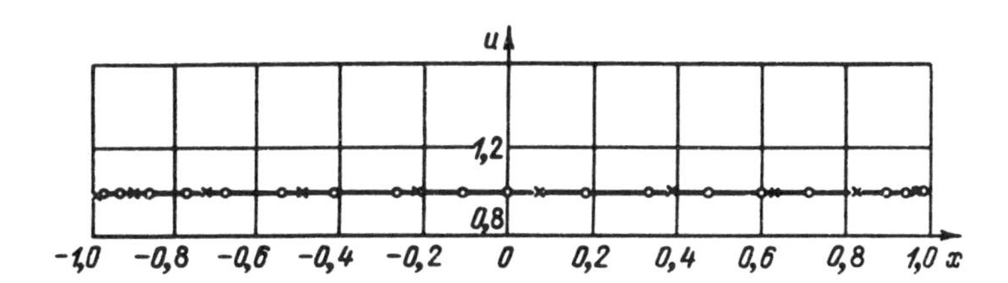

FIGURE 5.4. Index $\kappa = 0$ numerical solution to Equation (5.5.1) for a nonuniform division. The solid line corresponds to the exact value of the function $u(x)$, where $\gamma(x) = w(x)u(x)$ and $w(x) = \sqrt{(1-x)(1+x)}$; × × corresponds to $n = 10$ and ∘∘ to $n = 20$.

numerical solution (5.5.5) is found by considering the system (see Figure 5.5)

$$\sum_{i=1}^{n} \frac{u_n(x_i)a_i}{x_{0j} - x_i} = -\pi, \qquad j = 1, \dots, n-1,$$

$$\sum_{i=1}^{n} u_n(x_i)a_i = 0, \qquad j = n, \qquad x_i = \cos\frac{2i-1}{2n}\pi,$$

$$a_i = \frac{\pi}{n}, \qquad x_{0j} = \cos\frac{j}{n}\pi,$$

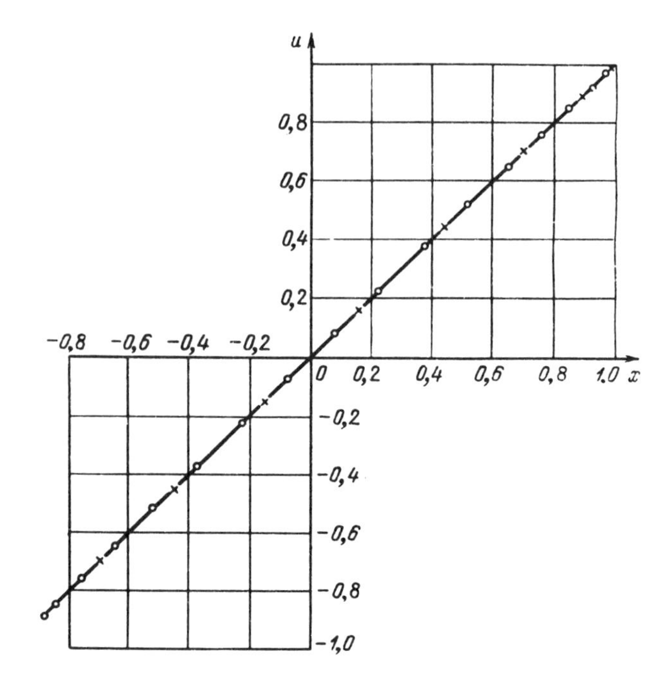

FIGURE 5.5. Index $\kappa = 1$ numerical solution to Equation (5.5.4) for a nonuniform division. The solid line corresponds to the exact value of the function $u(x)$, where $\gamma(x) = w(x)u(x)$ and $w(x) = (1 - x^2)^{-1/2}$; ×× corresponds to $n = 10$ and ∘∘ to $n = 20$.

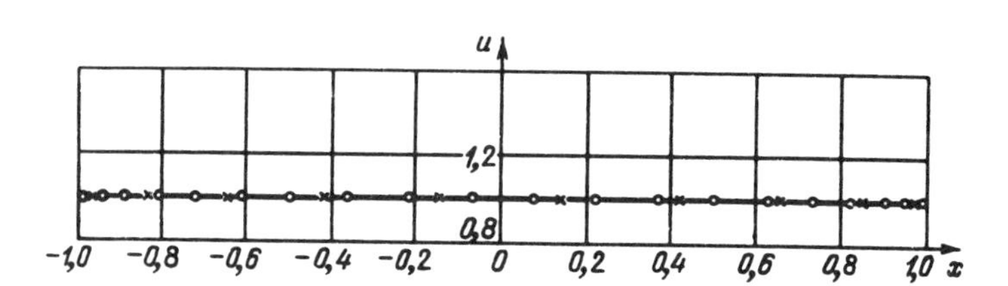

FIGURE 5.6. Index $\kappa = -1$ numerical solution to Equation (5.5.7) for a nonuniform division. The solid line corresponds to the exact value of the function $u(x)$, where $\gamma(x) = w(x)u(x)$ and $w(x) = \sqrt{1 - x^2}$; ×× corresponds to $n = 10$ and ∘∘ to $n = 20$.

and, finally, numerical solution (5.5.8) is found by considering the system (see Figure 5.6)

$$\gamma_{0n} + \sum_{i=1}^{n} \frac{u_n(x_i)a_i}{x_{0j} - x_i} = \pi x_{0j}, \qquad j = 1, \ldots, n + 1,$$

$$x_i = \cos\frac{i}{n+1}\pi, \qquad a_i = \frac{\pi}{n+1}\sin^2\frac{i}{n+1}\pi, \qquad x_{0j} = \cos\frac{2j-1}{2(n+1)}\pi.$$

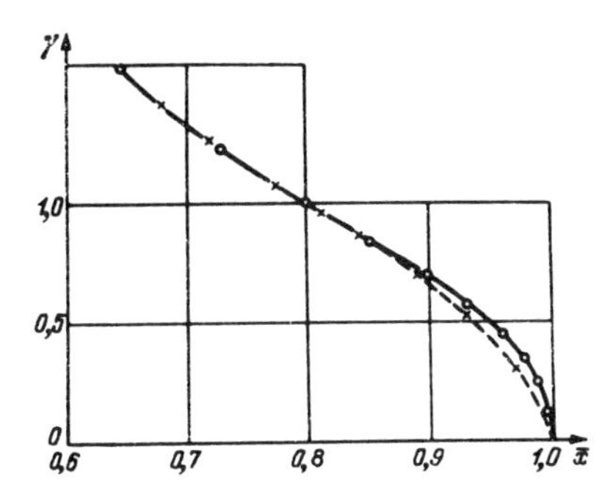

FIGURE 5.7. Index $\kappa = 0$ numerical solution to Equation (5.5.9) for a uniform($\times$) and nonuniform (●) divisions for $n = 30$ and the reference point coinciding with point $q = 0.8$.

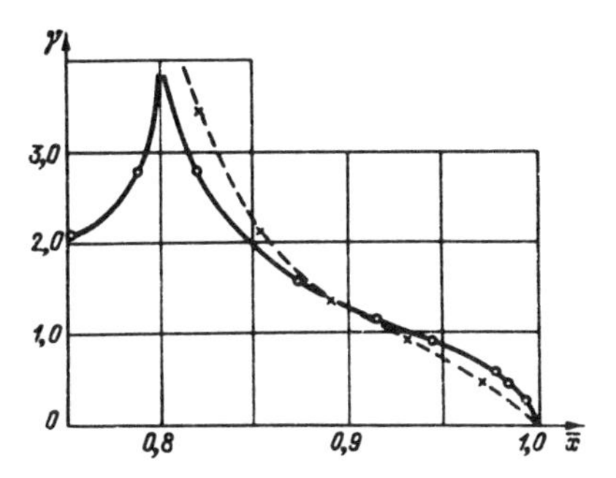

FIGURE 5.8. Index $\kappa = 0$ numerical solution to Equation (5.5.9) for $f(x_0) = 0, 0 < x < 0.8$, and $f(x_0) = -2\pi$, $0.8 < x < 1$ for a uniform ($\times$) and nonuniform (●) divisions for $n = 30$ and the reference point coinciding with point $q = 0.8$.

The calculations were carried out for $n = 10, 20, 30, 40$. It was found out that $|u(x_i) - u_n(x_i)| \leq 5 \times 10^{-6}$.

Figure 5.7 compares the results of numerical solution of the equation

$$\int_0^1 \frac{\gamma(x)\, dx}{x_0 - x} = f(x_0), \tag{5.5.9}$$

at equally ($\times$) and unequally ($\circ$) spaced grid points and $n = 30$. It was assumed that $f(x_0) = -2\pi$, $\kappa = 0$, and the reference point was placed at the given point $q = 0.8$ (the hinge point of the flap) over the domain of integration. At point q the right-hand side was put equal to $[f(q - 0) + f(q + 0)]/2$.

Figure 5.8 presents a comparison for the same grid points for

$$f(x) = \begin{cases} 0, & 0 < \times < 0.8, \\ -2\pi, & 0.8 < \times < 1. \end{cases} \tag{5.5.10}$$

6

Equations of the First Kind on a Circle Containing Hilbert's Kernel

6.1. EQUATION ON A CIRCLE

Consider the characteristic equation

$$\int_L \frac{\varphi(t)\,dt}{t_0 - t} = f(t_0), \tag{6.1.1}$$

where L is a unit-radius circle centered at the origin of coordinates. Let the sets $E = \{t_k,\ k = 1, \ldots, n\}$ and $E_0 = \{t_{0j},\ j = 0, 1, \ldots, n\}$ form a canonic division of the circle. The following theorem is true.

Theorem 6.1.1. *Let function $f(t) \in H$ on L. Then between a solution to the system of linear algebraic equations*

$$\sum_{k=1}^{n} \frac{\varphi_n(t_k) a_k}{t_{0j} - t_k} = f(t_{0j}), \qquad j = 1, \ldots, n, \tag{6.1.2}$$

where $a_k = t_{k+1} - t_k$ and $t_{n+1} = t_1$, and the solution to Equation (6.1.1),

$$\varphi(t) = -\frac{1}{\pi^2} \int_L \frac{f(t_0)\,dt_0}{t - t_0}, \tag{6.1.3}$$

there exists the relationship

$$|\varphi(t_k) - \varphi_n(t_k)| \leq \theta(t_k), \qquad k = 1, \ldots, n, \tag{6.1.4}$$

where $\theta(t_k)$ satisfies the inequality

$$\theta(t_k) \leq O(1/n^\lambda), \qquad 0 < \lambda \leq 1. \tag{6.1.5}$$

Proof. If $h = a_k$, then Systems (5.1.3) and (6.1.2) coincide, and hence,

$$\varphi_n(t_k) = \frac{1}{a_k} I_{0,k}^{(n)} \sum_{j=1}^{n} \frac{1}{b_j} I_{0,0j}^{(n)} \frac{f(t_{0j}) b_j}{t_k - t_{0j}}, \tag{6.1.6}$$

where $b_j = t_{0j+1} - t_j$, $k = 1, \ldots, n$, and $t_{0n+1} = t_{01}$.

As long as this time L is a circle, multipliers $I_{0,k}^{(n)}$ and $I_{0,0j}^{(n)}$ must be treated in a way different from Theorem 5.1.1. Remember that

$$t_k = \exp(i\theta_k), \qquad t_{0k} = \exp(i(\theta_k + \pi/n)) = \exp(i\theta_{0k}),$$

$$k = 1, \ldots, n.$$

Therefore one can write

$$I_{0,k}^{(n)} = \frac{\prod_{m=1}^{n}(t_{0m} - t_k)}{\prod_{m=1, m\neq k}^{n}(t_m - t_k)} = -t_k \frac{\prod_{m=1}^{n}[1 - \exp(i(\theta_{0m} - \theta_k))]}{\prod_{m=1, m\neq k}^{n}[1 - \exp(i(\theta_m - \theta_k))]}$$

$$= -t_k \frac{P_{2,k}^{(n)}}{P_{1,k}^{(n)}} \tag{6.1.7}$$

Because points t_k, $k = 1, \ldots, n$, divide circle L into equal parts, and t_{0k} is the center of the arc (t_{k+1}, t_k), $\theta_m - \theta_k = 2\pi(m-k)/n$ and $\theta_{0m} - \theta_k = 2\pi(m-k)/n + \pi/n$. By relabeling and taking into account periodicity of function $\exp(i\theta)$, we can write

$$P_{1,k}^{(n)} = \prod_{m=1}^{n-1} (1 - \exp(im \cdot 2\pi/n)),$$

$$P_{2,k}^{(n)} = \prod_{m=0}^{n-1} (1 - \exp(i(\pi/n + m \cdot 2\pi/n))).$$

In order to calculate $P_{1,k}^{(n)}$ we note that the numbers $\exp[im(2\pi/n)]$, $m = 0, 1, \ldots, n-1$, are the nth-power roots of the number $z = 1$. In other words,

$$z^n - 1 = \prod_{m=0}^{n-1} (z - \exp(im \cdot 2\pi/n))$$

$$= (z - 1) \prod_{m=1}^{n-1} (z - \exp(im \cdot 2\pi/n))$$

or

$$\frac{z^n - 1}{z - 1} = z^{n-1} + z^{n-2} + \cdots + 1 = \prod_{m=1}^{n-1} (z - \exp(im \cdot 2\pi/n)).$$

The latter equality is, in fact, an identity. Therefore, tending z to unity, one gets in the limit

$$\lim_{z \to 1} \frac{z^n - 1}{z - 1} = n = P_{1,k}^{(n)}. \tag{6.1.8}$$

In order to calculate $P_{2,k}^{(n)}$ we observe that the numbers $\exp[i(\pi/n + m \cdot 2\pi/n)]$, $m = 0, 1, \ldots, n - 1$, are the nth-power roots of the number $z = -1$, i.e.,

$$z^n + 1 = \prod_{m=0}^{n-1} [z - \exp(i(\pi/n + m \cdot 2\pi/n))].$$

Because the latter equality is valid for any z, for $z = 1$ we get

$$P_{2,k}^{(n)} = 2. \tag{6.1.9}$$

According to (6.1.7)–(6.1.9),

$$I_{0,k}^{(n)} = -t_k \frac{2}{n},$$

$$\frac{1}{a_k} I_{0,k}^{(n)} = \frac{1}{t_{k+1} - t_k} (-t_k) \frac{2}{n} = \frac{2}{n} \frac{1}{1 - e^{i2\pi/n}}$$

$$= \frac{1}{n \sin \pi/n} \left(\sin \frac{\pi}{n} + i \cos \frac{\pi}{n} \right) = i \frac{1}{\pi} + O\left(\frac{1}{n} \right). \tag{6.1.10}$$

In a similar way one can show that

$$I_{0,0j}^{(n)} = t_{0j}\frac{2}{n}, \qquad \frac{1}{b_j}I_{0,0j}^{(n)} = -i\frac{1}{\pi} + O\left(\frac{1}{n}\right). \tag{6.1.11}$$

From (6.1.6), (6.1.10), and (6.1.11) it follows that

$$\varphi_n(t_k) = -\frac{1}{\pi^2}\sum_{j=1}^{n}\frac{f(t_{0j})b_j}{t_k - t_{0j}} + \sum_{j=1}^{n}O\left(\frac{1}{n}\right)\frac{f(t_{0j})b_j}{t_k - t_{0j}}. \tag{6.1.12}$$

Together with formulas (1.2.7) and (1.2.10), this proves the validity of Theorem 6.1.1. ■

In a similar way the following theorem may be proved.

Theorem 6.1.2. *Let function $f(t)$ be of the form*

$$f(t) = \frac{\psi(t)}{|t - q|^{\nu}}, \tag{6.1.13}$$

where $\psi(t) \in H$ on L, $0 \le \nu < 1$. Also let the sets E and E_0 forming a canonic division of the circle L be chosen in such a way that the point q is the middle of the arc limited by the two nearest points from E and E_0. Then between a solution to the system of linear algebraic equations (6.1.2) *and the solution* (6.1.3) *to Equation* (6.1.1), *Relationship* (6.1.4) *holds in which quantity $\theta(t_k)$ satisfies the inequalities*:

1. *For all points $t_k \in L^*$,*

$$\theta(t_k) \le O\left(\frac{1}{n^{\lambda_1}}\right), \qquad 0 < \lambda_1 \le 1. \tag{6.1.14}$$

2. *For all points $t_k \in L$,*

$$\sum_{k=1}^{n}\theta(t_k)|a_k| \le O\left(\frac{1}{n^{\lambda_2}}\right), \qquad 0 < \lambda_2 \le 1. \tag{6.1.15}$$

To prove the theorem one has to employ Formulas (1.4.5) and (1.4.6) for a piecewise Lyapunov curve. Note that L^* mentioned in connection with (6.1.14) is the portion of L lying outside the δ neighborhood of point q:

$$L^* = L \setminus O(q, \delta). \tag{6.1.16}$$

Theorem 6.1.1 and Inequalities (1.2.12)–(1.2.14) may be used to prove the following theorem (in analogy to Theorem 5.1.5).

Theorem 6.1.3. *Let function $f(t) \in H$ on L and the sets E and E_0 be chosen in such a way that point $q \in E_0$ for $j = j_q$. Then between a solution to the system of linear algebraic equations*

$$\sum_{k=1}^{n} \frac{\varphi_n(t_k)a_k}{t_{0j} - t_k} = f(t_{0j}), \qquad j = 1, \ldots, n, \; j \neq j_q,$$

$$\sum_{k=1}^{n} \varphi_n(t_k)a_k = C, \qquad j = j_q, \tag{6.1.17}$$

and the class-Π *solution* $\varphi(t)$ *to Equation* (6.1.1) *on* L,

$$\varphi(t) = -\frac{1}{\pi^2(q-t)} \int_L \frac{(q-t_0)f(t_0)\,dt_0}{t-t_0} + \frac{i}{\pi}\frac{C}{q-t},$$

$$\int_L \varphi(t)\,dt = C, \tag{6.1.18}$$

Relationship (6.1.4) *holds, in which quantity* $\theta(t_k)$ *satisfies Inequalities* (6.1.14) *and* (6.1.15).

To prove the theorem it suffices to note that because L is a circle, one can always assume that $j_q = n$. ■

Note 6.1.1. If function $f(t)$ suffers a discontinuity of the first kind at a point $q \in L$ and belongs to the class H on the set $L \setminus q$, then the sets E and E_0 must be chosen in such a way that $q \in E_0$ for $j = j_q$, and the system of linear algebraic equations should be composed in the following way:

$$\sum_{k=1}^{n} \frac{\varphi_n(t_k)a_k}{t_{0j} - t_k} = \begin{cases} f(t_{0j}), & j = 1, \ldots, n, \; j \neq j_q, \\ \dfrac{f(q-0) + f(q+0)}{2}, & j = j_q. \end{cases} \tag{6.1.19}$$

Then it can be shown theoretically that $|\varphi(t_k) - \varphi_n(t_k)|$, where $\varphi(t)$ is a solution of Equation (6.1.1), behaves as in the case of Theorem 6.1.2; however, calculations demonstrate a better convergence.

Note 6.1.2. While considering Theorems 6.1.1–6.1.3, we used rectangle rule formulas for a singular integral over a circle (see Section 5.2). Now for constructing systems of linear algebraic equations in (6.1.2) and (6.1.15), we can use interpolation-type quadrature formulas for a singular integral derived in Section 2.2. In other words, we put $a_k = -i2\pi t_k/(2n+1)$, $k = 0, 1, \ldots, 2n$. Hence, the sets E and E_0 contain an odd number of points. From (6.1.10) and (6.1.11) it follows that

$$\frac{1}{a_k} I_{0,k}^{(n)} = \frac{1}{\pi i}, \qquad k = 0, 1, \ldots, 2n,$$

$$\frac{1}{b_j} I_{0,0j}^{(n)} = -\frac{1}{\pi i}, \qquad j = 0, 1, \ldots, 2n, \tag{6.1.20}$$

where $b_j = -i2\pi t_{0j}/(2n+1)$. According to the latter formula, in the case we have

$$\varphi_n(t_k) = -\frac{1}{\pi^2} \sum_{j=0}^{2n} \frac{f(t_{0j}) b_j}{t_k - t_{0j}}. \tag{6.1.21}$$

Thus, if $f(t) \in H_r(\alpha)$ on L, then by Formula (2.2.17) one gets

$$|\varphi(t_k) - \varphi_n(t_k)| \leq O\left(\frac{\ln n}{n^{r+\alpha}}\right). \tag{6.1.22}$$

It must be stressed that in the case under consideration the system of linear algebraic equations possesses the following property: a solution to the system is expressed through the right-hand side with the help of the same quadrature formula that was used for constructing the system.

By using the preceding interpolation quadrature formulas in Theorem 6.1.3 we deduce that the right-hand sides of Inequalities (6.1.14) and (6.1.15) must be replaced by $O(n^{-r-\alpha} \ln n)$ and $O(n^{-r-\alpha} \ln^2 n)$, respectively.

Next we consider the full equation of the first kind on a circle

$$\int_L \frac{\varphi(t)\, dt}{t_0 - t} + \int_L K(t_0, t) \varphi(t)\, dt = f(t_0), \tag{6.1.23}$$

where function $f(t) \in H$ on L and $K(t_0, t) \in H$ on $L \times L$, or the equation

$$\int_L \frac{K(t_0, t)}{t_0 - t} \varphi(t)\, dt = f(t_0), \tag{6.1.24}$$

where $K(t_0, t) \in H$ on $L \times L$ and $K(t_0, t_0) \neq 0$.

Equations (6.1.23) and (6.1.24) are equivalent to the corresponding Fredholm equations of the second kind. If these equations have unique solutions, then the corresponding Fredholm equations of the second kind also have unique solutions. To solve the equations numerically in the class of continuous functions, one has to consider the following systems of linear algebraic equations:

$$\sum_{k=1}^{n} \frac{\varphi_n(t_k) a_k}{t_{0j} - t_k} + \sum_{k=1}^{n} K(t_{0j}, t_k) \varphi_n(t_k) a_k = f(t_{0j}), \qquad j = 1, \ldots, n, \tag{6.1.25}$$

or

$$\sum_{k=1}^{n} \frac{K(t_{0j}, t_k)}{t_{0j} - t_k} \varphi_n(t_k) a_k = f(t_{0j}), \qquad j = 1, \ldots, n, \tag{6.1.26}$$

where points t_k, $k = 1, \ldots, n$, divide circle L into equal parts, t_{0k} divides arc $\overset{\frown}{t_k t_{k+1}}$ into two halves, and $a_k = -i2\pi t_k/(2n + 1)$, $k = 1, \ldots, n$, where n is an odd number (in the general case, $a_k = t_{k+1} - t_k$). It is supposed that either $f(t)$ and $K(t_0, t)$, or $K(t_0, t)$, $K(t_0, t_0) \neq 0$ on L belong to the class H on the corresponding sets.

If $f(t)$ suffers a discontinuity of the first kind at point q, then one has to use Note 6.1.1. However, if $f(t)$ suffers an integrable discontinuity at point q, then the points t_k and t_{0k} must be chosen in such a way that point q is the middle of the arc limited by the nearest points from the sets $E = \{t_k, k = 1, \ldots, n\}$ and $E_0 = \{t_{0k}, k = 1, \ldots, n\}$. In this case Systems (6.1.25) and (6.1.26) preserve their form.

Convergence of solutions of Systems (6.1.25) and (6.1.26) to those of Equations (6.1.23) and (6.1.24), respectively, may be proved in the same way as for the analogous equations on a segment, that is, one has to repeat in the discrete form the transfer to systems of linear algebraic equations for corresponding singular Fredholm equations of the second kind. Naturally, the systems will not be degenerate starting from a certain n_1.

If one has to find a solution to Equation (6.1.23) having at a fixed point q a singularity of the form $1/q - t)$, then the points t_k and t_{0k} must be chosen so that $q = t_{0k_q}$ and the following system of linear algebraic equations must be considered:

$$\sum_{k=1}^{n} \frac{\varphi_n(t_k)a_k}{t_{0j} - t_k} + \sum_{k=1}^{n} K(t_{0j}, t_k)\varphi_n(t_k)a_k = f(t_{0j}),$$

$$j = 1, \ldots, n, \; j \neq k_q,$$

$$\sum_{k=1}^{n} \varphi_n(t_k)a_k = C, \qquad j = k_q. \tag{6.1.27}$$

Note 6.1.3. All the results obtained in this section can be readily extended onto the case when L dealt with in Equations (6.1.1), (6.1.23), and (6.1.24) is a system of nonintersecting circles (Lifanov 1981).

6.2. EQUATIONS WITH HILBERT'S KERNEL

Consider the equation

$$\frac{1}{2\pi}\int_0^{2\pi} \cot\frac{\theta_0 - \theta}{2}\varphi(\theta)\,d\theta = f(\theta_0). \tag{6.2.1}$$

Let us choose on the interval $[0, 2\pi]$ points θ_k, $k = 1, \ldots, n$, that, being treated as points of unit circle L, divide it into n equal parts. Also let θ_{0k}, $k = 1, \ldots, n$, divide arc $\theta_k\theta_{k+1}$ into two halves.

Remember (Muskhelishvili 1952) that Equation (6.2.1) has a solution subject to the condition

$$\int_0^{2\pi} f(\theta)\,d\theta = 0. \tag{6.2.2}$$

In order to single out a unique solution, one has either to specify it at a point or to fix the integral of the solution (the latter situation is more frequent in applications). Therefore, if Equation (6.2.1) is solved numerically with the help of the approach used for solving the characteristic equation on a segment (see Chapter 5), then using the quadrature formulas for an integral containing Hilbert's kernel (see Sections 1.5 and 2.1), one must replace Equation (6.2.1) by the following system of linear algebraic equations:

$$\frac{1}{2\pi}\sum_{k=1}^{n}\cot\frac{\theta_{0m}-\theta_k}{2}\varphi_n(\theta_k)\frac{2\pi}{n}=f(\theta_{0m}),\qquad m=1,\ldots,n,$$

$$\frac{1}{2\pi}\sum_{k=1}^{n}\varphi_n(\theta_k)\frac{2\pi}{n}=C. \tag{6.2.3}$$

The number of equations in System (6.2.2) is more than that of unknowns. By the choice of points θ_k and θ_{0k} and Formula (1.2.5) we have

$$\sum_{m=1}^{n}\cot\frac{\theta_{0m}-\theta_k}{2}=0,\qquad k=1,\ldots,n. \tag{6.2.4}$$

Hence, after summing the first n equations entering System (6.2.3) and taking into account Equality (6.2.4), one gets

$$0=\sum_{m=1}^{n}f(\theta_{0m}). \tag{6.2.5}$$

Thus the system of the first n equations from System (6.2.3) is ill-conditioned and generally has no solution. The same may be said for the whole System (6.2.3).

Naturally, the idea of discarding one of the first equations of System (6.2.3) arises. Then, as shown in the following text, we arrive at a well-conditioned definite system that leads, however, to an unstable process of calculations. Therefore, one can apply the method of regularizing factors, used before for solving a characteristic singular integral equation on a segment for the case of a negative index (everywhere limited flow past an airfoil or flow past an airfoil with a sharp trailing edge).

The following theorem is true.

Theorem 6.2.1. *Let function $f(\theta)\in H$ on $[0,2\pi]$ and $f(0)=f(2\pi)$. Also, let Equality (6.2.2) hold for the function. Then, between a solution to the system of linear algebraic equations*

$$\gamma_{0n}+\frac{1}{2\pi}\sum_{k=1}^{n}\cot\frac{\theta_{0m}-\theta_k}{2}\varphi_n(\theta_k)\frac{2\pi}{n}=f(\theta_{0m}),\qquad m=1,\ldots,n,$$

$$\frac{1}{2\pi}\sum_{k=1}^{n}\varphi_n(\theta_k)\frac{2\pi}{n}=C \tag{6.2.6}$$

and the solution $\varphi(\theta)$ *to Equation* (6.2.1) *given by*

$$\varphi(\theta) = -\frac{1}{2\pi}\int_0^{2\pi}\cot\frac{\theta_0 - \theta}{2} f(\theta_0)\,d\theta_0 + C \tag{6.2.7}$$

(*Muskhelishvili* 1952) *and subject to the condition*

$$\frac{1}{2\pi}\int_0^{2\pi}\varphi(\theta)\,d\theta = C, \tag{6.2.8}$$

the relationship

$$|\varphi(\theta_k) - \varphi_n(\theta_k)| \le O(n^{-\lambda}\ln n) \tag{6.2.9}$$

holds where $\lambda = \alpha \in (0,1]$, *if* n *is an arbitrary number and* $f(\theta) \in H(\alpha)$, *and* $\lambda = r + \alpha$ *if* n *is odd and* $f^{(r)}(\theta) \in H(\alpha)$.

Proof. Let us sum the first n equations of System (6.2.6). Then taking into account (6.2.4) we get

$$\gamma_{0n} = \frac{1}{2\pi}\sum_{m=1}^{n} f(\theta_{0m})\frac{2\pi}{n}. \tag{6.2.10}$$

Hence, $\gamma_{0n} \to 0$ for $n \to \infty$, if and only if Equation (6.2.1) has a solution.

In what follows we will reduce System (6.2.6) to a system of the form of (6.2.2) for an equation on a circle by using Equality (2.2.15). Let us multiply the last equality of System (6.2.6) by $(-i)$ and sum the result with all the first n equations. Then, taking into account Equality (6.2.10) and multiplying both its sides by π, one gets

$$-\frac{i}{2}\sum_{k=1}^{n}\varphi_n(\theta_k)\frac{2\pi}{n} + \frac{1}{2}\sum_{k=1}^{n}\cot\frac{\theta_{0m}-\theta_k}{2}\varphi_n(\theta_k)\frac{2\pi}{n}$$

$$= \left[\pi f(\theta_{0m}) - \frac{1}{2}\sum_{k=1}^{n} f(\theta_{0k})\frac{2\pi}{n}\right] - i\pi C, \qquad m = 1,\dots,n,$$

or

$$\sum_{k=1}^{n}\frac{\hat{\theta}_n(t_k)a_k}{t_{0m} - t_k} = \hat{f}(t_{0m}), \qquad m = 1,\dots,n, \tag{6.2.11}$$

where $t_k = \exp(i\theta_k)$, $t_{0m} = \exp(i\theta_{0m})$, $\varphi_n(t_k) = \varphi_n(\theta_k)$, $a_k = 2\pi i t_k/n$, and

$$\hat{f}(t_{0m}) = \left[\pi f(\theta_{0m}) - \frac{1}{2}\sum_{k=1}^{n} f(\theta_{0k})\frac{2\pi}{n}\right] - i\pi C.$$

Because System (6.2.11) coincides with System (6.1.2), its solution is given by Equation (6.1.6). By Equalities (6.1.20) we deduce (see (6.1.21))

$$\hat{\varphi}_n(t_k) = -\frac{1}{\pi^2}\sum_{m=1}^{n}\frac{\hat{f}(t_{0m})b_m}{t_k - t_{0m}}, \tag{6.2.12}$$

where $b_m = 2\pi i t_{0m}/n$.

Thus, after using Equalities (2.2.15) and (6.2.4) again, one gets

$$\varphi_n(\theta_k) + \frac{1}{\pi^2}\sum_{m=1}^{n}\left(\frac{1}{2}\cot\frac{\theta_k - \theta_{0m}}{2} - \frac{i}{2}\right)$$
$$\times\left\{\left[\pi f(\theta_{0m}) - \frac{1}{2}\sum_{k=1}^{n} f(\theta_{0k})\frac{2\pi}{n}\right] - i\pi C\right\}\frac{2\pi}{n}$$
$$= -\frac{1}{2}\sum_{m=1}^{n}\cot\frac{\theta_k - \theta_{0m}}{2} f(\theta_{0m})\frac{2\pi}{n} + C. \tag{6.2.13}$$

By comparing (6.2.7) with (6.2.13) we terminate the proof of the theorem. ■

Note 6.2.1. If instead of considering (6.2.6), one considers the system

$$\gamma_{0n} + \frac{1}{2\pi}\sum_{k=1}^{n}\cot\frac{\theta_{0m} - \theta_k}{2}\varphi_n(\theta_k)\frac{2\pi}{n} = f(\theta_{0m}), \qquad m = 1,\dots,n,$$

$$\gamma_{0n} + \frac{1}{2\pi}\sum_{k=1}^{n}\varphi_n(\theta_k)\frac{2\pi}{n} + C, \tag{6.2.14}$$

then similar considerations result in

$$\varphi_n(\theta_k) = -\frac{1}{2\pi}\sum_{m=1}^{n}\exp\frac{\theta_k - \theta_{0m}}{2} f(\theta_{0m})\frac{2\pi}{n} - \frac{1}{2\pi}\sum_{m=1}^{n} f(\theta_{0m})\frac{2\pi}{n} + C. \tag{6.2.15}$$

The following theorems may be proved in a similar way.

Theorem 6.2.2. *Between a solution to the system of linear algebraic equations*

$$\frac{1}{2\pi}\sum_{k=1}^{n}\cot\frac{\theta_{0m}-\theta_k}{2}\varphi_n^-(\theta_k)\frac{2\pi}{n}=f(\theta_{0m}),\qquad m=1,\ldots,n-1,$$

$$\frac{1}{2\pi}\sum_{k=1}^{n}\varphi_n^-(\theta_k)\frac{2\pi}{n}=C \tag{6.2.16}$$

and the solution $\varphi^-(\theta)$ *of Equation* (6.2.1) *for any* $f(\theta)$,

$$\varphi^-(\theta)=-\frac{1}{2\pi}\int_0^{2\pi}\cot\frac{\theta-\theta_0}{2}f(\theta_0)\frac{2\pi}{n}$$
$$-\frac{1}{2\pi}\cot\frac{q-\theta}{2}\int_0^{2\pi}f(\theta_0)\,d\theta_0+C, \tag{6.2.17}$$

the relationship

$$|\varphi(\theta_k)-\varphi_n^-(\theta_k)|\le\eta(\theta_k),\qquad k=1,\ldots,n, \tag{6.2.18}$$

holds where $\eta(\theta_k)$ *satisfies Inequalities* (6.1.14) *and* (6.1.15) *in which the numbers* λ_1 *and* λ_2 *are defined by using the properties of function as was done in* (6.2.9). *Points* θ_k *and* θ_{0k}, $k=1,\ldots,n$, *are chosen in such a way that* $\theta_{0n}=q$ *for and* n.

Theorem 6.2.3. *Let* $f(\theta)\in H$ *on* $[0,2\pi]$ *and* $f(0)=f(2\pi)$. *Also, let points* θ_k *and* θ_{0k}, $k=1,\ldots,n$, *be chosen so that* $\theta_{k_q}=q$, $q\in[0,2\pi]$. *Then, between a solution to the system of linear algebraic equations*

$$\gamma_{0n}+\frac{1}{2\pi}\sum_{\substack{k=1\\k\neq k_q}}^{n}\cot\frac{\theta_{0m}-\theta_k}{2}\varphi_n^+(\theta_k)\frac{2\pi}{n}$$
$$=f(\theta_{0m})-\frac{1}{2\pi}\cot\frac{\theta_{0m}-q}{2}\varphi(q)\frac{2\pi}{n},\qquad m=1,\ldots,n, \tag{6.2.19}$$

where $\varphi(q)$ *is known, and the solution* $\varphi^+(\theta)$ *of Equation* (6.2.1),

$$\varphi^+(\theta)=-\frac{1}{2\pi}\int_0^{2\pi}\left(\cot\frac{\theta-\theta_0}{2}-\cot\frac{q-\theta_0}{2}\right)f(\theta_0)\,d\theta_0+\varphi(q), \tag{6.2.20}$$

inequality

$$|\varphi^+(\theta_k) - \varphi_n^+(\theta_k)| \le O\left(\frac{\ln n}{n^\alpha}\right), \qquad k = 1, \ldots, n, \qquad (6.2.21)$$

holds.

Note 6.2.2. Solutions to Systems (6.2.16) and (6.2.19) have, respectively, the form

$$\varphi_n^-(\theta_k) = -\frac{1}{2\pi}\sum_{m=1}^{n} \cot\frac{\theta_k - \theta_{0m}}{2} f(\theta_{0m})\frac{2\pi}{n}$$

$$-\frac{1}{2\pi}\cot\frac{\theta_{0n} - \theta_k}{2}\sum_{m=1}^{n} f(\theta_{0m})\frac{2\pi}{n} + C, \qquad (6.2.22)$$

$$\varphi_n^+(\theta_k) = \frac{1}{2\pi}\sum_{m=1}^{n}\left(-\cot\frac{\theta_k - \theta_{0m}}{2} + \cot\frac{\theta_{k_q} - \theta_{0m}}{2}\right) f(\theta_{0m})\frac{2\pi}{n}$$

$$-\frac{1}{2\pi}\sum_{m=1}^{n}\left(-\cot\frac{\theta_k - \theta_{0m}}{2} + \cot\frac{\theta_{k_q} - \theta_{0m}}{2}\right)$$

$$\times\frac{1}{2\pi}\cot\frac{\theta_{0m} - \theta_{k_q}}{2}\varphi(q)\left(\frac{2\pi}{n}\right)^2. \qquad (6.2.23)$$

From (6.2.22) it follows that the accuracy of calculations deteriorates in the neighborhood of point q whose equation was discarded, because, generally, $\Sigma_{m=1}^{n} f(\theta_{0m})(2\pi/n) \ne 0$.

In accordance with (6.2.23) we deduce that if $\varphi(q) = 0$, then for $f^{(r)}(\theta) \in H(\alpha)$ on $[0, 2\pi]$ and an odd n, the right-hand side of Equation (6.2.21) becomes $O(n^{-r-\alpha}\ln n)$. Therefore, if the value of $\varphi(q)$ is known, then it is advantageous to introduce a new function $\overline{\varphi}(\theta) = \varphi(\theta) - \varphi(q)$ for which $\overline{\varphi}(q) = 0$.

Note 6.2.3. If function $f(\theta)$ entering Equation (6.2.1) suffers an integrable discontinuity at point q, then, similarly to the equation on a segment, a system of linear algebraic equations may be taken in the form of (6.2.6). However, points θ_k and θ_{0k}, $k = 1, \ldots, n$, must be chosen in such a way that point q divides the segment into two equal parts limited by the nearest points belonging to the sets E and E_0.

If, however, function $f(\theta)$ suffers a discontinuity of the first kind at point $q \in [0, 2\pi]$, then the calculations show that the following strategy should be preferred: Points θ_k and θ_{0k} are chosen so that point q coincides with point θ_{0k_q}, and the right-hand sides of n first equations of system (6.2.6) must be taken as follows:

1. For equations whose number is equal to $m = 1, \ldots, n$, $m \neq k_q$, the right-hand sides are given by $f(\theta_{0m})$.
2. The right-hand side of the equation numbered $m = k_q$ is equal to $[f(q-0) + f(q+0)]/2$.

Consider the equation

$$\frac{1}{2\pi}\int_0^{2\pi} \cot\frac{\theta_0 - \theta}{2}\varphi(\theta)\,d\theta + \int_0^{2\pi} K(\theta_0, \theta)\varphi(\theta)\,d\theta = f(\theta_0) \quad (6.2.24)$$

or the equation

$$\frac{1}{2\pi}\int_0^{2\pi} K(\theta_0, \theta)\cot\frac{\theta_0 - \theta}{2}\varphi(\theta)\,d\theta = f(\theta_0), \quad (6.2.25)$$

where $K(\theta_0, \theta_0) \neq 0$ and $K(\theta_0, \theta) \in H$ on $[0, 2\pi] \times [0, 2\pi]$, and the functions $f(\theta_0)$ and $K(\theta_0, \theta)$ are periodic with respect to their coordinates.

Let us suppose that Equation (6.2.24) has a unique solution subject to the additional condition

$$\int_0^{2\pi} K_1(\theta)\varphi(\theta)\,d\theta = C, \quad (6.2.26)$$

where $K_1(\theta)$ is a nonzero function.

If kernel $K(\theta_0, \theta)$ satisfies the identity

$$\int_0^{2\pi} K(\theta_0, \theta)\,d\theta_0 = 0, \quad (6.2.27)$$

and the right-hand side satisfies Equality (6.2.2), then Equality (6.2.6) must be specified subject to some additional conditions demonstrated in succeeding text when considering problems of aerodynamics and elasticity.

However, if Equation (6.2.24) is uniquely solvable for any right-hand side, then the additional condition must be taken in the form:

$$\int_0^{2\pi} d\theta_0 \int_0^{2\pi} K(\theta_0, \theta)\varphi(\theta)\,d\theta = \int_0^{2\pi} f(\theta_0)\,d\theta_0. \quad (6.2.28)$$

These remarks must be taken into consideration when developing a numerical method for solving Equations (6.2.24) and (6.2.25).

Thus, Equation (6.2.24) may be represented by the system of linear algebraic equations:

$$\gamma_{0n} + \frac{1}{2\pi}\sum_{k=1}^{n} \cot\frac{\theta_{0m} - \theta_k}{2}\varphi_n(\theta_k)\frac{2\pi}{n}$$

$$+ \sum_{k=1}^{n} K(\theta_{0m}, \theta_k)\varphi_n(\theta_k)\frac{2\pi}{n} = f(\theta_{0m}), \qquad m = 1,\dots,n,$$

$$\sum_{k=1}^{n} K_1(\theta_k)\varphi_n(\theta_k)\frac{2\pi}{n} = C. \tag{6.2.29}$$

If a unique solution to Equation (6.2.24) is singled out with the help of condition $\varphi(q) = 0$, then the following system must be considered:

$$\gamma_{0n} + \frac{1}{2\pi}\sum_{\substack{k=1\\k\neq k_q}}^{n} \cot\frac{\theta_{0m} - \theta_k}{2}\varphi_n(\theta_k)\frac{2\pi}{n}$$

$$+ \sum_{\substack{k=1\\k\neq k_q}}^{n} K(\theta_{0m}, \theta_k)\varphi_n(\theta_k)\frac{2\pi}{n} = f(\theta_{0m}), \qquad m = 1,\dots,n, \tag{6.2.30}$$

where points θ_k and θ_{0m} are chosen as was done when considering System (6.2.19). The convenience of Systems (6.2.29) and (6.2.30) is associated with solvability of Systems (6.2.6) and (6.2.19), which allows us to transform the former systems into equivalent systems of linear algebraic equations for integral Fredholm equations of the second kind equivalent to Equation (6.2.24) in the class of continuous solutions. Thus, one may prove the convergence of solutions of Systems (6.2.29) and (6.2.30) to the accurate solution of Equation (6.2.24), with the estimates made previously for the characteristic equation.

If the right-hand side of Equation (6.2.24) suffers a discontinuity of the first kind or has an integrable discontinuity at point q, then System (6.2.29) must be composed into account Note 6.2.3.

The rationale for composing a system of the form of (6.2.29) for Equation (6.2.24) in the case when it has a unique solution for any right-hand side will be explained on the example of the equation

$$\frac{1}{2\pi}\int_0^{2\pi}\cot\frac{\theta - \theta_0}{2}\varphi(\theta)\,d\theta + \frac{1}{2\pi}\int_0^{2\pi}\varphi(\theta)\,d\theta = f(\theta_0), \tag{6.2.31}$$

which, according to Muskhelishvili (1952), has the solution

$$\varphi(\theta) = -\frac{1}{2\pi}\int_0^{2\pi}\cot\frac{\theta_0 - \theta}{2}f(\theta_0)\,d\theta_0 + \frac{1}{2\pi}\int_0^{2\pi}f(\theta_0)\,d\theta_0. \quad (6.2.32)$$

Let us, in fact, consider the system

$$\frac{1}{2\pi}\sum_{k=1}^{n}\cot\frac{\theta_{0m} - \theta_k}{2}\varphi_n(\theta_k)\frac{2\pi}{n} + \frac{1}{2\pi}\sum_{k=1}^{n}\varphi_n(\theta_k)\frac{2\pi}{n} = f(\theta_{0m}),$$

$$m = 1,\dots,n, \quad (6.2.33)$$

where points θ_k and θ_{0k}, $k = 1,\dots,n$, are chosen just as in the case of System (6.2.6). When we sum all the equations in (6.2.33), we deduce that the system is equivalent to

$$\frac{1}{2\pi}\sum_{k=1}^{n}\cot\frac{\theta_{0m} - \theta_k}{2}\varphi_n(\theta_k)\frac{2\pi}{n} = f(\theta_{0m}) - \frac{1}{2\pi}\sum_{p=1}^{n}f(\theta_{0p})\frac{2\pi}{n},$$

$$m = 1,\dots,n-1.$$

$$\frac{1}{2\pi}\sum_{k=1}^{n}\varphi_n(\theta_k)\frac{2\pi}{n} = \frac{1}{2\pi}\sum_{p=1}^{n}f(\theta_{0p})\frac{2\pi}{n}, \qquad m = n. \quad (6.2.34)$$

By comparing Systems (6.2.34) and (6.2.16), we see that a solution to System (6.2.34) is given by the formula

$$\begin{aligned}\varphi_n(\theta_k) = &-\frac{1}{2\pi}\sum_{m=1}^{n}\cot\frac{\theta_k - \theta_{0m}}{2}\left(f(\theta_{0m}) - \frac{1}{2\pi}\sum_{p=1}^{n}f(\theta_{0p})\right)\frac{2\pi}{n}\\ &-\frac{1}{2\pi}\cot\frac{\theta_{0n} - \theta_k}{2}\left[\sum_{m=1}^{n}\left(f(\theta_{0m}) - \frac{1}{2\pi}\sum_{p=1}^{n}f(\theta_{0p})\frac{2\pi}{n}\right)\frac{2\pi}{n}\right]\\ &+\frac{1}{2\pi}\sum_{p=1}^{n}f(\theta_{0p})\frac{2\pi}{n}. \end{aligned} \quad (6.2.35)$$

From Formula (6.2.35) it follows that in the presence of errors in calculated right-hand sides and the sum $\Sigma_{p=1}^{n} f(\theta_{0p})2\pi/n$, the error increases by the factor of $\cot(\theta_{0n} - \theta_k)/2$, which for $\theta_k \to \theta_{0n}$ has the order of n.

7

Singular Integral Equations of the Second Kind

7.1. EQUATION ON A SEGMENT

Consider the equation

$$a\varphi(t_0) + \frac{b}{\pi}\int_{-1}^{1}\frac{\varphi(t)\,dt}{t-t_0} = f(t_0), \tag{7.1.1}$$

where a and b are real numbers, $b \neq 0$, $a^2 + b^2 = 1$, and function $f(t)$ belongs to the class H on $[-1,1]$. Let us briefly recall some results obtained in Muskhelishvili (1952) for Equation (7.1.1).

The index κ of the equation is equal to $1, 0, -1$, and the corresponding solutions have the form

$$\varphi(t) = \omega(t)\psi(t),$$

$$\omega(t) = (1-t)^{\alpha}(1+t)^{\beta}, \qquad 0 < |\alpha|, |\beta| < 1,$$

$$\kappa = -(\alpha+\beta). \tag{7.1.2}$$

The number α is defined by the equality

$$a + b\cot\pi\alpha = 0. \tag{7.1.3}$$

Let us denote the left-hand side of Equation (7.1.1) by $I(t_0)$ and call the formula

$$I_n(t_0) = a\varphi_n(t_0) + \frac{b}{\pi}\int_{-1}^{1}\frac{\varphi_n(t)\,dt}{t - t_0},$$

$$\varphi_n(t) = \omega(t)\psi_n(t), \qquad \psi_n(t) = \sum_{k=1}^{n}\frac{\psi_n(t_k)P_n^{(\alpha,\beta)}(t)}{(t - t_k)P'^{(\alpha,\beta)}_n(t_k)}, \tag{7.1.4}$$

a quadrature-interpolation formula of index κ and order n (Lifanov and Saakyan 1982). Here $t_k, k = 1,\dots,n$, are the roots of the n-degree Jacobi polynomial $P_n^{(\alpha,\beta)}(t)$ corresponding to function $\omega(t)$, and $\psi_n(t_k) = \psi(t_k)$. According to Erdogan, Gupta, and Cook (1973), polynomial $P_n^{(\alpha,\beta)}(t)$ satisfies the relationship

$$a\omega(t_0)P_n^{(\alpha,\beta)}(t_0) + \frac{b}{\pi}\int_{-1}^{1}\frac{\omega(t)P_n^{(\alpha,\beta)}(t)}{t - t_0}\,dt$$

$$= -\frac{b}{2^{\kappa}\sin\alpha\pi}P_{n-\kappa}^{(-\alpha,-\beta)}(t_0). \tag{7.1.5}$$

Hence, the equality

$$I_n(t_0) = -\sum_{k=1}^{n}\frac{\psi_n(t_k)}{(t_0 - t_k)P'^{(\alpha,\beta)}_n(t_k)}$$

$$\times\left[\frac{b}{2^{\kappa}\sin\alpha\pi}P_{n-\kappa}^{(-\alpha,-\beta)}(t_0) + \frac{b}{\pi}\int_{-1}^{1}\frac{\omega(t)P_n^{(\alpha,\beta)}(t)}{t - t_k}\,dt\right], \tag{7.1.6}$$

is valid.

Function $\varphi_n(t)$ will be called an *approximate solution* to Equation (7.1.1). It will be found by equating functions $I_n(t_0)$ and $f_n(t_0)$, where $f_n(t_0)$ is an interpolation polynomial of the form of (2.3.2) for the function $f(t_0)$ constructed by using roots of the polynomial $P_{n-\kappa}^{(-\alpha,-\beta)}(t_0)$. Function $\varphi_n(t)$ is defined if the numbers $\psi_n(t_k), k = 1,\dots,n$, are known. By equating the functions $I_n(t_0)$ and $f_n(t_0)$ at the points $t_{0m}, m = 1,\dots,n-\kappa$, where t_{0m} are the roots of the polynomial $P_{n-\kappa}^{(-\alpha,-\beta)}(t)$, one arrives at the system of linear algebraic equations

$$-\frac{b}{\pi}\sum_{k=1}^{n}\frac{\psi_n(t_k)a_k}{t_{0m}-t_k}=f(t_{0m}),\qquad m=1,\ldots,n-\kappa$$

$$a_k=\frac{1}{P'^{(\alpha,\beta)}_n(t_k)}\int_{-1}^{1}\frac{\omega(t)P^{(\alpha,\beta)}_n(t)\,dt}{t-t_k}. \tag{7.1.7}$$

By putting $t_0=t_k$ in (7.1.5), we note that the coefficient a_k appearing in (7.1.7) may be written in the form

$$a_k=-\frac{\pi}{2^{\kappa}\sin\alpha\pi}\frac{P^{(-\alpha,-\beta)}_{n-\kappa}(t_k)}{P'^{(\alpha,\beta)}_n(t_k)},\qquad k=1,\ldots,n. \tag{7.1.8}$$

Let us consider different values of index κ. Let κ be equal to 0. Then, a unique solution to Equation (7.1.1) may be singled out by specifying the number α, $0<|\alpha|<1$, satisfying Equality (7.1.3). If one has to obtain a solution that is limited at point 1 and unlimited at point -1, then one has to choose a positive value of α. Because polynomials $P^{(\alpha,\beta)}_n(t)$ and $P^{(-\alpha,-\beta)}_{n-\kappa}(t)$ have the same number of roots, System (7.1.7) contains n unknowns as well as n equations.

Let κ be equal to 1. Then $-1<\alpha,\beta<0$ and Equation (7.1.1) has a nonunique solution. A required solution may be singled out by additionally employing Condition (5.1.46). In this case the degree of polynomial $P^{(-\alpha,-\beta)}_{n-\kappa}(t)$ is equal to $(n-1)$, and hence, System (7.1.7) contains n unknowns and $(n-1)$ equations. This may be remedied by digitizing Equation (5.1.46), i.e., by passing to the system

$$-\frac{b}{\pi}\sum_{k=1}^{n}\frac{\psi_n(t_k)a_k}{t_{0m}-t_k}=f(t_{0m}),\qquad m=1,\ldots,n-1,$$

$$\sum_{k=1}^{n}\psi_n(t_k)a_k=C,\qquad m=n. \tag{7.1.9}$$

Finally, let κ be equal to -1. In this case polynomial $P^{(-\alpha,-\beta)}_{n-\kappa}(t)$ has $(n+1)$ roots, and System (7.1.7) has more equations than unknowns and usually has no solutions. Therefore, similarly to the equation of the first kind, we introduce the regularizing factor γ_{0n} and consider the system

$$\gamma_{0n}-\frac{b}{\pi}\sum_{k=1}^{n}\frac{\psi_n(t_k)a_k}{t_{0m}-t_k}=f(t_{0m}),\qquad m=1,\ldots,n+1. \tag{7.1.10}$$

According to Muskhelishvili (1952) a solution exists only if the equality

$$\int_{-1}^{1} \frac{f(t)\,dt}{(1-t)^{\alpha}(1+t)^{\beta}} = 0, \qquad 0 < \alpha, \beta < 1,\ \alpha + \beta = 1,$$

holds.

Theorem 7.1.1. *Let function* $f(t) \in H_r(\alpha)$ *on* $[-1, 1]$. *Then, between solutions to the systems of linear algebraic equations* (7.1.7) *for* $\kappa = 0$, (7.1.9) *and* (7.1.10) *and the corresponding solutions to Equation* (7.1.1), *Inequality* (5.2.4) *holds, where* $R_n(t_k)$ *is an error of the quadrature-interpolation formula of index* κ *and order* $n^{-\kappa}$ *for the function*

$$\psi(t) = a\frac{f(t)}{\omega(t)} - \frac{b}{\pi}\int_{-1}^{1} \frac{f(t_0)\,dt_0}{\omega(t_0)(t_0 - t)} + T_k C, \tag{7.1.11}$$

which determines the index κ *solution in Equation* (7.1.2); $T_0 = T_{-1} = 0$, $T_1 = -\pi[\sin \alpha\pi]^{-1}$.

Proof. As shown when proving Theorem 5.1.1,

$$\psi_n(t_k) = \frac{1}{a_k} I_{\kappa,k}^{(n)} \left[\sum_{m=1}^{n-\kappa} I_{\kappa,0m}^{(n)} \frac{-b^{-1}\pi f(t_{0m})}{t_k - t_{0m}} + \nu_\kappa C \right], \qquad k = 1, \ldots, n,$$

$$I_{\kappa,k}^{(n)} = \prod_{p=1}^{n-\kappa} (t_k - t_{0p}) \Bigg/ \prod_{\substack{p=1 \\ p \neq k}}^{n} (t_k - t_p),$$

$$I_{\kappa,0m}^{(n)} = \prod_{p=1}^{n} (t_{0m} - t_p) \Bigg/ \prod_{\substack{p=1 \\ p \neq m}}^{n-\kappa} (t_{0m} - t_{0p}), \qquad \nu = 1,\ \nu_0 = \nu_{-1} = 0. \tag{7.1.12}$$

By representing the polynomials by products of linear multipliers and using Equation (7.1.8), one gets

$$I_{\kappa,k}^{(n)} = -\frac{2^{\kappa} a_k \sin \alpha\pi}{\pi} \frac{B_n^{(\alpha,\beta)}}{B_{n-\kappa}^{(-\alpha,-\beta)}}, \qquad k = 1, \ldots, n, \tag{7.1.13}$$

where $B_n^{(-\alpha,-\beta)}$ and $B_{n-\kappa}^{(-\alpha,-\beta)}$ are coefficients before the senior degrees of the variable in the corresponding Jacobi polynomials.

As previously noted, function $\psi(t)$, appearing in Equation (7.1.2) for the index κ solution, is defined by Equation (7.1.11). Let us consider this equality as an equation in the function $\xi(t) = [\omega(t)]^{-1}f(t)$. If $\varphi(t)$ is the index κ solution for Equation (7.1.1), then function $\xi(t)$ will be the index $-\kappa$ solution for Equation (7.1.11). If, in Equality (7.1.5), t is substituted by t_0, $\omega(t)$ by $1/\omega(t)$, and $P_n^{(\alpha,\beta)}(t)$ by $P_{n-\kappa}^{(-\alpha,-\beta)}(t)$, then the number b must be replaced by $-b$. Therefore, one gets

$$a\frac{P_{n-\kappa}^{(-\alpha,-\beta)}(t)}{\omega(t)} - \frac{b}{\pi}\int_{-1}^{1}\frac{P_{n-\kappa}^{(-\alpha,-\beta)}(t_0)\,dt_0}{t_0 - t} = \frac{2^{\kappa}b}{\sin\alpha\pi}P_n^{(\alpha,\beta)}(t). \quad (7.1.14)$$

Let us denote by $\Phi(t)$ the function obtained from $I(t_0)$ by replacing $\varphi(t)$ by $\xi(t)$, b by $-b$, and t_0 by t. If one denotes by $\Phi_{n-\kappa}(t)$ the function obtained from $\Phi(t)$ by substituting $\xi(t)$ by $\xi_{n-\kappa}(t)$, where $\xi_{n-\kappa}(t)$ is determined with the help of the roots $t_{0m}, m = 1,\ldots,n-\kappa$, of the polynomial $P_{n-\kappa}^{(-\alpha,-\beta)}(t)$, then at the points $t_k, k = 1,\ldots,n$, which are the roots of polynomial $P_n^{(\alpha,\beta)}(t)$, one gets

$$\Phi_{n-\kappa}(t_k) = \frac{b}{\pi}\sum_{m=1}^{n-\kappa}\frac{f(t_{0m})b_m}{t_k - t_{0m}}, \qquad k = 1,\ldots,n,$$

$$b_m = -\frac{2^{\kappa}\pi}{\sin\alpha\pi}\frac{P_n^{(\alpha,\beta)}(t_{0m})}{P'^{(-\alpha,-\beta)}_{n-\kappa}(t_{0m})}. \quad (7.1.15)$$

By representing the polynomials by products of linear multipliers again, one gets

$$I_{\kappa,0m}^{(n)} = -\frac{b_m\sin\alpha\pi}{2^{\kappa}\pi}\frac{B_{n-\kappa}^{(-\alpha,-\beta)}}{B_n^{(\alpha,\beta)}}, \qquad m = 1,\ldots,n-\kappa. \quad (7.1.16)$$

Finally, we observe that the equalities

$$\frac{B_n^{(\alpha,\beta)}}{B_{n-\kappa}^{(\alpha,\beta)}} = 2^{-\kappa}, \qquad \frac{\sin\alpha\pi\sin(-\alpha)\pi}{\pi^2} = -\frac{b^2}{\pi^2} \quad (7.1.17)$$

are valid. As a result, we get

$$\psi_n(t_k) = \frac{b}{\pi}\sum_{m=1}^{n-\kappa}\frac{f(t_{0m})b_m}{t_k - t_{0m}} - \frac{\nu_\kappa\sin\alpha\pi}{\pi}C, \qquad k = 1,\ldots,n. \quad (7.1.18)$$

Obviously,

$$-\nu_\kappa \frac{\sin \alpha\pi}{\pi} = T_\kappa$$

in Formula (7.1.11). Thus, Formulas (7.1.12) and (7.1.18) actually prove Theorem 7.1.1. The rate of convergence of an approximate solution to the accurate one at points $t_k, k = 1, \ldots, n$, is determined by the order of approximation attainable by using the quadrature-interpolation formula of the corresponding function $\psi(t)$. ■

Similarly to the equation of the first kind on a segment, we note that the regularizing factor γ_{0n} tends to zero for $n \to \infty$ only if the conditions of existence of the index $\kappa = -1$ solution are fulfilled. The results previously formulated for Equation (7.1.1) are also valid for the equation

$$a\varphi(t_0) + \frac{b}{\pi}\int_{-1}^{1} \frac{\varphi(t)\,dt}{t - t_0} + \int_{-1}^{1} K(t_0, t)\varphi(t)\,dt = f(t_0), \quad (7.1.19)$$

where it is supposed that the function $K(t_0, t) \in H$ on $[-1,1] \times [-1,1]$.

We will require that Equation (7.1.19) have a unique index $\kappa = 1, 0, -1$ solution subject to additional conditions that, for $\kappa = 1$, are imposed on the solution itself, and for $\kappa = -1$, on the functions $K(t_0, t)$ and $f(t_0)$.

The systems of linear algebraic equations for Equation (7.1.19) may be obtained from the corresponding systems for the characteristic equation by adding the term

$$\sum_{k=1}^{n} K(t_{0m}, t_k)\psi_n(t_k)a_k, \qquad m = 1, \ldots, n - \kappa.$$

7.2. EQUATION ON A CIRCLE

Consider the equation

$$a\varphi(t_0) + \frac{b}{\pi}\int_{L} \frac{\varphi(t)\,dt}{t - t_0} = f(t_0), \quad (7.2.1)$$

where a and b are real numbers, $b \neq 0$, $a^2 + b^2 = 1$, and the function $f(t) \in H$ on a unit-radius circle L centered at the origin of coordinates. As shown in Muskhelishvili (1952), Equation (7.2.1) has index $\kappa = 0$, i.e., is uniquely solvable for any right-hand side.

Let us denote the left-hand side of Equation (7.2.1) by $I(t_0)$ and call

$$I_n(t_0) = a\varphi_n(t_0) + \frac{b}{\pi}\int_L \frac{\varphi_n(t)\,dt}{t - t_0},$$

$$\varphi_n(t) = \frac{1}{2n+1}\sum_{k=0}^{2n} \varphi_n(t_k)\frac{t^{2n+1} - t_k^{2n+1}}{(t - t_k)t^n t_k^n}, \tag{7.2.2}$$

where points $t_k = \exp(i\theta_k), k = 1, \ldots, 2n$, divide L into $2n + 1$ equal parts, $\varphi_n(t_k) = \varphi(t_k)$, the *quadrature-interpolation formula* of the order n for the function $I(t_0)$.

By using Equality (2.2.6) the function $I_n(t_0)$ may be written in the form

$$I_n(t_0) = -\frac{b}{\pi}\sum_{k=0}^{2n} \frac{\varphi_n(t_k)a_k}{t_0 - t_k}$$

$$+ \sum_{k=0}^{2n} \frac{\varphi_n(t_k)(a + bi)(t_0^{2n+1} + (-a + bi)/(a + bi)t_k^{2n+1})}{(2n + 1)(t_0 - t_k)t_0^n t_k^n},$$

where $a_k = 2\pi i t_k/(2n + 1), k = 1, \ldots, 2n$.

Let us search for an approximate solution to Equation (7.2.1) in the class of functions of the form $\varphi_n(t)$. To do this it suffices to find the numbers $\varphi_n(t_k)$ from the system of $2n + 1$ linear algebraic equations

$$-\frac{b}{\pi}\sum_{k=0}^{2n} \frac{\varphi_n(t_k)a_k}{t_{0m} - t_k} = f(t_{0m}), \qquad m = 0, 1, \ldots, 2n, \tag{7.2.3}$$

obtained by equating functions $I_n(t_0)$ and $f_n(t_0)$ at points $t_{om}, m = 0, 1, \ldots, 2n$, where

$$f_n(t_0) = \frac{1}{2n+1}\sum_{m=0}^{2n} f(t_{0m})\frac{t_0^{2n+1} - t_{0m}^{2n+1}}{(t_0 - t_{0m})t_0^n t_m^n}, \tag{7.2.4}$$

and t_{0m} are the roots of the polynomial $t_0^{2n+1} + (-a + bi)/(a + bi)t_k^{2n+1}$.

Note that $f_n(t_{0m}) = f(t_{0m})$ and $t_{0m} = t_m \exp\{i(\pi + \psi)/(2n + 1)\}$, $m = 0, 1, \ldots, 2n$, where $\exp(i\psi) = (-a + bi)/(a + bi)$. Because $b \neq 0$, $-\pi < \psi < \pi$, and hence, $t_{0m} \neq t_k$ for any m and k. If $a = 0$, then $\psi = 0$, and point t_{0m} divides arc $\overset{\frown}{t_m t_{m+1}}$ into two halves.

The following theorem is true.

Theorem 7.2.1. *Let function $f(t) \in H_r(\alpha)$ on L. Then, between a solution to the system of linear algebraic equations* (7.2.3) *and the solution* $\varphi(t)$ *of Equation* (7.2.1), *the inequality*

$$|\varphi(t_k) - \varphi_n(t_k)| \leq O\left(\frac{\ln n}{n^{r+\alpha}}\right), \qquad k = 0, 1, \ldots, 2n, \tag{7.2.5}$$

holds.

Proof. Because System (7.2.3) is similar in structure to System (7.1.7), its solution is given by

$$\varphi_n(t_k) = \frac{\pi}{b}\frac{1}{a_k} I_{0,k}^{(n)} \sum_{m=0}^{2n} I_{0,0m}^{(n)} \frac{f(t_{0m})}{t_k - t_{0m}}, \qquad k = 0, 1, \ldots, 2n.$$

$$I_{0,k}^{(n)} = \prod_{p=0}^{2n} (t_{0p} - t_k) \Bigg/ \prod_{\substack{p=0 \\ p \neq k}}^{2n} (t_p - t_k),$$

$$I_{0,0m}^{(n)} = \prod_{p=0}^{2n} (t_{0m} - t_p) \Bigg/ \prod_{\substack{p=0 \\ p \neq m}}^{2n} (t_{0m} - t_{0p}). \tag{7.2.6}$$

By the choice of points t_k and t_{0k}, $k = 1, \ldots, 2n$, just as in the case of proving Theorem 6.1.1, one gets

$$I_{0,k}^{(n)} = -\frac{t_k}{2n+1}(1 + e^{i\psi}), \qquad I_{0,0m}^{(n)} = \frac{t_{0m}}{2n+1}(1 + e^{-i\psi}). \tag{7.2.7}$$

Because by the definition of $\exp(i\psi)$ we have $[1 + \exp(i\psi)][1 + \exp(-i\psi)] = 4b^2$, Formulas (7.2.6) and (7.2.7) result in

$$\varphi_n(t_k) = \frac{b}{\pi} \sum_{m=0}^{2n} \frac{f(t_{0m}) b_m}{t_k - t_{0m}}, \qquad k = 0, 1, \ldots, 2n, \tag{7.2.8}$$

where $b_m = i(2\pi t_{0m}/(2n+1))$.

According to Muskhelishvili (1952), the solution to Equation (7.2.1) has the form

$$\varphi(t) = af(t) - \frac{b}{\pi}\int_L \frac{f(t_0)\, dt_0}{t_0 - t}. \tag{7.2.9}$$

Let us denote the right-hand side of Equation (7.2.9) by $\Phi(t)$ and take for the function the quadrature-interpolation formula $\Phi_n(t)$ obtained from $\Phi(t)$ by substituting function $f(t)$ by $f_n(t)$ with the help of (7.2.4). By using Formula (2.2.6) again, one gets

$$\Phi_n(t) = \frac{b}{\pi} \sum_{m=0}^{2n} \frac{f(t_{0m}) b_m}{t - t_{0m}} + \sum_{m=0}^{2n} \frac{f(t_{0m})(a - bi)(t^{2n+1} - t_m^{2n+1})}{(2n+1)(t - t_{0m}) t^n t_{0m}^n}. \tag{7.2.10}$$

From (7.2.8) and (7.2.10), it follows that

$$\varphi_n(t_k) = \Phi_n(t_k), \qquad k = 0, 1, \ldots, 2n.$$

Then, by using (1) the results concerning approximation of periodic functions on a circle by polynomials of the preceding form (Ivanov 1968) and (2) the fact that a singular integral over a circle has the same differential properties that are characteristic of its density, we terminate the proof of Theorem 7.2.1. ■

Next, we consider the equation

$$a\varphi(t_0) + \frac{b}{\pi} \int_L \frac{\varphi(t)\, dt}{t - t_0} + \int_L K(t_0, t) \varphi(t)\, dt = f(t_0). \tag{7.2.11}$$

Supposing that the equation has a unique solution, we take for it the system of linear algebraic equations

$$-\frac{b}{\pi} \sum_{k=0}^{2n} \left[\frac{1}{t_{0m} - t_k} - \frac{\pi}{b} K(t_{0m}, t_k) \right] \varphi_n(t_k) a_k = f(t_{0m}),$$

$$m = 0, 1, \ldots, 2n. \tag{7.2.12}$$

If functions $f(t)$ and $K(t_0, t)$ are such that $f^{(r)}(t)$ and $K_{t_0}^{(r)}(t_0, t)$, $K_t^{(r)}(t_0, t)$ belong to the class $H(\alpha)$ on the sets L and $L \times L$, respectively, then Relationship (7.2.5) is also valid for a solution to the system of linear algebraic equations (7.2.12) and the solution $\varphi(t)$ of Equation (7.2.11).

7.3. EQUATION WITH HILBERT'S KERNEL

Consider the characteristic equation

$$a\varphi(\theta_0) + \frac{b}{2\pi} \int_0^{2\pi} \cot \frac{\theta - \theta_0}{2} \varphi(\theta)\, d\theta = f(\theta_0), \tag{7.3.1}$$

where it is supposed that $a \neq 0$ and $a^2 + b^2 = 1$. In this case the equation has a unique solution given by

$$\varphi(\theta) = af(\theta) - \frac{b}{2\pi}\int_0^{2\pi}\cot\frac{\theta_0 - \theta}{2}f(\theta_0)\,d\theta_0 + \frac{b^2}{2\pi}\int_0^{2\pi}f(\theta_0)\,d\theta_0.$$

Let us denote the left-hand side of (7.3.1) by $I(\theta_0)$ and call the function $I_n(\theta_0)$ obtained from $I(\theta_0)$ by replacing $\varphi(\theta)$ by

$$\varphi_n(\theta) = \frac{1}{2n+1}\sum_{k=0}^{2n}\varphi_n(\theta_k)\frac{\sin(2n+1)(\theta - \theta_k)/2}{\sin(\theta - \theta_k)/2}, \qquad (7.3.2)$$

the *quadrature-interpolation formula of the order n* for function $I(\theta_0)$. Here points $\theta_k, k = 1, \ldots, 2n$, are the points of a unit-radius circle dividing the latter into $2n + 1$ equal parts. By using Equalities (2.1.2) and (2.1.3), one gets

$$I_n(\theta_0) = \frac{b}{2\pi}\sum_{k=0}^{2n}\cot\frac{\theta_0 - \theta_k}{2}\varphi_n(\theta_k)\frac{2\pi}{2n+1}$$

$$+ \sum_{k=0}^{2n}\frac{\varphi_n(\theta_k)}{2n+1}\frac{a\sin(2n+1)(\theta_0 - \theta_k)/2 - b\cos(2n+1)(\theta_0 - \theta_k)/2}{\sin(\theta_0 - \theta_k)/2}. \qquad (7.3.3)$$

Let us look for an approximate solution to Equation (7.3.1) in the class of functions of the form $\varphi_n(\theta)$. This may be done by finding the numbers $\varphi_n(\theta_k)$, $k = 1, \ldots, 2n$, from the following system of linear algebraic equations:

$$\frac{b}{2\pi}\sum_{k=0}^{2n}\cot\frac{\theta_{0m} - \theta_k}{2}\varphi_n(\theta_k)\frac{2\pi}{2n+1} = f(\theta_{0m}), \qquad m = 0, 1, \ldots, 2n, \qquad (7.3.4)$$

obtained by equating the functions $I_n(\theta_0)$ and $f_n(\theta_0)$, where $f_n(\theta_0)$ is an interpolation polynomial over the points θ_{0m}, $m = 0, 1, \ldots, 2n$, for the function $f(\theta_0)$. These points are the roots of the function $a\sin(2n+1)(\theta_0 - \theta_k)/2 - b\cos(2n+1)(\theta_0 - \theta_k)/2$. Note that $\theta_{0m} = \theta_m + (\pi - 2\psi)/(2n+1)$, where $\exp(i\psi) = b + ai$, and because $b \neq 0$ and $\theta_{0m} \neq \theta_k$ for any m and k. If $a = 0$, then ψ is equal either to zero or π, and the point θ_{0m} divides the arc $\widetilde{\theta_m \theta_{m+1}}$ into two equal parts, but the systems of linear algebraic equations must be taken as indicated in Section 6.2.

The following is true.

Theorem 7.3.1. *Let function $f(\theta)$ entering Equation* (7.3.1) *belong to $H_r(\alpha)$ on* $[0, 2\pi]$, *and* $f(0) = f(2\pi)$. *Then* $|\varphi_n(\theta_k) - \varphi(\theta_k)|$, $k = 1, \ldots, 2n$, *satisfies an inequality of the form* (7.2.5), *where $\varphi(\theta)$ is a solution to Equation* (7.3.1), *and $\varphi_n(\theta_k)$ is a solution of System* (7.3.4).

Proof. By using the same approach with respect to System (7.3.4) as was used with respect to System (6.2.6), one gets

$$\begin{aligned}\varphi_n(\theta_k) = -\frac{b}{2\pi}\Bigg[&\sum_{m=0}^{2n} \cot\frac{\theta_k - \theta_{0m}}{2} f(\theta_{0m})\frac{2\pi}{2n+1}\\ &-\frac{bi}{2\pi}\sum_{p=0}^{2n}\varphi_n(\theta_p)\frac{2\pi}{2n+1}\sum_{m=0}^{2\pi}\cot\frac{\theta_k - \theta_{0m}}{2}\frac{2\pi}{2n+1}\\ &-i\sum_{m=0}^{2n} f(\theta_{0m})\frac{2\pi}{2n+1} - b\sum_{p=0}^{2n}\varphi_n(\theta_p)\frac{2\pi}{2n+1}\Bigg]. \end{aligned} \tag{7.3.5}$$

By the choice of points θ_k and θ_{0k} we have

$$\sum_{m=0}^{2n} \cot\frac{\theta_k - \theta_{0m}}{2}\frac{2\pi}{2n+1} = -\frac{2\pi a}{b}, \qquad k = 0, 1, \ldots, 2n, \tag{7.3.6}$$

$$\sum_{k=0}^{2n} \cot\frac{\theta_{0m} - \theta_k}{2}\frac{2\pi}{2n+1} = \frac{2\pi a}{b}, \qquad k = 0, 1, \ldots, 2n, \tag{7.3.7}$$

Let us demonstrate the validity of Equality (7.3.7). Let $\varphi(\theta)$ appearing in Equation (7.3.1) be identically equal to unity $[\varphi(\theta) \equiv 1]$. Then,

$$1 \cdot a + \frac{b}{2\pi}\int_0^{2\pi} \cot\frac{\theta_0 - \theta}{2} \cdot 1\, d\theta = a. \tag{7.3.8}$$

By substituting unity here with the help of formula (7.3.2) and putting $\theta_0 = \theta_{0m}$, one arrives at Equality (7.3.7).

Similarly, by using the formula for $\varphi(\theta)$, one can prove the validity of Equality (7.3.6).

Let us sum up all the equations comprising System (7.3.4). Then, taking into account (7.3.6), we get

$$\sum_{k=0}^{2n} \varphi_n(\theta_k) \frac{2\pi}{2n+1} = \frac{1}{a} \sum_{m=0}^{2n} f(\theta_{0m}) \frac{2\pi}{2n+1}. \tag{7.3.9}$$

From Equations (7.3.5), (7.3.6), and (7.3.9), we get

$$\varphi_n(\theta_k) = -\frac{b}{2\pi}\left[\sum_{m=0}^{2n} \cot\frac{\theta_k - \theta_{0m}}{2} f(\theta_{0m}) \frac{2\pi}{2n+1} - \frac{b}{a}\sum_{m=0}^{2n} f(\theta_{0m}) \frac{2\pi}{2n+1}\right], \qquad k = 0, 1, \ldots, 2n. \tag{7.3.10}$$

Let us next denote the right-hand side of the formula for $\varphi(\theta)$ by $\Phi(\theta)$ and take for the function the quadrature-interpolation formula $\Phi_n(\theta)$ of the order n, which is obtained from $\Phi(\theta)$ by replacing function $f(\theta_0)$ by $f_n(\theta_0)$. By using Equalities (2.1.2) and (2.1.3) again, we get

$$\Phi_n(\theta_k) = -\frac{b}{2\pi} \sum_{m=0}^{2n} \cot\frac{\theta_k - \theta_{0m}}{2} f(\theta_{0m}) \frac{2\pi}{2n+1} + \frac{b^2}{2\pi a} \sum_{m=0}^{2n} f(\theta_{0m}) \frac{2\pi}{2n+1}. \tag{7.3.11}$$

By comparing Formulas (7.3.10) and (7.3.11), applying the theory of approximation of periodic functions by trigonometric polynomials, and using the fact that a singular integral with Hilbert's kernel has the same differential properties as its density (Luzin 1951), we terminate proving Theorem 7.3.1. ■

Consider next the equation

$$a\varphi(\theta_0) + \frac{b}{2\pi}\int_0^{2\pi} \cot\frac{\theta - \theta_0}{2} \varphi(\theta)\, d\theta + \int_0^{2\pi} K(\theta_0, \theta)\varphi(\theta)\, d\theta = f(\theta_0). \tag{7.3.12}$$

We assume that the equation has a unique solution and represent it by the system of linear algebraic equations:

$$\frac{b}{2\pi}\sum_{k=0}^{2n}\left[\cot\frac{\theta_{0m}-\theta_k}{2}+\frac{2\pi}{b}K(\theta_{0m},\theta_k)\right]\varphi_n(\theta_k)\frac{2\pi}{2n+1}=f(\theta_{0m}),$$

$$m=0,1,\ldots,2n. \quad (7.3.13)$$

If functions $f(\theta_0)$ and $K(\theta_0,\theta)$ have the rth derivatives belonging to the class $H(\alpha)$ on $[0,2\pi]$, then Inequality (7.2.5) is valid for $|\varphi_n(\theta_k)-\varphi\theta_k)|$, where $\varphi_n(\theta_k)$ is a solution to System (7.3.13) and $\varphi(\theta)$ is a solution to Equation (7.3.12).

7.4. EQUATION ON A PIECEWISE SMOOTH CURVE WITH VARIABLE COEFFICIENTS

Consider the equation

$$a(t_0)\varphi(t_0)+\frac{b(t_0)}{\pi i}\int_L\frac{\varphi(t)\,dt}{t-t_0}=f(t_0), \quad (7.4.1)$$

where L is a piecewise smooth curve (Muskhelishvili 1952) with nodes $c_1,c_2,\ldots,c_n$ (see Figure 7.1), and the functions a, b, and f meet the Hölder condition on curve L, $a^2-b^2\neq 0$ on L.

In this case the index κ of the equation is defined as follows (Muskhelishvili 1952).

Let us take the function $\ln G(t)$, $G(t)=(a(t)-b(t))/(a(t)+b(t))$, keeping in mind its branch varying continuously on each of the smooth curves $L_1,L_2,\ldots,L_p$ constituting curve L. Denote by $\ln G_j(c_k)$ the limit to which the function $\ln G(t)$ tends as t approaches point c_k along the curve L_j, and take the function

$$\gamma(z)=\frac{1}{2\pi i}\int_L\frac{\ln G(t)\,dt}{t-z}. \quad (7.4.2)$$

Then, in the neighborhood of node c_k, $k=1,\ldots,n$, one has the representation

$$\gamma(z)=(\alpha_k+i\beta_k)\ln(z-c_k)+\gamma_0(z), \quad (7.4.3)$$

where $\gamma_0(z)$ is a function analytic within each of the sectors formed by curve L in the neighborhood of point c_k and tending to a certain limit for

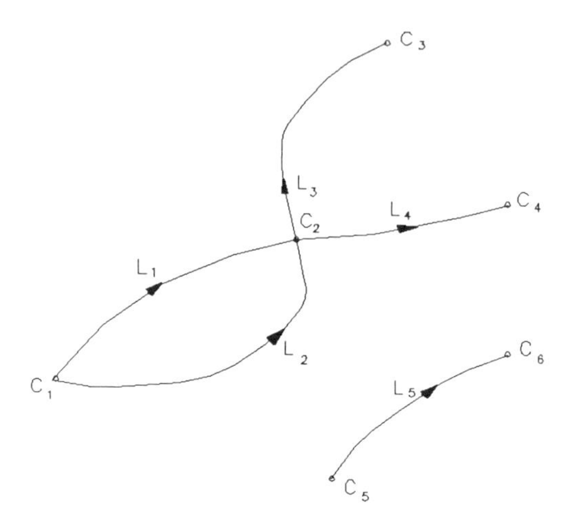

FIGURE 7.1. A piecewise smooth curve.

$z \to c_k$ along a route staying within a fixed sector. Further,

$$\alpha_k + i\beta_k = \sum_j \frac{\mp \ln G_j(c_k)}{2\pi i}, \tag{7.4.4}$$

where the sum encompasses al the numbers j of the arcs L_j converging to c_k. The upper and the lower signs correspond to issuing and incoming arcs, respectively. Hence, near a node c_k one has

$$l^{\gamma(z)} = (z - c_k)^{\alpha_k + i\beta_k}\Omega(z), \tag{7.4.5}$$

where $\Omega(z)$ has the same form as $\gamma_0(z)$ entering (7.4.3). Node c_k will be called singular if the corresponding number α_k is an integer; all the other nodes will be called nonsingular. Let $c_1, \dots, c_1$, $0 \le 1 \le n$, be nonsingular nodes and $c_{l+1}, \dots, c_n$ be singular nodes of the line L. A solution to the original Equation (7.4.1) will be sought in the class H^* on curve L. In other words, the solutions are either limited at the nodes or have integrable singularities. Therefore, we shall divide all the solutions into classes, assigning to the class h, $c_1, \dots, c_q$, $q = 0, 1, \dots, l$, all the solutions $\varphi(t) \in H^*$ on L that stay limited in the neighborhoods of nonsingular nodes $c_1, \dots, c_q$ and allow for integrable singularities at all the rest of the nonsingular nodes $c_{q+1}, \dots, c_1$. A fixed class h, $c_1, \dots, c_q$, is used to specify the canonic function $x(z)$ and the index κ of a solution

$$x(z) = l^{\gamma(z)} \cdot \prod_{k=1}^{n} (z - c_k)^{\lambda_k} = l^{\gamma(z)} \cdot \prod(z), \tag{7.4.6}$$

$$\kappa = -\sum_{k=1}^{n} \lambda_k, \tag{7.4.7}$$

where the integers λ_k are chosen, depending on the class $h, c_1, \ldots, c_q$, in the following way: $0 < \alpha_k + \lambda_k < 1$ for the nodes $c_1, \ldots, c_q$, $-1 < \alpha_k + \lambda_k < 0$, for the nodes $c_{q+1}, \ldots, c_l$, and $\alpha_k + \lambda_k = 0$ for the singular nodes $c_{l+1}, \ldots, c_n$.

Now the index κ solution to characteristic equation (7.4.1) is given by the formula (Muskhelishvili 1952)

$$\varphi(t) = \dot{a}^*(t)f(t) - \frac{b^*(t)z(t)}{\pi i} \int_L \frac{f(t_0)\,dt_0}{2(t_0)(t_0 - t)}$$

$$+ b^*(t)z(t)P_{\kappa-1}(t),$$

$$a^*(t) = \frac{a(t)}{a^2(t) - b^2(t)}, \qquad b^*(t) = \frac{b(t)}{a^2(t) - b^2(t)},$$

$$z(t) = [a(t) + b(t)]x^{(+)}(t) = \sqrt{a^2(t) - b^2(t)}\, l^{\gamma(t)}\Pi(t), \tag{7.4.8}$$

where for $\kappa > 0$, $P_{\kappa-1}(t_0)$ is a polynomial of the degree $\kappa - 1$, and for $\kappa \leq 0$, $P_{\kappa-1}(t_0) \equiv 0$.

Note that the function $z(t)$ may be presented in the form

$$z(t) = w(t) \cdot \prod_{k=1}^{n} (t - c_k)^{\gamma_k}, \tag{7.4.9}$$

where $w(t)$ is a function belonging to the class H_0 (Muskhelishvili 1952) and nonzero on L, and

$$\gamma_k = \alpha_k + \lambda_k + i\beta_k, \qquad k = 1, \ldots, n. \tag{7.4.10}$$

Also, it may readily be shown that

$$\lim_{z \to \infty} z^{\kappa} x(z) = 1, \tag{7.4.11}$$

i.e., the index κ canonic function $x(z)$ behaves at infinity as z^{κ}.

In order to develop numerical methods for solving Equation (7.4.1), let us first point out some of its properties. We start by writing it for the index κ solution, in the form

$$\int^{(+)}(w_\kappa^{(+)}\psi)(t_0) = f(t_0),$$

$$\int^{(+)}(w_\kappa^{(+)}\psi)(t_0) = a(t_0)w_\kappa^{(+)}(t_0)\psi(t_0) + \frac{b(t_0)}{\pi i}\int_L \frac{w_\kappa^{(+)}(t)\psi(t)\,dt}{t - t_0}. \tag{7.4.12}$$

Formula (7.4.8) will be presented in the form

$$S(t) = w_\kappa^{(+)}(t)\psi(t) = w_\kappa^{(+)}(t)\left[\int^{(-)}(w_\kappa^{(-)}f)(t) + \eta(\kappa)b(t)P_{\kappa-1}(t)\right], \tag{7.4.13}$$

where

$$w_\kappa^{(+)}(t) = z(t)/[a^2(t) - b^2(t)], \qquad w_\kappa^{(-)}(t) = z^{-1}(t),$$

$$\int^{(-)}(w_\kappa^{(-)}f)(t_0) = a(t_0)w_\kappa^{(-)}(t_0)f(t_0) - \frac{b(t_0)}{\pi i}\int_L \frac{w_\kappa^{(-)}(t)f(t)\,dt}{t - t_0},$$

$\eta(\kappa) = 1$ for $\kappa > 0$, and $\eta(\kappa) = 0$ for $\kappa \leq 0$.

For $\kappa < 0$ function $\varphi(t)$ given by (7.4.13) is a solution to the original characteristic equation (7.4.12) in the case when the conditions

$$\int_L w_\kappa^{(-)}(t)f(t)t^i\,dt = 0, \qquad j = 0, 1, \ldots, |\kappa| - 1, \tag{7.4.14}$$

hold. It should be noted that (Muskhelishvili 1952) the functions $f(t)t^j$, $j = 0, 1, \ldots, \kappa - 1$, are the proper functions of the operator $\int^{(+)}(w^{(+)}\kappa \cdot)(t)$ for $\kappa > 0$; the same functions are proper functions for the operator $\int^{(-)}(w^{(-)} \cdot)(t)$ for $\kappa < 0$ and $j = 0, 1, \ldots, -\kappa - 1$ [for the operator $\int^{(-)}(w^{(-)}\kappa \cdot)(t)$ its own index $\bar{\kappa} = -\kappa$].

For $\kappa \geq 1$ a solution to the equation

$$\int^{(-)}(w_\kappa^{(-)}f)(t_0) = \psi(t_0) \tag{7.4.15}$$

is given by the function

$$f(t_0) = \int^{(+)}(w_\kappa^{(+)}\psi)(t_0) \tag{7.4.16}$$

subject to the conditions

$$\int_L w_\kappa^{(+)}(t)\psi(t)t^j\,dt = 0, \qquad j = 0, 1, \ldots, \kappa - 1. \tag{7.4.17}$$

for $\kappa > 0$ the question of singling out the unique solution of Equation (7.4.12) arises. Because in this case the solution is not unique due to polynomial $P_{\kappa-1}(t)$ appearing in (7.4.13), one of the simplest ways of singling out a unique solution is by specifying the values of solution $\varphi(t)$ at κ different points, which do not coincide with nodes. However, more often one has to deal with singling out by means of specification of moments of the solution (Belotserkovskii et al. 1987, Zakharov and Pimenov 1982, Parton and Perlin 1982, Savruk 1981). In this connection we formulate the following result (Afendikova, Lifanov, and Matveev 1987).

Theorem 7.4.1. *Let curve L be piecewise smooth (in particular, it may be smooth). Then the system of equalities*

$$\int_L S(t)t^k\,dt = D_k, \qquad k = 0, 1, \ldots, \kappa - 1, \tag{7.4.18}$$

where D_k are fixed numbers (either real or complex), singles out a unique solution to the characteristic equation (7.4.12).

Proof. Let us assume that the statement is wrong. Then two different polynomials, $P_{\kappa-1,1}(t)$ and $P_{\kappa-1,2}(t)$, must exist, which satisfy Equality (7.4.13) for a function $\varphi(t)$. The difference of the polynomials, $P_{\kappa-1,3}(t) = P_{\kappa-1,1}(t) - P_{\kappa-1,2}(t)$, multiplied by $w_\kappa^{(+)}(t)b(t)$, satisfies Equalities (7.4.18) for $D_k = 0$, $k = 0, 1, \ldots, \kappa - 1$.

Consider next the equation

$$\int^{(-)}(w_\kappa^{(-)}f)(t_0) = b(t_0)P_{\kappa-1,3}(t_0). \tag{7.4.19}$$

If $\kappa > 0$ for the original characteristic equation, then for Equation (7.4.19) the index is negative ($\bar{\kappa} = -\kappa$), and the solution is unique for any Hölder right-hand side subject to the condition

$$\int_L \left[w_\kappa^{(+)}(t)b(t)P_{\kappa-1,3}(t)\right]t^j\,dt = 0, \qquad j = 0, 1, \ldots, \kappa - 1. \tag{7.4.20}$$

However, the right-hand side of Equation (7.4.19) incorporates a proper function for the operator $\int^{(+)}(w_\kappa^{(+)}\cdot)(t)$, and hence by Formula (7.4.16) one deduces that $f(t)\equiv 0$ on L, and $b(t)P_{\kappa-1,3}(t)\equiv 0$. Because $b(t)$ is a Hölder function on L and $b(t)\not\equiv 0$ on L, we deduce that $P_{\kappa-1,3}(t)\equiv 0$. ■

Note that the uniqueness conditions for a singular integral equation of the first kind on a system of segments of positive index $\kappa > 0$ in the form (7.4.18) were originally formulated in Lifanov (1979b); for an equation of the second kind with variable coefficients they were formulated in Lifanov and Matveev (1983). The preceding proof was proposed in Afendikova, Lifanov, and Matveev 1987).

For constructing a numerical method we will need a relationship for the operators $\int^{(+)}(w_\kappa^{(+)}\cdot)$. Let us denote $\xi_{n;m}(z) = P_n(z) + \lambda_{-m}(z-c)$ and $\xi_{n;0}(z) = P_n(z)$, where $P_n(z)$ is a polynomial of degree n, $\lambda_{-m}(z-c) = c_{-1}(z-c^{-1} + \cdots + c_{-m}(z-c)^{-m}$, $c\notin L$. The function $\xi_{n;m}$ will be called a *generalized polynomial of degree* $(n;m)$. The following theorems are true.

Theorem 7.4.2. *Let $a(t)$ and $b(t)$ be arbitrary functions belonging to the class H on L. Then the relationship*

$$a(t_0)w_\kappa^{(\pm)}(t_0)\xi_{n;m}(t_0) \pm \frac{1}{\pi i}\int_L \frac{b(t)w_\kappa^{(\pm)}(t)\xi_{n;m}(t)}{t-(t_0)}dt = \bar{\xi}_{n\mp\kappa;m}(t_0). \tag{7.4.21}$$

holds for any function $\xi_{n;m}(z)$.

Proof. Let us prove the theorem for $w_\kappa^{(+)}$. Consider function $x(z)\cdot\xi_{n;m}(z)$, where $x(z)$ is the index κ canonic function (7.4.6) for Equation (7.4.12). Because $x(z)$ is analytic outside L, when $x^{(\pm)}(t)\in H^*$ on L, and at infinity behaves as $z^{-\kappa}$, there exists such a function $\bar{\xi}_{n-\kappa;m}(z)$ that the function $\Phi(z) = x(z)\xi_{n;m}(z) - \bar{\xi}_{n-\kappa;m}(z)$ is analytic outside L, $\Phi^{(\pm)}(t)\in H^*$ on L, and $\Phi(z)\to 0$ for $|z|\to\infty$. Then by the Sokhotsky–Plemeli formulas (Muskhelishvili 1952) one gets

$$\frac{1}{\pi i}\int_L \frac{\phi(\tau)-\phi^{(-)}(\tau)}{\tau - t}d\tau = \phi^{(+)}(t) + \phi^{(-)}(t). \tag{7.4.22}$$

Taking into account that $\xi_{n;m}^{(+)}(t) = \xi_{n;m}^{(-)}(t) = \xi_{n;m}(t)\bar{\xi}_{n-\kappa;m}^{(+)}(t) = \bar{\xi}_{n-\kappa;m}^{(-)}(t) = \bar{\xi}_{n-\kappa;m}(t)$, and $x^{(\pm)}(t) = [a(t)\mp b(t)]\cdot z(t)/(a^2(t)-b^2(t))$, one gets from Formula (7.4.22) Relationship (7.4.21) for $w_\kappa^{(+)}$. In order to get the same relationship for $w_\kappa^{(-)}$ one has to consider the function $x^{-1}(z)\cdot\xi_{n;m}(z)$. Then, the following theorem may be formulated.

Theorem 7.4.3. *Let $b(t)$ be a polynomial of degree l. Then the following relationship is valid for operators $\int^{(\pm)}(w_\kappa^{(\pm)}\cdot)$:*

$$\int_{(\pm)}(w_\kappa^{(\pm)}\xi_{n;m})(t_0) = \xi^*_{n\mp\kappa;m}(t_0), \qquad n \mp \kappa \geq l - 1. \quad (7.4.23_O)$$

For a singular integral operator $\int_\kappa^{(\pm)}(w_\kappa^{(\pm)}\cdot)$ of the second kind with variable real coefficients of polynomials $P_n(t)$ and L being a segment, Relationship (7.4.23$_O$) was originally derived by Elliott (1980). Elliott also proved that a system of polynomials $\{P_n(t)\}$ orthogonal with the weight $w_\kappa^{(\pm)}(t)$ transforms into a system of polynomials $\{Q_m(t)\}$ orthogonal with the weight $w_\kappa^{(\mp)}(t)$. Thus, he generalized the known relationships for Chebyshev and Jacobi polynomials. Theorems 7.4.2 and 7.4.3 were originally formulated and proved in Afendikova, Lifanov, and Matveev (1986, 1987). ■

By using (7.4.23$_O$) one may construct a numerical method for the characteristic equation supposing that $b(t)$ is a polynomial of degree l. Matveev (1988) has also done this for the case $b(t) = b_{r_1}(t)\cdot b_1(t)$, where $b_{r_1}(t)$ is a polynomial of degree $r_1 \geq 0$, and $b_1(t)$ may vanish at a finite number of points belonging to the line L.

For a fixed index κ and a natural number n we choose two systems of points $E_0 = \{t_{01}, t_{02}, \ldots, t_{0n-\kappa}\}$ and $E = \{t_1, t_2, \ldots, t_n\}$ different in pairs. For $\kappa < 0$ we get $n - \kappa \geq l + |\kappa|$ and E_0 does not contain any roots of $b(t_0)$.

For function $f(t)$ we choose an approximating generalized polynomial $f_{n_1-\kappa-1;n_2}(t)$ of degree $n_1 + n_2 - \kappa - 1$, where $n_1 + n_2 = n$. In particular, the polynomial may satisfy the equality $f_{n_1-\kappa-1;n_2}(t_{0m}) = f(t_{0m})$. Then the index κ solution to the characteristic equation (7.4.12) is given by the function $w_\kappa^{(+)}(t)\psi_{n_1-1;n_2}(t)$, where $\psi_{n_1-1;n_2}(t)$ is a generalized polynomial of degree $n_1 + n_2 - 1$, and the following equality holds:

$$\int^{(+)}\left(w_\kappa^{(+)}\psi_{n_1-1;n_2}\right)(t_0) = f_{n_1-\kappa_1-1;n_2}(t_0). \qquad (7.4.23)$$

As an equality of generalized polynomials of degree $n_1 + n_2 - \kappa - 1$, the latter equality is equivalent to the following system of $n - \kappa$ linear algebraic equations:

$$\int^{(+)}\left(w_\kappa^{(+)}\psi_{n_1-1;n_2}\right)(t_{0m}) = f_{n_1-\kappa-1;n_2}(t_{0m}),$$

$$m = 1, 2, \ldots, n - \kappa, \quad (7.4.24)$$

in the coefficients n of the generalized polynomial $\psi_{n_1-1;n_2}$.

If $\kappa = 0$, then Equation (7.4.23) has a unique solution, and hence, System (7.4.24) also has a unique solution.

If $\kappa > 0$, then by Theorem 7.4.1, Equation (7.4.23) has a unique solution only if the κ conditions (7.4.18) are met. Thus, in this case the system of linear algebraic equations (7.4.18) and (7.4.24), where $\varphi(t)$ is to be replaced by $w_\kappa^{(+)}(t)\psi_{n_1-1,n_2}(t)$, has a unique solution.

If $\kappa < 0$, then the function $w_\kappa^{(+)}\psi_{n_1-1;n_2}(t)$ obtained from (7.4.13), where $f(t)$ is replaced by $f_{n_1-\kappa-1;n_2}(t)$, is a solution to Equation (7.4.23) only if Conditions (7.4.14) for $f_{n_1-\kappa-1;n_2}(t)$ are met. However, as far as the function $f_{n_1-\kappa-1;n_2}(t)$ is an arbitrary function, the conditions will not necessarily be met. Therefore, the function $f_{n_1-\kappa_1-1;n_2}(t)$ appearing in Equation (7.4.23) must be replaced by

$$f^*_{n_1-\kappa-1;n_2}(t) = f_{n_1-\kappa-1;n_2}(t) - b(t)\sum_{\nu=1}^{|\kappa|-1} \gamma_{n,\nu} t^\nu,$$

where the numbers $\gamma_{n,\nu}$, $\nu = 0, 1, \ldots, |\kappa| - 1$, must be such that all the conditions (7.4.14) are met. Numbers $\gamma_{n,\nu}$ exist and are uniquely defined by Conditions (7.4.14) for the function $f^*_{n_1-\kappa-1;n_2}(t)$, and if $\|f(t) - f_{n_1-\kappa-1;n_2}(t)\|_c \to 0$, then $\gamma_{n,\nu} \to 0$ for $n \to \infty$, $\nu = 0, 1, \ldots, |\kappa| - 1$. Thus, for $\kappa < 0$, instead of Equation (7.4.23), one must consider the equation

$$b(t_0)\sum_{\nu=0}^{|\kappa|=1} \gamma_{n,\nu} t_0^\nu + \int^{(+)}\left(w_\kappa^{(+)}\psi_{n_1-1;n_2}\right)(t_0) = f_{n_1-\kappa-1;n_2}(t_0). \quad (7.4.25)$$

For the chosen numbers this equation has a unique solution, and, hence, the equivalent system of linear algebraic equations

$$b(t_{0m})\sum_{\nu=0}^{|\kappa|-1} \gamma_{n,\nu} t_{0m}^\nu + \int^{(+)}\left(w_\kappa^{(+)}\psi_{n_1-1;n_2}\right)(t_{0m}) = f_{n_1-\kappa-1;n_2}(t_{0m}),$$

$$m = 1, \ldots, n - \kappa, \quad (7.4.26)$$

also has a unique solution.

From the preceding statements, the next theorem follows.

Theorem 7.4.4. *Let κ be an index of the solution $\varphi_\kappa(t) = w_\kappa^{(+)}(t)\psi(t)$ to Equation* (7.4.12). *Then, the system of linear algebraic equations*

$$\eta(-\kappa)b(t_{0m})\sum_{\nu=0}^{|\kappa|-1} \gamma_{n,\nu} t_{0m}^\nu + \int^{(+)}\left(w_\kappa^{(+)}\psi_{n_1-1;n_2}\right)(t_{0m})$$

$$= f_{n_1-\kappa-1;n_2}(t_{0m}), \qquad m = 1,2,\ldots,n-\kappa,$$

$$\eta(\kappa)\int_L w_\kappa^{(+)}(t)\psi_{n_1-1;n_2}(t)t^j\,dt = \eta(\kappa)D_j, \quad j = 0,1,\ldots,\kappa-1,$$

$$(7.4.27)$$

in the coefficients of the generalized interpolation polynomial $\psi_{n_1-1;n_2}(t)$ *for* $\kappa \geq 0$ *and the set of the coefficients and the coefficients* $\gamma_{n;\,\nu}$, $\nu = 0,1,\ldots,|\kappa| - 1$, *for* $\kappa < 0$, *is well-conditioned for any* n, $n - \kappa \geq l - 1$. *For* $\kappa < 0$ *the last* κ *equalities must be discarded, because* $\eta(\kappa) = 0$, $\kappa \leq 0$, *and* $\eta(\kappa) = 1$, $\kappa > 0$. *The following relationship holds between the accurate solution* $\varphi_\kappa(t)$ *and an approximate solution* $\varphi_{n,\,\kappa}(t) = w_\kappa^{(+)}(t)\psi_{n_1-1;n_2}(t)$:

$$S_\kappa(t) - S_{n,\,\kappa}(t) = w_\kappa^{(+)}(t)\int^{(-)}\left[w_\kappa^{(-)}(f - f_{n_1-\kappa-1;n_2})\right](t). \quad (7.4.28)$$

In analogy to the preceding sections, the numbers $\gamma_{n,\,\nu}$ will be called *regularizing variables*.

Note 7.4.1. From (7.4.28) it follows that an estimate of the difference $\varphi_\kappa(t) - \varphi_{n,\,\kappa}(t)$ in a certain metric of the space of functions on L is equal to an estimate of the function on the right-hand side of (7.4.28) in the same metric. Thus, if $w_\kappa^{(+)}(t)$ is limited on L (this is so if L is a piecewise smooth curve, but a solution is sought in the class of functions limited at all the nodes), and $f(t)$ is a function of the class H and L, then, according to Muskhelishvili (1952), the estimate may be obtained in a uniform metric.

Note 7.4.2. Instead of System (7.4.27) one may consider a system of equalities of coefficients before equal degrees of the variable t_0 in generalized polynomials entering Equality (7.4.26). In this case, Formula (7.4.28) preserves its form.

Note 7.4.3. A detailed description of the system of linear algebraic equations (7.4.27) corresponding to Equation (7.4.12) in interval $[-1,1]$ with real coefficients and a right-hand side is presented in works by Matveev (Matveev and Molyakov 1988, Matveev 1988) together with instructions for calculating all the elements of the system.

The results obtained in this section may be transferred in a natural way to the full singular integral equation of the second kind with variable coefficients:

$$a(t_0)S(t_0) + \frac{b(t_0)}{\pi i}\int_L \frac{S(t)\,dt}{t - t_0} + \int_L k(t_0,t)S(t)\,dt = f(t_0). \quad (7.4.29)$$

7.5. EQUATION WITH HILBERT'S KERNEL AND VARIABLE COEFFICIENTS

Considerations similar to those for Equation (7.4.12) may be applied to the equation

$$a(\theta_0)S(\theta_0) - \frac{b(\theta_0)}{2\pi}\int_0^{2\pi}\cot\frac{\theta-\theta_0}{2}S(\theta)\,d\theta = f(\theta_0), \quad (7.5.1)$$

where $a(\theta), b(\theta), f(\theta) \in H$ on $[0, 2\pi]$ and are periodic. Therefore, we shall present only the necessary relationships and a corresponding system of linear algebraic equations for a fixed index κ. A detailed theory of the equation is presented in Gakhov (1958).

Let us introduce the following notation:

$$\Gamma^{(\pm)}(w_\kappa^{(\pm)}\psi)(\theta_0) = a(\theta_0)w_\kappa^{(\pm)}(\theta_0)\psi(\theta_0)$$

$$\mp\frac{b(\theta_0)}{2\pi}\int_0^{2\pi} w_\kappa^{(\pm)}(\theta)\psi(\theta)\cot\frac{\theta-\theta_0}{2}\,d\theta, \quad (7.5.2)$$

where

$$w_\kappa^{(\pm)}(\theta) = \exp\frac{(\mp\mu(\theta))}{z(\theta)}, \qquad z(\theta) = \sqrt{a^2(\theta)+b^2(\theta)},$$

$$\mu(\theta_0) = -\frac{1}{2\pi}\int_0^{2\pi}[\arg(a(\theta)+ib(\theta)) - \kappa\theta]\cot\frac{\theta-\theta_0}{2}\,d\theta.$$

Then Equation (7.5.1) for the index κ solution and its solution may be written in the form (Gakhov 1958)

$$\Gamma^{(+)}(w_\kappa^{(+)}\psi)(\theta_0) = f(\theta_0),$$

$$S(\theta_0) = w_\kappa^{(+)}(\theta_0)\psi(\theta_0) = w_\kappa^{(+)}(\theta_0)\left[\Gamma^{(-)}(w_\kappa^{(-)}f)(\theta_0) + b(\theta_0)P_\kappa(\theta_0)\right], \quad (7.5.3)$$

where for $\kappa \geq 0$,

$$P_\kappa(\theta) = \beta_0 + 2\eta(\kappa)\left[\sum_{k=1}^{\kappa=1}(\alpha_k\sin k\theta + \beta_k\cos k\theta)\right.$$

$$\left. + \alpha_\kappa(A_\kappa\sin\kappa\theta - B_\kappa\cos\kappa\theta)\right],$$

$$A_\kappa = \frac{1}{\pi}\int_0^{2\pi}\sigma(\theta)\sin\kappa\theta\,d\theta,$$

$$B_\kappa = \frac{1}{\pi}\int_0^{2\pi} \sigma(\theta)\cos \kappa\theta\, d\theta,$$

$$\sigma(\theta) = a(\theta)\exp(-\mu(\theta))/z(\theta),$$

and for $\kappa < 0$, $P_\kappa(\theta) \equiv 0$. For $\kappa > 0$ the coefficients β_0, α_k and β_k, $k = 1, 2, \dots, \kappa - 1$, α_κ, are uniquely defined by specifying 2κ relationships

$$\int_0^{2\pi} w_\kappa^{(+)}(\theta)\psi(\theta)\cos k\theta\, d\theta = c_k, \qquad k = 0, 1, \dots, \kappa - 1,$$

$$\int_0^{2\pi} w_\kappa^{(+)}(\theta)\psi(\theta)\sin k\theta\, d\theta = c_{k+\kappa-1}, \quad k = 1, \dots, \kappa - 1,$$

$$\int_0^{2\pi}\int_0^{2\pi} w_\kappa^{(+)}(\theta)\psi(\theta)\rho(\theta_0)\sin \kappa(\theta - \theta_0)\, d\theta\, d\theta_0 = c_{2\kappa-1},$$

$$\rho(\theta) = b(\theta)\exp(\mu(\theta))/z(\theta). \tag{7.5.4}$$

For $\kappa = 0$, the number β_0 is fully determined if $\int_0^{2\pi}\sigma(\theta)\, d\theta \neq 0$; otherwise, β_0 has an arbitrary value, and the function $\varphi(\theta)$ is a solution to Equation (7.5.3) only if the condition

$$\int_0^{2\pi} w_\kappa^{(-)} f(\theta)\, d\theta = 0 \tag{7.5.5}$$

is met.

For $\kappa < 0$ the function $\varphi(\theta)$ as defined by (7.5.3) gives a unique solution to Equation (7.5.3) if the following -2κ conditions are met:

$$\int_0^{2\pi} w_\kappa^{(-)}(\theta) f(\theta)\cos k\theta\, d\theta = 0, \qquad k = 0, 1, \dots, -\kappa - 1,$$

$$\int_0^{2\pi} w_\kappa^{(-)}(\theta) f(\theta)\sin k\theta\, d\theta = 0, \qquad k = 1, 2, \dots, -\kappa - 1,$$

$$\int_0^{2\pi}\int_0^{2\pi} w_\kappa^{(-)}(\theta) f(\theta)\delta(\theta_0)\sin \kappa(\theta - \theta_0)\, d\theta\, d\theta_0 = 0,$$

$$\delta(\theta) = -b(\theta)\exp(-\mu(\theta))/r(\theta). \tag{7.5.6}$$

Similarly to the preceding section, the following theorem may be proved.

Theorem 7.5.1. *If $b(\theta)$ is a trigonometric polynomial of degree l, then the operator $\Gamma^{(\pm)}(w_\kappa^{(\pm)}\cdot)$ transforms an arbitrary trigonometric polynomial of* *degree* $n > \max(l \pm \kappa; 2\kappa)$ *into a trigonometric polynomial of degree* $n \mp \kappa$.

Let the full equation

$$\Gamma^{(+)}(w_\kappa^{(+)}\psi)(\theta_0) + \int_0^{2\pi} k(\theta_0, \theta) w_\kappa^{(+)}(\theta)\psi(\theta)\, d\theta = f(\theta_0) \quad (7.5.7)$$

be given, where $k(\theta_0, \theta)$ and $f(\theta_0)$ are Hölder functions.

By using the inversion formula (7.5.3) for the characteristic equation, Equation (7.5.7) may be transformed into a Fredholm equation of the second kind in the function $w_\kappa^{(+)}(\theta)\psi(\theta)$. Because function $w_\kappa^{(\pm)}(\theta)$ belongs to the class H, if $a(\theta)$ and $b(\theta)$ also d_0, and the Hilbert operator stays in the class H too, then the kernel and right-hand side of the derived Fredholm equation of the second kind are periodic and also belong to the class H on $[0, 2\pi]$. If $\kappa < 0$, then the Fredholm equation solution is a solution to the original equation, subject to Conditions (7.5.5) and (7.5.6), where $f(\theta)$ must be replaced by $f(\theta) - \int_0^{2\pi} k(\theta, \tau) w_\kappa^{(+)}(\tau)\psi(\tau)\, d\tau$.

Next we suppose that Equation (7.5.7) has a unique solution if Conditions (7.5.4) for the required solution for $\kappa > 0$ are fulfilled. Then the corresponding Fredholm equation of the second kind also has a unique solution. Suppose that for $\kappa \le 0$ the Fredholm equation has a unique solution that is a solution to the original equation subject to the preceding conditions. Under the assumptions, a numerical method for solving Equation (7.5.7) is constructed as follows.

Let $E = \{\theta_k,\ k = 1, 2, \ldots, 2n + 1\}$ and $E_0 = \{\theta_{0j},\ j = 1, 2, \ldots, 2(n - \kappa) + 1\}$ be a pair of nonintersecting systems of points on a unit circle centered at the origin of coordinates. An approximate solution $\varphi_{n,\kappa}(\theta)$ will be sought in the form $w_\kappa^{(+)}(\theta)\psi_n(\theta)$, where $\psi_n(\theta)$ is a trigonometric polynomial approximating a function $\psi(\theta)$, for example, a truncated sum of Fourier series for function $\psi(\theta)$ most often represented by a polynomial of the form

$$\psi_n(\theta) = \sum_{k=1}^{2n+1} \psi_n(\theta_k) T_{n,k}(\theta), \quad (7.5.8)$$

where $T_{n,k}(\theta)$ is a trigonometric polynomial of degree n such that $T_{n,k}(\theta_k) = 1$, $k = 1, \ldots, 2n + 1$. The numbers $\psi_n(\theta_k)$, $k = 1, \ldots, 2n + 1$, will be calculated from the following system of linear algebraic equations:

$$\eta(-\kappa) b(\theta_{0j}) y_{|\kappa|}(\theta_{0j}) + \sum_{k=1}^{2n+1} \psi_n(\theta_k) q_{kj}$$

$$+ \sum_{k=1}^{2n+1} k(\theta_{0j}, \theta_k)\psi_n(\theta_k) h_k = f(\theta_{0j}), \qquad j = 1, \ldots, 2(n - \kappa)$$

$$+1,$$

$$\eta(\kappa)\sum_{k=1}^{2n+1}\psi_n(\theta_k)\cos(j\theta_k)h_k = \eta(\kappa)c_j, \qquad j = 0,1,\ldots,\kappa-1,$$

$$\eta(\kappa)\sum_{k=1}^{2n+1}\psi_n(\theta_k)\sin(j\theta_k)h_k = \eta(\kappa)c_{j+\kappa-1}, \qquad j = 1,\ldots,\kappa-1,$$

$$\eta(\kappa)\sum_{k=1}^{2n+1}\psi_n(\theta_k)\int_0^{2\pi}\int_0^{2\pi}w_\kappa^{(+)}(\theta)T_{n,k}(\theta)\sin$$

$$(\kappa(\theta-\theta_0))\rho(\theta_0)\,d\theta\,d\theta_0 = c_{2\kappa-1}, \qquad (7.5.9)$$

where $y_{|\kappa|}(\theta)$ for $\kappa < 0$ is a trigonometric polynomial of degree $|\kappa|$ and of the same form as in the case of $P_\kappa(\theta)$ in (7.5.3) for $\kappa > 0$. The coefficients of the polynomial are regularizing variables chosen under the solvability condition for System (7.5.9). Similarly to the preceding section, they may be shown to be defined uniquely. The coefficients g_{kj} and h_k are determined by the relationships

$$g_{kj} = \Gamma^{(+)}(w_\kappa^{(+)}T_{n,k})(\theta_{0j}), \qquad k = 1,\ldots,2n+1,$$

$$j = 1,\ldots,2(n-\kappa)+1,$$

$$h_k = \int_0^{2\pi}w_\kappa^{(+)}(\theta)T_{n,k}(\theta)\,d\theta, \quad k = 1,\ldots,2n+1. \qquad (7.5.10)$$

System (7.5.9) corresponding to the characteristic equation (7.5.3) ($k \equiv 0$) and has a unique solution for any $n > \max(l+\kappa, 2\kappa)$. The following theorem is valid for System (7.5.9) corresponding to Equation (7.5.7).

Theorem 7.5.2. *Let Equation* (7.5.7) *have a unique solution subject to the corresponding assumptions of the form* (7.5.4) *for* $\kappa > 0$ *or of the form* (7.5.6) *if* $f(\theta)$ *is replaced by* $f(\theta) - \int_0^{2\pi}k(\theta,\tau)\varphi(\tau)\,dt$ *for* $\kappa < 0$. *Then starting from a certain n, System* (7.5.9) *has a unique solution, and the rate of convergence of an approximate solution* $\varphi_{n,\kappa}(\theta)$ *to the accurate solution* $\varphi_\kappa(\theta)$ *for a given metric of the space of periodic functions on* $[0,2\pi]$ *will be the same as the rate of convergence (in the same metric) of quadrature formulas for a singular integral with Hilbert's kernel and a regular integral, if densities of the latter integrals are approximated by interpolation trigonometric polynomials of the form* (7.5.8) *on the sets of points* E *and* E_0 *for the variables* θ *and* θ_0, *respectively.*

8

Singular Integral Equations with Multiple Cauchy Integrals

8.1. ANALYTICAL SOLUTION TO A CLASS OF CHARACTERISTIC INTEGRAL EQUATIONS

Consider a one-dimensional characteristic integral equation of the first kind, depending on a parameter

$$\int_{L_1} \frac{\varphi(t^1, t_0^2)\, dt^1}{t^1 - t_0^1} = f(t_0^1, t_0^2) \tag{8.1.1}$$

where L_1 is a curve lying in the plane of complex variable t^1, and $f(t_0^1, t_0^2)$ belongs to the class H on the set $L_1 \times L_2$, where L_2 is a curve lying in the plane of complex variable t^2. In what follows, L_1 and L_2 are supposed to be limited. Under the constraints imposed on $f(t_0^1, t_0^2)$, the sought solution $\varphi(t^1, t_0^2)$ must meet the condition H with respect to t_0^2 for any $t^1 \in L_1$.

Let L_1 be a smooth closed curve or a system of nonintersecting smooth closed curves. The unique solution to this equation, as well as to Equation (6.1.1), is the function

$$\varphi(t^1, t_0^2) = -\frac{1}{\pi^2} \int_{L_1} \frac{f(t_0^1, t_0^2)\, dt_0^1}{t_0^1 - t^2}. \tag{8.1.2}$$

This is a consequence of the fact that if the function $\varphi(t^1, t_0^1)$ satisfies the identity

$$\int_{L_1} \frac{\varphi(t^1, t_0^2)\, dt^1}{t^1 - t_0^1} \equiv 0, \tag{8.1.3}$$

then $\varphi(t^1, t_0^2) \equiv 0$.

Now let L_1 be the integral $[a, b]$. Then the solution to the above equation has the form (see (5.1.2))

$$\varphi(t^1, t_0^2) = -\frac{1}{\pi^2} R_\kappa(t^1) \int_a^b \frac{f(t_0^1, t_0^2)\, dt_0^1}{R_\kappa(t_0^1)(t_0^1 - t^1)} + \frac{1}{\pi} R_\kappa(t^1) \nu_\kappa C(t_0^2). \tag{8.1.4}$$

For $\kappa = 1$, the condition

$$\int_a^b \varphi(t^1, t_0^2)\, dt^1 = C(t_0^2) \tag{8.1.5}$$

holds, and for $\kappa = -1$, the function $\varphi(t^1, t_0^2)$ is a solution if the identity

$$\int_a^b \frac{f(t^1, t_0^2)\, dt^1}{\sqrt{(t^1 - a)(b - t^1)}} \equiv 0 \tag{8.1.6}$$

is true.

If L_1 is a system of nonintersecting segments, then the index $\kappa = m$ (where m is the number of segments) general solution has the form

$$\varphi(t^1, t_0^2) = -\frac{1}{\pi^2} R_m(t^1) \int_{L_1} \frac{f(t_0^1, t_0^2)\, dt_0^1}{R_m(t_0^1)(t_0^1 - t^1)}$$

$$+ R_m(t^1) P_{m-1}(t^1, t_0^2), \tag{8.1.7}$$

where $P_{m-1}(t^1, t_0^1) = P_1(t_0^2)(t^1)^{m-1} + P_2(t_0^2)(t^1)^{m-2} + \cdots + P_m(t_0^2)$ and $P_k(t_0^2)$, $k = 1, \ldots, m$, are arbitrary functions of the class H on L_2, expressible both uniquely and linearly through the functions

$$C_k(t_0^2) \equiv \int_{L_1} \varphi(t^1, t_0^2)(t^1)^{k-1}\, dt^1, \qquad k = 1, \ldots, m. \tag{8.1.8}$$

A similar procedure is applicable to a characteristic singular integral equation of the second kind,

$$a(t_0^1)\varphi(t_0^1,t_0^2) + \frac{b(t_0^1)}{\pi}\int_{L_1}\frac{\varphi(t^1,t_0^2)\,dt^1}{t^1-t_0^1} = f(t_0^1,t_0^2), \qquad (8.1.9)$$

where L_1 is a piecewise smooth curve. It is obvious that arbitrary functions of the variable t_0^2 entering the general solution of a given positive index κ must have the same singularities as the function $f(t_0^1, t_0^2)$ or the variable t_0^2.

In a similar way one can consider the procedure of finding solutions to Equation (8.1.1) which have the form $\varphi(t^1, t_0^2) = \psi(t^1, t^2)/(q - t^1)$, with q being an internal point of the curve L_1 differing from the end points of the curve.

By supposing that $a(t_0^1)$ and $b(t_0^1)$ entering Equation (8.1.9) are constants, one can write the equation in the operator form

$$A_{L_1,t_0^1}(\varphi(t^1,t_0^2) = f(t_0^1,t_0^2), \qquad (8.1.10)$$

where A_{L_1}, t_0^1 designates the operators applied to the function $\varphi(t^1, t_0^2)$ entering the left-hand side of (8.1.9). A_{L_1}, t_0^1 will be called a characteristic singular operator of the second kind.

Let us next consider characteristic singular integral equations of the second kind with double Cauchy integrals whose operators may be represented by products of the corresponding one-dimensional operators with respect to each of the variables, i.e., equations of the form

$$A_{L_1,t_0^1}A_{L_2,t_0^2}\varphi(t^1,t^2) = f(t_0^1,t_0^2) \qquad (8.1.11)$$

or

$$A_{L_1,t_0^1}\big(A_{L_2,t_0^2}\varphi(t^1,t^2)\big) \equiv A_{L_2,t_0^2}\big(A_{L_1,t_0^1}\varphi(t^1,t^2)\big) = f(t_0^1,t_0^2),$$

meaning

$$a_1a_2(\varphi(t_0^1,t_0^2)) + \frac{a_2b_1}{\pi}\int_{L_1}\frac{\varphi(t^1,t_0^2)\,dt^1}{t^1-t_0^1} + \frac{a_1b_2}{\pi}\int_{L_2}\frac{\varphi(t_0^1,t^2)\,dt^2}{t^2-t_0^2}$$

$$+\frac{b_1b_2}{\pi^2}\iint_{L_1\times L_2}\frac{\varphi(t^1,t^2)}{(t^1-t_0^1)(t^2-t_0^2)}\,dt^1\,dt^2 = f(t_0^1,t_0^2). \qquad (8.1.12)$$

where $a_k^2 + b_k^2 = 1$, a_k, and b_k are real numbers, $k = 1, 2$, $f(t_0^1, t_0^2) \in H$ on $L_1 \times L_2$, and L_1 and L_2 are plane curves.

1. Let L_1 and L_2 be unit-radius circles centered at the origin of coordinates. By using Formula (7.2.9) repeatedly, we conclude that the function

$$\varphi(t^1, t^2) = A^{-1}_{L_2, t^2}\left(A^{-1}_{1, t^2} f(t_0^1, t_0^2)\right)$$

$$\equiv a_1 a_2 f(t^1, t^2) - \frac{a_2 b_1}{\pi} \int_{L_1} \frac{f(t_0^1, t^2)\, dt_0^1}{t_0^1 - t^1}$$

$$- \frac{a_1 b_2}{\pi} \int_{L_2} \frac{f(t^1, t_0^2)\, dt_0^2}{t_0^2 - t^2} + \frac{b_1 b_2}{\pi^2}$$

$$\times \iint_{L_1 \times L_2} \frac{f(t_0^1, t_0^2)\, dt_0^1\, dt_0^2}{(t_0^1 - t^1)(t_0^2 - t^2)} \tag{8.1.13}$$

is a unique solution of Equation (8.1.12). Hereinafter $A^{-1}_{L_k, t_k}$ denotes operations performed on the function $f(t_0^1, t_0^2)$ with respect to a corresponding variable either in Formula (7.2.9) or Formulas (5.1.2) and (7.1.2) for the corresponding indices of the operators under consideration.

2. Let now L_1 be a circle and L_2 be the interval $[-1, 1]$. We start by seeking a solution whose index with respect to the coordinate t^2 is equal to 1. In this case operator A_{L_1, t_0^1} is uniquely invertible for any right-hand side. To invert the operator A_{L_2, t_0^2} uniquely one has to know the function $\int_{L_2} \varphi(t^1, t^2)\, dt^2$, which will be defined uniquely if the function

$$A_{L_1 t_0^1} \int_{L_2} \varphi(t^1, t^2)\, dt^2 = C_1(t_0^1). \tag{8.1.14}$$

supposed to belong to the class H on L_1, is specified.

Thus, to find a unique solution to Equation (8.1.12) or Equation (8.1.11), one has to consider the system

$$A_{L_1, t_0^1}\left(A_{L_2, t_0^2} \varphi(t^1, t^2)\right) = f(t_0^1, t_0^2),$$

$$A_{L_1, t_0^1}\left(\int_{L_2} \varphi(t^1, t^2)\, dt^2\right) = C_1(t_0^1). \tag{8.1.15}$$

By applying Formulas (7.2.9) or (7.1.2) for $\kappa = 1$, one gets

$$\varphi(t^1,t^2) = A^{-1}_{L_2,t^2}\left(A^{-1}_{L_1,t^1} f(t^1_0,t^2_0)\right) + T_1(\alpha_2)\omega_1^{\alpha_2}(t^2) A^{-1}_{L_1,t^1} C_1(t^1_0)$$

$$= a_1 a_2 f(t^1,t^2) - \frac{a_2 b_1}{\pi}\int_{L_1} \frac{f(t^1_0,t^2)\,dt^1_0}{t^1_0 - t^1} - \frac{a_1 b_2}{\pi}\omega_1^{\alpha_2}(t^2)$$

$$\times \int_{L_2} \frac{f(t^1,t^2_0)\,dt^2_0}{\omega_1^{\alpha_2}(t^2_0)(t^2_0 - t^2)} + \frac{b_1 b_2}{\pi^2}\omega_1^{\alpha_2}(t_2)$$

$$\times \iint_{L_1\times L_2} \frac{f(t^1_0,t^2_0)\,dt^1_0\,d^2_0}{\omega_1^{\alpha_2}(t^2_0)(t^1_0 - t^1)(t^2_0 - t^2)}$$

$$+ T_1(\alpha_2)\omega_1^{\alpha_2}(t^2)\left[a_1 C_1(t^1) - \frac{b_1}{\pi}\int_{L_1}\frac{C_1(t^1_0)\,dt^1_0}{t^1_0 - t^1}\right], \qquad (8.1.16)$$

where (see (3.4.35)) $T_1(\alpha_2) = -\sin \alpha\pi)/\pi$, $-1 < \alpha_2 < 0$, $a_2 + b_2 \cot \pi\alpha_2 = 0$, $\omega_1^{\alpha_2}(t^2) = (1-t^2)^{\alpha_2}(1+t^2)^{\beta_2}$, and $-(\alpha_2 + \beta_2) = 1$.

If for a solution to Equation (8.1.11), $\kappa = 0$ or -1 in the variable t^2, then it is given by

$$\varphi(t^1,t^2) = A^{-1}_{L_2,t^2}\left(A^{-1}_{L_1,t^1} f(t^1_0,t^2_0)\right), \qquad (8.1.17)$$

where $\omega_1^{\alpha_2}(t^2)$ must be replaced by $\omega_\kappa^{\alpha_2}(t^2)$, $\kappa = 0, -1$. If the index is equal to -1 in the variable t^2, then the function $\varphi(t^1,t^2)$ given by (8.1.17) is a solution if the condition

$$\int_{L_1} \frac{f(t^1,t^2_0)\,dt^2_0}{\omega_{-1}^{\alpha_2}(t^2_0)} = 0 \qquad (8.1.18)$$

is met. Note that the possibility of obtaining the general solution to Equation (8.1.11) from System (8.1.15) allows us to change the order of operations performed with respect to the variables t^1 and t^2. The form of the result will be the same. However, if the result is found by inverting the operators in Equation (8.1.11), then it does depend on the order in which the operators are applied [because the arbitrary function $C_1(t^1_0)$ enters in different ways]. The convenience of System (8.1.15) will become clear when we construct direct numerical methods for solving Equation (8.1.11).

3. Let L_1 and L_2 be parts of the interval $[-1,1]$. We will seek the index 1 solution to both variables, i.e., $\kappa = (1,1)$. In this case opera-

tor A_{L_1, t_0^1} is uniquely solvable if the function

$$C_2(t_0^2) = \int_{L_1} \left(A_{L_2, t_0^2} \varphi(t^1, t^2) \right) dt^1$$

is known.

For operator A_{L_2, t_0^2} to be uniquely solvable one has to know the function $\int_{L_2} \varphi(t^1, t^2)\, dt^2$, which may be determined uniquely with the help of the operator A_{L_1, t_0^1} if the function $C_1(t_0^1) = A_{L_1, t_0^1}$ $(\int_{L_2} \varphi(t^1, t^2)\, dt^2)$ and the number $C = \int_{L_1}(\int_{L_2} \varphi(t^1, t^2)\, dt^2)\, dt^1$ are known.

Thus, in the case under consideration, a unique solution will be found from the system

$$A_{L_1, t_0^1}\left(A_{L_2, t_0^1} \varphi(t^1, t^2) \right) = f(t_0^1, t_0^2), \quad \int_{L_1} dt^1 \left(A_{L_2, t_0^2} \varphi(t^1, t^2) \right) = C_2(t_0^2),$$

$$A_{L_1, t_0^2}\left(\int_{L_2} \varphi(t^1, t^2)\, dt^2 \right) = C_1(t_0^1), \quad \int_{L_1} dt^1 \left(\int_{L_2} \varphi(t^1, t^2)\, dt^2 \right) = C.$$

(8.1.19)

By applying Formula (7.1.2) for the index 1, one gets

$$\varphi(t^1, t^2) = A^{-1}_{L_2, t^2}\left(A^{-1}_{L_1, t^1} f(t_0^1, t_0^2) + T_1(\alpha_1)\, \omega_1^{\alpha_1}(t^1) C_2(t_0^2) \right)$$

$$+ T_1(\alpha_2)\, \omega_1^{\alpha_2}(t^2)\left(A^{-1}_{L_1, t^1} C_1(t_0^1) + T_1(\alpha_1)\, \omega_1^{\alpha_1}(t^1) C \right).$$

(8.1.20)

Let the solution now have index 0 in the variable t^1 and index 1 in the variable t^2, i.e., $\kappa = (0, 1)$. Then the operator A_{L_1, t_0^1} is uniquely invertible for any right-hand side; to make the same true for the operator A_{L_2, t_0^2}, one has to know the function $\int_{L_2} \varphi(t^1, t^2)\, dt^2$, and hence, the index $\kappa = (0, 1)$ solution may be found uniquely from the system

$$A_{L_1, t_0^1}\left(A_{L_2, t_0^2} \varphi(t^1, t^2) \right) = f(t_0^1, t_0^2),$$

$$A_{L_1, t_0^1}\left(\int_{L_2} \varphi(t^1, t^2)\, dt^2 \right) = C_1(t_0^1). \qquad (8.1.21)$$

By again using Formula (7.1.2) for $\kappa = 0$ with respect to the variable t^1 and $\kappa = 1$ with respect to the variable t^2, one gets

$$\varphi(t^1,t^2) = A_{L_2,t^2}^{-1}\left(A_{L_1,t^1}^{-1} f(t_0^1,t_0^2)\right) + T_1(\alpha_2)\,\omega_1^{\alpha_2}(t^2)\,A_{L_1,t^1}^{-1}C_1(t_0^1)$$

$$= a_1 a_2 f(t^1,t^2) - \frac{a_2 b_1}{\pi}\,\omega_0^{\alpha_1}(t^1)\int_{L_1}\frac{f(t_0^1,t^2)\,dt_0^1}{\omega_0^{\alpha_1}(t_0^1)(t_0^1 - t^1)}$$

$$- \frac{a_1 b_2}{\pi}\,\omega_1^{\alpha_2}(t^2)\int_{L_2}\frac{f(t^1,t_0^2)\,dt_0^2}{\omega_1^{\alpha_2}(t_0^2)(t_0^2 - t^2)}$$

$$+ \frac{b_1 b_2}{\pi^2}\,\omega_0^{\alpha_1}(t^1)\,\omega_1^{\alpha_2}(t^2)$$

$$\times \iint_{L_1\times L_2}\frac{f(t_0^1,t_0^2)\,dt_0^1\,dt_0^2}{\omega_0^{\alpha_1}(t_0^1)\,\omega_1^{\alpha_2}(t_0^2)(t_0^1 - t^1)(t_0^2 - t^2)}$$

$$+ T_1(\alpha_2)\,\omega_1^{\alpha_2}(t^2)$$

$$\times \left(a_1 C_1(t^1) - \frac{b_1}{\pi}\,\omega_0^{\alpha_1}(t^1)\int_{L_1}\frac{C_1(t_0^1)\,dt_0^1}{\omega_0^{\alpha_1}(t_0^1)(t_0^1 - t^1)}\right). \qquad (8.1.22)$$

Let us stress once again the functions $C_1(t_0^1)$ and $C_2(t_0^2)$ meet condition H on the corresponding curves.

Let us indicate the solutions and the systems they can be obtained from, for all the other combinations of indices. If $\kappa = (-1,1)$, then subject to the condition

$$\int_{L_1}\frac{f(t_0^1,t_0^2)\,dt_0^1}{\omega_{-1}^{\alpha_1}(t_0^1)} \equiv \int_{L_1}\frac{C_1(t_0^1)\,dt_0^1}{\omega_{-1}^{\alpha_1}(t_0^1)} = 0, \qquad (8.1.23)$$

the system has the form (8.1.21), and the solution has the form (8.1.22).

Unique solutions of the indices $\kappa = (0,0)$, $(0,-1)$, $(-1,0)$, $(-1,-1)$ may be found from Equation (8.1.21). However, for the variable corresponding to a negative index, the right-hand side must satisfy identity (8.1.23). Solutions of the indices $\kappa = (1,0)$, $(1,-1)$ are found in a similar way.

Thus, if L_1 and L_2 are portions of the segment $[-1,1]$, then the index $\kappa = (\kappa_1, \kappa_2)$ solution $\varphi_{(\kappa_1, \kappa_2)}(t^1, t^2)$ is given by the formula

$$\varphi_{(\kappa_1, \kappa_2)}(t^1, t^2) = A^{-1}_{L_2, t^2}\Big(A^{-1}_{L_1, t^1} f(t_0^1, t_0^2) + T_{\kappa_1}(\alpha_1)\,\omega_{\kappa_1}^{\alpha_1}(t^1) C_2(t_0^2)\Big)$$

$$+ T_{\kappa_2}(\alpha_2)\,\omega_{\kappa_2}^{\alpha_2}(t^2)\Big(A^{-1}_{L_1, t^1} C_1(t_0^1)$$

$$+ T_{\kappa_1}(\alpha_1)\,\omega_{\kappa_1}^{\alpha_1}(t^1) C\Big). \tag{8.1.24}$$

Note that for the index $\kappa = (1, -1)$ the function $\varphi(t^1, t^2)$, given by Formula (8.1.24), is a solution to Equation (8.1.11) if the condition

$$\int_{L_2} \frac{f(t_0^1, t_0^2)\, dt_0^2}{\omega_{-1}^{\alpha_2}(t_0^2)} \equiv \int_{L_2} \frac{C_2(t_0^2)\, dt_0^2}{\omega_{-1}^{\alpha_2}(t_0^2)} = 0 \tag{8.1.25}$$

is fulfilled.

4. If $a_1 = a_2 = 0$ in Equation (8.1.11), then the preceding formulas for solutions of the equation provide inversion formulas for multiple singular Cauchy integrals on products of plane curves (Lifanov 1978a, 1978b, 1979).
5. If a full singular integral equation with multiple Cauchy integrals, that is, an equation of the form

$$A_{L_1, t_0^1} A_{L_2, t_0^2} \varphi(t^1, t^2) + \iint_{L_1 \times L_2} K(t_0^1, t_0^2, t^1, t^2)\,\varphi(t^1, t^2)\, dt^1\, dt^2$$

$$= f(t_0^1, t_0^2), \tag{8.1.26}$$

is considered, then by using Vekua's approach (Muskhelishvili 1952) (i.e., by solving the equation with respect to the characteristic part for a given index) it may be reduced to an equivalent (in the sense of finding solutions) integral Fredholm equation of the second kind.

8.2. NUMERICAL SOLUTION OF SINGULAR INTEGRAL EQUATIONS WITH MULTIPLE CAUCHY INTEGRALS

Consider equation

$$\iint_{L_1 \times L_2} \frac{\varphi(t^1, t^2)\, dt^1\, dt^2}{(t_0^1 - t^1)(t_0^2 - t^2)} = f(t_0^1, t_0^2), \tag{8.2.1}$$

where L_1 and L_2 are unit-radius circles. Let the sets $E^i = \{t^i_{k_i},\ k_i =$

$1, \ldots, n^i\}$ and $E_0^i = \{t_{0j_i}^i,\ j_i = 1, \ldots, n^i\}$ form a canonic division of the circle L_i, $i = 1, 2$. The following theorem is true (Lifanov 1978a).

Theorem 8.2.1. *Let the function $f(t^1, t^2) \in H$ on the torus $L_1 \times L_2$. Then between a solution to the system of linear algebraic equations,*

$$\sum_{k_1=1}^{n^1} \sum_{k_2=1}^{n^2} \frac{\varphi_{n^1 n^2}\left(t_{k_1}^1, t_{k_2}^2\right) a_{k_1}^1 a_{k_2}^2}{\left(t_{0j_1}^1 - t_{k_1}^1\right)\left(t_{0j_2}^2 - t_{k_2}^2\right)} = f\left(t_{0j_1}^1, t_{0j_2}^2\right),$$

$$j_i = 1, \ldots, n^i, i = 1, 2, \quad (8.2.2)$$

and a solution to Equation (8.2.1),

$$\varphi(t^1, t^2) = \frac{1}{\pi^4} \iint_{L_1 \times L_2} \frac{f(t_0^1, t_0^2)\, dt_0^1\, dt_0^2}{(t^1 - t_0^1)(t^2 - t_0^2)}, \quad (8.2.3)$$

the relationship

$$\varphi\left(t_{k_1}^1, t_{k_2}^2\right) - \varphi_{n^1 n^2}\left(t_{k_1}^1, t_{k_2}^2\right) \leq \theta\left(t_{k_1}^1, t_{k_2}^2\right),$$

$$k_i = 1, \ldots, n^i, i = 1, 2, \quad (8.2.4)$$

holds where the quantity $\theta(t_{k_1}^1, t_{k_2}^2)$ satisfies the inequality

$$\theta\left(t_{k_1}^1, t_{k_2}^2\right) \leq O(1/N^\lambda), \qquad \lambda > 0,\ N = \min(n^1, n^2), \quad (8.2.5)$$

and $n^i/N \leq R < +\infty$.

For the number λ we have:

1. $\lambda = \alpha - \epsilon$ *(where ϵ is however small), if $f \in H(\alpha)$ on $L_1 \times L_2$ and* $a_{k_p}^p = t_{k_p+1}^p - t_{k_p}^p$.
2. $\lambda = r + \alpha - \epsilon$, *if $f_{t^k}^{(r)} \in H(\alpha)$, $k = 1, 2$ and* $a_{k_p}^p = i[2\pi t_{k_p}^p / (2m^p + 1)]$, $n^p = 2m^p + 1$, $p - 1, 2$.

Proof. Let $\tilde{M}_{0,t^1}^{(n^1)}$ be a matrix obtained from the matrix whose determinant is $\Delta_0^{(n)}$ in System (5.1.3), by substituting t_0^1 and t^1 for t_0 and t, respectively. Then the determinant of System (8.2.2) may be written in the form

$$D_0^{(n^1, n^2)} = \prod_{m_1=1}^{n^1} \prod_{m_2=1}^{n^2} \left(a_{m_1}^1\right)^{n^2} \left(a_{m_2}^2\right)^{n^1} \Delta_0^{(n^1, n^2)}, \quad (8.2.6)$$

where $a^i_{m_i}$ is in degree n^i, $i = 1, 2$,

$$\Delta_0^{(n^1,n^2)} = \begin{vmatrix} \tilde{M}_{0,t^1}^{(n^1)} \dfrac{1}{t_{01}^2 - t_1^2} & \cdots & \tilde{M}_{0,t^1}^{(n^1)} \dfrac{1}{t_{01}^2 - t_{n^2}^2} \\ \cdots & \cdots & \cdots \\ \tilde{M}_{0,t^1}^{(n^1)} \dfrac{1}{t_{0n^2}^2 - t_1^2} & \cdots & \tilde{M}_{0,t^1}^{(n^1)} \dfrac{1}{t_{0n^2}^2 - t_{n^2}^2} \end{vmatrix}.$$

By calculating determinant $\Delta_0^{(n^1,n^2)}$ with the help of the approach used for calculating determinant $\Delta_0^{(n)}$, one arrives at $\Delta_0^{(n^1,n^2-1)}$.

Continuing the process, we get

$$D_0^{(n^1,n^2)} = \prod_{m_1,m_2=1}^{n^1,n^2} \left(a^1_{m_1}\Delta_0^{(n^1)}\right)^{n^2} \left(a^2_{m_2}\Delta_0^{(n^2)}\right)^{n^1}, \tag{8.2.7}$$

where $\Delta_0^{(n^i)}$ is obtained from $\Delta_0^{(n)}$ by substituting n^i for n and for t_k, t_{0k}, $t^i_{k_i}, t^i_{0k_i}$ respectively, $i = 1, 2$.

It is evident that $D_0^{(n^1,n^2)} \neq 0$ for any n^1 and n^2. Hence, by applying the Cramér rule to System (8.2.2), one gets

$$\varphi_{n^1n^2}\left(t^1_{k_1}, t^2_{k_2}\right) = \frac{D^{(n^1,n^2)}_{0,(k_1,k_2)}}{D_0^{(n^1,n^2)}}$$

$$= \frac{1}{a^1_{k_1}a^2_{k_2}} \sum_{j_1=1}^{n^1} \sum_{j_2=1}^{n^2} (-1)^{(k_2-1)n^1+(j_2-1)n^1+k_1+j_1}$$

$$\times \frac{\Delta^{(n^1,n^2)}_{0,(k_1,k_2),(j_1,j_2)}}{\Delta_0^{(n^1,n^2)}} f\left(t^1_{0j_1}, t^2_{0j_2}\right),$$

$$k_i = 1, \ldots, n^i, i = 1, 2, \tag{8.2.8}$$

where $D^{(n^1,n^2)}_{0,(k_1,k_2)}$ is obtained from $D_0^{(n^1,n^2)}$ by substituting the free terms* column for the column number (k_1, k_2); $\Delta^{(n^1,n^2)}_{0,(k_1,k_2),(j_1,j_2)}$ is obtained from $\Delta_0^{(n^1,n^2)}$ by striking out the column number (k_1, k_2) and the

* $(k_1, k_2) > (\tilde{k}_1, \tilde{k}_2)$ if $k_2 > \tilde{k}_2$; or $k_2 = \tilde{k}_2$, but $k_1 > \tilde{k}_1$; $(k_1, k_2) = (\tilde{k}_1, \tilde{k}_2)$ if $k_1 = \tilde{k}_1$, $k_2 = \tilde{k}_2$.

row number (j_1, j_2). Let us denote by $\tilde{M}^{(n^1)}_{0,t^1,(k_1,j_1)}$ $(\tilde{M}^{(n^1)}_{0,t^1,(k_1,0)}$; $\tilde{M}^{(n^1)}_{0,t^1,(0,j_1)})$ the matrix obtained from matrix $M^{(n^1)}_{0,t^1}$ by striking out the column number k_1 and the row number j_1 (the column number k_1 only; the row number j_1 only, respectively). Then one gets

$$\Delta^{(n^1,n^2)}_{0,(k_1,k_2),(j_1,j_2)}$$

$$= \begin{vmatrix} \tilde{M}^{(n^1)}_{0,t^1}\dfrac{1}{t^2_{01}-t^2_1} & \cdots & \tilde{M}^{(n^1)}_{0,t^1},(k_1,0)\dfrac{1}{t^2_{01}-t^2_{k_2}} & \cdots & \tilde{M}^{(n^1)}_{0,t^1}\dfrac{1}{t^2_{01}-t^2_{n^2}} \\ \cdots & \cdots & \cdots & \cdots & \cdots \\ \tilde{M}^{(n^1)}_{0,t^1,(0,j_1)}\dfrac{1}{t^2_{0j_2}-t^2_1} & \cdots & \tilde{M}^{(n^1)}_{0,t^1,(k_1,j_1)}\dfrac{1}{t^2_{0j_2}-t^2_{k_2}} & \cdots & \tilde{M}^{(n^1)}_{0,t^1,(0,j_1)}\dfrac{1}{t^2_{0j_2}-t^2_{n^2}} \\ \cdots & \cdots & \cdots & \cdots & \cdots \\ \tilde{M}^{(n^1)}_{0,t^1}\dfrac{1}{t^2_{0,n^2}-t^2_1} & \cdots & \tilde{M}^{(n^1)}_{0,t^1,(k_1,0)}\dfrac{1}{t^2_{0n^2}-t^2_{k_2}} & \cdots & \tilde{M}^{(n^1)}_{0,t^1}\dfrac{1}{t^2_{0n^2}-t^2_{n^2}} \end{vmatrix}. \tag{8.2.9}$$

Next we move the matrix column with $\tilde{M}^{(n^1)}_{0,t^1,(k_1,0)}$ and the matrix row with $\tilde{M}^{(n^1)}_{0,t^1,(0,j_1)}$ to the first positions in the columns and rows, respectively. The rearrangement does not result in a change in the sign of determinant (8.2.9). In fact, the movement of a column of determinant (8.2.9) from the said matrix column to the first position requires $n^1(k_2 - 1)$ rearrangements, whereas the movement of the entire matrix column requires $(n^1 - 1)n^1(k_2 - 1)$ such rearrangements, because no rearrangements are performed in the matrix column itself. The number thus obtained is evidently even. The transformed determinant (8.2.9) will be denoted by $\tilde{\Delta}^{(n^1,n^2)}_{0,(k_1,k_2),(j_1,j_2)}$. It can be calculated by using the following approach. First, the last matrix row is subtracted from all the preceding matrix rows save the first one. Next the last matrix row from which the j_1th row is preliminarily struck out is subtracted from the first matrix row, and common multipliers appearing in the matrix rows and columns are taken out of the determinant sign. Then the last matrix row becomes

$$\left(\tilde{M}^{(n^1)}_{0,t^1,(k_1,0)} \quad \tilde{M}^{(n^1)}_{0,t^1} \quad \cdots \quad \tilde{M}^{(n^1)}_{0,t^1}\right).$$

A similar operation will be performed in the last determinant with respect to the matrix columns. The resulting common multipliers will be taken out of the determinant sign again. Then the last row in the last determinant will have the form

$$\left(0 \quad 0 \quad \cdots \quad \tilde{M}^{(n^1)}_{0,t^1}\right).$$

By applying the rule of expanding a determinant in a matrix row, we come to the block determinant whose order is smaller by 1 as compared with the original. By repeating the preceding operations n^2 times, one gets

$$\Delta^{(n^1,n^2)}_{0,(k_1,k_2),(j_1,j_2)} = \left(\Delta_0^{(n^1)}\right)^{n^2-1}\left(\Delta^{(n^2)}_{0j_2,k_2}\right)^{n^1-1}\Delta^{(n^1)}_{0,(j_1,k_1)}\Delta^{(n^2)}_{0,(j_2,k_2)},$$

$$\Delta^{(n^2)}_{0,j_2,k_2} = \prod\prod_{\substack{1\le m<p\le n^2\\ p\ne j_2}}\left(t_{0p}^2 - t_{0m}^2\right)\prod\prod_{\substack{1\le m<p\le n^2\\ p\ne k^2}}\left(t_m^2 - t_p^2\right)\prod_{m=1}^{j_2-1}\left(t_{0m}^2 - t_{0j_2}^2\right)$$

$$\times\prod_{n=1}^{k_2-1}\left(t_{k_2}^2 - t_m^2\right)\Big/\prod_{m,p=1}^{n^2}\left(t_{0m}^2 - t_p^2\right)$$

$$= (-1)^{j_2+k_1}\,\Delta_0^{(n^2)}.$$

Thus, we have

$$\Delta^{(n^1,n^2)}_{0,(k_1,k_2),(j_1,j_2)}$$

$$= (-1)^{(n^1-1)(k_2+j_2)}\left(\Delta_0^{(n^1)}\right)^{n^2-1}\left(\Delta_0^{(n^2)}\right)^{n^1-1}\Delta^{(n^1)}_{0,(j_1,k_1)}\Delta^{(n_2)}_{0,(j_2,k_2)}. \quad (8.2.10)$$

By substituting this formula into (8.2.8), we get

$$\varphi_{n^1n^2}\left(t_{k_1}^1, t_{k_2}^2\right) = \frac{1}{a_{k_1}^1 a_{k_2}^2}\sum_{j_1=1}^{n^2}\sum_{j_2=1}^{n^2}(-1)^{k_1+j_1}\frac{\Delta^{(n^1)}_{0,(j_1,k_1)}}{\Delta_0^{(n^1)}}$$

$$\times(-1)^{k_2+j_2}\frac{\Delta^{(n^2)}_{0,(j_2,k_2)}}{\Delta_0^{(n^2)}}f\left(t_{0j_1}^1, t_{0j_2}^2\right). \quad (8.2.11)$$

The use of Formula (5.1.13) for $\kappa = 0$ results in

$$\varphi_{n^1,n^2}\left(t_{k_1}^1, t_{k_2}^2\right) = \frac{1}{a_{k_1}^1}I^{(n^1)}_{0,k_1}\frac{1}{a_{k_2}^2}I^{(n^2)}_{0,k_2}$$

$$\times\sum_{j_1=1}^{n^1}\sum_{j_2=1}^{n^2}\frac{1}{b_{0j_1}^1}I^{(n^1)}_{0,0j_1}\frac{1}{b_{0j_2}^2}I_{0,0j_2}\frac{f\left(t_{0j_1}^1, t_{0j_2}^2\right)b_{0j_1}^1 b_{0j_2}^2}{\left(t_{k_1}^1 - t_{0j_1}^1\right)\left(t_{k_2}^2 - t_{0j_2}\right)}, \quad (8.2.12)$$

where $b_{0j_k}^k = t_{0j_k+1}^k - t_{0j_k}^k$ or $b_{0j_k}^k = 2\pi i t_{0j_k}^k/(2m^k+1)$ (if $a_{k_p}^p$ have the corresponding form).

Hence, by using Formulas (6.1.10), (6.1.11), and (6.1.20), Formula (8.2.12) may be reduced to

$$\varphi_{n^1,n^2}\left(t_{k_1}^1, t_{k_2}^2\right) = \frac{1}{\pi^4} \sum_{j_1=1}^{n^1} \sum_{j_2=1}^{n^2} \frac{f\left(t_{0j_1}^1, t_{0j_2}^2\right) b_{0j_1}^1 b_{0j_2}^2}{\left(t_{k_1}^1 - t_{0j_1}^1\right)\left(t_{k_2}^2 - t_{0j_2}^2\right)} + O(N^{-1} \ln^2 N), \tag{8.2.13}$$

because if one uses Formulas (6.1.10) and (6.1.11), then

$$\sum_{j_1=1}^{n^1} \sum_{j_2=1}^{n^2} \frac{|f(t_{0j_1}, t_{0j_2}||b_{0j_1}^1||b_{0j_2}^2|}{|t_{k_1}^1 - t_{0j_1}^1||t_{k_2}^2 - t_{0j_2}^2|} \leq O(\ln^2 N),$$

and if Formulas (6.1.20) are used, then the quantity $O(N^{-1} \ln^2 N)$ must be discarded in Equation (8.2.13).

The use of the quadrature formula for multiple singular integrals terminates the proof of Theorem 8.2.1. ■

Theorem 8.2.2. *Let function $f(t_0^1, t_0^2)$ have the form*

$$f(t_0^1, t_0^2) = \frac{\psi(t_0^1, t_0^2)}{|q_1 - t_0^1|^{\nu_1}|q_2 - t_0^2|^{\nu_2}}, \tag{8.2.14}$$

where $\psi(t_0^1, t_0^2) \in H$ on $L_1 \times L_2$, $q_i \in L_i$, $\nu_i \in [0,1)$, $i = 1,2$, and the sets E^i and E_0^i form a canonic division L_i such as in the case of Theorem 6.1.2.

Then between a solution to the system of linear algebraic equations (8.2.2) and the solution (8.2.3) to Equation (8.2.1), Relationship (8.2.4) holds where $\theta(t_{k_1}^1, t_{k_2}^2)$, $k_i = 1, \ldots, n^i$, $i = 1,2$ satisfies the inequalities:

1. *For all points $(t_{k_1}^1, t_{k_2}^2) \in [L_1 \setminus O(q_1, \delta)] \times [L_2 \setminus O(q_2, \delta)]$, δ is however small:*

$$\theta\left(t_{k_1}^1, t_{k_2}^2\right) \leq O\left(\frac{1}{N^{\lambda_1}}\right), \qquad \lambda_1 > 0. \tag{8.2.15}$$

2. *For all points $(t_{k_1}^1, t_{k_2}^2)$,*

$$\sum_{k_1=1}^{n^1} \sum_{k_2=1}^{n^2} \theta\left(t_{k_1}^1, t_{k_2}^2\right)|\Delta_{k_1}^1||\Delta t_{k_2}^2| \leq O\left(\frac{1}{N^{\lambda_2}}\right), \qquad \lambda_2 > 0. \tag{8.2.16}$$

Note that if $\nu_2 = 0$, then Inequality (8.2.15) is fulfilled for all points $(t^1_{k_1}, t^2_{k_2})$ belonging to $[L_1 \setminus O(q_1, \delta)] \times L_2$. Similarly, if $\nu_1 = 0$, then $\nu_2 > 0$.
Theorem 8.2.1 may be generalized for the case when L_i in Equation (8.2.1) is a set of L_i nonintersecting circles $L_1, \ldots, L_i^{l_i}$.

Theorem 6.1.3 may also be generalized onto Equation (8.2.1); however, a singularity of the form $1/(q - t)$ may exist either for one of the coordinates or for two coordinates simultaneously. The cases must be considered separately.

Definition 8.2.1. *A function $\varphi(t^1, t^2)$ is said to belong to the class Π_i on $L_1 \times L_2$ where L_1 and L_2 are unit-radius circles if it has the form*

$$\varphi(t^1, t^2) = \frac{\psi(t^1, t^2)}{q_i - t^i}, \qquad q_i \in L_i,\ i = 1, 2,\ \psi(t^1, t^2) \in H \text{ on } L_1 \times L_2.$$

A function $\varphi(t^1, t^2)$ belongs to the class $\Pi_{1\,2}$ on $L_1 \times L_2$ if, other conditions being equal, it has the form

$$\varphi(t^1, t^2) = \frac{\psi(t^1, t^2)}{(q_1 - t^1)(q_2 - t^2)}.$$

The following theorem is true.

Theorem 8.2.3 (Lifanov 1978a). *Let function $f(t^1, t^2) \in H$ on $L_1 \times L_2$, where L_1 and L_2 are unit-radius circles. Then between solutions to the systems of linear algebraic equations,*

$$\sum_{k_1=1}^{n^1} \sum_{k_2=1}^{n^2} \frac{\varphi_{n^1 n^2}\left(t^1_{k_1}, t^2_{k_2}\right) a^1_{k_1} a^2_{k_2}}{\left(t^1_{0j_1} - t^1_{k_1}\right)\left(t^2_{0j_2} - t^2_{k_2}\right)} = f\left(t^1_{0j_1}, t^2_{0j_2}\right),$$

$$j_p = 1, \ldots, n^p,\ p = 1, 2,\ j_1 \neq j_{q_1},$$

$$\sum_{k_1=1}^{n^1} \sum_{k_2=1}^{n^2} \frac{\varphi_{n^1 n^2}\left(t_{k_1}, t_{k_2}\right) a^1_{k_1} a^2_{k_2}}{t^2_{0j_2} - t^2_{k_2}} = C_2\left(t^2_{0j_2}\right),$$

$$j_1 = j_{q_1},\ j_2 = 1, \ldots, n^2; \qquad (8.2.17)$$

$$\sum_{k_1=1}^{n^1} \sum_{k_2=1}^{n^2} \frac{\varphi_{n^1n^2}\left(t_{k_1}^1, t_{k_2}^2\right) a_{k_1}^1 a_{k_2}^2}{\left(t_{0j_1}^1 - t_{k_1}^1\right)\left(t_{0j_2}^2 - t_{k_2}^2\right)} = f\left(t_{0j_1}^1, t_{0j_2}^2\right),$$

$$j_p = 1, \dots, n^p,\ j_p \neq j_{q_p},\ p = 1, 2,$$

$$\sum_{k_1=1}^{n^1} \sum_{k_2=1}^{n^2} \frac{\varphi_{n^1n^2}\left(t_{k_1}^1, t_{k_2}^2\right) a_{k_1}^1 a_{k_2}^2}{t_{0j_1}^1 - t_{k_1}^1} = C_1\left(t_{0j_1}^1\right),$$

$$j_1 = 1, \dots, n^1,\ j_1 \neq j_{q_1},\ j_2 = j_{q_2},$$

$$\sum_{k_1=1}^{n^1} \sum_{k_2=1}^{n^2} \frac{\varphi_{n^1n^2}\left(t_{k_1}^1, t_{k_2}^2\right) a_{k_1}^1 a_{k_2}^2}{t_{0j_2}^2 - t_{k_2}^2} = C_2\left(t_{0j_2}^2\right),$$

$$j_2 = 1, \dots, n^2,\ j_2 \neq j_{q_2},\ j_1 = j_{q_1},$$

$$\sum_{k_1=1}^{n^1} \sum_{k_2=1}^{n^2} \varphi_{n^1n^2}\left(t_{k_1}^1, t_{k_2}^2\right) a_{k_1}^1 a_{k_2}^2 = C,$$

$$j_1 = j_{q_1},\ j_2 = j_{q_2}, \quad (8.2.18)$$

where the sets E^p *and* E_0^p *form a canonic division* L_p, $p = 1, 2$ (*for* (8.2.17), $q_1 = t_{0j_{q_1}}^1$, *and for* (8.2.18), $q_p = t_{0j_{q_p}}^p$, $p = 1, 2$), *and the solutions* $\varphi(t^1, t^2)$ *to Equation* (8.2.1),

$$\varphi(t^1, t^2) = \frac{1}{\pi^4} \frac{1}{q - t^1} \left[\iint_{L_1 \times L_2} \frac{(q_1 - t_0^1) f(t_0^1, t_0^2)\, dt_0^1\, dt_0^2}{(t^1 - t_0^1)(t^2 - t_0^2)} - \pi i \int_{L_2} \frac{C_2(t_0^2)\, dt_0^2}{t^2 - t_0^2} \right], \quad (8.2.19)$$

$$\iint_{L_1 \times L_2} \frac{\varphi(t^1, t^2)\, dt^1\, dt^2}{t_0^2 - t^2} = C_2(t_0^2), \qquad C_2(t_0^2) \in H \text{ on } L_2,$$

$$\varphi(t^1, t^2) = \frac{1}{\pi^4 (q_1 - t^1)(q_2 - t^2)} \times \left[\iint_{L_1 \times L_2} \frac{(q_1 - t_0^1)(q_2 - t_0^2) f(t_0^1, t_0^2)\, dt_0^1\, dt_0^2}{(t^1 - t_0^1)(t^2 - t_0^2)} \right.$$

$$- \pi i \int_{L_1} \frac{(q_1 - t_0^1) C_1(t_0^1)\, dt_0^1}{t^1 - t_0^1}$$

$$\left. - \pi i \int_{L_2} \frac{(q_2 - t_0^2) C_2(t_0^2)\, dt_0^2}{t^2 - t_0^2} - \pi^2 C \right],$$

$$\iint_{L_1 \times L_2} \frac{\varphi(t^1, t^2)\, dt^1\, dt^2}{t_0^i - t^i} = C_i(t_0^i), \qquad C_i(t_0^i) \in H \text{ on } L_i,\ i = 1, 2,$$

$$\iint_{L_1 \times L_2} \varphi(t^1, t^2)\, dt^1\, dt^2 = C, \tag{8.2.20}$$

respectively, Relationship (8.2.4) *holds in which the quantity* $\theta(t_{k_1}^2, t_{k_2}^2)$ *satisfies Inequalities* (8.2.16) *and* (8.2.15) *at all points* $(t_{k_1}^1, t_{k_2}^2) \in L_2 \times [L_1 \setminus O(q_1, \delta)]$ *in the former case and at all points* $(t_{k_1}^1, t_{k_2}^2) \in [L_1 \setminus O(q_1, \delta)] \times [L_2 \setminus O(q_2, \delta)]$ *in the latter case. Note that the numbers* λ_1 *and* λ_2 *depend on the differential properties of the functions* $f(t^1, t^2)$, $C_1(t_0^1)$, *and* $C_2(t_0^2)$ *as in Theorem* 8.2.1 (*if* $a_{k_1}^1, a_{k_2}^2$ *are chosen in the same manner*).

Note 8.2.1. Theorems 8.2.1–8.2.3 may be readily generalized onto the equation

$$\int \cdots \int_{L_1 \times \cdots \times L_l} \frac{\varphi(t)\, dt}{((t_0 - t))} = f(t_0), \tag{8.2.21}$$

where $L_1, \ldots, L_l$ are circles whose solution in the class of functions H on $L_1 \times \cdots \times L_l$ [if $f(t_0)$ belongs to the same class] is given by the function (see (2.1.12))

$$\varphi(t) = \frac{(-1)^l}{\pi^{2l}} \int \cdots \int_{L_1 \times \cdots \times L_l} \frac{f(t_0)\, dt_0}{((t - t_0))}. \tag{8.2.22}$$

In order to write the subsequent results in a compact form and to emphasize the special structure of the matrices of Systems (8.2.2), (8.2.17), and (8.2.18) of linear algebraic equations, one should be reminded of the following definitions (Proskuryakov 1967).

The *right-hand direct product* of matrix A of the order of $m \times m$ and matrix B of the order $n \times n$ is an array $\tilde{C} = A \dot{\times} B$ whose element posed at the intersection of the jth row and the ith column is defined by $\tilde{C}_{ji} = a_{ji}B$.

The *left-hand direct product* of the same matrices is an array $D = A \dot{\times} B$, where $D_{ji} = Ab_{ji}$.

The products possess the properties

$$A \times\!\cdot\, B = B \dot{\times} A, \qquad A \times\!\cdot\, (B \times\!\cdot\, \tilde{C}) = (A \times\!\cdot\, B) \times\!\cdot\, \tilde{C},$$

$$(A \times\!\cdot\, B)^{-1} = A^{-1} \times\!\cdot\, B^{-1}.$$

Similar properties are also valid for the left-hand product.

For any direct product,

$$|A \times B| = |A|^n |b|^m.$$

Let us write in matrix form the systems of linear algebraic equations for characteristic equations on a segment and a circle for the case of uniform grids:

$$1. \quad M_{0,t}^{(n)} \varphi_t = f_{t_0}^0,$$

$$2. \quad M_{1,t}^{(n)} \varphi_t = \begin{pmatrix} f_{t_0}^1 \\ C \end{pmatrix},$$

$$3. \quad M_{-1,t}^{(n)} \begin{pmatrix} \lambda_{0n} \\ \varphi_t \end{pmatrix} = f_{t_0}^{-1},$$

where

$$M_{0,t}^{(n)} = \left(a_{j,k}^0\right)_{j,k=1}^n, \qquad a_{j,k}^0 = \frac{a_k}{t_{0j} - t_k}, \qquad j,k = 1,\ldots,n,$$

$$M_{1,t}^{(n)} = \left(a_{j,k}^1\right)_{j,k=1}^n, \qquad a_{j,k}^1 = \frac{a_k}{t_{0j} - t_k},$$

$$j,k = 1,\ldots,n,\ j \neq n,\ a_{n,k}^1 = a_k,\ j = n;\ k = 1,\ldots,n,$$

$$M_{-1,t}^{(n)} = \left(a_{j,k}^{-1}\right)_{j=1,k=0}^{n+1,n}, \qquad a_{j,k}^{-1} = \frac{a_k}{t_{0j} - t_k},$$

$$k = 1,\ldots,n,\ j = 1,\ldots,n+1,\ a_{j,0}^{-1} = 1,\ k = 0,\ j = 1,\ldots,n+1,$$

$$\varphi_t = (\varphi_n(t_k))_{k=1}^n, \qquad \begin{pmatrix} \lambda_{0n} \\ \varphi_t \end{pmatrix} = \begin{pmatrix} \lambda_{0n} \\ \varphi_n(t_k) \end{pmatrix}_{k=1}^n,$$

$$f_{t_0}^0 = (f(t_{0j}))_{j=1}^n \qquad \begin{pmatrix} f_{t_0}^1 \\ C \end{pmatrix} = \begin{pmatrix} f(t_{0j}) \\ C \end{pmatrix}_{j=1}^{n-1}, \qquad f_{t_0}^{-1} = (f(t_{0j}))_{j=1}^{n+1},$$

Here the points numbered earlier in j starting from zero are numbered starting from 1. If the grids are nonuniform, then φ_t must be replaced by ψ_t $[\varphi_t = \omega(t)\psi_t]$.

In light of the latter definitions, the systems of linear algebraic equations (8.2.2), (8.2.17), and (8.2.18) for a torus may be presented in the form

$$\left(M_{0,t^1}^{(n^1)} \times M_{0,t^2}^{(n^2)}\right)\varphi_{t^1,t^2} = f_{t^1,t^2}^{0,0}, \tag{8.2.2$'$}$$

$$\left(M_{1,t^1}^{(n^1)} \times M_{0,t^2}^{(n^2)}\right)\varphi_{t^1,t^2} = \begin{pmatrix} f_{t_0^1,t_0^2}^{1,0} \\ C_{n^1,t_0^2} \end{pmatrix}, \tag{8.2.17$'$}$$

$$\left(M_{1,t^1}^{(n^1)} \times M_{1,t^2}^{(n^2)}\right)\varphi_{t^1,t^2} = \begin{pmatrix} f_{t_0^1,t_0^2}^{1,1} \\ C_{n^1,t_0^2} \\ C_{t_0^1,n^2} \\ C_{n^1,n^2} \end{pmatrix}, \tag{8.2.18$'$}$$

$$\varphi_{t^1,t^2} = \left(\varphi_{n^1,n^2}\left(t_{k_1}^1, t_{k_2}^2\right)\right)_{k_1,k_2=1}^{n^1,n^2},$$

where $j_{q_i} = n^i$, $i = 1, 2$, and the pairs (k_1, k_2) are arranged in lexicographic order.

In a similar sense, we treat the right-hand side matrix columns appearing in Systems (8.2.2′), (8.2.17′), and (8.2.18′).

Let us next consider a characteristic integral equation of the second kind with multiple Cauchy singular integrals (see Equation (8.1.12) for the case when both L_1 and L_2 are circles). The following theorem is true.

Theorem 8.2.4. *Let the function $f(t_0^1, t_0^2) \in H_r(\alpha)$ on the torus $L_1 \times L_2$. Then between a solution to the system of linear algebraic equations,*

$$\frac{b_1 b_2}{\pi^2}\left(M_{0,t^1}^{(n^1)} \times M_{0,t^2}^{(n^2)}\right)\varphi_{t^1,t^2} = f_{t_0^1,t_0^2}^{0,0}, \tag{8.2.23}$$

where $M_{0,t^1}^{(n^1)}$ and $M_{0,t^2}^{(n^2)}$ are matrices obtained from the matrix of System (7.2.3) *by subscribing the letters or, in other words,*

$$t^p_{0m_p} = t^p_{m_p} \exp\left\{i \frac{\pi + \psi_p}{2n^p + 1}\right\}, \qquad e^{i\psi_p} = \frac{(-a_p + b_p i)}{(a_p + b_p i)},$$

$$t^p_{m_p}, m_p = 0, 1, \ldots, n^p,$$

divide the circle L_p, $p = 1, 2$, *into equal parts, and* $a_{k_p} = 2\pi i t^p_{k_p}/(2n^p + 1)$, *and the solution* $\varphi(t^1, t^2)$ (*see Equation* (8.2.13) *to Equation* (8.2.12), *an inequality of the form* (8.2.4) *holds where the quantity* $\theta(t^1_{k_1}, t^2_{k_2})$ *satisfies Inequality* (8.2.5) *with* $\lambda = r + \alpha - \epsilon$ *and* $\epsilon > 0$ *being however small.*

Now let L_1 and L_2 associated with Equation (8.1.12) be a circle and a segment, respectively. Then the following theorem is true.

Theorem 8.2.5. *Let* $f(t^1, t^2)$ *appearing in Equation* (8.1.12) *belong to* H *on* $L_1 \times L_2$, *and* $a_1 = a_2 = 0$, $b_1 = b_2 = 1$. *Then Relationship* (8.2.4) *holds, on the one hand, between solutions to the systems of linear algebraic equations,*

$$\left(M^{(n^1)\cdot}_{0,t^1} \times M^{(n^2)}_{0,t^2}\right)\varphi_{t^1,t^2} = f^{0,0}_{t^1_0,t^2_0}, \tag{8.2.24}$$

$$\left(M^{(n^1)\cdot}_{0,t^1} \times M^{(n^2)}_{1,t^2}\right)\varphi_{t^1,t^2} = \begin{pmatrix} f^{0,1}_{t^1_0,t^2_0} \\ C_{t^1_0,n^2} \end{pmatrix}, \tag{8.2.25}$$

$$\left(M^{(n^1)\cdot}_{0,t^1} \times M^{(n^2)}_{-1,t^2}\right)\begin{pmatrix} \gamma_{t^1,0n^2} \\ \varphi_{t^1,t^2} \end{pmatrix} = f^{0,-1}_{t^1_0,t^2_0}, \tag{8.2.26}$$

where $M^{(n^1)}_{0,t^1}$ *is the matrix of System* (7.2.3), *and the matrices* $M^{(n^2)}_{\kappa,t^2}$, $\kappa = 1, -1$, *are the matrices of Systems* (5.1.3)–(5.1.5) *multiplied by* $1/\pi$, *and the solutions* $\varphi(t^1, t^2)$ *of an equation under consideration,*

$$\varphi(t^1, t^2) = \frac{1}{\pi^2}\sqrt{\frac{1 - t^2}{1 + t^2}} \iint_{L_1 \times L_2} \sqrt{\frac{1 + t^2_0}{1 - t^2_0}} \frac{f(t^1_0, t^2_0)\, dt^1_0\, dt^2_0}{(t^1 - t^1_0)(t^2 - t^2_0)}, \tag{8.2.27}$$

$$\varphi(t^1, t^2) = \frac{1}{\pi^2\sqrt{1 - (t^2)^2}}$$

$$\times\left[\iint_{L_1 \times L_2} \frac{\sqrt{1 - (t^2_0)^2}\, f(t^1_0, t^2_0)\, dt^1_0\, dt^2_0}{(t^1 - t^2_0)(t^2 - t^2_0)}\right.$$

$$- \pi \int_{L_1} \frac{C_1(t_0^1)\, dt_0^1}{(t^1 - t_0^1)} \Bigg], \tag{8.2.28}$$

$$\varphi(t^1, t^2) = \frac{1}{\pi^2} \sqrt{1 - (t^2)^2} \iint_{L_1 \times L_2} \frac{f(t_0^1, t_0^2)\, dt_0^1\, dt_0^2}{\sqrt{1 - (t_0^2)^2}\,(t^1 - t_0^1)(t^2 - t_0^2)}, \tag{8.2.29}$$

where in the case of Equation (8.2.28) *function* $\varphi(t^1, t^2)$ *meets the condition*

$$\frac{1}{\pi^2} \iint_{L_1 \times L_2} \frac{\varphi(t^1, t^2)\, dt^1\, dt^2}{t_0^1 - t^1} = C_1(t_0^1), \tag{8.2.30}$$

and in the case of Equation (8.2.29) *function* $\varphi(t^1, t^2)$ *is a solution to the original equation subject to condition* (8.1.18). *In Equation* (8.2.4) *the quantity* $\theta(t_{k_1}^1, t_{k_2}^2)$ *satisfies both Inequality* (8.2.15) *for the points* $(t_{k_1}^1, t_{k_2}^2) \in L_1 \times [-1 + \delta, 1 - \delta]$ *and Inequality* (8.2.16).

Note that System (8.2.25) has the form

$$\frac{1}{\pi^2} \sum_{k_1=1}^{n^1} \sum_{k_2=1}^{n^2} \frac{\varphi_{n^1 n^2}\left(t_{k_1}^1, t_{k_2}^2\right) a_{k_1} h}{\left(t_{0j_1}^1 - t_{k_1}^1\right)\left(t_{0j_2}^2 - t_{k_2}^2\right)} = f\left(t_{0j_1}^1, t_{0j_2}^2\right),$$

$$j_1 = 1, \ldots, n^1,\ j_2 = 1, \ldots, n^2 - 1,$$

$$\frac{1}{\pi^2} \sum_{k_1=1}^{n^1} \sum_{k_2=1}^{n^2} \frac{\varphi_{n^1 n^2}\left(t_{k_1}^1, t_{k_2}^2\right) a_{k_1} h}{t_{0j_1}^1 - t_{k_1}^1} = C_1\left(t_{0j_1}^1\right),$$

$$j_1 = 1, \ldots, n^1,\ j_2 = n^2. \tag{8.2.31}$$

In order to ensure a higher-order convergence of an approximate solution to the accurate one, one has to account for the singularity. For example, in the case under consideration with L_1 and L_2 being a circle and a segment respectively, function $\varphi(t^1, t^2)$ may be represented in the form

$$\varphi(t^1, t^2) + \omega_\kappa(t^2)\psi(t^1, t^2), \tag{8.2.32}$$

where $\omega_\kappa(t^2)$ is of the same form as in (5.2.1).

Then, $\psi_{t^1,t^2} = (\psi_{n^1,n^2}(t^1_{k_1}, t^2_{k_2}))^{n^1,n^2}_{k_1,k_2=1}$, for φ_{t^1,t^2} entering Systems (8.2.24)–(8.2.26) must be substituted and the matrices $m^{(n^2)}_{\kappa,t^2}$, $\kappa = 0, 1, -1$, should be replaced by the matrices of the systems of linear algebraic equations (5.2.16), (5.2.2), and (5.2.15) multipied by $1/\pi$. The rate of convergence of an approximate solution to the accurate one, i.e., the numbers λ_1 and λ_2 in formulas of the form (8.2.15) and (8.2.16) for all points $(t^1_{k_1}, t^2_{k_2}) \in L_1 \times L_2$, will be determined by the order of approximation of Cauchy double integrals by the corresponding quadrature sums.

Next let L_1 and L_2 in Equation (8.2.12) be segments, and $a_1 = a_2 = 0$, $b_1 = b_2 = 1$. Then, for uniform grids, the index $\kappa = (0, 0)$, $(0, 1)$, and $(0, -1)$ solutions must be sought from systems of linear algebraic equations of the form (8.2.24)–(8.2.26), where $M^{(n^1)}_{0,t^1}$ is the System (5.1.3) matrix multiplied by $1/\pi$. For index $\kappa = (1, 1)$, $(1, -1)$, and $(-1, -1)$ the systems of linear algebraic equations have the form

$$\left(M^{(n^1)}_{1,t^1} \times M^{(n^2)}_{1,t^2}\right)\varphi_{t^1,t^2} = \begin{pmatrix} f^{1,1}_{t^1_0,t^2_0} \\ C_{n^1,t^2_0} \\ C_{t^1_0,n^2} \\ C_{n^1,n^2} \end{pmatrix}, \tag{8.2.33}$$

$$\left(M^{(n^1)}_{1,t^1} \times M^{(n^2)}_{-1,t^2}\right)\begin{pmatrix} \gamma_{t^1,0n^2} \\ \varphi_{t^1,t^2} \end{pmatrix} = \begin{pmatrix} f^{1,-1}_{t^1_0,t^2_0} \\ C_{n^1,t^2_0} \end{pmatrix}, \tag{8.2.34}$$

$$\left(M^{(n^1)}_{-1,t^1} \times M^{(n^2)}_{-1,t^2}\right)\begin{pmatrix} \gamma_{0n^1,0n^2} \\ \gamma_{t^1,0n^2} \\ \gamma_{0n^1,t^2} \\ \varphi_{t^1,t^2} \end{pmatrix} = f^{-1,-1}_{t^1_0,t^2_0}, \tag{8.2.35}$$

where M_{κ,t^p}, $\kappa = 0, 1, -1$, $p = 1, 2$, are the Systems (5.1.3)–(5.1.5) matrices multiplied by $1/\pi$. Between solutions to these systems and the corresponding solutions of the characteristic singular integral equation of the first kind with a double Cauchy integral, Relationship (8.2.4) holds where the quantity $\theta(t^1_{k_1}, t^2_{k_2})$ satisfies both Inequality (8.2.15) for all points $(t^1_{k_1}, t^2_{k_2}) \in [-1 + \delta, 1 - \delta] \times [-1 + \delta, 1 - \delta]$ and Inequality (8.2.16).

Let us consider in greater detail System (8.2.35):

$$\gamma_{0n^1,0n^2} + \sum_{k_1=1}^{n^1} \frac{\gamma_{t^1_{k_1},0n^2} h_1}{t^1_{0j_1} - t^1_{k_1}} + \sum_{k_2=1}^{n^2} \frac{\gamma_{0n^1,t^2_{k_2}} h_2}{t^2_{0j_2} - t^2_{k_2}}$$

$$+\sum_{k_1=1}^{n^1}\sum_{k_2=1}^{n^2}\frac{\varphi_{n^1n^2}\left(t_{k_1}^1,t_{k_2}^2\right)h_1h_2}{\left(t_{0j_1}^1-t_{k_1}^1\right)\left(t_{0j^2}^2-t_{k_2}^2\right)}=f\left(t_{0j_1}^1,t_{0j_2}^2\right),$$

$$j_p=0,1,\ldots,n^p,\ p=1,2,\quad (8.2.36)$$

where $\gamma_{0n^1,0n^2}$, $\gamma_{t_{k_1}^1}$, and $\gamma_{0n^1,t_{k_2}^2}$ are the regularizing factors.

Similar systems may be constructed for indices $\kappa=(1,0)$, $(-1,0)$, and $(-1,1)$.

If nonuniform grids are used, then the solution must be presented in the form

$$\varphi(t^1,t^2)=\omega_{\kappa_1}(t^1)\,\omega_{\kappa_2}(t^2)\,\psi(t^1,t^2),\quad (8.2.37)$$

where $\kappa_1,\kappa_2=0,1,-1$. In Systems (8.2.24)–(8.2.26) and (8.2.33)–(8.2.35), ψ_{t^1,t^2} must be substituted for φ_{t^1,t^2}, and the matrices of Systems (5.2.16), (5.2.2), and (5.2.15) must be substituted for the matrices $M_{\kappa,t^p}^{(n^p)}$, $p=1,2$; $\kappa=0,1,-1$.

Now let a_p and b_p entering Equation (8.1.12) be real, $a_p^2+b_p^2=1$, $p=1,2$, and L_1 and L_2 be either a segment or a circle. Because the equation was constructed by using a product of one-dimensional singular operators, the corresponding systems of linear algebraic equations must be constructed with the help of the left-hand-side direct product of matrices of corresponding systems of linear algebraic equations for one-dimensional characteristic singular integral equations. Thus, the systems of linear algebraic equations will have the form of Systems (8.2.24)–(8.2.26) and (8.2.33)–(8.2.35), where ψ_{t^1,t^2} is substituted for φ_{t^1,t^2}, and the matrices $M_{\kappa,t^p}^{(n^p)}$, $p=1,2$; $\kappa=0,1,-1$ are the matrices of Systems (7.1.7), (7.1.9), (7.1.10), or (7.2.3).

Next we consider full equation (8.1.26) with a regular kernel, for which the systems of linear algebraic equations may be obtained from the corresponding systems for characteristic equation (8.1.12) by adding the terms

$$\sum_{k_1=1}^{n^1}\sum_{k_2=1}^{n^2}K\left(t_{0m_1}^1,t_{0m_2}^2,t_{k_1}^1,t_{k_2}^2\right)\psi_{n^1n^2}\left(t_{k_1}^1,t_{k_2}^2\right)a_{k_1}a_{k_2},$$

$$m_p=1,\ldots,n^p-\kappa_p.$$

We will require that Equation (8.1.26) have a unique solution subject to corresponding conditions imposed onto the characteristic part. In this case, the system of linear algebraic equations for the equation is equivalent to a system of linear algebraic equations approximating the corresponding

Fredholm integral equation of the second kind, to which the preceding equation is equivalent and that has a unique solution. Thus, an approximate solution to Equation (8.1.26) is seen to converge to the corresponding accurate solution.

8.3. ABOUT AN INTEGRODIFFERENTIAL EQUATION

The following two-dimensional integrodifferential equation is used in aerodynamics to calculate approximately aerodynamic characteristics of a thin rectangular finite-span wing in stationary flow (Bisplinghoff et al. 1955):

$$\int_{-b}^{b}\int_{-l}^{l}\gamma_z'(x,z)\frac{x_0-x+\sqrt{b^2+(z_0-z)^2}}{(x_0-x)(z_0-z)}\,dx\,dz=f(x_0,z_0), \quad (8.3.1)$$

where $x_0\in(-b,b)$, $z_0\in(-l,l)$, and $f(x,z)\in H$ on $[-b,b]\times[-l,l]$.

In the class of functions $\gamma(x,z)$ meeting the conditions

$$\gamma(x,l)\equiv\gamma(x,-l)\equiv\gamma(b,z)\equiv 0,$$

$$\gamma_z'(x,l)=\infty, \qquad \gamma_z'(x,-l)=\infty, \qquad \gamma(-b,z)=\infty, \quad (8.3.2)$$

characteristic of aerodynamics, the equation is equivalent to the following Fredholm equation of the second kind:

$$\gamma(t,\xi)+\frac{1}{\pi l}\sqrt{\frac{b-t}{b+t}}\int_{-b}^{b}\gamma(x,\xi)\,dx-\frac{1}{\pi^2}\int_{-l}^{l}\gamma(t,z)K(\xi,z)\,dz$$

$$=\frac{1}{\pi^4 b}\sqrt{\frac{b-t}{b+t}}\int_{-b}^{b}\int_{-l}^{l}\sqrt{\frac{b+x_0}{b-x_0}}\,\frac{\psi_l(\xi,z_0)f(x_0,z_0)\,dx_0\,dz_0}{t-x_0},$$

$$K(\xi,z)=\int_{-l}^{l}\frac{\psi_l(\xi,z_0)\,dz_0}{\left(b+\left[b^2+(z_0-z)^2\right]^{1/2}\right)\sqrt{b^2+(z_0-z)^2}},$$

$$\psi_l(\xi,z_0)=\sqrt{1-z_0^2}\int_{-l}^{\xi}\frac{d\tau}{\sqrt{1-\tau^2}\,(\tau-z_0)},$$

$$\psi_l(-l,z_0)\equiv\psi_l(l,z_0)\equiv 0. \quad (8.3.3)$$

Note 5.3.4 implies that the following theorem is true.

Theorem 8.3.1. *If Equation* (8.3.3), *and hence, Equation* (8.3.1), *has a unique solution within the required class of functions, then between a solution to the system of linear algebraic equations*

$$\sum_{i=1}^{n}\sum_{k=1}^{N}\gamma_{n,N}(x_i,z_{0k})\left[\frac{x_{0j}-x_i+\sqrt{b^2+(z_{0m}-z_k)^2}}{(x_{0j}-x_i)(z_{0m}-z_k)}\right.$$

$$\left.-\frac{x_{0j}-x_i+\sqrt{b^2+(z_{0m}-z_{k+1})^2}}{(x_{0j}-x_i)(z_{0m}-z_{k+1})}\right]h_1=f(x_{0j},z_{0m}),$$

$$j=1,\ldots,n,\ m=1,\ldots,N, \quad (8.3.4)$$

where $x_i=-b+ih_1$, $h_1=2b/(n+1)$, $x_{0i}=x_i+h_1/2$, $i=1,\ldots,n$, $z_k=-l+(k-1)h_2$, $h_2=2l/N$, $z_{0k}=z_k+h_2/2$, $k=1,\ldots,N$, $z_{N+1}=l$, *and the solution* $\gamma(x,z)$ *to Equation* (8.3.1) *subject to Conditions* (8.3.2), *the relationship*

$$|\gamma(x_i,z_{0k})-\gamma_{n,N}(x_i,z_{0k})|\le\theta(x_i,z_{0k}),$$

$$i=1,\ldots,n,\ k=1,\ldots,N, \quad (8.3.5)$$

holds where quantity $\theta(x_i,z_{0k})$ *meets the following conditions*:

1. *For all points* $(x_i,z_{0k})\in[-b+\delta,b-\delta]\times[-l,l]$,

$$\theta(x_i,z_{0k})=O_\delta(h^{\lambda_1}),\qquad \lambda_1>0. \quad (8.3.6)$$

2. *For all points* $(x_i,z_{0k})\in[-b,b]\times[-l,l]$,

$$\sum_{i=1}^{n}\sum_{k=1}^{N}\theta(x_i,z_{0k})h_1h_2=O(h^{\lambda_2}),\qquad \lambda_2>0, \quad (8.3.7)$$

where $h=\max(h_1,h_2)$, $h/h_i\le R<+\infty$, $i=1,2$ *for* $h\to 0$.

Proof. It suffices to prove that System (8.3.4) is equivalent to the system of linear algebraic equations approximating Equation (8.3.3). To do this we write System (8.3.4) in the form

$$\sum_{i=1}^{n} \sum_{k=1}^{N} \gamma_{n,N}(x_i, z_{0k}) \frac{bh_1}{x_{0j} - x_i} \left(\frac{1}{z_{0m} - z_k} - \frac{1}{z_{0m} - z_{k+1}} \right)$$

$$= f(x_{0j}, z_{0m}) - \sum_{\nu=1}^{n} \sum_{\mu=1}^{N} \gamma_{n,N}(x_\nu, z_{0\mu})$$

$$\times \left[\frac{x_{0j} - x_\nu + \sqrt{b^2 + (z_{0m} - z_\mu)^2} - b}{(x_{0j} - x_\nu)(z_{0m} - z_k)} \right.$$

$$\left. - \frac{x_{0j} - x_\nu + \sqrt{b^2 + (z_{0m} - z_{\mu+1})^2} - b}{(x_{0j} - x_\nu)(z_{0m} - z_{k+1})} \right] h_1,$$

$$j = 1, \ldots, n, \; m = 1, \ldots, N. \quad (8.3.8)$$

The matrix of System (8.3.8) with respect to the unknowns on the left-hand side is a left-hand side direct product of the well-conditioned matrices $M_{0,x}^{(n)}$ and $M_{1,z}^{(N)}$ of Systems (5.1.3) and (5.1.55), respectively. By solving the system with respect to the left-hand-side unknowns and performing appropriate transformations, one arrives at the system of linear algebraic equations approximating Equation (8.3.3). ■

This is a Fredholm equation of the second kind. Because it is physically clear that Equation (8.3.1) has a unique solution, Equation (8.3.3) also has this feature. Thus, starting from certain sufficiently large values of n and N, the system of linear algebraic equations (8.3.4) becomes solvable, the solution converging to the corresponding solution of Equation (8.3.1).

Part III: Application of the Method of Discrete Vortices to Aerodynamics: Verification of the Method

9

Formulation of Aerodynamic Problems and Discrete Vortex Systems

9.1. FORMULATION OF AERODYNAMIC PROBLEMS

Here we consider steady and unsteady motion of a wing of an arbitrary plan form in an inviscid incompressible fluid (see Figure 9.1).

In aerodynamic problems both the form of a body and the law of its motion are supposed to be known. If a body under study is elastic, then the law of deformation is also supposed to be specified.* The conditions under which a motion (flight) proceeds are also assumed to be known. As a rule, the medium through which a body travels is supposed to be boundless and initially undisturbed. However, one may consider the motion of a body through a medium disturbed by wind, currents, atmospheric turbulence, etc. In this case, one has to determine disturbed flow velocity $\mathbf{V}(x, y, z, t) = \{V_x, V_y, V_z\}$ and pressure $p(x, y, z, t)$. For calculating the four required functions V_x, V_y, V_z, and p, one has three Euler equations as well as the continuity equation (Golubev 1949).

Of paramount importance are boundary conditions at the surface of a body. If a medium is viscous, then the "no-slip" condition must be employed, i.e., at the surface of a body $\mathbf{V} = 0$. For an inviscid fluid the only no-penetration condition is used according to which the normal component of the flow velocity is equal to zero at the surface of a body:

$$\mathbf{V}_{\text{rel}} \cdot \mathbf{n} = 0, \qquad (x, y, z) \in \sigma, \tag{9.1.1}$$

* Aeroelasticity studies problems where both the deformations of a body and the flow of a fluid have to be determined (G. Ch.).

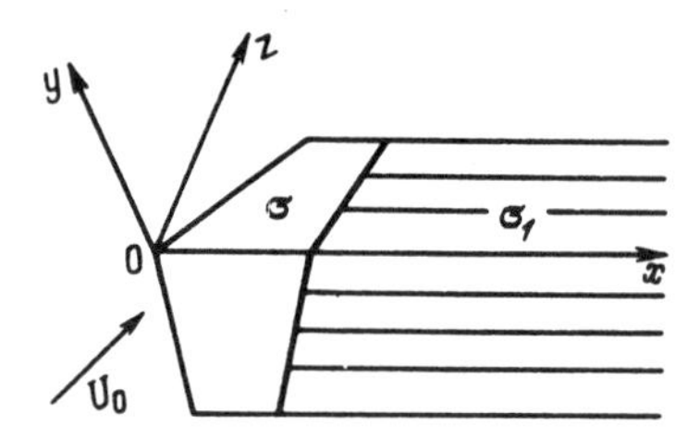

FIGURE 9.1. Schematic representation of a finite-span wing. The circulatory problem.

where **n** is a unit vector of the normal to the body surface σ at a point (x, y, z).

It should be pointed out that there exists a class of problems associated with flows past parachutes and paragliders whose surfaces are formed by thin fabrics. In this case, the surface is permeable, but the law of cross flow is known. It is determined experimentally and is actually a relationship characterizing a given material. Usually, the relationship connects the cross-flow velocity and the pressure drop across a fabric.

Further simplification of the formulation of the general problem is closely related to the following experimental evidence. Flow past a body is accompanied by formation of an aerodynamic vortex wake within which the fluid participates not only in translational and deformational motions, but in rotational motion as well. Outside the region flow is irrotational. In this case the perturbed flow outside a body and the wake is potential, i.e., may be characterized not by the three unknown functions V_x, V_y, and V_z, but by only one function $\Phi(x, y, z, t)$, called the flow velocity potential. In this case,

$$V_x = \partial\Phi/\partial x, \qquad V_y = \partial\Phi/\partial y, \qquad V_z = \partial\Phi/\partial z, \tag{9.1.2}$$

and the continuity equation becomes the Laplace equation

$$\frac{\partial^2\Phi}{\partial x^2} + \frac{\partial^2\Phi}{\partial y^2} + \frac{\partial^2\Phi}{\partial z^2} = 0 \tag{9.1.3}$$

outside σ and σ_1, where σ_1 is the wake formed behind a moving body. Because it is natural to suppose that far from the surface of a body σ and its wake σ_1 flow disturbances die away, a solution $\Phi(x, y, z, t)$ to Equation (9.1.3) must meet the condition:

$$\lim_{r\to\infty} \nabla\Phi = 0, \qquad r = \sqrt{x^2 + y^2 + z^2}, \tag{9.1.4}$$

for points (x, y, z) located infinitely far from the body σ and its wake σ_1.

The Euler equations of motion may be integrated, resulting in the well-known relationship between pressure p and derivatives of the velocity potential Φ with respect to coordinates and time—the Cauchy–Lagrange integral

$$\frac{\partial \Phi}{\partial t} + \frac{\mathbf{V}_{\text{rel}}^2}{2} - \frac{\mathbf{V}^{*2}}{2} + \frac{p - p_\infty}{\rho} = 0, \qquad (9.1.4')$$

where $\mathbf{V}_{\text{rel}}$ and $\mathbf{V}^*$ are the relative and translational velocity components, respectively, ρ is the density of the fluid (assumed to be constant), and p_∞ is the known pressure at infinity. Thus, the unknown function $p(x,y,z,t)$ may also be excluded from the general problem formulation.

The following major peculiarity of the general problem formulation should be mentioned. When constructing the vortex wake behind a body, one has to take into account general theorems of hydrodynamics that have been ignored until now, which follow from the general properties of flow velocity fields and the corresponding equations.

The properties are:

a. In steady flows vortices are directed along streamlines.
b. In unsteady flows vortices shed from a body (the so-called *free vortices*) move along the paths of liquid particles and together with the latter.
c. Velocity circulation over any closed contour is independent of time.
d. A change in the circulation of a bound vortex is accompanied by shedding a free vortex.
e. Because the vortex wake σ_1 does not develop any lift, the pressure across it must stay continuous:

$$p_- = p_+, \qquad (x, y, z) \in \sigma_1. \qquad (9.1.5)$$

Here the subscripts plus $(+)$ and minus $(-)$ refer to opposite sides of the surface σ_1.

From the Joukowski theorem "in the small" (Belotserkovsky 1967), it follows that the relative velocity of free vortices is equal to zero; in other words, they travel together with the particles of a fluid.

The physical nature of the problem under consideration and the required accuracy level determine one more, extremely important stage of the problem formulation, namely, the choice of the flow mode. The major modes are:

a. Noncirculatory flow past a body for which the vortex wake is ignored. This mode is used mostly for analyzing flows past highly elongated bodies as well as stationary oscillating bodies.

b. Circulatory separated flow satisfying all physically evident conditions, including the requirement of the flow velocity and pressure being finite throughout the flow domain. What was previously said has a decisive impact on the choice of the flow mode. Thus, when studying flow past a thin wing with sharp edges (either leading or trailing, or side edges), one has to admit shedding of free vortices from all of them and impose the Chaplygin–Joukowski condition concerning the velocity finiteness at all the edges. Otherwise, the flow velocity at the edges tends to infinity. Note that in this case even a steady motion of a wing results in pulsating (unsteady) flows.
c. Some simplified modes of circulatory flow past a body for which certain restriction may be lifted. Most popular are the modes that do not require the flow velocity and pressure at the leading and side edges of a thin wing, as well as at a cusp of the surface of a wing to be finite. As a result, the problem may be solved as a steady one, shedding free vortices only from the trailing edge of a wing.

Extensive experience in applying the preceding modes has been accumulated. All of them provide satisfactory data with respect to summary quantities; however, they generate locally incorrect data in the vicinity of sharp edges and cusps.

9.2. FUNDAMENTAL CONCEPTS OF THE METHOD OF DISCRETE VORTICES

The formulated aerodynamic problems will be solved with the help of the vortex method, whose major features are described in succeeding text.

Both the wing σ and the wake σ_1 are replaced by a continuous vortex surface: the wing σ by a surface formed of bound summary and free vortices stationary with respect to the wing, and the wake σ_1 by a surface of free vortices traveling together with the fluid particles along the streamlines. This continuous vortex layer will be used for calculating the velocity field $\mathbf{V}$ whose potential is given by the function $\Phi(x, y, z, t)$ and that meets all the requirements mentioned in the preceding section.

Thus, when solved by the vortex method, the problem of aerodynamics is reduced to calculating the strength of a vortex layer modeling the wing σ and the wake σ_1, whose induced velocity field satisfies all the boundary conditions mentioned in Section 9.1.

In the framework of the method of discrete vortices, numerical realization of all the conditions of boundary problems of aerohydrodynamics is implemented as follows.

A system of discrete vortices is substituted for a continuous vortex layer in such a way that in the limit, when the number of vortices tend to infinity, one arrives at the required vortex layer. The following vortex

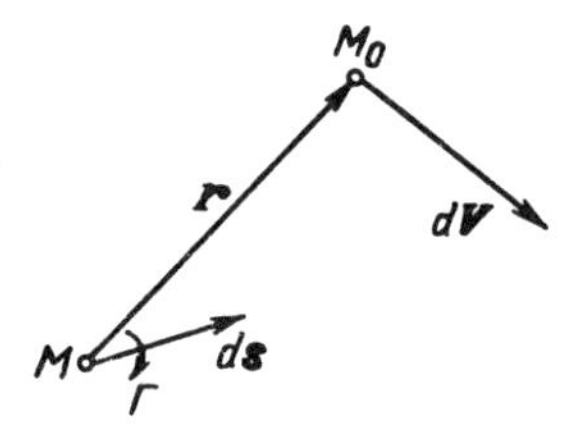

FIGURE 9.2. Illustration of the Biot–Savart law.

models are employed as primary elements in problems under consideration. In plane problems, this is a vortex filament of infinite-span. In problems of flow past thin wings at small angles of attack, these are oblique horseshoe vortices. Rectilinear vortex segments–closed vortex polygons (as a rule, triangles or quadrangles) are used in plane problems, general nonlinear three-dimensional problems.

The flow velocity fields induced by vortices are calculated with the help of the Biot–Savart formula according to which a segment of a vortex line $d\mathbf{s}$ long (whose circulation is equal to Γ) induces at a point M_0 (whose distance from $d\mathbf{s}$ is defined by the radius-vector $\mathbf{r}$) the flow velocity given by

$$d\mathbf{V} = \frac{\Gamma}{4\pi} \frac{d\mathbf{s} \times \mathbf{r}}{r^3}. \tag{9.2.1}$$

The direction of $d\mathbf{s}$ must be chosen in such a way that the circulation around it, in accordance with the right-hand rule, is positive (see Figure 9.2). If the element $d\mathbf{s}$ begins at point M, then $\mathbf{r} = \mathbf{MM}_0$. Let us denote by Π a given discrete vortex entity along which the strength Γ stays constant due to the known theorems of hydrodynamics concerning vortex filaments (Golubev 1949). Then the velocity field induced by the vortex is given by the formula

$$\mathbf{V} = \frac{\Gamma}{4\pi} \int_{\Pi} \frac{d\mathbf{s} \times \mathbf{r}}{r^3}. \tag{9.2.2}$$

The resulting velocity field satisfies the continuity equation throughout the space (with the exception of the discrete vortex itself). Additionally, the flow velocities attenuate as the distance from vortex elements increases, tending to zero at infinity.

In steady problems a wake trails into infinity. However, in unsteady problems it changes all the time: older vortices are carried away by the stream, while newly born ones are shed from the body. Vortex systems must be constructed in accordance with all the theorems about vortex conservation in the framework of a chosen flow mode.

At each instant the surface of a body is replaced by summary vortices that are not divided into bound and free ones and are fixed with respect to the body. Curvilinear free vortices are approximated by systems of rectilinear segments.

Although the number of discrete vortices is finite, it can increase without limit. This is done with the help of a fixed algorithm that ensures fulfillment of the following requirements underlying the efficiency of the method:

a. Near a body the dimensions of vortex grids must be approximately equal in all directions.
b. The points at which the no-penetration boundary conditions are met must be located approximately at the centers of vortex polygons (the so-called *collocation points*). This ensures Cauchy principal values of singular integrals.
c. At the boundaries of surfaces as well as in the vicinity of cusps where the flow velocity may become infinitely large, positions of vortices and boundary points are chosen in accordance with the employed scheme. If the requirement is formulated that the flow velocity is finite at edges and cusps (the *Chaplygin–Joukowski* (C + Z) *condition*), then collocation points are placed at and/or near the edges or cusps, i.e., the discrete analog of the fitness condition is employed. Otherwise, vortices are placed here. In what follows, this rule of positioning discrete vortices and reference* points at the edges is called the *B-condition of the method of discrete vortices*, which was originally formulated in Belotserkovsky (1955c).

The construction of vortex wake pattern is an important stage of solving a problem. In linear steady and unsteady problems this is specified in a most natural manner. However, in the case of nonlinear problems a wake must be constructed gradually: for steady flows, by the method of iterations; for unsteady flows, with the help of time steps. Circulations of the first discrete free vortices shed from a body are calculated; their strengths remain constant as they are carried away by the flow.

The preceding method allows all conditions of an aerohydrodynamic problem to be satisfied. Generally, the procedure of solution reduces to implementing (a) the solution of systems of linear algebraic equations satisfying boundary conditions at a body surface and (b) the construction of a vortex wake downstream of a body.

*They are called reference points in the present book.

9.3. FUNDAMENTAL DISCRETE VORTEX SYSTEMS

Consider the velocity field induced by a vortex segment A_1A_2 whose strength Γ is constant and directed from A_1 to A_2. An arbitrary point A of the segment may be presented in the form

$$\mathbf{r}_A = r_1 + tr_{1\,2}, \qquad 0 \leq t \leq 1, \tag{9.3.1}$$

where $\mathbf{r}_A = \mathbf{OA}$, $\mathbf{r}_1 = \mathbf{OA}_2$, $\mathbf{r}_{1\,2} = \mathbf{A}_1\mathbf{A}_2$, and O is a certain point of the space. Hence, for the arc element we have

$$d\mathbf{s} = \mathbf{r}_{1\,2}\,dt. \tag{9.3.2}$$

Let us next put the element $d\mathbf{s}$ at the point A. Then for the vector $\mathbf{r}$ one gets

$$\mathbf{r} = \mathbf{r}_0 - \mathbf{r}_A = \mathbf{r}_{1\,0} - t\mathbf{r}_{1\,2}, \tag{9.3.3}$$

where $\mathbf{r}_0 = \mathbf{OM}_0$, $\mathbf{r}_{1\,0} = \mathbf{A}_1\mathbf{M}_0$. Note also that

$$r = \left[\left(tr_{1\,2} - \frac{\mathbf{r}_{1\,0}\cdot\mathbf{r}_{1\,2}}{r_{1\,2}}\right)^2 + \alpha\right]^{1/2}, \qquad \alpha = r_{1\,0}^2 - \left(\frac{\mathbf{r}_{1\,0}\cdot\mathbf{r}_{1\,2}}{r_{1\,2}}\right)^2. \tag{9.3.4}$$

From Formulas (9.2.1) and (9.3.2)–(9.3.4) one gets the following formula for the velocity $\mathbf{V}$ at point M_0 induced by the vortex segment A_1A_2:

$$\mathbf{V} = \frac{\Gamma}{4\pi}\int_0^1 \frac{(\mathbf{r}_{1\,2}\times\mathbf{r}_{1\,0})\,dt}{\left[\left(tr_{1\,2} - \dfrac{\mathbf{r}_{1\,0}\cdot\mathbf{r}_{1\,2}}{r_{1\,2}}\right)^2 + r_{1\,0}^2 - \left(\dfrac{\mathbf{r}_{1\,0}\cdot\mathbf{r}_{1\,2}}{r_{1\,2}}\right)^2\right]^{3/2}}. \tag{9.3.5}$$

The primitive of the integrand in the latter equation is readily obtained:

$$\mathbf{V} = \frac{\Gamma}{4\pi}\frac{\mathbf{r}_{1\,2}\times\mathbf{r}_{1\,0}}{r_{1\,2}\alpha}\left.\frac{tr_{1\,2} - \dfrac{\mathbf{r}_{1\,0}\cdot\mathbf{r}_{1\,2}}{r_{1\,2}}}{\sqrt{\left(tr_{1\,2} - \dfrac{\mathbf{r}_{1\,0}\cdot\mathbf{r}_{1\,2}}{r_{1\,2}}\right)^2 + \alpha}}\right|_{t=0}^{1} \tag{9.3.6}$$

or

$$\mathbf{V} = \frac{\Gamma}{4\pi}\frac{\mathbf{r}_{12}\times\mathbf{r}_{10}}{r_{12}\alpha}\left[\frac{r_{12} - \dfrac{\mathbf{r}_{10}\cdot\mathbf{r}_{12}}{r_{12}}}{\sqrt{r_{12}^2 - 2\mathbf{r}_{10}\cdot\mathbf{r}_{12} + r_{10}^2}} + \frac{\mathbf{r}_{10}\cdot\mathbf{r}_{12}}{r_{10}r_{12}}\right]$$

$$= \frac{\Gamma}{4\pi}\frac{\mathbf{r}_{12}\times\mathbf{r}_{10}}{r_{10}^2 r_{12}^2 - (\mathbf{r}_{12}\cdot\mathbf{r}_{10})^2}\left(\frac{-\mathbf{r}_{12}\cdot\mathbf{r}_{20}}{r_{20}} + \frac{\mathbf{r}_{12}\cdot\mathbf{r}_{10}}{r_{10}}\right). \quad (9.3.7)$$

Let us next introduce a three-dimensional Cartesian system of coordinates $OXYZ$ as shown in Figure 9.1; in other words, let the triplet $\mathbf{i}, \mathbf{j}, \mathbf{k}$ be a right-hand system. Consider a special case when both the point M_0 and the vortex segment lie in the plane OXZ, and the segment is parallel to one of the coordinate axes, say, $\mathbf{A}_1\mathbf{A}_2 \| \mathbf{OZ}$. Then

$$M_0 = (x_0, z_0, 0), \qquad A_1 = (x_1, z_1, 0), \qquad A_2 = (x_1, z_2, 0),$$

$$\mathbf{r}_{12} = (z_2 - z_1)\mathbf{k}, \qquad \mathbf{r}_{10} = (x_0 - x_1)\mathbf{i} + (z_0 - z_1)\mathbf{k},$$

$$\mathbf{r}_{12} \times r_{10} = -\mathbf{j}(x_0 - x_1)(z_2 - z_1),$$

$$r_{12}^2 = (z_2 - z_1)^2,$$

$$\mathbf{r}_{20} = (x_0 - x_2)\mathbf{i} + (z_0 - z_2)\mathbf{k},$$

$$\mathbf{r}_{12}\cdot\mathbf{r}_{20} = (z_2 - z_1)(z_0 - z_2),$$

$$\mathbf{r}_{10}\cdot\mathbf{r}_{12} = (z_2 - z_1)(z_0 - z_1), \qquad r_{10}^2 = (x_0 - x_1)^2 + (z_0 - z_1)^2,$$

and from (9.3.7) one gets

$$\mathbf{V} = -\mathbf{j}\frac{\Gamma}{4\pi}\frac{1}{x_0 - x_1}\left(\frac{-(z_0 - z_2)}{\sqrt{(x_0 - x_1)^2 + (z_0 - z_2)^2}} + \frac{z_0 - z_1}{\sqrt{(x_0 - x_1)^2 + (z_0 - z_1)^2}}\right). \quad (9.3.8)$$

Next we consider a rectilinear vortex of a semi-infinite span that originates at the point $A_1 = (x_1, y_1, z_1)$ and tends to infinity by passing through the point $A_2 = (x_2, y_2, z_2)$.

Then the parameter t appearing in Formula (9.3.1) varies between 0 and $+\infty$. Hence, the upper limit in (9.3.6) is equal to $+\infty$, and upon substituting the limits one gets

$$\mathbf{V} = \frac{\Gamma}{4\pi} \frac{\mathbf{r}_{12} \times \mathbf{r}_{10}}{r_{12}\alpha} \left[1 + \frac{\mathbf{r}_{10} \cdot \mathbf{r}_{12}}{r_{10} r_{12}} \right]. \tag{9.3.9}$$

However, if the direction at a vortex proceeds from infinity through point A_2 to point A_1, then the minus sign is to be used in Formula (9.3.2) for $d\mathbf{s}$, and one gets

$$\bar{\mathbf{V}} = -\frac{\Gamma}{4\pi} \frac{\mathbf{r}_{12} \times \mathbf{r}_{10}}{r_{12}\alpha} \left[1 + \frac{\mathbf{r}_{10} \cdot \mathbf{r}_{12}}{r_{10} r_{12}} \right]. \tag{9.3.10}$$

Let us consider an example when a vortex originates at point $A_1 = (x_1, z_1, 0)$ and tends to infinity passing through the point $A_2 = (x_2, z_1, 0)$. Point $M_0 = (x_0, z_0, 0)$ also lies in the plane OXZ. Then we have $\mathbf{r}_{12} = (x_2 - x_1)\mathbf{i}$, $\mathbf{r}_{10} = (x_0 - x_1)\mathbf{i} + (z_0 - z_1)\mathbf{k}$, $\mathbf{r}_{12} \times \mathbf{r}_{10} = \mathbf{j}(x_2 - x_1)(z_0 - z_1)$, $r_{12} = x_2 - x_1$ (it is assumed that $x_2 > x_1$), $\mathbf{r}_{12} \cdot \mathbf{r}_{10} = (x_2 - x_1)(x_0 - x_1)$, $\alpha = (z_0 - z_1)^2$, and

$$\mathbf{V} = \mathbf{j}\frac{\Gamma}{4\pi} \frac{1}{z_0 - z_1} \left[1 + \frac{x_0 - x_1}{\sqrt{(x_0 - x_1)^2 + (z_0 - z_1)^2}} \right]. \tag{9.3.11}$$

Next we consider a rectilinear vortex of an infinite span. It will be assumed that the direction at the vortex is determined by the parameter appearing in (9.3.1), i.e., that Formula (9.3.2) preserves its form. Then the limits of integration in (9.3.5) become $-\infty$ and $+\infty$. Hence, by substituting $t = -\infty$ and $t = +\infty$ into Formula (9.3.6) one gets

$$\mathbf{V} = \frac{\Gamma}{2\pi} \frac{\mathbf{r}_{12} \times \mathbf{r}_{10}}{r_{12}\alpha}. \tag{9.3.12}$$

In Formulas (9.3.1) and (9.3.3), $\mathbf{r}_{12}$ may be assumed to be an arbitrary vector parallel to the direction at the vortex.

Let, for example, a vortex pass through point $A_1 = (x_1, z_1, 0)$ and be parallel to axis OZ (and directed in the same direction). Let the directing vector $\mathbf{r}_{1\,2}$ coincide with vector k. The flow velocity induced by the vector will be calculated at point $M_0 = (x_1, z_0, 0)$. The $\mathbf{r}_{1\,2} \times \mathbf{r}_{1\,0} = -\mathbf{j}(x_0 - x_1)$, $\alpha = (x_0 - x_1)^2$, and $r_{1\,2} = 1$. Thus, from Formula (9.3.12) one has

$$\mathbf{V} = -j\frac{\Gamma}{2\pi}\frac{1}{x_0 - x_1}. \qquad (9.3.13)$$

When we consider problems involving nonzero-thickness airfoils, we will come across a situation when $A_1 = (x, 0, y_1)$, $\mathbf{r}_{1\,2} = \mathbf{k}$, and $M_0 = (x_0, 0, y_0)$. Then $\mathbf{r}_{1\,0} = (x_0 - x_1)\mathbf{i} + (y_0 - y_1)\mathbf{j}$, $\mathbf{r}_{1\,2} \times \mathbf{r}_{1\,0} = -\mathbf{j}(x_0 - x_1) + \mathbf{i}(y_0 - y_1)$, and $\mathbf{r}_{1\,2} \cdot \mathbf{r}_{1\,0} = 0$. Thus, we have

$$\mathbf{V} = \frac{\Gamma}{2\pi}\frac{-j(x_0 - x_1) + i(y_0 - y_1)}{(x_0 - x_1)^2 + (y_0 - y_1)^2}. \qquad (9.3.14)$$

Let us proceed by considering usual and oblique horseshoe vortices.

According to Belotserkovsky (1967), a usual horseshoe vortex (see Figure 9.3a) is a vortex of a constant strength Γ composed of segment $[A_1(x_1, z_1, 0), A_2(x_1, z_2, 0)]$ and two semi-infinite rectilinear vortices $(A_1, +\infty)$ and $(A_2, +\infty)$ parallel to axis OX. The direction at the vortex will be specified by vector $\mathbf{r}_{1\,2}$. The vortex will be denoted by $\Pi(A_1, A_2)$. Let us calculate the velocity $\mathbf{V}$ induced by the vortex at point $M_0 = (x_0, z_0, 0)$ lying outside the vortex. Flow velocities induced at point M_0 by the vortices (A_1, A_2), $(A_1, +\infty)$, and $(A_2, +\infty)$ will be denoted by $\mathbf{V}_{1\,2}$, $\mathbf{V}_1$, and $\mathbf{V}_2$, respectively. Using Formulas (9.3.8)–(9.3.10), one gets

$$\mathbf{V}_{1\,2} = \mathbf{j}\frac{\Gamma}{4\pi}\frac{1}{x_0 - x_1}\left(\frac{z_0 - z_2}{\sqrt{(x_0 - x_1)^2 + (z_0 - z_2)^2}} - \frac{z_0 - z_1}{\sqrt{(x_0 - x_1)^2 + (z_0 - z_1)^2}}\right),$$

$$\mathbf{V}_2 = \mathbf{j}\frac{\Gamma}{4\pi}\frac{1}{z_0 - z_2}\left(1 + \frac{x_0 - x_2}{\sqrt{(x_0 - x_2)^2 + (z_0 - z_2)^2}}\right).$$

$$\mathbf{V}_1 = -\mathbf{j}\frac{\Gamma}{4\pi}\frac{1}{z_0 - z_1}\left(1 + \frac{x_0 - x_1}{\sqrt{(x_0 - x_1)^2 + (z_0 - z_1)^2}}\right). \qquad (9.3.15)$$

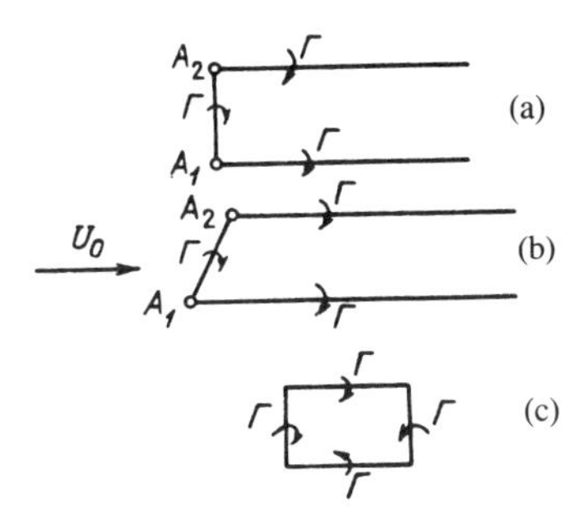

FIGURE 9.3. (a) A straight horseshoe vortex; (b) an oblique horseshoe vortex; (c) a closed rectangular vortex frame.

Because

$$\mathbf{V} = \mathbf{V}_{1\,2} + \mathbf{V}_1 + \mathbf{V}_2, \tag{9.3.16}$$

$$\mathbf{V} = \mathbf{j}\frac{\Gamma}{4\pi}\left[\frac{1}{z_0 - z_2} - \frac{1}{z_0 - z_1} + \frac{\sqrt{(x_0 - x_1)^2 + (z_0 - z_2)^2}}{(x_0 - x_1)(z_0 - z_2)} - \frac{\sqrt{(x_0 - x_1)^2 + (z_0 - z_1)^2}}{(x_0 - x_1)(z_0 - z_1)}\right]$$

$$= \mathbf{j}\frac{\Gamma}{4\pi}\int_{z_1}^{z_2}\frac{1}{(z_0 - z)^2}\left(1 + \frac{x_0 - x_1}{\sqrt{(x_0 - x_1)^2 + (z_0 - z)^2}}\right)dz$$

$$= \mathbf{j}\frac{\Gamma}{4\pi}\int_{z_1}^{z_2}\int_{x_1}^{+\infty}\frac{dx\,dz}{\left[(x_0 - x)^2 + (z_0 - z)^2\right]^{3/2}}. \tag{9.3.17}$$

An oblique horseshoe vortex (see Figure 9.3b) differs from the usual one in that the segment $[A_1, A_2]$ is not parallel to the OZ axis, i.e., $A_1 = (x_1, z_1, 0)$, $A_2 = (x_2, z_2, 0)$, and $x_2 \neq x_1$. Let the equation of the vortex line (A_1, A_2) be

$$x(z) = a + zb, \tag{9.3.18}$$

i.e., $x_1 = x(z_1) = a + z_1 b$ and $x_2 = x(z_2) = a + z_2 b$. Then for the same point $M_0(x_0, z_0, 0)$ we have

$$\mathbf{r}_{1\,2} = (x_2 - x_1)\mathbf{i} + (z_2 - z_1)\mathbf{k} = (z_2 - z_1)b\mathbf{i} + (z_2 - z_1)\mathbf{k}$$

$$= (z_2 - z_1)(b\mathbf{i} + \mathbf{k}),$$

$$\mathbf{r}_{10} = (x_0 - x_1)\mathbf{i} + (z_0 - z_1)\mathbf{k},$$

$$\mathbf{r}_{12} \times \mathbf{r}_{10} = \mathbf{j}[(x_2 - x_1)(z_0 - z_1) - (x_0 - x_1)(z_2 - z_1)]$$

$$= \mathbf{j}(z_2 - z_1)[b(z_0 - z_1) - (x_0 - x_1)],$$

$$\mathbf{r}_{12} = (z_2 - z_1)\sqrt{1 + b^2},$$

$$\mathbf{r}_{12} \cdot \mathbf{r}_{10} = (z_2 - z_1)[b(x_0 - x_1) + (z_0 - z_1)],$$

$$\mathbf{r}_{12} \cdot \mathbf{r}_{20} = (z_2 - z_1)[b(x_0 - x_2) + (z_0 - z_2)].$$

Hence, by (9.3.7),

$$\mathbf{V}_{12} = \mathbf{j}\frac{\Gamma}{4\pi}\frac{1}{\lambda(x_0, z_0)}\left[\frac{(x_0 - x_2)b + (z_0 - z_2)}{\sqrt{(x_0 - x_2)^2 + (z_0 - z_2)^2}} - \frac{(x_0 - x_1)b + (z_0 - z_1)}{\sqrt{(x_0 - x_1)^2 + (z_0 - z_1)^2}}\right], \tag{9.3.19}$$

where $\lambda(x_0, z_0) \equiv x_0 - a - z_0 b \equiv x_0 - x(z_0)$.

Because for the vortices $(A_1, +\infty)$ and $(A_2, +\infty)$ the same formulas are valid, we have

$$\mathbf{V} = \mathbf{j}\frac{\Gamma}{4\pi}\left[\frac{1}{z_0 - z_2} - \frac{1}{z_0 - z_1} + \frac{\sqrt{(x_0 - x_2)^2 + (z_0 - z_2)^2}}{\lambda(x_0, z_0)(z_0 - z_2)} - \frac{\sqrt{(x_0 - x_1)^2 + (z_0 - z_1)^2}}{\lambda(x_0, z_0)(z_0 - z_1)}\right]$$

$$= \mathbf{j}\frac{\Gamma}{4\pi}\int_{z_1}^{z_2}\frac{1}{(z_0 - z_1)^2}\left(1 + \frac{x_0 - x(z)}{\sqrt{(x_0 - x(z))^2 + (z_0 - z)^2}}\right)dz$$

$$= \mathbf{j}\frac{\Gamma}{4\pi}\int_{z_1}^{z_2}\int_{x(z)}^{+\infty}\frac{dx\,dz}{\left[(x_0 - x)^2 + (z_0 - z)^2\right]^{3/2}}. \tag{9.3.20}$$

Note that if $b = 0$, i.e., $\Pi(A_1, A_2)$ is a usual horseshoe vortex, then (9.3.20) coincides with (9.3.17), because in this case $x(z) = x_1$ and $\lambda = x_0 - x(z_0) \equiv x_0 - x_1$.

Note also the following circumstance. If point M_0 lies at the segment $[A_1, A_2]$, then Formulas (9.3.17) and (9.3.20) must be used in the integral form, and one gets

$$\mathbf{V} = \mathbf{j}\frac{\Gamma}{4\pi}\left[\frac{1}{z_0 - z_2} - \frac{1}{z_0 - z_2}\right], \tag{9.3.21}$$

because in this case either $x_0 - x_1 = 0$ or $x_0 - x(z) = 0$.

Finally, let us consider a vortex of constant strength Γ having the form of a rectangle in the plane OXZ, whose sides are parallel to the coordinate axes (see Figure 9.3c). Let the corner points of the vortex coincide with the points $A_1 = (x_1, z_1, 0)$, $A = (x_1, z_2, 0)$ (where $z_2 > z_1$), $A_3 = (x_2, z_2, 0)$ (where $x_2 > x_1$), and $A_4 = (x_2, z_1, 0)$. The direction at the vortex will be defined by the vector $\mathbf{r}_{1\,2}$. By $\mathbf{V}_{1\,2}$, $\mathbf{V}_{2\,3}$, $\mathbf{V}_{3\,4}$, and $\mathbf{V}_{4\,1}$ we denote the flow velocities induced at point $M_0 = (x_0, z_0, 0)$ by the vortices (A_1, A_2), (A_2, A_3), (A_3, A_4), and (A_4, A_1), respectively. Taking into account the directions of the vortices, one gets from (9.3.7) (see also (9.3.8)),

$$\mathbf{V}_{1\,2} = \mathbf{j}\frac{\Gamma}{4\pi}\frac{1}{x_0 - x_1}\left[\frac{z_0 - z_2}{\sqrt{(x_0 - x_1)^2 + (z_0 - z_2)^2}} - \frac{z_0 - z_1}{\sqrt{(x_0 - x_1)^2 + (z_0 - z_1)^2}}\right],$$

$$\mathbf{V}_{2\,3} = -\mathbf{j}\frac{\Gamma}{4\pi}\frac{1}{z_0 - z_2}\left[\frac{x_0 - x_2}{\sqrt{(x_0 - x_2)^2 + (z_0 - z_2)^2}} - \frac{x_0 - x_1}{\sqrt{(x_0 - x_1)^2 + (z_0 - z_2)^2}}\right],$$

$$\mathbf{V}_{3\,4} = \mathbf{j}\frac{\Gamma}{4\pi}\frac{1}{x_0 - x_2}\left[\frac{-(z_0 - z_2)}{\sqrt{(x_0 - x_2)^2 + (z_0 - z_2)^2}} + \frac{z_0 - z_1}{\sqrt{(x_0 - x_2)^2 + (z_0 - z_1)^2}}\right],$$

$$\mathbf{V}_{4\,1} = -\mathbf{j}\frac{\Gamma}{4\pi}\frac{1}{z_0 - z_1}\left[\frac{-(x_0 - x_2)}{\sqrt{(x_0 - x_2)^2 + (z_0 - z_1)^2}}\right.$$
$$\left. + \frac{x_0 - x_1}{\sqrt{(x_0 - x_1)^2 + (z_0 - z_1)^2}}\right]. \tag{9.3.22}$$

Hence, for the flow velocity **V** induced at point M_0 by the whole of the vortex, we have

$$\mathbf{V} = \mathbf{V}_{1\,2} + \mathbf{V}_{2\,3} + \mathbf{V}_{3\,4} + \mathbf{V}_{4\,1}$$
$$= \mathbf{j}\frac{\Gamma}{4\pi}\left[\frac{\sqrt{(x_0 - x_1)^2 + (z_0 - z_2)^2}}{(x_0 - x_1)(z_0 - z_2)} - \frac{\sqrt{(x_0 - x_2)^2 + (z_0 - z_2)^2}}{(x_0 - x_2)(z_0 - z_2)}\right.$$
$$\left. + \frac{\sqrt{(x_0 - x_2)^2 + (z_0 - z_1)^2}}{(x_0 - x_2)(z_0 - z_1)} + \frac{\sqrt{(x_0 - x_1)^2 + (z_0 - z_1)^2}}{(x_0 - x_1)(z_0 - z_1)}\right]. \tag{9.3.23}$$

This formula may also be presented in the form:

$$\mathbf{V} = -\mathbf{j}\frac{\Gamma}{4\pi}\int_{z_1}^{z_2}\frac{1}{(z_0 - z)^2}\left[\frac{x_0 - x_2}{\sqrt{(x_0 - x_2)^2 + (z_0 - z)^2}}\right.$$
$$\left. - \frac{x_0 - x_1}{\sqrt{(x_0 - x_1)^2 + (z_0 - z)^2}}\right]$$
$$= \mathbf{j}\frac{\Gamma}{4\pi}\int_{z_1}^{z_2}\int_{x_1}^{x_2}\frac{dx\,dz}{\left[(x_0 - x)^2 + (z_0 - z)^2\right]^{3/2}}. \tag{9.3.24}$$

10

Two-Dimensional Problems for Airfoils

10.1. STEADY FLOW PAST A THIN AIRFOIL

Let us start by considering two-dimensional steady flow past a thin, slightly curved isolated airfoil (see Figure 10.1). Let the flow velocity be defined by the formula $\mathbf{U}_0 = u_x\mathbf{i} + u_y\mathbf{j}$. We also assume that the projection of the airfoil onto the plane OXZ occupies the strip $-b \leq x \leq b$. Because the airfoil is only slightly curved, the no-penetration boundary condition may be transferred onto the strip $-b \leq x \leq b$. This means that the vortex sheet representing the airfoil is located within the strip. The strength of the sheet is independent of z and will be denoted by $\gamma(x)$. The no-penetration condition is met at all the points of the strip; in other words, the sum of normal components of the flow velocities induced by the vortex sheet at a point of the strip and of the corresponding component of the oncoming flow velocity is equal to zero.

In the case under consideration the method of discrete vortices reduces to the following (Belotserkovsky 1967). The vortex sheet representing an airfoil is modeled by infinitely long vortex filaments of constant strength Γ_k described by the equation $x = x_k$, $x_k = -b + kh$, $h = 2b/(n+1)$, $k = 1, \ldots, n$, and the no-penetration condition (9.1.1) is met at the points $x_{0m} = x_m + h/2$, $m = 0, 1, \ldots, n$. The normal component V_{km} of the velocity induced by the kth vortex at the mth reference point is equal to (see (9.3.13))

$$V_{km} = \Gamma_k \omega_{km} = -\frac{\Gamma_k}{2\pi(x_{0m} - x_k)}. \qquad (10.1.1)$$

At the mth reference point the normal component V_m of the velocity

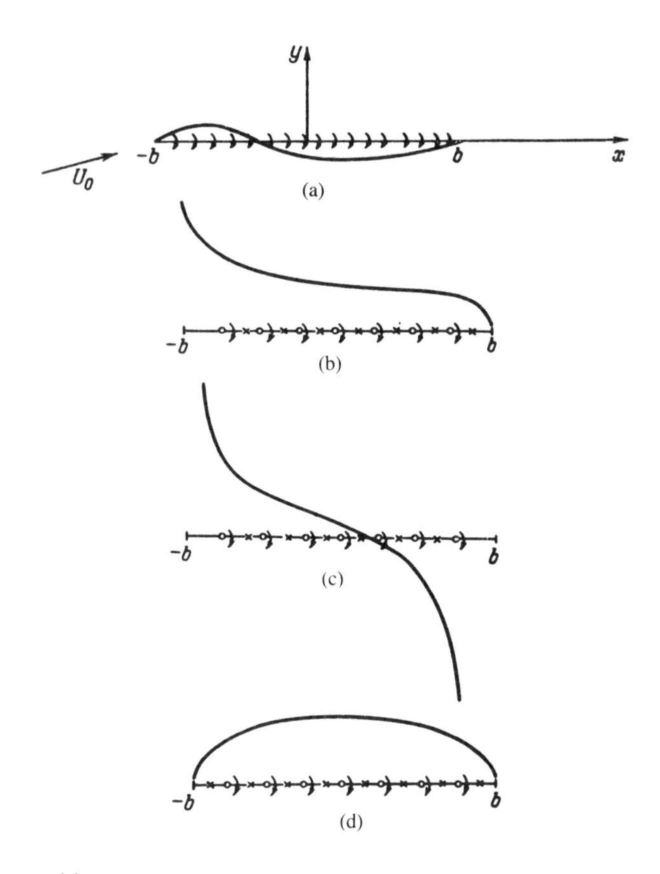

FIGURE 10.1. (a) Simulation of a thin, slightly curved airfoil by a vortex sheet in the interval $[-b, b]$ of the OX axis; (b), (c), and (d) are positions of discrete vortices and reference points in $[-b, b]$ for circulatory, noncirculatory, and finite-velocity airfoil problems, respectively.

induced by a system of discrete vortices is equal to

$$V_m = \sum_{k=1}^{n} \Gamma_k \omega_{km}. \qquad (10.1.2)$$

Consider circulatory flow past an airfoil (Figure 10.1b). According to experimental evidence (Belotserkovsky 1967), the vortex strength must be unlimited at the leading edge (i.e., at the point $-b$) and limited at the trailing edge (at the point b). Hence, by the B condition of the method of discrete vortices, a vortex must be located near a leading edge and a reference point located near a trailing edge; therefore, reference points should be numbered by $m = 1, \ldots, n$. If the no-penetration condition (9.1.1) is met at the points, then one gets the following system of equalities:

$$\sum_{k=1}^{n} \Gamma_k \omega_{km} = V_m^*, \qquad V_m^* = -\mathbf{U}_0 \mathbf{n}(x_{0m}) \qquad m = 1, \dots, n, \quad (10.1.3)$$

or, taking into account Formula (10.1.1),

$$-\frac{1}{2\pi} \sum_{k=1}^{n} \frac{\Gamma_k}{x_{0m} - x_k} = V_m^*, \qquad m = 1, \dots, n. \qquad (10.1.4)$$

From Section 1.3 it follows that if $\Gamma_k = \gamma_n(x_k)h$ (where γ_n is an approximate value of γ), then the system of linear algebraic equations (10.1.4) approximates the singular integral equation

$$-\frac{1}{2\pi} \int_{-b}^{b} \frac{\gamma(x)\, dx}{x_0 - x} = V^*(x_0), \qquad x_0 \in (-b, b). \qquad (10.1.5)$$

By Theorem 5.1.1, Relationships (5.1.6)–(5.1.8) hold between a solution to System (10.1.4) and the exact value of the vortex strength that is unlimited at point $-b$ and limited at point b.

For noncirculatory flow (see Figure 10.1c) with the strength $\gamma(x)$ being unlimited at both edges, the B condition requires that discrete vortices be located near both edges; in other words, in this case $m = 1, \dots, n-1$ for reference points. By employing the no-penetration condition at the points, one gets $n-1$ equations for n unknown circulations of discrete vortices. Hence, the number of unknowns is more than that of equations; however, by adding the no-circulation flow condition we have n equations and n unknowns. As a result, one gets the following system of linear algebraic equations:

$$\sum_{k=1}^{n} \Gamma_k \omega_{km} = V_m^*, \qquad m = 1, \dots, n-1,$$

$$\sum_{k=1}^{n} \Gamma_k = 0, \qquad (10.1.6)$$

and, hence, Theorem 5.1.1 is applicable for $C = 0$.

For limited flow speeds (see Figure 10.1d) when the vortex strength is limited at both edges, the B condition requires that reference points lie nearest to the edges, i.e., in this case, $m = 0, 1, \dots, n$ for the reference points. By using the no-penetration condition at the points, one gets $n+1$

equations for n unknown circulations of discrete vortices (i.e., the number of equations exceeds that of unknowns). However, it can be made the same by introducing a regularizing term γ_{0n} (which is, in fact, a new additional unknown). In other words, we consider the system

$$\gamma_{0n} + \sum_{k=1}^{n} \Gamma_k \omega_{km} = V_m^*, \qquad m = 0, 1, \ldots, n, \qquad (10.1.7)$$

and, hence, Theorem 5.1.1 is applicable in this case too.

Thus, Theorem 5.1.1 provides a mathematical foundation for steady unbounded flow past a thin airfoil. The same theorems verify the B condition of the method of discrete vortices. (The B condition is a discrete analog of the Chaplygin–Joukowski hypothesis). Examples of numerical calculations for certain problems were considered previously (see Figures 5.1–5.3).

Note that according to Inequality (5.1.30), the method of discrete vortices allows us to calculate summary aerodynamic characteristics to any specified accuracy.

The method of discrete vortices has the following important peculiarity: the function $\gamma(x)$ obtained by solving a system of linear algebraic equations is defined by relative positions of discrete vortices and reference points only and is not specified *a priori*. This is of special importance when one has to solve new problems of aerodynamics that are not yet analyzed from the mathematical point of view.

Next we consider a thin airfoil with a flap (see Figure 10.2). In this case the normal component of the oncoming flow velocity suffers a discontinuity of the first kind at the point where the flap is deflected (point q). Hence, at point q the right-hand side of Equation (10.1.5) suffers a discontinuity of the first kind too, and the function $\gamma(x)$ has a logarithmic singularity at that point. By using Note 5.1.4 we see that for the problem under consideration the strength of discrete vortices may be found as follows:

1. The discrete vortices and reference points must be positioned so that point q lies midway between the nearest discrete vortex and reference point. Then, depending on the type of problem, one has to consider one of the systems (10.1.3), (10.1.6), or (10.1.7). Foundation of the numerical scheme is provided by Note 5.1.4.
2. Calculations show that more accurate results in the neighborhood of q may be obtained is reference points are chosen in such a way that q is one of them and the normal component of the oncoming flow velocity at the point is assumed to be equal to $(V^*(q - 0) + V^*(q + 0))/2$.

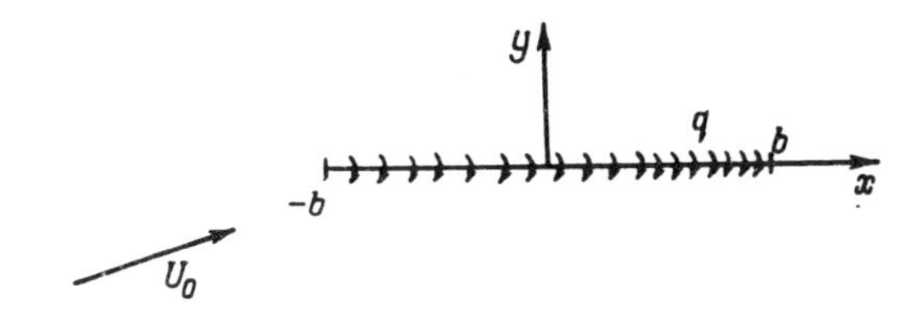

FIGURE 10.2. Simulation of a thin airfoil with a flap by a vortex sheet. q is the leading edge of the flap.

3. If it is required that an equal number of vortices be distributed over the flap and the rest of an airfoil (as in the case of calculating hinge moments due to the flap), then one should make use of Note 5.3.3.

Let us continue by considering steady two-dimensional flow past an airfoil near a solid surface (see Figure 10.3). As earlier, the airfoil is positioned at straight line $y = 0$, and the nearby solid surface is described by the equation $y = -H$. In this case the method of discrete vortices is applied as follows (Belotserkovsky 1967): Both discrete vortices and reference points are chosen at the airfoil as was done before; in addition, discrete vortices are positioned at points $A_{kH} = (x_k, -2H)$ of the plane OXY, and their strengths are specified as $\Gamma_{kH} = -\Gamma_k$ in order to ensure no penetration of the line $y = -H$. From (9.3.14) it follows that the normal flow velocity component V^{*H}_{km} induced by the vortex Γ_{kH} at the reference point x_{0m} is given by

$$V^{*H}_{km} = -\frac{\Gamma_{kH}}{2\pi}\frac{x_{0m} - x_k}{(x_{0m} - x_k)^2 + 4H^2} = \Gamma_{kH}\,\omega^H_{km} = -\Gamma_k\,\omega^H_{km}. \quad (10.1.8)$$

Thus, for circulatory flow one gets the following system of linear algebraic equations:

$$\sum_{k=1}^{n} \Gamma_k\,\omega_{km} - \sum_{k=1}^{n} \Gamma_k\,\omega^H_{km} = V^*_m, \qquad m = 1, \ldots, n. \quad (10.1.9)$$

In accordance with what was said in Section 1.3, the latter system approximates the equation

$$-\frac{1}{2\pi}\int_{-b}^{b} \gamma(x)\left[\frac{1}{x_0 - x} - \frac{x_0 - x}{(x_0 - x)^2 + 4H^2}\right] dx = V^*(x_0). \quad (10.1.10)$$

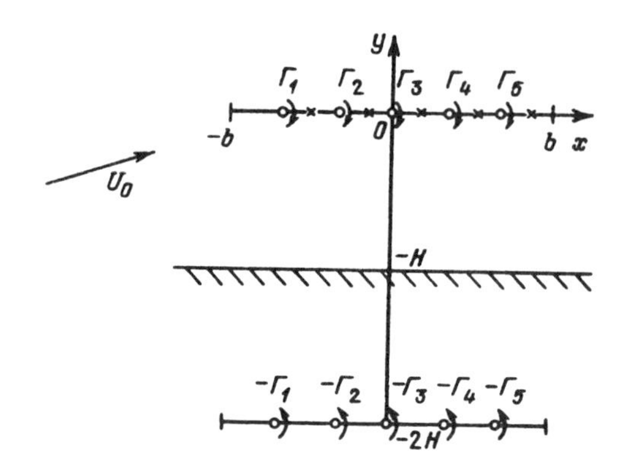

FIGURE 10.3. A system of discrete vortices and reference points for an airfoil near a solid surface.

According to Theorem 5.3.1, System (10.1.9) yields a solution converging to the solution $\gamma(x)$ of Equation (10.1.10), which vanishes at point b and tends to infinity at point $-b$.

Because Equation (10.1.10) has the form of Equation (5.3.1), for noncirculatory flows with finite velocities one gets systems analogous to (5.3.6) and (5.3.7). Thus, Theorem 5.3.1 provides mathematical foundation of the method of discrete vortices for steady flow past an airfoil near a solid surface.

If more accurate calculated results are needed for the immediate neighborhood of airfoil edges, then one has to use unequally spaced discrete vortices and reference points. In this case it is advisable to position the airfoil in the interval $[-1, 1]$ (by putting $x = bt$) and to make use of Theorems 5.2.1 and 5.2.2 for an isolated airfoil or Theorem 5.3.2 for an airfoil near a solid surface.

10.2. AIRFOIL CASCADES

Consider a cascade of thin airfoils (Belotserkovsky 1967) presented at the cross section $z = 0$ by a system of segments $[-b, b] \times y_k$, where $y_k = kl$, $k = 0, \pm 1, \pm 2, \ldots$; l is a fixed positive number and $[-b, b]$ is an interval of the axis OX. Let us consider steady two-dimensional unbounded flow described in the preceding section. Because the flow past any of the airfoils is subject to the same conditions, the strength of the bound vortex sheet at an airfoil depends on the coordinate x only and is independent of both z and y_k, $k = 0, \pm 1, \pm 2, \ldots$. Therefore, the method of discrete vortices is used as follows.

Let us replace the continuous vortex sheet simulating airfoil $[-b, b] \times y_k$ by a system of discrete vortex filaments parallel to the axis OZ and crossing the plane $z = 0$ at points $(x_i, y_k, 0)$, $i = 1, \dots, n$, $k = 0, \pm 1, \pm 2, \dots$, whose strength is equal to $\Gamma_i = \gamma_n(x_i)h$ (points x_i on $[-b, b]$ were chosen as in the preceding section).

In what follows a system of discrete vortex filaments (x_i, y_i), $k = 0, \pm 1, \dots$, is called the ith vortex chain. Flow velocities induced by the chains will be calculated at points $(x_{0j}, 0)$, $j = 0, 1, \dots, n$. According to (9.3.14), velocity components induced at point $(x_{0j}, 0)$ by a discrete vortex filament (x_i, y_k) of strength Γ_i are given by

$$V_{x,kij} = \frac{\Gamma_i}{2\pi} \frac{-y_k}{(x_{0j} - x_i)^2 + y_k^2} = \Gamma_k \omega_{x,kij},$$

$$V_{y,kij} = \frac{\Gamma_i}{2\pi} \frac{-(x_{0j} - x_i)}{(x_{0j} - x_i)^2 + y_k^2} = \Gamma_i \omega_{y,kij}. \tag{10.2.1}$$

Hence, the ith chain of discrete vortices induces at point $(x_{0j}, 0)$ flow velocity whose components are given by

$$V_{x,ij} = \Gamma_i \sum_{k=-\infty}^{\infty} \omega_{x,kij}, \qquad V_{y,ij} = \Gamma_i \sum_{k=-\infty}^{\infty} \omega_{y,kij}. \tag{10.2.2}$$

By employing the notion of complex potential (Golubev 1949), it can be readily shown that the normal velocity component $V_{y,ij}$ is equal to

$$V_{y,ij} = \Gamma_i \omega_{ij} = \frac{\Gamma_i}{2l} \coth \frac{\pi}{l} (x_{0j} - x_i). \tag{10.2.3}$$

Thus, the normal velocity component at point $(x_{0j}, 0)$, due to all vortex chains forming a cascade of airfoils, is given by

$$V_{yj} = \sum_{i=1}^{n} \Gamma_k \omega_{ij} = \frac{1}{2l} \sum_{i=1}^{n} \Gamma_i \coth \frac{\pi}{l} (x_{0j} - x_i). \tag{10.2.4}$$

Let us denote the normal flow velocity component at reference points of airfoils by V_j^*. The no-penetration condition gives:

for circulatory flow,

$$\frac{1}{2l} \sum_{i=1}^{n} \Gamma_1 \coth \frac{\pi}{l} (x_{0j} - x_i) = V_j^*, \qquad j = 1, \dots, n; \tag{10.2.5}$$

for noncirculatory flow,

$$\frac{1}{2l}\sum_{i=1}^{n}\Gamma_i \coth\frac{\pi}{l}(x_{0j}-x_i) = V_j^*, \qquad j = 1,\ldots,n-1,$$

$$\sum_{i=1}^{n}\Gamma_i = 0, \qquad j = n; \tag{10.2.6}$$

for flow with finite velocities,

$$\gamma_{0n} + \frac{1}{2l}\sum_{i=1}^{n}\Gamma_i \coth\frac{\pi}{l}(x_{0j}-x_i) = V_j^*, \qquad j = 0,1,\ldots,n. \tag{10.2.7}$$

According to Section 1.3, for $\Gamma_i = \gamma_n(x_i)h$, Systems (10.2.5)–(10.2.7) approximate the singular integral equation

$$\frac{1}{2l}\int_{-b}^{b}\gamma(x)\coth\frac{\pi}{l}(x_0-x)\,dx = V^*(x_0), \qquad x_0 \in (-b,b), \tag{10.2.8}$$

which may be reduced to (5.3.9). In fact, Equation (10.2.8) may be written in the form

$$\frac{1}{2l}\int_{-b}^{b}\frac{(x_0-x)\coth(\pi(x_0-x)/l)}{x_0-x}\gamma(x)\,dx = V^*(x_0). \tag{10.2.9}$$

Hence, it suffices to denote $K(x_0,x) = (x_0-x)\coth(\pi(x_0-x)/l)$ and $K(x_0,x_0) = l/\pi$, where the kernel is an analytic function. Because it is physically obvious that, subject to corresponding additional conditions, Equation (10.2.8) has a unique solution for each index $\kappa = 0, 1, -1$, solutions to Systems (10.2.5)–(10.2.7) converge to the corresponding exact solutions of the equation, and Relationships (5.1.6)–(5.1.8) are valid for them.

Thus, the method of discrete vortices is also fully verified for cascades of airfoils.

10.3. THIN AIRFOIL WITH EJECTION

The effects of both air ejection and injection (suction) are used to increase lift of aircraft wings during take-off and landing. In order to calculate an increase in lift due to ejection, we simulate an airfoil by a vortex sheet whose strength is to be found.

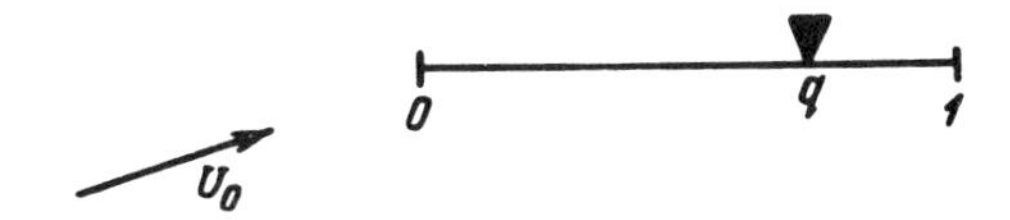

FIGURE 10.4. An airfoil with a sink at point q. The discrete vortices and reference points are distributed in such a way that point q is one of the reference points.

Thus, consider a thin airfoil (a plate) located at the segment $[-b, b]$ of axis OX with air ejected at point $q \in (-b, b)$ (see Figure 10.4). If fluid is ejected on the upper surface of an airfoil, then, due to the no-penetration condition, there are no singularities in the flow velocity distribution at the lower surface of the profile. The flow velocity pattern at the upper surface of an airfoil in the neighborhood of point q is determined by the expression

$$V_{rs} = Q/(2\pi r). \tag{10.3.1}$$

where V_{rs} are the radial velocity components induced by a sink, r is the distance between a point under consideration and the sink, and Q is the flow rate of the sink. Because the sink lies at the axis OX, the flow velocity induced by it at the axis is parallel to the latter and, taking into account the flow direction, is determined by the formula

$$V_{xs} = Q/[2\pi(q - x)]. \tag{10.3.2}$$

Let $\gamma(x)$ be the strength of the vortex sheet at point $x \in [-b, b]$. Then, at the upper and lower surface of a plate the vortex sheet induces tangential velocities (Nekrasov 1947)

$$V_{x\gamma}^{+,-} = \pm\gamma(x)/2. \tag{10.3.3}$$

where the plus $(+)$ and minus $(-)$ signs refer to the upper and lower surface, respectively. Tangential flow velocities due to a sink are given by Formula (10.3.2), i.e.,

$$V_{xs} = V_{xs}^{+} = V_{xs}^{-}. \tag{10.3.4}$$

For the lower surface of an airfoil, one has

$$V_{x\gamma}^{-} + V_{xs}^{-} \equiv \varphi(x), \tag{10.3.5}$$

where, according to the problem formulation, $\varphi(x)$ is a smooth function in the neighborhood of the sink.

From (10.3.3)–(10.3.5) one gets for the value of $\gamma(x)$ in the neighborhood of a sink,

$$\gamma(x) = \frac{Q}{\pi(q-x)} - 2\varphi(x) = \frac{\psi(x)}{q-x}. \qquad (10.3.6)$$

where $\psi(q) = Q/\pi$. At the upper airfoil surface, the summary tangential flow velocity due to the vortices and the sink has a singularity of the form of (10.3.6).

The strength $\gamma(x)$ of the layer under consideration must satisfy both Equation (10.1.5) of the theory of airfoils and the conditions at the edges of an airfoil. The function $\gamma(x)$ will be found with the help of a numerical method developed on the basis of the method of discrete vortices.

Let us choose the points where discrete vortices, $\{x_i, i = 1, \dots, n\}$, and the reference points, $\{x_{j0}, j = 0, 1, \dots, n\}$, are located, as was done when proving Theorem 5.1.4, i.e., so that point q is one of the reference points and the relative positions of discrete vortices and reference points repeat the ones used in the preceding sections. While constructing the system of linear algebraic equations for determining the strength Γ_i of the discrete vortices, the no-penetration boundary condition is met for all reference points x_{0j} except point $x_{0j_q} = q$, which coincides with the sink. In fact, it is nonsensical to speak about a normal velocity component at the point where a sink is located. Thus, we deduce that in the case of a circulatory problem when $\gamma(x)$ is unlimited at the leading edge and limited at the trailing edge, there are n discrete vortices and $n-1$ linear algebraic equations (defined by reference points x_{0j}, $j = 0, 1, \dots, n$, $j \neq j_q$) for calculating the latter. The value of circulation over the whole of the airfoil, $\int_{-b}^{b} \gamma(x)\,dx$, is unknown. Therefore, to make the problem well-posed we note that the strength of a discrete vortex located near a sink may be assumed to be known and equal to $\Gamma_{j_q} = \psi(q)h/(q - x_{j_q})$ or $\Gamma_{j_q+1} = \psi(q)h/(q - x_{j_q+1})$, because function $\psi(x)$ entering (10.3.6) is smooth in the neighborhood of point q (by all means, it belongs to the class H).

Thus, for a circulatory problem one has to consider the system of linear algebraic equations

$$\sum_{\substack{i=1 \\ i \neq j_q}}^{n} \Gamma_i \omega_{ij} = V_j^* - \Gamma_{j_q} \omega_{j_q j}, \qquad j = 1, \dots, n, \; j \neq j_q, \qquad (10.3.6')$$

where $\Gamma_i = \gamma_n(x_i)h$.

In this case mathematical foundation of the chosen numerical scheme is provided by Theorem 5.1.4. Calculations carried out with the help of the proposed mode demonstrate good convergence of an approximate solution to the exact one (see Figure 10.5).

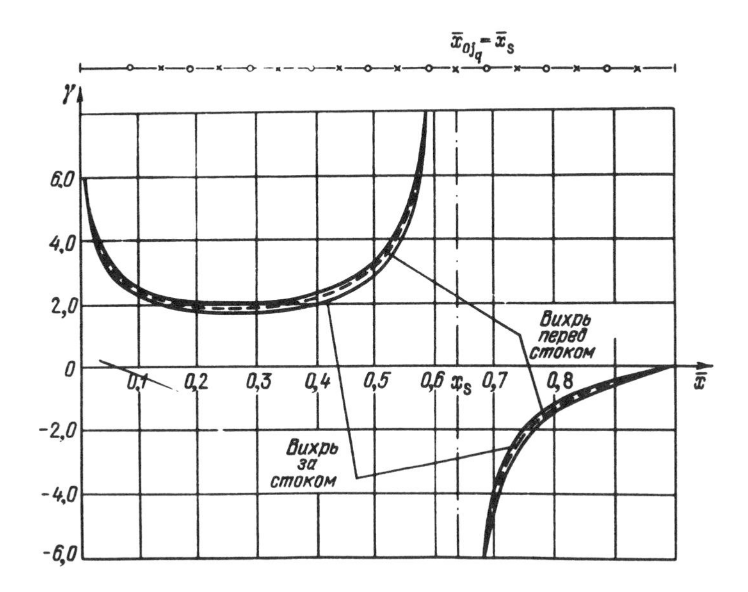

FIGURE 10.5. The vortex sheet strength distribution along an airfoil with a sink at point x_1 for a circulatory problem. The dashed and solid lines correspond to the exact and numerical solutions, respectively.

Consider next the problem of flow with finite velocities past an airfoil. For a given oncoming flow one has to find such a sink flow rate Q at point q for which the strength $\gamma(x)$ of a vortex sheet is limited at both the leading and the trailing edge. By the B condition of the method of discrete vortices, in this case one has to choose $j = 0, 1, \ldots, n$, omitting, however, the reference point number j_q corresponding to the sink. Thus, we arrive at a system of n equations with n unknowns:

$$\sum_{i=1}^{n} \Gamma_i \omega_{ij} = V_j^*, \qquad j = 0, 1, \ldots, n, j \neq j_q. \tag{10.3.7}$$

By Theorem 5.1.5 the approximate solution $\gamma_n(x_i) = \Gamma_i/h$ converges to the solution of Equation (10.1.5):

$$\gamma(x) = -\frac{2}{\pi} \frac{\sqrt{b^2 - x^2}}{q - x} \int_{-b}^{b} \frac{(q - x_0) V^*(x_0)\, dx_0}{\sqrt{b^2 - x_0^2}\,(x - x_0)}. \tag{10.3.8}$$

Hence, the sink flow rate ensuring flow with finite velocities past an airfoil is given by the formula (see (10.3.8))

$$Q = \pi\psi(q) = \pi \lim_{x\to q} \gamma(x)(q-x) = -2\sqrt{b^2-q^2}\int_{-b}^{b} \frac{V^*(x_0)\,dx_0}{\sqrt{b^2-x_0^2}}. \tag{10.3.9}$$

Finally, of interest is a vortex sheet corresponding to noncirculatory flow past an airfoil at a fixed sink flow rate Q at point q. In this case the B condition requires that the no-penetration condition be met at points x_{0j}, $j = 1,\ldots,n-1$, $j \neq j_q$. Assuming that the discrete vortex strength Γ_{j_q} is known, one gets $n-2$ equations for $n-1$ unknowns. The system of equations may be augmented by digitizing the equation

$$\int_{-b}^{b} \gamma(x)\,dx = 0, \tag{10.3.10}$$

which is, in fact, the condition of no-circulation flow past an airfoil. Thus, we have to consider the system

$$\sum_{\substack{i=1\\ i\neq j_q}}^{n} \Gamma_i\omega_{ij} = V_j^* - \Gamma_{j_q}\omega_{j_q j}, \qquad j = 1,\ldots,n-1, j\neq j_q,$$

$$\sum_{\substack{i=1\\ i\neq j_q}}^{n} \Gamma_i = -\Gamma_{j_q}, \qquad j = j_q. \tag{10.3.11}$$

Similarly to Theorems 5.1.4 and 5.1.5, it can be shown that a solution to the latter system of equations converges, with the same estimates as in the case of the theorems, to the solution of Equation (10.1.5):

$$\gamma(x) = -\frac{2}{\pi}\frac{1}{\sqrt{b^2-x^2}}\int_{-b}^{b} \frac{\sqrt{b^2-x_0^2}\,V^*(x_0)\,dx_0}{x-x_0}$$

$$+\frac{1}{q-x}\sqrt{\frac{b^2-q^2}{b^2-x^2}}\,\frac{Q}{\pi}. \tag{10.3.12}$$

If an airfoil contains several sinks, then one has to employ Theorem 5.1.6 and the corresponding notes. In order to use unequally spaced points when considering problems of flow past an airfoil with a sink, one has to use the results of Section 5.2.

10.4. FINITE-THICKNESS AIRFOIL WITH A SMOOTH CONTOUR

In the three preceding sections of this chapter we dealt only with very thin airfoils. However, of great practical importance are airfoils having a finite thickness (Kotovskii et al. 1980, Matveev and Molyakov 1988).

Let an airfoil contour L be described by a plane simple curve $x = x(t)$ and $y = y(t)$, $t \in [0, l]$, on the plane OXY, which may be either unclosed or closed (Muskhelishvili 1952). In the former case, it is supposed that at point $M(t)$ lying on the curve (whose radius-vector $\mathbf{r}_M = x\mathbf{i} - y\mathbf{j}$), $r'_M = \sqrt{x'^2(t) + y'^2(t)} \neq 0$ on $[0, l]$, both $x'(t)$ and $y'(t)$ meet the Hölder condition of degree α, i.e., $x'(t), y'(t) \in H(\alpha)$ (Muskhelishvili 1952). In the latter case, if point $M(x(0), y(0)) = M(x(l), y(l))$ is not a cusp (i.e., if the airfoil has no sharp edges), the $x'(0+0) = x'(l-0), y'(0+0) = y'(l-0)$. However, if the point is a cusp, then there exist one-sided unit vectors tangential at the point; otherwise, there exist $\boldsymbol{\tau}(0+0) = [x'(0+0)\mathbf{i} + y'(0+0)\mathbf{j}]/r'_{M(0+0)}$ and $\boldsymbol{\tau}(l-0) = [x'(l-0)\mathbf{i} + y'(l-0)\mathbf{j}]/r'_{M(l-0)}$. When L is a closed curve, then we put $l = 2\pi$.

Now let the airfoil be immersed into steady ideal incompressible flow characterized by the free-stream velocity vector $\mathbf{U}_0 = \mathbf{i}U_x + \mathbf{j}U_y$. Let the surface of the airfoil be modeled by a vortex sheet whose density at point $M(t) = M(x(t), y(t))$ of the curve L is equal to $\gamma(t)$ (Belotserkovsky 1967). Then the condition that the normal flow velocity component be equal to zero at point M_0 of the airfoil L (the condition of airfoil no-penetration at point M_0), i.e., the condition according to which the sum of the normal velocity components $\mathbf{U}_0$ and the velocity $\mathbf{V}$ induced by the airfoil vortex layer is equal to zero, may be represented in the form

$$\mathbf{V} \cdot \mathbf{n}_{M_0} = -\mathbf{U}_0 \cdot \mathbf{n}_{M_0}, \tag{10.4.1}$$

where

$$\mathbf{V} = \frac{1}{2\pi} \int_L \frac{\mathbf{i}y_1(t_0, t) - \mathbf{j}x_1(t_0, t)}{r^2_{MM_0}} \gamma(t)\, ds_t,$$

$$x_1(t_0, t) = x(t_0) - x(t_1), \qquad y_1(t_0, t) = y(t_0) - y(t),$$

$$r_{MM_0} = |\mathbf{r}_{MM_0}| = |\mathbf{i}x_1(t_0, t) + \mathbf{j}y_1(t_0, t)|,$$

$$\mathbf{n}_{M_0} = [\mathbf{i}y'(t_0) - \mathbf{j}x'(t_0)]/r'_{M_0},$$

Hence, Equation (10.4.1) may be written in the form

$$\frac{1}{2\pi}\int_L \frac{y'(t_0)y_1(t_0,t) + x'(t_0)x_1(t_0,t)}{r_{MM_0}^2 r'_{M_0}} \gamma(t) r'_M \, dt = f(t_0), \quad (10.4.2)$$

where $t_0 \in (0, l)$ and $f(t_0) = -\mathbf{U}_0 \cdot \mathbf{n}_{M_0}$.

The latter equation is valid for both unclosed and closed curves L.

If L is an unclosed curve, then Equation (10.4.2) may be written in the form

$$\frac{1}{2\pi}\int_0^l \frac{K(t_0,t)}{t_0 - t} \gamma(t) \, dt = f(t_0), \qquad t_0 \in (0, l), \qquad (10.4.3)$$

where

$$K(t_0,t) = \frac{y'(t_0)y_2(t_0,t) + x'(t_0)x_2(t_0,t)}{r_{2,MM_0}^2 r'_{M_0}} r'_M,$$

$$x_2(t_0,t) = \frac{x_1(t_0,t)}{t_0 - t}, \qquad y_2(t_0,t) = \frac{y_1(t_0,t)}{t_0 - t},$$

$$r_{2,MM_0}^2 = x_2^2(t_0,t) + y_2^2(t_0,t).$$

If functions $x'(t)$ and $y'(t)$ belong to the class $H(\alpha)$, then

$$K(t_0,t) \equiv 1, \qquad t_0 \in (0, l). \qquad (10.4.4)$$

From this equation it follows that Equation (10.4.3) may be presented in the form

$$\frac{1}{2\pi}\int_0^l \frac{\gamma(t) \, dt}{t_0 - t} + \int_0^l K_1(t_0,t)\gamma(t) \, dt = f(t_0), \qquad t_0 \in (0, l), \quad (10.4.5)$$

where

$$K_1(t_0,t) = \frac{k(t_0,t) - 1}{t_0 - 1}.$$

Let $x''(t_0), y''(t_0) \in H(\alpha)$ on $[0, l]$. Then according to Muskhelishvili (1952), $K(t_0, t)$ and $K_1(t_0, t)$ also belong to $H(\alpha)$ on $[0, l]$, uniformly with respect to each coordinate to the other.

Let now L be a closed curve ($l = 2\pi$). Then Equation (10.4.2) cannot be presented in the form (10.4.3) with a continuous kernel. $K(t_0, t)$ for x'', $y'' \dot{\in} H$ on $[0, l]$, because the kernel in Equation (10.4.2) tends to infinity for $t = t_0$ and $t = t_0 + l$. Therefore, first Equation (10.4.2) is written in the form

$$\frac{1}{2\pi}\int_0^{2\pi} \frac{\tilde{K}(t_0, t)}{\sin((t_0 - t)/2)} \gamma(t)\, dt = f(t_0), \qquad t_0 \in [0, 2\pi], \quad (10.4.6)$$

where

$$\tilde{K}(t_0, t) = \frac{y'(t_0)\tilde{y}_2(t_0, t) + x'(t_0)\tilde{x}_2(t_0, t)}{\tilde{r}^2_{2, MM_0} r'_{M_0}} r'_M,$$

$$\tilde{x}_2(t_0, t) = \frac{x_1(t_0, t)}{\sin((t_0 - t)/2)}, \qquad \tilde{y}_2(t_0, t) = \frac{y_1(t_0, t)}{\sin((t_0 - t)/2)},$$

$$\tilde{r}^2_{2, MM_0} = \tilde{x}^2_2(t_0, t) + \tilde{y}^2_2(t_0, t).$$

Functions $\tilde{x}_2(t_0, t)$ and $\tilde{y}_2(t_0, t)$ suffer discontinuities of the first kind, because $\tilde{x}_2(t_0, t) \to 2x'(t_0)$ for $t \to t_0$ and $\tilde{x}_2(t_0, t) \to -2x'(t_0)$ for $t \to t_0 + 2\pi$; however, function $\tilde{r}^2_{2, MM_0}$ is continuous at point t_0 and does not vanish on $[0, 2\pi]$.

Consider the functions

$$x_3(t_0, t) = \tilde{x}_2(t_0, t) - 2\cos\frac{t_0 - t}{2} x'(t_0),$$

$$y_3(t_0, t) = \tilde{y}_2(t_0, t) - 2\cos\frac{t_0 - t}{2} y'(t_0). \qquad (10.4.7)$$

Because $2\cos(\frac{1}{2})(t_0 - t)x'(t_0) \to 2x'(t_0)$ for $t \to t_0$ and $2\cos(\frac{1}{2})$ $(t_0 - t)x'(t_0) \to -2x'(t_0)$ for $t \to t_0 + 2\pi$, the functions $x_3(t_0, t)$ and $y_3(t_0, t)$ are continuous at point t_0. Thus, for a continuous curve, Equation (10.4.2) may be written in the form

$$\frac{1}{2\pi}\int_0^{2\pi}\cot\frac{t_0-t}{2}A(t_0,t)\gamma(t)\,dt$$

$$+\int_0^{2\pi}B(t_0,t)\gamma(t)\,dt = f(t_0),$$

$$A(t_0,t) = \frac{2r'_{M_0}r'_M}{\tilde{r}^2_{2,MM_0}},$$

$$B(t_0,t) = \frac{y'(t_0)\tilde{y}_3(t_0,t)+x'(t_0)\tilde{x}_3(t_0,t)}{\tilde{r}^2_{2,MM_0}r'_{M_0}}r'_M,$$

$$\tilde{x}_3(t_0,t) = \frac{x_3(t_0,t)}{\sin((t_0-t)/2)},$$

$$\tilde{y}_3(t_0,t) = \frac{y_3(t_0,t)}{\sin((t_0-t)/2)}. \qquad (10.4.8)$$

Now we note that $A(t_0,t_0)=\frac{1}{2}$, $t_0\in[0,2\pi]$. Hence, Equation (10.4.2) for a closed curve L finally may be written in the form

$$\frac{1}{4\pi}\int_0^{2\pi}\frac{t_0-t}{2}\gamma(t)\,dt+\int_0^{2\pi}\tilde{K}_1(t_0,t)\gamma(t)\,dt = f(t_0), \qquad (10.4.9)$$

where $t_0\in[0,2\pi]$,

$$\tilde{K}_1(t_0,t) = A(t_0,t)-\tfrac{1}{2}+B(t_0,t).$$

If $x''(t), y''(t)\in H(\alpha)$ on $[0,2\pi]$, then by using the results of Muskhelishvili (1952) again one can show that the kernel of Equation (10.4.9), $\tilde{K}_1(t_0,t)\in H(\alpha)$ on $[0,2\pi]$.

Now we can proceed to using the method of discrete vortices to solve numerically problems of steady flow past finite-thickness airfoils.

If L is an unclosed curve, then Equation (10.4.5) is a singular integral equation of the first kind with a Cauchy kernel on $[0,l]$, and, hence, discrete vortices and reference points must be distributed along $[0,l]$ as was done in Section 10.1 (see Systems (10.1.3), (10.1.6), and (10.1.7). If the first and second derivatives of the functions $y(t)$ and $x(t)$ are small in absolute values, then Equation (10.4.5) may be replaced by the equation

$$\frac{1}{2\pi}\int_0^{l}\frac{\gamma(t)\,dt}{t_0-t} = f(t_0). \qquad (10.4.10)$$

However, if an airfoil contour is described by $y = y(x)$, $x \in [-b, b]$, and $y'(x)$ and $y''(x)$ are sufficiently small in absolute values, then Equation (10.4.10) may be written in the form (to an accuracy of second-order infinitesimal quantities)

$$\frac{1}{2\pi}\int_{-b}^{b} \frac{\gamma(x)\,dx}{x_0 - x} = -U_x y'(x_0) + U_y. \tag{10.4.11}$$

This equation describes a thin, slightly curved airfoil.

Some examples of choosing vortex modes for such airfoils are presented in Figures 10.6 and 10.7.

Next we consider the situation when L is a closed contour. In this case discrete vortices are placed at points $(x(t_i), y(t_i))$ and reference points are placed at points $(x(t_{0i}), y(t_{0i}))$, $i = 1, 2, \ldots, n$, where $t_i \in [0, 2\pi]$ (interpreted as points of a unit-radius circle) divide the latter into n equal parts and t_{0i} are the middles of the parts. The flow conditions are formulated as follows:

1. If flow is known to be noncirculatory, then one has to consider the system of linear algebraic equations

$$\gamma_{0n} + \sum_{i=1}^{n} \Gamma_i w_{ij} = -V_j^*, \qquad j = 1, \ldots, n,$$

$$\sum_{i=1}^{n} \Gamma_i = 0, \tag{10.4.12}$$

where γ_{0n} is a regularizing variable. The results of using this scheme to calculate flow past a circular cylinder are presented in Figures 10.8 and 10.9.

2. If a point of the contour is known where the flow strength $\gamma(t)$ vanishes, then the positions of discrete vortices are chosen in such a way that the point is one of them. Let the number of the point be equal to n. Then the system

$$\gamma_{0n} + \sum_{i=1}^{n-1} \Gamma_i w_{ij} = -V_j^*, \qquad j = 1, \ldots, n, \tag{10.4.13}$$

is to be considered, and the distributed strength of the vortex layer is found from the formula

$$\gamma(t_i) = \frac{\Gamma_i}{\sqrt{x'^2(t_i) + y'^2(t_i)}\, 2\pi/n}, \qquad i = 1, \ldots, n. \tag{10.4.14}$$

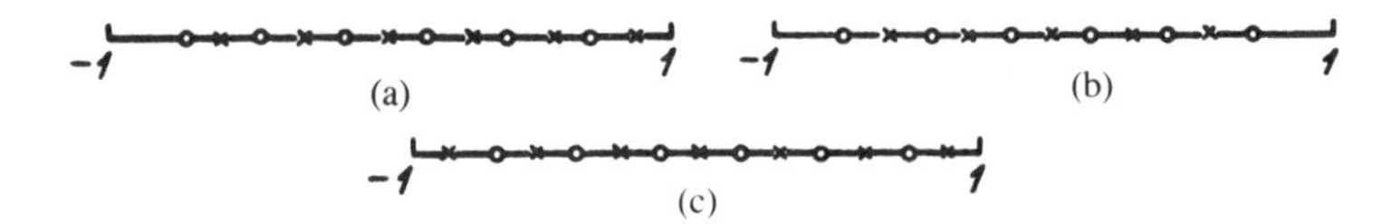

FIGURE 10.6. Distributions of discrete vortices (∘ ∘ ∘) and reference points (× × ×) for a thin, slightly curved airfoil considered for equally spaced grid points. (a), (b), and (c) correspond to the circulatory, noncirculatory, and finite-velocity problems, respectively.

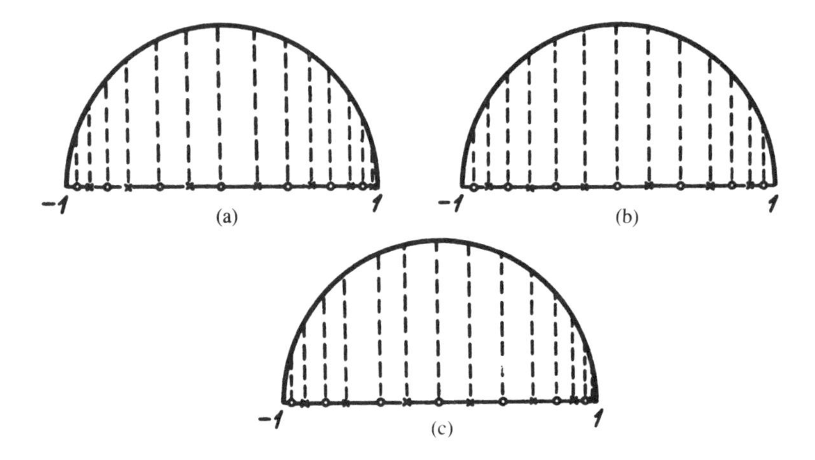

FIGURE 10.7. Distributions of discrete vortices (∘ ∘ ∘) and reference points (× × ×) for a thin, slightly curved airfoil considered for unequally spaced grid points. (a), (b), and (c) correspond to the circulatory, noncirculatory, and finite-velocity problems, respectively.

10.5. METHOD OF A "SLANTING NORMAL"

In the preceding section Equation (10.4.2) was derived appropriate for the case when the airfoil contour L is a smooth closed curve and the nondimensional thickness is sufficiently large (no less than 6–10%). However, for smaller nondimensional thicknesses, one has to retain a comparatively large number of discrete vortices. Even more complicated is the situation when a contour contains angular points or sharp edges. In the latter case the solution contains singularities due to the points, and the regular kernels of Equations (10.4.5) and (10.4.9) have nonintegrable singularities at the points. Although the method of discrete vortices allows us to solve the problem numerically, it is necessary to deal with a rather large number of discrete vortices. Therefore, the following approach is often used. It is especially useful for the cases when an airfoil has sharp edges or when its nondimensional thickness is small (Shipilov 1986).

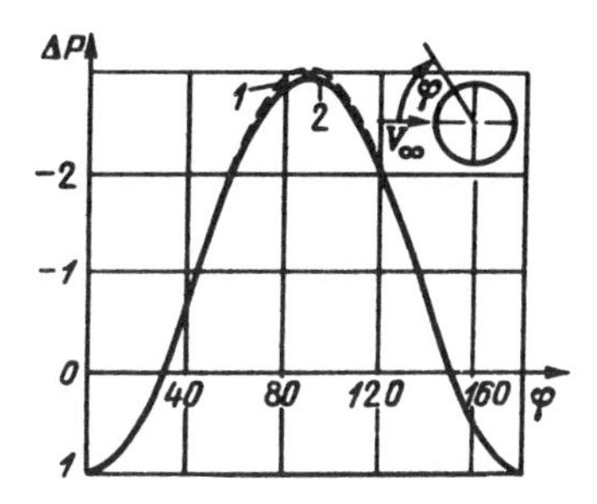

FIGURE 10.8. Distribution of Δp along a circle. The dashed and solid lines correspond to the exact and numerical solutions, respectively.

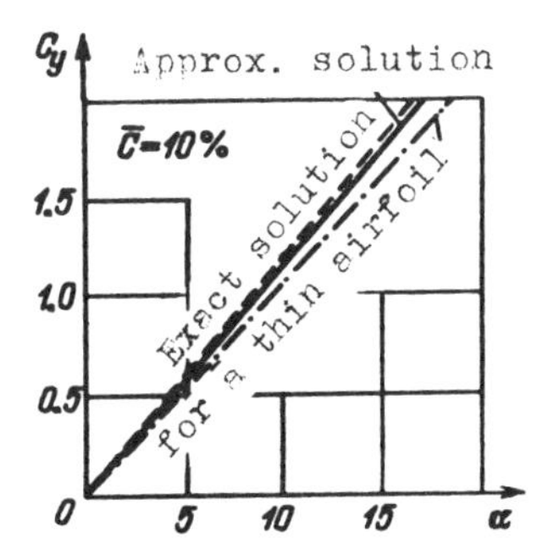

FIGURE 10.9. The dependence of coefficient C_y on the angle of attack α for a 10% symmetric airfoil.

An unclosed curve L_1 is chosen within an airfoil, with respect to which the closed contour L is divided into two unclosed smooth contours L^+ and L^- placed above and below the curve L_1, respectively (see Figure 10.10). If contour L contains a sharp edge, then it is a common end of the curves L^+, L^-, and L_1. Curve L_1 is such that the normal to it at each point M_0 crosses the curves L^+ and L^- at only one point, M_0^+ and M_0^-, respectively, thus establishing a one-to-one correspondence between the points of the curves L_1, L^+ and L_1, L^-. Let the parametric equations of the curves $L_1, L^\pm$ have the form $x = x_1(t)$, $y = y_1(t)$ and $x = x^\pm(t)$, $y = y^\pm(t)$, respectively, where $t \in [-1, 1]$. Let us denote the unit vector of the outer normal to the curve L on $L^\pm$ by $\mathbf{n}^\pm_{M_0^\pm}$. Then the no-penetration condition (10.4.1) for the contour L may be written in the form

$$\mathbf{V}(M_0^+) \cdot \mathbf{n}^+_{M_0^+} = -\mathbf{U}_0 \cdot \mathbf{n}^+_{M_0^+}, \qquad M_0^+ \in L^+,$$

$$\mathbf{V}(M_0^-) \cdot \mathbf{n}^-_{M_0^-} = -\mathbf{U}_0 \cdot \mathbf{n}^-_{M_0^-}, \qquad M_0^- \in L^-, \tag{10.5.1}$$

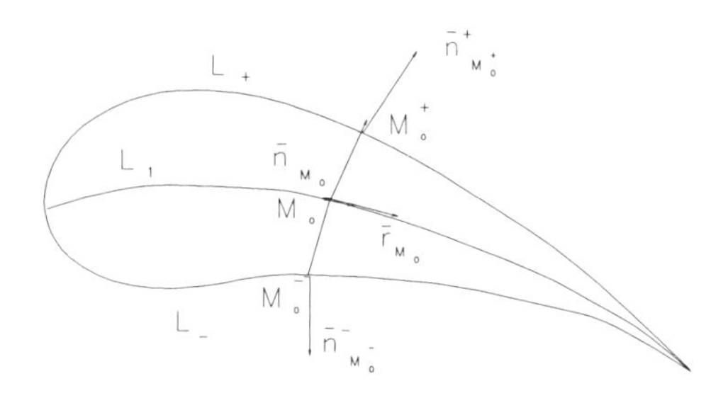

FIGURE 10.10. The principle of transference of boundary conditions from the surface of an airfoil onto the middle line. At point M_0 the boundary condition is determined by vectors $\mathbf{n}^+_{M_0^+}$ and $\mathbf{n}^-_{M_0^-}$, respectively.

where points $M_0^\pm \in L^\pm$ are taken at one and the same value of the parameter t_0. Let point $M_0 \in L_1$ correspond to points M_0^+ and M_0^-, and let $\mathbf{n}_{M_0}$ and $\boldsymbol{\tau}_{M_0}$ be unit vectors of the normal and the tangent to curve L_1 at point M_0 ($\mathbf{n}_{M_0^+}$ and $\mathbf{c}^+_{M_0^+}$ are directed to one side from L_1). Then one can write

$$\mathbf{n}^+_{M_0^+} = a^+(t_0)\mathbf{n}_{M_0} + b^+(t_0)\boldsymbol{\tau}_{M_0},$$

$$\mathbf{n}^-_{M_0^-} = a^-(t_0)\mathbf{n}_{M_0} + b^-(t_0)\boldsymbol{\tau}_{M_0}. \tag{10.5.2}$$

Instead of the velocity field $\mathbf{V}$ disturbed by the airfoil L, we shall consider the velocity field $\mathbf{V}_1$ induced by a vortex sheet of strength $\gamma(t)$ and a sheet of sources of strength $\mu(t)$ distributed along the curve L_1. We shall also require that Equalities (10.5.1) hold for the velocity field $\mathbf{V}_1$ at the points of the curve L_1. Thus, at the points outside L_1 one has

$$\mathbf{V}_1(M_0) = \frac{1}{2\pi}\int_{-1}^{1} \mathbf{w}(M_0, t)\gamma(t) r_M \, dt + \frac{1}{2\pi}\nabla\int_{-1}^{1} \mu(t) \ln\frac{1}{r_{MM_0}} \times r'_M \, dt, \tag{10.5.3}$$

where $M_0 \not\subset L_i$, $\mathbf{w}(t_0, t)$ is a factor before $\gamma(t)$ in (10.4.1), and ∇ is the gradient sign; also,

$$\mathbf{V}_1^+(M_0) \cdot \mathbf{n}^+_{M_0^+} = -\mathbf{U}_{M_0} \cdot \mathbf{n}^+_{M_0^+}, \qquad M_0 \in L_1,$$

$$\mathbf{V}_1^-(M_0) \cdot \mathbf{n}^-_{M_0^-} = -\mathbf{U}_{M_0} \cdot \mathbf{n}^-_{M_0^-}, \qquad M_0 \in L_1, \tag{10.5.4}$$

where $\mathbf{V}_1^{\pm}(M_0)$ is the flow field velocity $\mathbf{V}_1$ at point M_0 on the side of $\mathbf{n}^{\pm}_{M_0^{\pm}}$. As far as the vector field $\mathbf{V}_1$ is continuous, it approximates sufficiently well Equation (10.5.1) on curves L^+ and L^- and, hence, on curve L; in other words, the no-penetration condition may be approximately satisfied for the contour L. Calculated data confirm this conclusion (Shipilov 1986).

Let us rcplace the vectors $\mathbf{n}^{\pm}_{M_0^{\pm}}$ entering Equalities (10.5.4), according to (10.5.2). By using the properties of velocity fields induced by both the vortex and the source layer, one gets for points of L_1 (Belotserkovsky 1967, Tikhonov and Samarsky 1966)

$$\frac{a^{\pm}(t_0)}{2\pi}\int_{-1}^{1}\mathbf{w}(M_0,t)\cdot\mathbf{n}_{M_0}\cdot\gamma(t)r'_M\,dt \pm \frac{b^{\pm}(t_0)}{2}\gamma(t_0) + \frac{b^{\pm}(t_0)}{2}$$

$$\times\int_{-1}^{1}\mathbf{w}(M_0,t)\cdot\boldsymbol{\tau}_{M_0}\gamma(t)r'_M\,dt \pm \frac{a^{\pm}(t_0)}{2}\mu(t_0)$$

$$+\frac{a^{\pm}(t_0)}{2\pi}\int_{-1}^{1}\mu(t)$$

$$\times\ln\frac{1}{r_{MM_0}}\cdot r'_M\,dt + \frac{b^{\pm}(t_0)}{2\pi}\int_{-1}^{1}\mu(t)\frac{\partial}{\partial\tau_{M_0}}\ln\frac{1}{r_{MM_0}}\cdot r'_m\,dt$$

$$= f_1^{\pm}(t_0), \qquad t_0\in(-1,1). \tag{10.5.5}$$

Unlike the tangential component, the normal velocity component induced by the vortex layer has a singularity at the points of curve L_1, and an opposite statement is true for the components of a simple layer potential. Hence, Equations (10.5.5) form a system of two singular integral equations in the functions $\gamma(t)$ and $\mu(t)$. However, it is more convenient to pass from System (10.5.5) to the system

$$\frac{a^+t_0 \pm a^-(t_0)}{2\pi}\int_{-1}^{1}\mathbf{w}(M_0,t)\cdot\mathbf{n}_{M_0}\cdot\gamma(t)\,dt + \frac{b^+(t_0)\mp b^-(t_0)}{2}\cdot\gamma(t_0)$$

$$+\frac{b^+(t_0)\pm b^-(t_0)}{2\pi}\int_{-1}^{1}\mathbf{w}(M_0,t)\cdot\boldsymbol{\tau}_{M_0}\cdot\gamma(t)\,dt$$

$$+\frac{a^+(t_0)\mp a^-(t_0)}{2}\mu(t_0) + \frac{a^+(t_0)\pm a^-(t_0)}{2\pi}$$

$$\times \int_{-1}^{1} \mu(t) \frac{\partial}{\partial n_{M_0}} \left(\ln \frac{1}{r_{MM_0}} \right)$$

$$\times r'_M \, dt + \frac{b^+(t_0) \pm b^-(t_0)}{2\pi} \int_{-1}^{1} \mu(t) \frac{\partial}{\partial \tau_{M_0}} \left(\ln \frac{1}{r_{MM_0}} \right) r'_M \, dt$$

$$= f_1^+(t_0) \pm f_1^-(t_0). \tag{10.5.6}$$

Next consider the case when the contour L is symmetric with respect to L_1, and M_0^+ and M_0^- are the points where the normal to L_1 at point M_0 crosses the curves L^+ and L^- respectively (see Figure 10.10). Then, it can be assumed that in Equalities (10.5.2),

$$a^+(t) = -a^-(t), \qquad b^+(t_0) = b^-(t_0), \tag{10.5.7}$$

and hence, System (10.5.6) becomes

$$\frac{b^+(t_0)}{\pi} \int_{-1}^{1} \mathbf{w}(M_0, t) \cdot \boldsymbol{\tau}_{M_0} \gamma(t) r'_m \, dt + a^+(t_0) \mu(t_0)$$

$$+ \frac{b^+(t_0)}{\pi} \int_{-1}^{1} \mu(t) \frac{\partial}{\partial \tau_{M_0}} \left(\ln \frac{1}{r_{MM_0}} \right) r'_M \, dt = f_1^+(t_0) + f_1^-(t_0),$$

$$\frac{a^+(t_0)}{\pi} \int_{-1}^{1} \mathbf{w}(M_0, t) \cdot \mathbf{n}_{M_0} \gamma(t) r'_M \, dt + b^+(t_0) \gamma(t_0)$$

$$+ \frac{a^+(t_0)}{\pi} \int_{-1}^{1} \mu(t) \frac{\partial}{\partial n_{M_0}} \left(\ln \frac{1}{r_{MM_0}} \right) r'_M \, dt = f_1^+(t_0) - f_1^-(t_0).$$

$$\tag{10.5.8}$$

This system is a diagonal singular integral system because the first equation is singular with respect to one function, and the second equation is singular with respect to the other function. If under the circumstances L_1 is a segment of a straight line, then $\mathbf{w}(M_0, t) \cdot \boldsymbol{\tau}_{M_0} = \partial / \partial_{n_{M_0}}$ $(\ln(1/r_{MM_0}) \equiv 0$ for $M_0, M \in L_1$, and, hence, System (10.5.8) splits into two independent equations

$$a^+(t_0) \mu(t_0) + \frac{b^+(t_0)}{\pi} \int_{-1}^{1} \mu(t) \frac{\partial}{\partial \tau_{M_0}} \left(\ln \frac{1}{r_{MM_0}} \right) r'_M \, dt$$

$$= f_1^+(t_0) + f_1^-(t_0),$$

$$b^+(t_0)\gamma(t_0) + \frac{a^+(t_0)}{\pi}\int_{-1}^{1} \mathbf{w}(M_0, t) \cdot \mathbf{n}_{M_0}\gamma(t)\,dt$$

$$= f_1^+(t_0) - f_1^-(t_0), \tag{10.5.9}$$

each of which is a singular integral equation of the second kind on a segment with variable coefficients.

The solution to System (10.5.9) must be chosen subject to the condition that the disturbed flow velocity vanishes at a sharp edge and is limited at all the other points. An example of both analytic and numerical solutions to system (10.5.9) for the case when L is an ellipse is presented in Shipilov (1986).

10.6. A PERMEABLE AIRFOIL

Recently, much attention has been paid to calculating aerodynamic characteristics of parachutes and delta-wing planes (Belotserkovsky et al. 1987) whose surfaces are partially permeable. Therefore, of practical interest is the problem of flow past an unclosed contour L when the corresponding airfoil is permeable. In this case the boundary condition (10.4.1) for the velocity field is presented in the form

$$\mathbf{V}(M_0) \cdot \mathbf{n}_{M_0} + \mathbf{U}_0 \cdot \mathbf{n}_{M_0} = W(M_0), \tag{10.6.1}$$

where $W(M_0)$ is the cross-flow velocity at point M_0 of the curve L. It is supposed that $\mathbf{W}(M_0) = W(M_0) = \mathbf{n}_{M_0}$. Usually, the function $W(M_0)$ is assumed to depend on the pressure drop, i.e., $W(M_0) = F(\Delta p(M_0))$, and the pressure drop is expressible through the vortex sheet strength $\gamma(t_0)$ at point M_0 of an airfoil, which corresponds to the value of the parameter t_0 (Belotserkovsky et al. 1987). Thus, by modeling the contour L by a vortex layer of strength $\gamma(t)$ we arrive at a solution to the singular integral equation (see Equation (10.4.5))

$$F_1(\gamma(t_0)) + \frac{1}{2\pi}\int_{-1}^{1}\frac{\gamma(t)\,dt}{t_0 - t} + \int_{-1}^{1} K_1(t_0, t)\gamma(t)\,dt = f(t_0),$$

$$t_0 \in (-1, 1). \tag{10.6.2}$$

As a rule, the function $F(\Delta p(M_0))$ is obtained experimentally and depends on the material used to manufacture an airfoil. The concrete method for solving Equation (10.6.2) numerically depends to a large extent on the form of the function $F_1(\gamma(t_0))$.

If $F_1(\gamma(t_0)) \equiv 0$ (an airfoil is fully impermeable), then one arrives at a singular integral equation of the first kind for which a sufficiently large number of methods of numerical solution were previously presented.

However, if $F_1(\gamma(t_0)) = a(t_0)\gamma(t_0)$, i.e., is a linear function of $\gamma(t_0)$, then we arrive at a singular integral equation of the second kind with variable coefficients (see Chapter 7) for the methods of numerical solution). Some examples of numerical solutions of such equations may be found in Matveev and Molyakov (1988).

Finally, if $F_1(\gamma(t_0))$ is a nonlinear function of $\gamma(t_0)$, then Equation (10.6.2) may be solved numerically be employing the method of discrete vortices and an iterative procedure (Belotserkovsky et al. 1987). In this book we present some calculated data; however, the questions of verifying the numerical schemes remain open.

11

Three-Dimensional Problems

11.1. FLOW WITH CIRCULATION PAST A RECTANGULAR WING

Consider a rectangular wing, i.e., a plane plate lying in the plane OXZ within rectangle $\sigma = [-b, b] \times [-l, l]$. Let the flow past the wing be steady and characterized by the velocity vector $\mathbf{U}_0 = U_x \mathbf{i} + U_y \mathbf{j}$. According to Belotserkovsky (1967), the wing and the wake occupying the strip $[b, +\infty] \times [-l, l]$ may be modeled by a continuous vortex sheet; the latter may be modeled by discrete straight horseshoe vortices $\Pi_{ik} = \Pi(A_{ik}, A_{ik+1})$ of strength $\Gamma_{ik} = \gamma_z(x_i, z_{0k})h_1$, where $A_{ik}, = A(x_i, z_k)$, $x_i = -b + ih_1$, $h_1 = 2b/(n + 1)$, $i = 1, \ldots, n$, $x_{0i} = x_i + h_1/2$, $z_k = -l + (k - 1)h_2$, $h_2 = 2l/N$, $k = 1, \ldots, N + 1$, and $z_{0k} = z_k + h_2/2$. The subscript on γ indicates that we seek the components of the vortex sheet of the wing depending on the z coordinate; in what follows, the subscript z is omitted. In aerodynamics simulation of continuous vortex sheet of a wing and the vortex sheet downstream of it by discrete horseshoe vortices is justified from the physical point of view. Due to physical reasons, the aerodynamic problem under consideration is described by the function $\gamma(x, z)$, which becomes unlimited as it nears the leading edge $-b \times [-l, l]$ of a wing and vanishes as it nears all the other edges. The fact that the function vanishes at the trailing edge $b \times [-l, l]$ agrees with the Chaplygin–Joukowski condition of the flow shedding smoothly from the trailing edge. The problem thus formulated is called the circulatory problem of flow past a wing.

Let us denote by $V_{ik}^{jm} = \Gamma_{ik}\,\omega_{ik}^{jm}$ the normal velocity component induced by vortex Π_{ik} at the reference point $P_{jm} = (x_{0j}, z_{0m})$ and by V^{jm} the normal velocity component induced at the same point by the system of

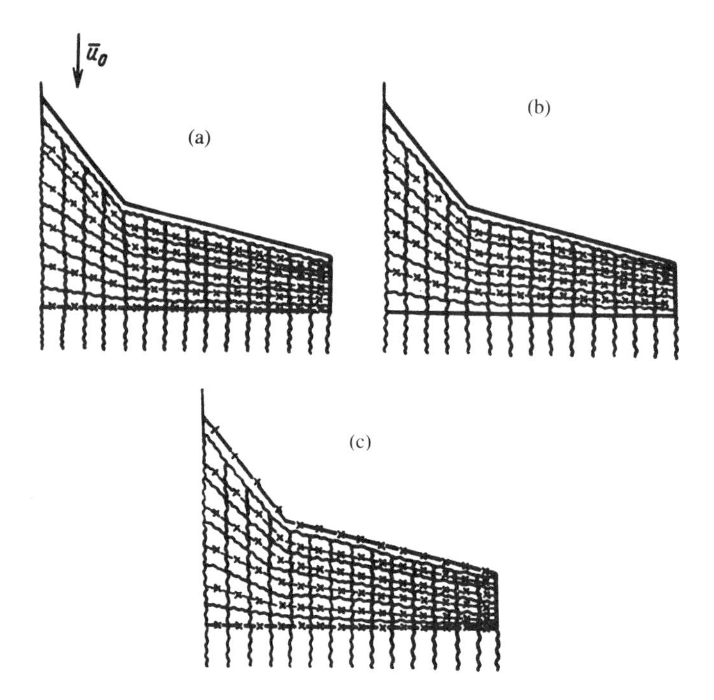

FIGURE 11.1. Schematic distribution of discrete horseshoe vortices (wavy lines) and reference points ($\times\times\times$) for a three-dimensional wing of a complicated plan form. (a), (b), and (c) correspond to the circulatory, noncirculatory, and finite-velocity problems, respectively.

straight discrete horseshoe vortices. By the B condition, the discrete vortices and reference points must be located near the leading and the trailing edges, respectively (see Figure 11.1a). Therefore, in Belotserkovsky (1967) it was proposed to find the values of $\gamma(x_i, z_{0k})$ corresponding, in the limit, to the preceding problem of aerodynamics, from a system of linear algebraic equations obtained by fulfilling the no-penetration condition at the reference points P_{jm}, $j = 1, \ldots, n$, $m = 1, \ldots, N$, i.e., from the system

$$\sum_{i=1}^{n} \sum_{k=1}^{N} \Gamma_{ik} \omega_{ik}^{jm} = -U_y, \qquad j = 1, \ldots, n, m = 1, \ldots, N, \quad (11.1.1)$$

or, in a detailed form (see (9.3.17) and (3.4.1), from

$$\frac{1}{4\pi} \sum_{i=1}^{n} \sum_{k=1}^{N} \gamma(x_i, z_{0k}) h_1 \left[\frac{K_2(x_i, z_{k+1}, x_{0j}, z_{0m})}{(x_{0j} - x_i)(z_{0m} - z_{k+1})} \right.$$

$$\left. - \frac{k_2(x_i, z_k, x_{0j}, z_{0m})}{(x_{0j} - x_i)(z_{0m} - z_k)} \right] = -U_y,$$

$$j = 1, \ldots, n, \; m = 1, \ldots, N,$$

$$K_2(x, z, x_0, z_0) = x_0 - x + \sqrt{(x_0 - x)^2 + (z_0 - z)^2}. \quad (11.1.2)$$

By employing the results of Section 3.4, we deduce that the latter system of linear algebraic equations approximates the integral equation

$$\frac{1}{4\pi} \int_{-b}^{b} \int_{-l}^{l} \frac{\gamma(x, z)}{(z_0 - z)^2} \left(1 + \frac{x_0 - x}{\sqrt{(x_0 - x)^2 + (z_0 - z)^2}} \right) dx \, dz = -U_y. \quad (11.1.3)$$

System (11.1.2) may be obtained from Equation (11.1.3) as follows. First, Equation (11.1.3) is written for each reference point P_{jm}, $j = 1, \ldots, n$, $m = 1, \ldots, N$, and then the quadrature formula (3.4.1) is applied to the integral entering the equation.

Next, upon adjusting System (11.1.2) to straight horseshoe vortices, the system can be shown to approximate:

1. An integral equation any of whose solutions vanish at the side edges.
2. An integral equation any of whose solutions vanish at the trailing edge and tend to infinity at the leading edge of a wing.

We start by writing System (11.1.2) in the form

$$\sum_{i=1}^{n} \sum_{k=1}^{N} \gamma(x_i, z_{0k}) h_1 F(z_{0m}, z_k, z_{k+1}) (1 + \operatorname{sign}(x_{0j} - x_i))$$

$$= -\sum_{\nu=1}^{n} \sum_{\mu=1}^{N} \gamma(x_\nu, z_{0\mu}) h_1 \left[\frac{r_{\nu, \mu+1}(x_{0j}, z_{0m}) - |x_{0j} - x_\nu|}{(x_{0j} - x_\nu)(z_{0m} - z_{\mu+1})} \right.$$

$$\left. + \frac{r_{\nu, \mu}(x_{0j}, z_{0m}) - |x_{0j} - x_\nu|}{(x_{0j} - x_\nu)(z_{0m} - z_\mu)} \right] - 4\pi U_y,$$

$$j = 1, \ldots, n, \; m = 1, \ldots, N, \quad (11.1.4)$$

$$F(z_{0m}, z_k, z_{k+1}) = \frac{1}{z_{0m} - z_{k+1}} - \frac{1}{z_{0m} - z_k},$$

$$r_{\nu, \mu}(x_{0j}, z_{0m}) = \sqrt{(x_{0j} - x_\nu)^2 + (z_{0m} - z_\mu)^2}.$$

Because

$$\operatorname{sign}(x_{0j} - x_i) = \begin{cases} 1, & i \leq j, \\ -1, & i > j, \end{cases} \tag{11.1.5}$$

System (11.1.4) becomes

$$\sum_{k=1}^{N} 2\left(\sum_{i=1}^{j} \gamma(x_i, z_{0k}) h_1\right)\left(\frac{1}{z_{0m} - z_{k+1}} - \frac{1}{z_{0m} - z_k}\right) = \Omega(x_{0j}, z_{0m}),$$

$$j = 1, \ldots, n,\ m = 1, \ldots, N, \tag{11.1.6}$$

where $\Omega(x_{0j}, z_{0m})$ denotes the right-hand side of System (11.1.4).

By using Theorem 5.1.8 one gets

$$2\sum_{i=1}^{j} \gamma(x_i, z_{0k}) h_1 = -\frac{1}{\pi^2} \sum_{m=1}^{N} \left(\sqrt{l^2 - z_{0m}^2} \sum_{\lambda=1}^{k} \frac{h_2}{\sqrt{l^2 - z_\lambda^2}\,(z_\lambda - z_{0m})}\right)$$

$$\times \Omega(x_{0j}\, z_{0m}) h_2 + \alpha(j, k),$$

$$j = 1, \ldots, n,\ k = 1, \ldots, N, \tag{11.1.7}$$

where $|\alpha(j, k)|$ coincides with the quantity $\theta(x_i, z_{0k})$ used in Theorem 8.3.1.

Now we see that System (11.1.2) does approximate the integral equation

$$2\int_{-b}^{x_0} \gamma(x, z)\, dx = \frac{1}{\pi^2} \int_{-l}^{l} \psi_l(z, z_0) \left[\int_{-b}^{b} \int_{-l}^{l} \frac{\gamma(x, \tau)}{(z_0 - \tau)^2}\right.$$

$$\times \left(\frac{x_0 - x}{\sqrt{(x_0 - x)^2 + (z_0 - \tau)^2}}\right.$$

$$\left.\left. - \operatorname{sign}(x_0 - x)\right) dx\, d\tau + 4\pi U_y\right] dz_0, \tag{11.1.8}$$

where the function $\psi_l(z, z_0)$ is defined by Formula (8.3.3).

From Equation (11.1.8) and the properties of the function $\psi_l(z, z_0)$ we deduce that

$$\int_{-b}^{x_0} \gamma(x, l)\, dx \equiv \int_{-b}^{x_0} \gamma(x, -l)\, dx \equiv 0,$$

which is equivalent to the identities

$$\gamma(x, -l) \equiv \gamma(x, l) \equiv 0. \tag{11.1.9}$$

Note that Equation (11.1.3) is equivalent to Equation (11.1.8) for functions $\gamma(x, z)$ meeting Condition (11.1.9).

On the other hand, System (11.1.2) is equivalent to the system

$$\sum_{i=1}^{n} \sum_{k=1}^{N} \gamma(x_i, z_{0k}) \frac{h_1}{x_{0j} - x_i} \Lambda(z_{0m}, z_{k+1}, z_k)$$

$$= - \sum_{\nu=1}^{n} \sum_{\mu=1}^{N} \gamma(x_\nu, z_{0\mu}) \big[K(x_{0j}, z_{0m}, x_\nu, z_{k+1})$$

$$-K(x_{0j}, z_{0m}, x_\nu, z_k) \big] - 4\pi U_y,$$

$$j = 1, \dots, n, \; m = 1, \dots, N,$$

$$\Lambda(z_{0m}, z_{k+1}, z_k) = \operatorname{sign}(z_{0m} + z_{k+1}) - \operatorname{sign}(z_{0m} - z_k)$$

$$= \begin{cases} 0, & k \neq m, \\ -2, & k = 1, \end{cases}$$

$$K(x_0, z_0, x, z) = \frac{x_0 - x + \sqrt{(x_0 - x)^2 + (z_0 - z)^2} - |z_0 - z|}{(x_0 - x)(z_0 - z)}. \tag{11.1.10}$$

Hence, by denoting the right-hand side of the latter system by $\Omega_1(x_{0j}, z_{0m})$ and applying Theorem 5.1.1, one gets

$$\gamma(x_i, z_{0m}) = \frac{1}{2\pi^2} \sqrt{\frac{b - x_i}{b + x_i}} \sum_{j=1}^{n} \sqrt{\frac{b + x_{0j}}{b - x_{0j}}}\, \Omega_1(x_{0j}, z_{0m})$$

$$\times \frac{h_1}{x_l - x_{0j}} + \alpha_1(i, m),$$

$$i = 1, \ldots, n, \; m = 1, \ldots, N, \quad (11.1.11)$$

where the quantity $|\alpha_1(i, m)|$ is the same as the quantity $\theta(t_{k_1}, t_{k_2})$ for System (8.2.33)–(8.2.35).

System (11.1.11) approximates an integral equation of the second kind whose solution (if it exists) meets the condition

$$\gamma(b, z) \equiv 0. \qquad \gamma(-b, z) \equiv \infty. \quad (11.1.12)$$

The integral equation has the form

$$\gamma(x, z_0) = \frac{1}{2\pi^2}\sqrt{\frac{b - x}{b + x}} \int_{-b}^{b} \int_{-l}^{l} \frac{1}{z_0 - z}$$

$$\times \left(-\pi + \int_{-b}^{b} \sqrt{\frac{b + x_0}{b - x_0}} \frac{\sqrt{(x_0 - \tau)^2 + (z_0 - z)^2}}{(x - x_0)(x_0 - \tau)} \, dx_0 \right) \gamma_z'$$

$$\times (\tau, z) \, d\tau \, dz - 2\sqrt{\frac{b - x}{b + x}} U_y.$$

Thus, if System (11.1.2) for straight horseshoe vortices is solvable and the sequence of its solutions is convergent, then in the limit the solutions result in the function $\gamma(x, z)$ possessing the required properties at the edges of the wing. In the following text, the system is shown to be solvable.

11.2. FLOW WITHOUT CIRCULATION PAST A RECTANGULAR WING

Consider the problem of noncirculatory flow past a wing. The problem arises, for example, when studying an interaction of flow with a stationary oscillating wing by employing the so-called virtual inertia. In this case all the edges function under the same conditions, and, hence, the z component $\gamma_z(x, z)$ of the continuous vortex sheet simulating a wing must tend to infinity when approaching either the leading or the trailing edge. Hence, by the B condition, discrete vortices must be located near the leading and the trailing edge (see Figure 11.1b). In this case the problem is modeled as follows. Oncoming flow is supposed to have the needed normal velocity component U_y; the vortex layer of the wing is modeled by straight discrete horseshoe vortices Π_{ik} with bounded vortices (A_{ik}, A_{ik+1}) parallel to the OZ axis and the free vortices $(A_{ik}, (+\infty, z_k))$ and $(A_{ik+1}, (+\infty, z_{k+1}))$ parallel to the OX axis. Because there must be no wake downstream of the wing (i.e., the strength of the vortex wake is equal to zero), the summary circulation due to the discrete vortices along each chord of the wing must be equal to zero:

$$\sum_{i=1}^{n} \Gamma_{ik} = 0, \qquad k = 1, \ldots, N.$$

Thus, to find the circulations Γ_{ik}, $i = 1, \ldots, n$, $k = 1, \ldots, N$, one has to consider the system of linear algebraic equations (Belotserkovsky 1967)

$$\sum_{i=1}^{n} \sum_{k=1}^{N} \Lambda_{ik}\, \omega_{ik}^{jm} = -U_y, \qquad j = 1, \ldots, n-1,\ m = 1, \ldots, N,$$

$$\sum_{i=1}^{n} \Gamma_{im} = 0, \qquad j = n,\ m = 1, \ldots, N. \tag{11.2.1}$$

By reasoning in the same way as in the case of circulatory flow past a wing, System (11.2.1) may be easily shown to approximate the integral equation (11.2.3) supplemented by the condition

$$\int_{-b}^{b} \gamma_z(x, z_0)\, dx \equiv 0. \tag{11.2.2}$$

Similarly, we can show that System (11.2.1) approximates the integral equation (11.2.8) supplemented by the additional condition (11.2.2), and, hence, the solution of this equation again satisfies Conditions (11.1.9). Thus, instead of System (11.1.10), one gets

$$\sum_{i=1}^{n} \gamma_z(x_i, z_{0m}) \frac{h_1}{x_{0j} - x_i} = -\frac{1}{2}\Omega_1(x_{0j}, z_{0m}),$$

$$j = 1, \ldots, n-1,\ m = 1, \ldots, N,$$

$$\sum_{i=1}^{n} \gamma_z(x_i, z_{0m}) h_1 = 0, \qquad j = n,\ m = 1, \ldots, N. \tag{11.2.3}$$

From Theorem 5.1.1 it follows that

$$\gamma_z(x_i, z_{0m}) = \frac{1}{2\pi^2\sqrt{b^1 - x_i^2}} \sum_{j=1}^{n} \sqrt{b^2 - x_{0j}^2} \frac{\Omega_1(x_{0j} z_{0m})h_1}{x_i - x_{0j}}$$

$$+ \alpha_2(i, m),$$

$$i = 1, \ldots, n, \; m = 1, \ldots, N, \quad (11.2.4)$$

where the quantity $|\alpha_2(i, m)|$ is of the same type as in Equation (11.1.11).

The latter system approximates an integral equation of the second kind whose solution (if it exists) tends to infinity when approaching the leading and the trailing edges.

In order to find the vortex sheet component parallel to the OX axis $\gamma_x(x, z)$, one can employ the equality (Bisplinghoff, Ashley, and Halfman 1955)

$$\frac{\partial \gamma_z(x, z)}{\partial z} = \frac{\partial \gamma_x(x, z)}{\partial x} \quad (11.2.5)$$

or consider the system of equations obtained from (11.2.1) by substituting z and x on the left-hand side for x and z, respectively, i.e., by considering straight horseshoe vortices $\Pi_{ik} = \Pi(A_{ik}, A_{i+1,k})$ of strength $\Gamma_{ik} = \gamma_x(x_{0i}, z_k)h_2$ composed of the vortex segment $(A_{ik}, A_{i+1,k})$ and a pair of semi-infinite vortices $(A_{ik}, (x_i, +\infty))$ and $(A_{i+1,k}, (x_{i+1}, +\infty))$ and putting $j = 1, \ldots, n$, $m = 1, \ldots, N - 1$.

When considering the noncirculatory problem, the following way of modeling a continuous vortex sheet by discrete vortices may also be readily employed. Because the vortex sheet exists only within the area occupied by the wing, it may be conveniently approximated by discrete closed vortices $\Pi_{0,ik} = \Pi(A_{ik}, A_{i,k+1}, A_{i+1,k+1}, A_{i+1,k})$ whose circulation is equal to $Q_{ik} = Q(x_{0i}, z_{0k})$ and that are composed of vortex segments $(A_{ik}, A_{i,k+1})$, $(A_{i,k+1}, A_{i+1,k+1})$, $(A_{i+1,k+1}, A_{i+1,k})$, and $(A_{i+1,k}, A_{ik})$, $i = 0, 1, \ldots, n$, $k = 1, \ldots, N$. From Figure 11.1b it is seen that

$$\gamma_z(x_i, z_{0k})h_1 = \frac{Q(x_{0i}, z_{0k}) - Q(x_{0i-1}, z_{0k})}{h_1} h_1 \quad (11.2.6)$$

because the segment $[A_{ik}, A_{i,k+1}]$ is a part of the discrete vortices $\Pi_{0,i-1,k}, \Pi_{0,ik}$ characterized, in the former case, by circulation—in the negative direction (with respect to the OZ axis) and in the latter case by circulation in the positive direction. Analogously, we have

$$\gamma_x(x_{0i}, z_k)h_2 = \frac{Q(x_{0i}, z_{0k}) - Q(x_{0i}, z_{0k-1})}{h_2}h_2. \qquad (11.2.7)$$

Here $Q(x_{0i-1}, z_{0k}) = 0$ for $i = 0$, $k = 1, \ldots, N$ and $Q(x_{0i}, z_{0k-1}) = 0$ for $k = 1$, $i = 0, 1, \ldots, n$.

From Relationships (11.2.6) and (11.2.7), it follows that

$$Q(x_{0j}, z_{0m}) = \sum_{i=0}^{j} \gamma_z(x_i, z_{0m})h_1 = \sum_{k=1}^{m} \gamma_x(x_{0j}, z_k)h_2, \qquad (11.2.8)$$

where $j = 0, 1, \ldots, n$, $m = 1, \ldots, N$.

Because the functions $\gamma_z(x, z)$ and $\gamma_x(x, z)$ may have at the edges only integrable power singularities of the order of $\rho^{-1/2}$, where ρ is the distance from an edge, and because the flow is noncirculatory, we deduce that the function $Q(x, z)$ must vanish at all the edges of the wing as $\rho^{1/2}$.

The no-penetration boundary condition must be fulfilled at the reference points (x_{0j}, z_{0m}), $j = 0, 1, \ldots, n$, $m = 1, \ldots, N$, and this results in the system of linear algebraic equations

$$\sum_{i=1}^{n} \sum_{k=1}^{N} Q_{ik}\, \omega_{ik}^{jm} = -U_y, \qquad j = 0, 1, \ldots, n,\ m = 1, \ldots, N, \qquad (11.2.9)$$

where ω_{ik}^{jm} is the factor before $j\Gamma$ in Formula (9.3.23) for $x = x_{0j}$, $x_1 = x_i$, $x_2 = x_{i+1}$, $z_0 = z_{0m}$, $z_1 = z_k$, and $z_2 = z_{k+1}$.

By using Formula (9.3.24) one gets

$$\sum_{i=0}^{n} \sum_{k=1}^{N} Q(x_{0i}, z_{0k}) \int_{z_k}^{z_{k+1}} \int_{x_i}^{x_{i+1}} \frac{dx\,dz}{\left[(x_{0j} - x)^2 + (z_{0m} - z)^2\right]^{3/2}} = -U_y,$$

$$j = 0, 1, \ldots, n,\ m = 1, \ldots, N. \qquad (11.2.10)$$

From Section 3.4 it follows (see Formula (3.4.17)) that System (11.2.10) approximates the integral equation

$$\int_{-b}^{b} \int_{-l}^{l} \frac{Q(x, z)\,dx\,dz}{\left[(x_0 - x)^2 + (z_0 - z)^2\right]^{3/2}} = -U_y. \qquad (11.2.11)$$

In the class of functions $Q(x, z)$ vanishing at the edges of a wing, Equation (11.2.11) is equivalent to Equation (11.1.3) supplemented by Condition

(11.2.2), i.e., to the equation

$$\int_{-b}^{b}\int_{-l}^{l}\frac{\gamma_z(x,z)(x_0-x)\,dx\,dz}{(z_0-z)^2\sqrt{(x_0-x)^2+(z_0-z)^2}}=-U_y \quad (11.2.12)$$

where $\gamma_z(x,z)=\frac{\partial}{\partial x}Q(x,z)$. Equation (11.2.12) is consistent with Equation (11.2.6). This statement may be easily proved by integrating the left-hand side of Equation (11.2.11) by parts in x and taking into account the condition $Q(b,z)\equiv Q(-b,z)\equiv 0$.

Finally, we observe that System (11.2.10) is well-conditioned because its matrix corresponds to the Hadamard criterion. In fact, if $(x_i, z_k)\neq(x_j, x_m)$, then

$$a_{ik}^{jm}=\int_{z_k}^{z_{k+1}}\int_{x_i}^{x_{i+1}}\frac{dx\,dz}{\left[(x_{0j}-x)^2+(z_{0m}-2)^2\right]^{3/2}}>0, \quad (11.2.13)$$

and if $i=j$, $k=m$, then from (3.4.17) it follows that

$$a_{jm}^{jm}=\int_{z_m}^{z_{m+1}}\int_{x_j}^{x_{j+1}}\frac{dx\,dz}{\left[(x_{0j}-x)^2+(x_{0m}-z)^2\right]^{3/2}}<0. \quad (11.2.14)$$

The sum of all elements a_{ik}^{jm} of the row (j,m) in the matrix of System (11.2.10) is equal to

$$\sum_{i=1}^{n}\sum_{k=1}^{N}a_{ik}^{jm}=\int_{-b}^{b}\int_{-l}^{l}\frac{dx\,dz}{\left[(x_{0j}-x)^2+(z_{0m}-z)^2\right]^{3/2}}<0. \quad (11.2.15)$$

Thus, from Formulas (11.2.13)–(11.2.15) one gets

$$\sum_{\substack{i=0\\ i\neq j}}^{n}\sum_{\substack{k=1\\ k\neq m}}^{N}a_{ik}^{jm}<-a_{jm}^{jm}=\left|a_{jm}^{jm}\right|. \quad (11.2.16)$$

This inequality expresses the Hadamard criterion.

Because for any numbers a_i and b_i the equality

$$\sum_{i=1}^{n}a_ib_i=-\sum_{i=1}^{n}\left(\sum_{k=1}^{i}a_k\right)(b_{i+1}-b_i) \quad (11.2.17)$$

holds where $b_{n+1} = 0$), System (11.2.1) is equivalent to a system of the form (11.2.10) and, hence, is also well-conditioned.

Note 11.2.1. By employing Formula (9.3.17) it also can be shown that the system of linear algebraic equations (11.1.2) for the problem of circulatory flow past a rectangular wing is well-conditioned.

11.3. A WING OF AN ARBITRARY PLAN FORM

In this section we consider wings of complicated plan forms as well as a schematic flying vehicle as a whole.

Let us start by considering a plane wing in the form of a canonical trapezoid (a trapezoid lying in the plane OXZ whose two sides are parallel to the axis OX; see Figure 11.1). We will show that the results obtained previously for a rectangular wing are also valid for this case. Then, we will continue by considering a complicated lifting surface divided into canonical trapezoids (see Figure 11.2).

Thus, we will consider the linear problem of steady flow past a canonical trapezoid σ whose sides lie on the straight lines $z = 0$ and $z = l$, and the leading and the trailing edges are specified by the equations

$$x_-(z) = a_- + zb_-, \qquad x_+(z) = a_+ + zb_+. \tag{11.3.1}$$

The oncoming flow velocity will be specified as $\mathbf{U}_0 = U_x\mathbf{i} + U_y\mathbf{j}$, $U_x > 0$.

In Belotserkovsky (1967) it was proposed to solve the problem of finding the strength of a continuous vortex sheet modeling trapezoid σ by using the method of slanting horseshoe vortices constructed in the following way. Each side edge of trapezoid σ is divided into $n + 1$ equal parts. The grid points are connected by straight lines on which the bound vortices (the vortex segments of slanting horseshoe vortices) are located. Segment $[0, l]$ of the OZ axis is divided into N equal parts, and straight lines parallel to the OX axis are drawn through the grid points (including the end points $z = 0$ and $z = l$). The points of intersection of the straight lines parallel to the OX axis and passing through the middles of the segments formed by the division of segment $[0, l]$ of the OZ axis and of the straight lines connecting the middles of the corresponding segments of the side edges of trapezoid σ are assumed to be the reference points.

The following approach will be used for describing discrete vortices and reference points. Let $D = [0, 1] \times [0, l]$ be a rectangle lying in the plane OX^1Z and F be the mapping of the rectangle onto the trapezoid described by Formula (3.2.16). Let us next choose points $x_i^1 = ih_1$, $x_{0i}^1 = x_i^1 + h_1/2$, $h_1 = 1/(n + 1)$, $i = 0, 1, \ldots, n$, $z_k = (k - 1)h_2$, $z_{0k} = z_k + h_2/2$, $h_2 = l/N$, $k = 1, \ldots, N + 1$. In what follows it is assumed that $h/h_k \leq R < +\infty$,

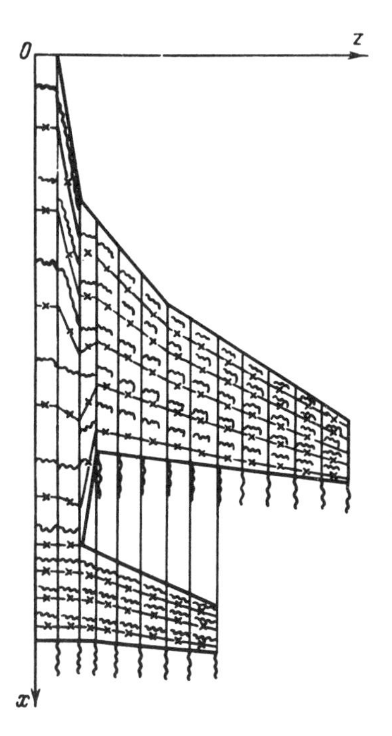

FIGURE 11.2. The vortex system for a schematic aircraft in the framework of the circulatory problem.

$h = \max(h_1, h_2)$, $k = 1, 2$, for $h \to 0$. By $A_{i,k} = (x_{i,k}, z_k)$, $A_{i,0k} = (x_{i,0k}, z_{0k})$, and $A_{0i,0k} = (x_{0i,0k}, z_{0k})$ we denote points of the canonical trapezoid σ that are, respectively, images of point $A^1_{i,k} = (x^1_i, z_k)$, $A^1_{i,0k} = (x^1_i, z_{0k})$, and $A^1_{0i,0k} = (x^1_{0i}, z_{0k})$ belonging to rectangle D, obtained by using the mapping F. By Π_{ik} we denote the slanting horseshoe vortex incorporating the bound vortex $(A_{i,k}, A_{i,k+1})$, whose strength is $\Gamma_{ik} = \gamma(x_{i,0k}, z_{0k})(x_{i+1,0k} + x_{i,0k})$. Note that by the definition of the mapping F,

$$x_{i+1,0k} - x_{i,0k} = J(z_{0k})h_1, \tag{11.3.2}$$

where $J(z_{0k})$ is the Jacobian of the mapping F. Hence, the strength of vortex Π_{ik} is given by

$$\Gamma_{ik} = \varphi_{ik} h_1, \tag{11.3.3}$$

where $\varphi_{ik} = \gamma(x_{i,0k}, z_{0k})J(z_{0k})$, $i = 1, \ldots, n$, $k = 1, \ldots, N$.

Flow velocities induced by the slanting horseshoe vortices will be computed at reference points $P_{jm} = (x_{0j0m}, z_{0m})$, which are images of the points $P^1_{jm} = (x^1_{0j}, z_{0m})$. Note that the line of the bound vortices

$(A_{i,k}, A_{i,k+1})$, $k = 1, \ldots, N$, is the image of the straight line $x^1 = x_i^1$ for the mapping F, and, hence, the equation of the line is given by

$$x_i(z) = x(x_i^1, z) = a(x_i^1) + zb(x_i^1),$$

$$a(x_i^1) = a_- + x_i^1(a_+ - a_-), \qquad b(x_i^1) = b_- + x_i^1(b_+ - b_-). \quad (11.3.4)$$

Let V_{ik}^{jm} be the normal velocity component induced by the slanting horseshoe vortex Π_{ik} at point P_{jm}, and let V_{jm} be the normal velocity component induced at the point by the entire system of slanting horseshoe vortices. Then, by using Formula (9.3.20) one gets

$$V_{jm} = \sum_{i=1}^{n} \sum_{k=1}^{N} V_{ik}^{jm} = \sum_{i=1}^{n} \sum_{k=1}^{N} \Gamma_{ik} \omega_{ik}^{jm},$$

$$\omega_{ik}^{jm} = \frac{1}{4\pi}\left[\frac{1}{z_{0m} - z_{k+1}} - \frac{1}{z_{0m} - z_k} + \frac{r(A_{i,k+1}, P_{j,m})}{\lambda_{0j,0m,i}(z_{0m} - z_{k+1})} - \frac{r(A_{i,k}, P_{j,m})}{\lambda_{0j,0m,i}(z_{0m} - z_k)}\right],$$

$$r(A_{i,k}, P_{j,m}) = \sqrt{(x_{0j,0m} - x_{i,k})^2 + (z_{0m} - z_k)^2},$$

$$\lambda_{0j,0m,i} = x\left(x_{0j}^1, z_{0m}\right) - x(x_i^1, z_{0m}) = \left(x_{0j}^1 - x_i^1\right)J(z_{0m}). \quad (11.3.5)$$

From Formula (11.3.5) and the results of Section 3.4 it follows that if function $\varphi(x^1, z) = \gamma(x(x^1, z)z)J(z)$ is such that $\partial\varphi/\partial z \in H^*$ on D, then

$$V_{jm} = \frac{1}{4\pi}\iint_\sigma \frac{\gamma(x,z)}{(z_{0m} - z)^2}\left(1 + \frac{x_{0j} - x}{\sqrt{(x_{0j} - x)^2 + (z_{0m} - z)^2}}\right) dx\, dz + \alpha(x_{0j}, z_{0m}), \quad (11.3.6)$$

where the quantity $|\alpha(x_{0j}, z_{0m})|$ is the same as $\theta(x_{0j}\, z_{0m})$ entering Equation (3.4.3).

Because we consider circulatory flow past a canonical trapezoid σ, or by the B condition discrete vortices must be located near the leading edge, and reference points must be located near the trailing edge. Therefore the work of Belotserkovsky (1967) proposed to calculate circulations Γ_{ik} of the discrete vortices from the following system of linear algebraic equations

$$\sum_{i=1}^{n} \sum_{k=1}^{N} \Gamma_{ik} \omega_{ik}^{jm} = V_{jm}^{*}, \qquad j = 1, \ldots, n, \; m = 1, \ldots, N, \quad (11.3.7)$$

where V_{jm}^{*} is the normal component of the oncoming flow velocity at the reference point P_{jm}.

Now we see that taking into account (11.3.6), the latter system of equations approximates the integral equation

$$\frac{1}{4\pi} \iint_{\sigma} \frac{\gamma(x, z)}{(z_0 - z)^2} \left(1 + \frac{x_0 - x}{\sqrt{(x_0 - x)^2 + (z_0 - z)^2}} \right) dx\, dz$$

$$= -V^{*}(x_0, z_0). \qquad (11.3.8)$$

Next, following the procedure of Sections 11.1. and 11.2, we will show that in the framework of the method of slanting horseshoe vortices for circulatory flow, System (11.3.7) (1) is well-conditioned, (2) approximates an integral equation any solution of which vanishes at the side edges, and (3) approximates an integral equation any solution of which vanishes at the trailing edge and tends to infinity at the leading edge of the trapezoid σ. Thus, the method of slanting horseshoe vortices will be shown to single out the solution to Equation (11.3.8) that corresponds to the physical flow pattern of the problem under consideration.

To prove the first statement, we have to apply again the transform (11.2.17) to the summands $\Sigma_{i=1}^{n} \Gamma_{ik} \omega_{ik}^{jm}$ for a fixed k, having denoted the quantities $\gamma(x_{i,0k}, z_{0k}) J(z_{0k})$ and ω_{ik}^{jm} by a_k and b_i, respectively. Then we have to employ Formula (9.3.20) and note that

$$\omega_{i+1,k}^{jm} - \omega_{ik}^{jm} = \int_{z_k}^{z_{k+1}} \int_{x(x_i^1, z)}^{x(x_{i+1}^1, z)} \frac{dx\, dz}{\left[(x_0 - x)^2 + (z_0 - z)^2 \right]^2}, \quad (11.3.9)$$

where $x(x_i^1, z)$ and $x(x_{i+1}^1, z)$ are equations describing the lines on which the bound vortices $(A_{i,k}, A_{i,k+1})$ and $(A_{i+1,k}, A_{i+1,k+1})$ lie. Finally, it suffices to show that the Hadamard criterion holds for the transformed system (as was done before for System (11.2.10)).

In order to prove the second and the third statements, we note that

$$r(A_{i,k}, P_{j,m}) = \sqrt{\begin{matrix} \lambda_{0j,0m,i}^2 + 2\lambda_{0j,0m\,i} b(x_i^1)(z_{0m} - z_k) \\ +(z_{0m} - z_k)^2 \theta^2(x_i^1) \end{matrix}},$$

$$\theta^2(x_i^1) = 1 + b^2(x_i^1). \tag{11.3.10}$$

Therefore, System (11.3.7) may be written similarly to System (11.1.6):

$$2\sum_{k=1}^{N}\left(\sum_{i=1}^{j}\varphi_{ik}h_1\right)\left(\frac{1}{z_{0m} - z_{k+1}} - \frac{1}{z_{0m} - z_k}\right)$$

$$= -4\pi V_{jm}^* - \sum_{\nu=1}^{n}\sum_{\mu=1}^{N}\varphi_{\nu\mu}h_1\left(\frac{r(A_{\nu,\mu+1}, P_{jm}) - |\lambda_{0j,0m,i}|}{\lambda_{0j,0m,i}(z_{0m} - z_{\mu+1})}\right.$$

$$\left. - \frac{r(A_{\nu\mu}, P_{jm}) - |\lambda_{0j,0m,i}|}{\lambda_{0j,0m,i}(z_{0m} - z_\mu)}\right),$$

$$j = 1, \dots, n,\ m = 1, \dots, N. \tag{11.3.11}$$

By solving System (11.3.11) with respect to $(\Sigma_{i=1}^{j}\varphi_{ik}h_1)$ and passing to the limit, we deduce that System (11.3.11) approximates the integral equation

$$2\int_0^{x_0^1}\varphi(x^1, z)\,dx^1 = -\frac{1}{\pi^2}\int_0^l \psi(z, z_0)\Omega_2(x_0, z_0)\,dz_0, \tag{11.3.12}$$

where

$$\psi(z, z_0) = \sqrt{z_0(l - z_0)}\int_0^z \frac{d\nu}{\sqrt{\nu(l - \nu)}\,(\upsilon - z_0)},$$

and $\Omega_2(x_0, z_0)$ is the limit of the right-hand side of System (11.3.11) for n, $N \to \infty$. From Equation (11.3.12) it follows that

$$\varphi(x^1, 0) \equiv \varphi(x^1, l) \equiv 0. \tag{11.3.13}$$

As far as for the trapezoid under consideration, $J(z) > 0$ on $[0, l]$, the following identity follows from Identities (11.3.13):

$$\gamma\big(x(^1, 0), 0\big) \equiv \gamma(x(x^1, l), l) \equiv 0, \tag{11.3.14}$$

according to which function $\gamma(x, z)$ vanishes at the side edges of the trapezoid σ.

Note 11.3.1. If one of the side edges of the trapezoid σ degenerates into a point (i.e., σ is a triangle), then the Jacobian $J(z)$ vanishes at the point, and hence, the preceding analysis does not allow us to determine the value of $\gamma(x, z)$ at the point. This issue still awaits analysis.

Let us again use Formula (11.3.10) for transforming System (11.3.7) as follows:

$$\sum_{i=1}^{n} \sum_{k=1}^{N} \varphi_{ik} \Lambda(z_{0m}, z_{k+1}, z_k) \frac{\theta(x_i^1)}{J(z_{0m})} \frac{h_1}{x_{0j}^1 - x_i^1}$$

$$= -4\pi V_{jm}^* + \Omega_3\left(x_{0j}^1, z_{0m}\right), \qquad j = 1, \ldots, n,\ m = 1, \ldots, N, \tag{11.3.15}$$

where $\Omega_3(x_{0j}^1, z_{0m})$ denotes the terms transferred from the left-hand to the right-hand side. Then, by applying to System (11.3.15) considerations similar to those applied to System (11.3.10) and passing to the limit, one gets

$$2\gamma\left(x(x^1, z_0), z_0\right)\sqrt{1 + b^2(x^1)} = \frac{4}{\pi}\sqrt{\frac{1 - x^1}{x^1}} \int_0^1 \sqrt{\frac{x_0^1}{1 - x_0^1}}$$

$$\times\left(-4\pi U_y + \Omega_3(x_0^1, z_0)\right)\frac{dx_0^1}{x^1 - x_0^1}. \tag{11.3.16}$$

According to Equation (11.3.16),

$$\gamma(x(1, z_0), z_0) \equiv \gamma(x_+(z_0), z_0) \equiv 0,$$

$$\gamma(x(0, z_0), z_0) \equiv \gamma(x_-(z_0), z_0) = \infty. \tag{11.3.17}$$

Next we will consider noncirculatory flow past a canonical trapezoid σ. In analogy to the noncirculatory flow problem, simulation of the vortex sheet by slanting horseshoe vortices in the framework of the present problem allows us to conclude that circulations Γ_{ik} of the vortices may be found by considering the system of linear algebraic equations

$$\sum_{i=1}^{n}\sum_{k=1}^{N}\Gamma_{ik}\,\omega_{ik}^{jm} = -U_y, \qquad j = 1,\ldots,n-1,\, m = 1,\ldots,N,$$

$$\sum_{i=1}^{n}\Gamma_{im} = 0, \qquad j = n,\, m = 1,\ldots,N. \tag{11.3.18}$$

Similarly to the circulatory problem, we deduce that in the present case the strength $\gamma(x,z)$ of the continuous vortex sheet on σ satisfies Equation (11.3.8) supplemented by the condition

$$\int_{x_-(z)}^{x_+(z)}\gamma(x,z)\,dx \equiv 0, \qquad z \in [0,l], \tag{11.3.19}$$

according to which the vortex sheet circulation along any chord of the wing is zero.

Note that by introducing the function

$$Q(x,z) = \int_{x_-(z)}^{x}\gamma(x,z)\,dx, \tag{11.3.20}$$

one can prove that Equation (11.3.8) supplemented with Condition (11.3.19) is equivalent to the equation

$$\int_{-l}^{l}\int_{x_-(z)}^{x_+(z)}\frac{Q(x,z)}{\left[(x_0-x)^2+(z_0-z)^2\right]^{3/2}}\,dx\,dx = -U_y, \tag{11.3.21}$$

because function $Q(x,z)$ vanishes at all the edges of a wing. In fact, similarly to the circulatory problem, function $\gamma(x,z)$ may be shown to vanish at the side edges because system (11.3.18) approximates an integral equation whose solutions possess the preceding property. Similarly to the previous section, function $\gamma(x,z)$ may be shown to tend to infinity at both the leading and trailing edge of the canonical trapezoid σ.

Finally, note that System (11.3.18) is well-conditioned.

Similarly to the preceding section, we deduce that $Q(x_{i,k}, z_k)$ is the strength of a closed discrete vortex composed of the segments $[A(_{i,k}, z_k)$, $A(x_{i,k+1}, z_{k+1})]$, $[A(x_{i,k+1}, z_{k+1})$, $A(x_{i+1,k+1}, z_{k+1})]$, $[A(x_{i+1,k+1}, z_{k+1})$, $A(x_{i+1,k}, z_k)]$, and $[A(x_{i+1,k}, z_k)$, $A(x_{i,k}, z_k)]$.

Note 11.3.2. The results obtained in this section are fully applicable to a wing equipped with a flap, i.e., when the right-hand side of Equation

(11.3.8) suffers a discontinuity of the first kind on the straight line $x = x(q^1, z)$ where $x(x^1, z) = x^1[x_+(z) - x_-(z)] + x_-(z)$. However, in this case point q^1 must be placed midway between the nearest points x_i^1 and x_{0j}^1.

Next we consider steady flow past a finite-span wing of a complicated plan form and a schematic flying vehicle (see Figure 11.2). The lifting surface σ lies in the plane OXZ and its contour is composed of segments. By drawing through the angular points of the contour straight lines parallel to the OX axis, we divide the surface σ into canonical trapezoids σ_ϵ, $\epsilon = 1, \ldots, p$, which can intersect each other along side edges only. Let the side edges σ_ϵ be specified by the equations $z = l_\epsilon^1$, $z = l_\epsilon^2$, $l_\epsilon^1 < l_\epsilon^2$, and the leading and the trailing edge by equations

$$x_-^\epsilon(z) = a_-^\epsilon + zb_-^\epsilon, \qquad x_+^\epsilon(z) = a_+^\epsilon + zb_+^\epsilon, \qquad (11.3.22)$$

respectively, where $\epsilon = 1, \ldots, p$ and a_+^ϵ, a_-^ϵ, b_+^ϵ, and b_-^ϵ are constants.

Let us consider p planes $OX^\epsilon Z$ and choose on each of them a rectangle $D_\epsilon = [0, 1] \times [l_\epsilon^1, l_\epsilon^2]$. Next consider mapping F_ϵ of the rectangle onto σ_ϵ, specified by the formula

$$x(x^\epsilon, z) = x^\epsilon[x_+^\epsilon(z) - x_-^\epsilon(z)] + x_-^\epsilon(z), \qquad z = z. \quad (11.3.23)$$

The Jacobian of the mapping is equal to $J_\epsilon(z) = x_+^\epsilon(z) - x_-^\epsilon(z)$. Similarly to the way it was done before for a canonical trapezoid, we choose points $x_{i_\epsilon}^\epsilon$ and $x_{0i_\epsilon}^\epsilon$, $i = 1, \ldots, n_\epsilon$, on segment $[0, 1]$ of the axis OX^ϵ separated by distances h_1^ϵ and points z_{k_ϵ} and z_{0k_ϵ}, $k = 1, \ldots, N_\epsilon$, on segment $[l_\epsilon^1, l_\epsilon^2]$ separated by distances h_2^ϵ. If trapezoids σ_ϵ and σ_ν are located one after the other in the oncoming flow direction (i.e., if $l_\epsilon^k = l_\nu^k$, $k = 1, 2$), then $h_2^\epsilon = h_2^\nu$, and, hence, $N_\epsilon = N_\nu$. Thus, coincidence of the lines of free semi-infinite vortices on σ_ϵ and σ is ensured.

Next, the mapping F_ϵ is used to specify the slanting horseshoe vortices $\Pi_{i_\epsilon k_\epsilon}$ on the trapezoid σ_ϵ as well as the reference points $P_{j_\epsilon m_\epsilon}$. Let us denote by

$$V_{i_\nu k_\nu}^{j_\epsilon m_\epsilon} = \Gamma_{i_\nu k_\nu} \omega_{i_\nu k_\nu}^{j_\epsilon m_\epsilon}$$

the normal velocity component induced by the slanting horseshoe vortex $\Pi_{i_\nu k_\nu}$ of trapezoid σ_ν with intensity $\Gamma_{i_\nu k_\nu}$ at point $P_{j_\epsilon m_\epsilon}$ of trapezoid σ_ϵ, by $v_{j_\epsilon m_\epsilon}^\nu$ denote the normal velocity component induced by the entire system of slanting horseshoe vortices on σ_ν, and by $V_{j_\epsilon m_\epsilon}$ denote the normal velocity component induced by the entire system of vortices on σ at the same point $P_{j_\epsilon m_\epsilon}$. Then,

$$V_{j_\epsilon m_\epsilon} = \sum_{\nu=1}^{p} V^{\nu}_{j_\epsilon m_\epsilon} = \sum_{\nu=1}^{p} \sum_{i_\nu=1}^{n_\nu} \sum_{k_\nu=1}^{N_\nu} V^{j_\epsilon m_\epsilon}_{i_\nu k_\nu}. \tag{11.3.24}$$

Because we consider a circulatory problem (i.e., the solution being sought must be unlimited at the leading edges of trapezoids and limited at their trailing edges), by the B condition the following system of linear algebraic equations must be considered for obtaining unknown circulations of discrete vortices (Belotserkovsky and Skripach 1975):

$$\sum_{\nu=1}^{p} \sum_{i_\nu=1}^{n_\nu} \sum_{k_\nu=1}^{N_\nu} \Gamma_{i_\nu k_\nu} \omega^{j_\epsilon m_\epsilon}_{i_\nu k_\nu} = -V^{*}_{j_\epsilon m_\epsilon},$$

$$j_\epsilon = 1, \ldots, n_\epsilon,\ m_\epsilon = 1, \ldots, N_\epsilon,\ \epsilon = 1, \ldots, p, \tag{11.3.25}$$

where $V^{*}_{j_\epsilon m_\epsilon}$ is the oncoming flow normal velocity component at the reference point $P_{j_\epsilon m_\epsilon}$.

However, if a noncirculatory problem is considered for which the solution must be unlimited at both the leading and trailing edges, then by the B condition one has to consider the system

$$\sum_{\nu=1}^{p} \sum_{i_\nu=1}^{n_\nu} \sum_{k_\nu=1}^{N_\nu} \Gamma_{i_\nu k_\nu} \omega^{j_\epsilon m_\epsilon}_{i_\nu k_\nu} = -V^{*}_{j_\epsilon, m_\epsilon},$$

$$j_\epsilon = 1, \ldots, n_\epsilon - 1,\ m_\epsilon = 1, \ldots, N_\epsilon,\ \epsilon = 1, \ldots, p,$$

$$\sum_{i_\epsilon=1}^{n_\nu} \Gamma_{i_\epsilon m_\epsilon} = 0, \qquad j_\epsilon = n_\epsilon,\ M_\epsilon = 1, \ldots, N_\epsilon,\ \epsilon = 1, \ldots, p. \tag{11.3.26}$$

By isolating the summands corresponding to trapezoid σ_ϵ, on the left-hand sides of Systems (11.3.25) and (11.3.26), we deduce that the sought functions $\gamma(x, z)$ behave as needed at both the leading and the trailing edges.

12

Unsteady Linear and Nonlinear Problems

The ability of the method of discrete vortices to provide simultaneously a discrete model of the physical phenomenon under consideration and a method for solving numerically the corresponding mathematical problem is fully displayed in the analysis of unsteady linear and nonlinear, ideal incompressible flows past lifting surfaces. However, the systems of integral and differential equations describing the problems remain virtually unstudied. Here we shall try to fill the gap.

12.1. LINEAR UNSTEADY PROBLEM FOR A THIN AIRFOIL

Consider a thin airfoil traveling through an ideal incompressible fluid with average translational velocity $\mathbf{U}_0$ at an angle of attack α. Let us introduce a frame of reference $OXZY$ fixed at the airfoil, i.e., in fact we shall consider unsteady flow past the airfoil. In the case of unsteady flow, circulation around an airfoil varies with time. By the theorem about the constancy of circulation along a material loop (which does not come across singularities), free vortices that are shed from the airfoil continue moving with the flow and their circulation is time-independent (Golubev 1949). Free vortices remain parallel to the bound vortices at the airfoil and, similarly to the latter, are parallel to the OZ axis.

As far as the problem is considered in the linear formulation and for small angles of attack only, free vortices may be assumed to move in the plane of the airfoil (i.e., in the plane OXZ) with speed U_0 (Belotserkovsky, Skripach, and Tabachnikov 1971). Hence, a free vortex shed from the airfoil at instant τ travels, by instant t, through distance $U_0(t - \tau)$.

It is required to calculate the strengths of the bound and free vortex sheets. This will be done by employing the method of discrete vortices (Belotserkovsky and Nisht 1978). Let the airfoil occupy the segment $[-1, 1]$ of the axis, OX. Because a free vortex sheet sheds from the trailing edge of the airfoil, the flow velocity at the latter must be finite, and hence, a reference point and a discrete vortex must be located near the trailing and the leading edges, respectively. Thus, discrete vortices and reference points may be conveniently distributed in the following way. The discrete vortices are at points $x_i = -1 + (i - 3/4)h$, $h = 2/n$, $i = 1, \ldots, n$, and the reference points are at points $x_{0_j} = x_j + h/2 = -1 + (j - \frac{1}{4})h$, $j = 1, \ldots, n$. The discrete time spacing, Δt, will be chosen in accordance with the formula

$$U_0 \, \Delta t = h. \tag{12.1.1}$$

For simplicity, we put $U_0 = 1$. Therefore, the coordinate of a free vortex shed from an airfoil at instant t_s, at instant t_r is equal to

$$\xi_{sr} = x_n + (t_r - t_s + \Delta t) = x_n + (r - s + 1)\,\Delta t. \tag{12.1.2}$$

Let the circulation of the discrete vortex located at instant r at point x_i be equal to r_{ir}, and let the circulation of free vortices shed prior to the instant from the airfoil be Λ_s, $s = 1, \ldots, r$. By employing the no-penetration condition at reference points x_{0_j}, $j = 1, \ldots, n$, we arrive at the system of equations

$$\sum_{i=1}^{n} \Gamma_{ir}\,\omega_{ij} + \sum_{s=1}^{r} \Lambda_s\,\omega_{sjr} = -V_j^*, \qquad j = 1, \ldots, n,\ r = 1, 2, \ldots, \tag{12.1.3}$$

where ω_{ij} is the normal velocity component at reference point x_j induced by a unit-strength vortex placed at point x_i, and ω_{sjr} is the normal velocity component induced at the same point by a unit-strength vortex located at point ξ_{sr}.

In the latter system of equations r_{ir}, $i = 1, \ldots, n$, and Λ_r are unknown at the reference instant r, while $\Lambda_1, \ldots, \Lambda_{r-1}$ are determined at the preceding reference instants and remain unchanged as time passes (these are circulations of the free discrete vortices). Hence, the system contains $n + 1$ unknowns and n equations. Let us supplement the system under consideration by an equation for the constancy of circulation around a material loop encompassing both the airfoil and the wake. If an airfoil starts moving from the state of rest, then the equation has the form

$$\sum_{i=1}^{n} \Gamma_{ir} + \sum_{s=1}^{r} \Lambda_s = 0, \qquad r = 1, 2, \ldots. \tag{12.1.4}$$

Thus, to solve the problem, one has to consider the system of linear algebraic equations

$$\sum_{i=1}^{n} \Gamma_{ir}\,\omega_{ij} + \sum_{s=1}^{r} \Lambda_s\,\omega_{sjr} = -U_{yr}, \qquad j = 1,\ldots,n,\ r = 1,2,\ldots,$$

$$\sum_{i=1}^{n} \Gamma_{ir} + \sum_{s=1}^{r} \Lambda_s = 0, \qquad r = 1,2,\ldots. \tag{12.1.5}$$

By putting $\Gamma_{ir} = \gamma(x_i, t_r)$, $\Lambda_s = \delta(t_s)\,\Delta t$, and taking into account Formula (9.3.13), the latter system of equations may be rewritten in the form

$$\sum_{i=1}^{n} \frac{\gamma(x_i, t_r)h}{x_{0_j} - x_i} + \sum_{s=1}^{r} \frac{\delta(t_s)\,\Delta t}{x_{0_j} - x_n - (t_r - t_s + \Delta t)} = 2\pi f(t_r),$$

$$j = 1,\ldots,n,\ r = 1,2,\ldots,$$

$$\sum_{i=1}^{n} \gamma(x_i, t_r)h + \sum_{s=1}^{r} \delta(t_s)\,\Delta t = 0, \qquad r = 1,2,\ldots,\ f(t_r) = U_{yr}. \tag{12.1.6}$$

This system of equations must be solved step by step for $r = 1$, $r = 2$, etc.; hence, it is convenient to rewrite it in the form

$$\sum_{i=1}^{n} \frac{\gamma(x_i, t_r)h}{x_{0_j} - x_i} + \frac{\delta(t_r)\,\Delta t}{x_{0_j} - x_n - \Delta t} = 2\pi f(t_r)$$

$$- \sum_{s=1}^{r-1} \frac{\delta(t_s)\,\Delta t}{x_{0_j} - x_n - (t_r - t_s + \Delta t)},$$

$$j = 1,\ldots,n,\ r = 1,2,\ldots,$$

$$\sum_{i=1}^{n} \gamma(x_i, t_r)h + \delta(t_r)\,\Delta t = -\sum_{s=1}^{r-1} \delta(t_s)\,\Delta t,$$

$$r = 1,2,\ldots. \tag{12.1.7}$$

The matrix of the latter system has the same sign as that of System (5.1.4); hence, it is both well-conditioned and solvable, and by Formula

(5.1.13) one has

$$\gamma(x_i, t_r) = -\frac{1}{h} I_{1i}^{(n+1)} \left[\sum_{j=1}^{n} \frac{1}{h} I_{1,0j}^{(n+1)} \frac{h}{x_i - x_{0j}} \right.$$

$$\times \left(2\pi f(t_r) - \sum_{s=1}^{r-1} \frac{\delta(t_s)\,\Delta t}{x_{0j} - x_n - (t_r - t_s + \Delta t)} \right)$$

$$\left. + \sum_{s=1}^{r-1} \delta(t_s)\,\Delta t \right], \qquad i = 1, \ldots, n,\ r = 1, 2, \ldots,$$

$$\delta(t_r) = -\frac{1}{\Delta t} I_{1,n+1}^{(n+1)} \left[\sum_{j=1}^{n} \frac{1}{h} L_{1,0j}^{(n+1)} \frac{h}{x_{n+1} - x_{0j}} \right.$$

$$\times \left(2\pi f(t_r) - \sum_{s=1}^{r-1} \frac{\delta(t_s)\,\Delta t}{x_{0j} - x_n - (t_r - t_s + \Delta t)} \right)$$

$$\left. + \sum_{s=1}^{r-1} \delta(t_s)\,\Delta t \right], \qquad i = n + 1,\ r = 1, 2, \ldots,$$

(12.1.8)

where

$$I_{1i}^{(n+1)} = \prod_{m=1}^{n} (x_i - x_{0m}) \Bigg/ \prod_{\substack{m=1 \\ m \neq i}}^{n+1} (x_i - x_m),$$

$$I_{1,0j}^{(n+1)} = -\prod_{m=1}^{n+1} (x_{0j} - x_m) \Bigg/ \prod_{\substack{m=1 \\ m \neq j}}^{n} (x_{0j} - x_{0m}), \qquad x_{n+1} = x_n + \Delta t.$$

On the other hand, by the results obtained in Section 1.3 and supposing that the functions $\gamma(x, t)$ and $\delta(t)$ belong to class H^* at the corresponding sets, we deduce that System (12.1.6) approximates the following system of integral equations:

$$\int_{-1}^{1} \frac{\gamma(x,t)\,dx}{x_0 - x} + \int_0^t \frac{\delta(\tau)\,d\tau}{x_0 - 1 - (t-\tau)} = 2\pi f(t),$$

$$x_0 \in (-1,1),\, t \geq 0,$$

$$\int_{-1}^{1} \gamma(x,t),dx + \int_0^t \delta(\tau)\,d\tau = 0. \tag{12.1.9}$$

Let us show that the latter system of equations has a unique solution meeting the required boundary conditions. It is physically obvious that the first equation of the system must be considered with respect to the function $\gamma(x,t)$ unlimited at the point $x = -1$, as a singular integral equation of index $\kappa = 0$ on $[-1,1]$. Therefore, by solving the equation with respect to $\gamma(x,t)$, one gets

$$\gamma(x,t) = f_1(x,t) + \frac{1}{\pi^2}\sqrt{\frac{1-x}{1+x}} \int_0^t \delta(\tau)\,d\tau$$

$$\times \int_{-1}^{1} \sqrt{\frac{1+x_0}{1-x_0}} \frac{dx_0}{(x-x_0)(x_0 - 1 - (t-\tau))},$$

$$f_1(x,t) = -\frac{1}{\pi^2}\sqrt{\frac{1-x}{1+x}} \int_{-1}^{1} \sqrt{\frac{1+x_0}{1-x_0}} \frac{2\pi f(t)\,dx_0}{x - x_0} \equiv 2\sqrt{\frac{1-x}{1+x}} f(t). \tag{12.1.10}$$

Let us recall some formulas presented in Gakhov (1977), Muskhelishvili (1952), and Prudnikov, Brychkov, and Marichev 1983):

$$\int_{-1}^{1} \sqrt{\frac{1+x}{1-x}} \frac{dx}{x_0 - x} \equiv -\pi, \qquad |x_0| < 1,$$

$$\int_{-1}^{1} \sqrt{\frac{1-x}{1+x}} \frac{dx}{x_0 - x} \equiv \pi, \qquad |x_0| < 1,$$

$$\int_{-1}^{1} \frac{dx}{(x-b)\sqrt{1-x^2}} = \pm \frac{1}{\sqrt{b^2-1}} \arcsin \frac{-1+xb}{x-b}\bigg|_{-1}^{1} = \mp \frac{\pi}{\sqrt{b^2-1}}, \tag{12.1.11}$$

where the upper and the lower signs correspond to $b > 1$ and $b < -1$, respectively. From the latter integral we get

$$\int_{-1}^{1} \sqrt{\frac{1+x_0}{1-x_0}} \frac{dx_0}{b-x_0} = -\pi + \pi\sqrt{\frac{b+1}{b-1}}, \quad b > 1,$$

$$\int_{-1}^{1} \sqrt{\frac{1-x}{1+x}} \frac{dx}{b-x} = \pi - \pi\sqrt{\frac{b-1}{b+1}}, \quad b > 1. \quad (12.1.12)$$

Formulas (12.1.11) and (12.1.12) yield

$$\int_{-1}^{1} \sqrt{\frac{1+x_0}{1-x_0}} \frac{dx_0}{(x-x_0)(x_0-1-(t-\tau))}$$

$$= \sqrt{\frac{2+(t-\tau)}{t-\tau}} \frac{\pi}{1+(t-\tau)-x}. \quad (12.1.13)$$

Hence, Formula (12.1.10) becomes

$$\gamma(x,t) = f_1(x,t) + \frac{1}{\pi}\sqrt{\frac{1-x}{1+x}} \int_0^t \sqrt{\frac{2+(t-\tau)}{t-\tau}} \frac{\delta(\tau)\, d\tau}{1+(t-\tau)-x}. \quad (12.1.14)$$

By substituting this expression for $\gamma(x,t)$ into the second equation of System (12.1.9) and using the second equation of System (12.1.12), we get

$$\int_0^1 \sqrt{\frac{2+(t-\tau)}{t-\tau}} \delta(\tau)\, d\tau = -f_2(t),$$

$$f_2(t) = \int_{-1}^{1} f_1(x,t)\, dx = 2\pi f(t). \quad (12.1.15)$$

This is the Abel equation (Goursat 1934). Let us rewrite it in the form

$$\int_0^t \frac{\delta(\tau)\, d\tau}{\sqrt{t-\tau}} = \frac{-f_2(t)}{\sqrt{2}} - \frac{1}{\sqrt{2}} \int_0^t \frac{\sqrt{t-\tau}}{\sqrt{(t-\tau)+2}+\sqrt{2}} \delta(\tau)\, d\tau. \quad (12.1.16)$$

Then, by applying the inversion formula for the Abel integral (Goursat 1934), we get

$$\delta(t) + \frac{1}{\sqrt{2}\,\pi}\int_0^t \frac{d\tau}{\sqrt{t-\tau}} \int_0^\tau \frac{\partial}{\partial \tau}\left(\frac{\sqrt{\tau - s}}{\sqrt{2+\tau-s} + \sqrt{2}} \right)\delta(s)\, ds$$

$$= \frac{1}{\sqrt{2}\,\pi}\left[\frac{-f_2(0)}{\sqrt{t}} - \int_0^t \frac{f_2'(\tau)\, d\tau}{\sqrt{t-\tau}} \right]. \tag{12.1.17}$$

From Formulas (12.1.16) and (12.1.17) we deduce that Equation (12.1.17) has the form

$$\delta(t) + \frac{1}{2\pi}\int_0^t K(t,s)\,\delta(s)\, ds = -\sqrt{2}\,\frac{U_y(0)}{\sqrt{t}} - \sqrt{2}\int_0^t \frac{U_y'(\tau)\, d\tau}{\sqrt{t-\tau}},$$

$$K(t,s) = \int_s^t \frac{d\tau}{[\sqrt{2+\tau-s} + \sqrt{2}\,]\sqrt{t-\tau}\;\sqrt{\tau-s}\,\sqrt{2+\tau-s}}. \tag{12.1.18}$$

The kernel $K(t,s)$ is continuous because it always may be written in the form

$$K(t,s) = \int_0^1 \frac{dz}{\sqrt{z(1-z)}\,\sqrt{2+z(t-s)}\left[\sqrt{2+z(t-s)} + \sqrt{2}\right]}. \tag{12.1.19}$$

Hence, Equation (12.1.18) is the Volterra equation of the second kind having a unique solution for any integrable right-hand side (Goursat 1934). Thus we deduce that the system of equations (12.1.9), which describes the problem under consideration, has a unique solution obtainable from Equation (12.1.18) and Formula (12.1.14). Figure 12.1 presents the types of vortex systems.

12.2. NUMERICAL SOLUTION OF THE ABEL EQUATION

It is obvious that by solving the preceding linear unsteady problem by the method of discrete vortices, we develop at the same time a numerical method for solving the Abel equation (12.1.15). An analogous numerical scheme for solving the equation was proposed by L. N. Poltavsky, and

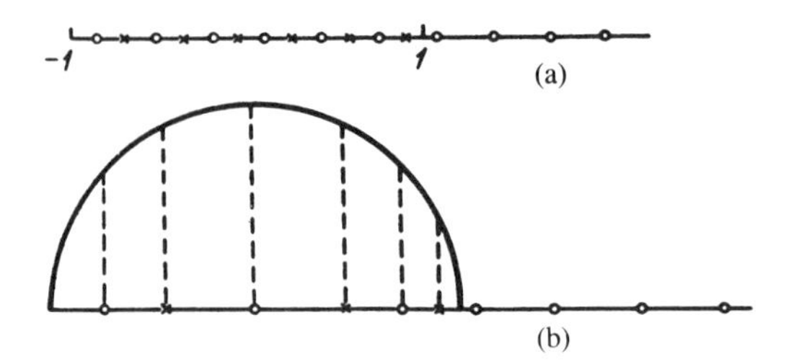

FIGURE 12.1. Distributions of discrete vortices (○ ○ ○) and calculation points (× × ×) over an airfoil (the segment $[-1, 1]$) and in the wake. (a) and (b) correspond to equally and unequally spaced points over the airfoil, respectively.

calculations carried out by E. B. Rodin demonstrated rapid convergence of the process. Rodin was also the first to extend the method onto the two-dimensional Abel equation.

Consider the equation

$$\int_0^t \frac{\varphi(s)\,ds}{(t-s)^\alpha} = f(t), \qquad 0 < \alpha < 1, \tag{12.2.1}$$

which has the solution

$$\varphi(t) = \frac{\sin \alpha\pi}{\pi}\left[\frac{f(0)}{t^{1-\alpha}} + \int_0^t \frac{f'(s)\,ds}{(t-s)^{1-\alpha}}\right] \tag{12.2.2}$$

for any $f(t)$ having a continuous derivative function $f(t)$ (Privalov 1935).

The numerical method for solving this equation is constructed as follows. We start by considering conjugated sequences of numbers. Let us consider the sequence of numbers $\{\lambda_{ik}\}$, $k = 1, 2, 3, \ldots,$ $i = 1, \ldots, k$, subject to the conditions

$$1. \quad \lambda_{ik} = \lambda_{i-j,\,k-j}, \qquad i \leq j \leq i-1,$$

$$2. \quad \lambda_{kk} \neq 0. \tag{12.2.3}$$

For the sequence $\{\lambda_{ik}\}$ and the number $\kappa \neq 0$ we define the sequence $\{\lambda^*_{ik}\}$, where $k = 1, 2, \ldots,$ $i = 1, \ldots,$ k, in the following way:

a. Quantity λ^*_{ik} satisfies the system of equations

$$\sum_{j=1}^{k} \lambda^*_{1j}\lambda_{jk} = \kappa, \qquad k = 1, 2, \ldots, \tag{12.2.4}$$

which always has a solution, because $\lambda_{kk} \neq 0$.

b. By definition,

$$\lambda_{kk}^{*} = \lambda_{11}^{*}, \qquad \lambda_{k-1,k}^{*} = \lambda_{12}^{*}, \ldots, \lambda_{k-i,k}^{*} = \lambda_{1,i+1,\ldots}^{*}. \quad (12.2.5)$$

Definition 12.2.1. *Sequences* $\{\lambda_{ik}^{*}\}$ *and* $\{\lambda_{ik}\}$ *are called conjugated sequences with the conjugation index* κ.

Consider the sum

$$\lambda_{ss}^{*}\lambda_{sk} + \lambda_{s,s+1}^{*}\lambda_{s+1,k} + \cdots + \lambda_{sk}^{*}\lambda_{kk} \equiv \sum_{p=s}^{k} \lambda_{sp}^{*}\lambda_{pk}. \quad (12.2.6)$$

By making use of Properties (12.2.3)–(12.2.5) we get from (12.2.6),

$$\sum_{p=s}^{k} \lambda_{sp}^{*}\lambda_{pk} = \sum_{j=1}^{k-s+1} \lambda_{1j}^{*}\lambda_{j,k-s+1} = \kappa. \quad (12.2.7)$$

This identity will be called the *fundamental identity* for conjugated sequences of index κ.

Next consider the system of linear algebraic equations

$$\sum_{i=1}^{k} \lambda_{ik}^{*} x_i = A_k, \qquad k = 1, \ldots, n. \quad (12.2.8)$$

Let us multiply both sides of the latter system by λ_{kn} and sum the result with respect to k from 1 to n:

$$\sum_{k=1}^{n} \lambda_{kn} \sum_{i=1}^{k} \lambda_{ik}^{*} x_i = \sum_{k=1}^{n} \lambda_{kn} A_k.$$

Next we invert the order of summation on the left-hand side of the equality:

$$\sum_{i=1}^{n} x_i \sum_{k=i}^{n} \lambda_{ik}^{*}\lambda_{kn} = \sum_{k=1}^{n} \lambda_{kn} A_k. \quad (12.2.9)$$

By the fundamental identity (12.2.7) we get the equality

$$\sum_{i=1}^{n} x_i = \frac{1}{\kappa} \sum_{k=1}^{n} \lambda_{kn} A_k, \quad (12.2.10)$$

valid for any n. Therefore, substituting $n - 1$ for n and subtracting, we obtain

$$x_n = \frac{1}{\kappa}\left(\sum_{k=1}^{n} \lambda_{kn} A_k - \sum_{k=1}^{n-1} \lambda_{k,n-1} A_k\right). \tag{12.2.11}$$

Now we will consider Equation (12.2.1). Let us divide the semi-axis $[0, \infty)$ into semi-intervals $E_i = [(i-1)h, ih)$, $i = 1, 2, \ldots,$ and choose in E_i an arbitrary point ξ_i. Consider the sequence $\{\lambda_{ik} = [\{(k - i + 1)h\}^\alpha - \{(k-i)h\}^\alpha]/\alpha\}$ that satisfies Equalities (12.2.3). Let us construct the conjugated sequence $\{\lambda_{ik}^*\}$ (with index $\kappa = \pi/\sin \alpha\pi$).

Consider the system of linear algebraic equations

$$\sum_{i=1}^{k} \lambda_{ik}^* \varphi_h(\xi_i) h = f(kh), \qquad k = 1, 2, \ldots . \tag{12.2.12}$$

By Formula (12.2.11) we have

$$\varphi_n(\xi_n) h = \frac{\sin \alpha\pi}{\pi}\left[\sum_{k=1}^{n} \lambda_{kn} f(kh) - \sum_{k=1}^{n-1} \lambda_{k,n-1} f(kh)\right].$$

Using Relationship (12.2.3), we get

$$\varphi_n(\xi_n) = \frac{\sin \pi\alpha}{\pi h}\left\{\lambda_{1n} f(h) + \sum_{j=2}^{n} \lambda_{jn}[f(jh) - f((j-1)h)]\right\}. \tag{12.2.13}$$

Let us start by supposing that $f(0) = 0$ and $f''(t)$ is continuous. Then, Formula (12.2.13) becomes

$$\varphi_n(\xi_n) = \frac{\sin \pi\alpha}{\pi} \sum_{j=1}^{n} \int_{(j-1)h}^{jh} \frac{f'\left[(j-1)h + \theta_j h\right] ds}{(nh - s)^{1-\alpha}}, \qquad 0 < \theta_j < 1, \tag{12.2.14}$$

because

$$\lambda_{ik} = \int_{(i-1)h}^{ih} \frac{ds}{(kh - s)^{1-\alpha}}.$$

Now it may be readily shown that

$$|\varphi(nh) - \varphi_h(\xi_n)| \le h\frac{\sin \alpha\pi}{\pi} f''_{\max}\int_0^{nh} \frac{ds}{(nh-s)^{1-\alpha}}$$

$$\le \frac{\sin \alpha\pi}{\pi} f''_{\max}(nh)^{\alpha} h = O(h). \qquad (12.2.15)$$

where

$$f''_{\max} = \max_{t\in[0,nh]} f''(t), \qquad nh \le T < +\infty.$$

Next let $f(0) \neq 0$. Then,

$$\varphi_h(\xi_n) = \frac{\sin \alpha\pi}{\pi h}\lambda_{1n} f(0) + \frac{\sin \alpha\pi}{\pi}\sum_{j=1}^{n} \lambda_{jn}[f(jh) - f(j-1)h]. \qquad (12.2.16)$$

and, hence,

$$|\varphi(nh) - \varphi_h(\xi_n)| \le O(h) + \frac{\sin \alpha\pi}{\pi}|f(0)|\left|\frac{\lambda_{1n}}{h} - \frac{1}{(nh)^{1-\alpha}}\right|. \qquad (12.2.17)$$

In accordance with the previous definition of λ_{1n}, we have

$$\left|\frac{\lambda_{1n}}{h} - \frac{1}{(nh)^{1-\alpha}}\right| \le \frac{(1-\theta_n)h^{1-\alpha}}{[(n-1)h + \theta_n h]^{1-\alpha}(nh)^{1-\alpha}}, \qquad (12.2.18)$$

where $0 < \theta_n < 1$, $n = 1, 2, \ldots$.

With analogous estimates, these results may be transferred onto the equation

$$\int_0^t \frac{\varphi(s)\,ds}{(t-s)^{\alpha}} + \int_0^t K(t,s)\varphi(s)\,ds = f(s), \qquad (12.2.19)$$

for which one has to consider the system

$$\sum_{i=1}^{k} \lambda^*_{i,k}\varphi_h(\xi_i)h + \sum_{i=1}^{k} K(kh,\xi_i)\varphi(\xi_i)h = f(kh), \qquad k = 1, 2, \ldots .$$

Next consider the two-dimensional Abel equation

$$\int_0^x \int_0^y \frac{\varphi(\xi, \eta)\, d\xi\, d\eta}{(x-\xi)^{\beta}(y-\eta)^{\alpha}} = f(x, y), \qquad (12.2.20)$$

where $0 < \alpha, \beta < 1$. Representing the integral entering (12.2.20) as an iterated one and solving the resulting Abel equations sequentially, we get

$$\varphi(x, y) = \frac{\sin \alpha\pi}{\pi} \frac{\sin \beta\pi}{\pi} \left[\frac{f(0,0)}{x^{1-\beta} y^{1-\alpha}} + \frac{1}{y^{1-\alpha}} \int_0^x \frac{f'_{\xi}(\xi, 0)\, d\xi}{(x-\xi)^{1-\beta}} + \frac{1}{x^{1-\beta}} \int_0^y \frac{f'_{\eta}(0, \eta)\, d\eta}{(y-\eta)^{1-\alpha}} + \int_0^x \int_0^y \frac{f''_{\xi\eta}(\xi, \eta)\, d\xi\, d\eta}{(x-\xi)^{1-\beta}(y-\eta)^{1-\alpha}} \right].$$

(12.2.21)

Let us divide the semi-axis OY into semi-intervals $E_i^{\xi} = [(i-1)h_y, ih_y)$, $i = 1, 2, \ldots,$ and divide the axis OX into the sets $E_j' = [(j-1)h_x, jh_x)$, $j = 1, 2, \ldots$. Consider the sequence of numbers $\{\lambda_{ik}, \{\lambda_{jk}'\}$, where

$$\lambda_{ik} = \int_{(i-1)h}^{ih} \frac{ds}{(kh - s)^{1-\alpha}}, \qquad \lambda_{jm}' = \int_{(j-1)h}^{jh} \frac{ds}{(mh - s)^{1-\beta}},$$

$k = 1, 2, \ldots,$ $i = 1, \ldots, k,$ $m = 1, 2, \ldots,$ $j = 1, \ldots, m$. It is obvious that the sequences satisfy Conditions (12.2.3). Let us construct conjugated sequences $\{\lambda_{ik}^*\}, \{\lambda_{ik}\}$ with the conjugation index $\kappa = \pi/\sin \alpha\pi$ and $\{\lambda_{jm}^*\}, \{\lambda_{jm}'\}$ with $\kappa = \pi/\sin \beta\pi$.

Next consider the system of linear algebraic equations

$$\sum_{i=1}^{N} \sum_{j=1}^{M} \lambda_{iN}^* \lambda_{jM}^* \varphi_{NM}(\xi_i, \eta_j) h_y h_x' = f(Mh_x, Nh_y), \qquad M, N = 1, 2, \ldots.$$

(12.2.22)

This system is solvable, and the estimated difference between its solution and the exact solution (12.2.21) has a form analogous to (12.2.17).

Note 12.2.1. An analogous numerical method may be applied to the equation

$$\int_{x_0}^x \int_{\psi(\xi)}^y \frac{\varphi(\xi, \eta)\, d\xi\, d\eta}{(x-\xi)^{\beta}(y-\eta)^{\alpha}} = f(x, y), \qquad (12.2.23)$$

where $0 < \alpha, \beta < 1$. It should be mentioned that the case $\alpha = \beta = \frac{1}{2}$, $f(x_0, y) \equiv 0$ is of paramount significance for supersonic aerodynamics.

12.3. SOME EXAMPLES OF NUMERICAL SOLUTION OF THE ABEL EQUATION

Consider the equation

$$\int_0^t \frac{\varphi(s)\, ds}{\sqrt{t-s}} = t \tag{12.3.1}$$

for $t \in [0, 1]$. By using Formula (12.2.2) we deduce that the exact solution is given by the function $\varphi(t) = (2/\pi)\sqrt{t}$.

Equation (12.3.1) was solved numerically with the step $h = 0.1$ by using Formula (12.2.8). Figure 12.2 compares the exact (solid lines) and approximate (open circles) results. Calculations have shown that the inequality $|\varphi(t_k) - \varphi_n(t_k)| \le 10^{-5}$, $k = 1, \ldots, 10$, is valid.

If the right-hand side of Equation (12.3.1) is put equal to t^2, then the exact solution is given by $\varphi(t) = (8/(3\pi))t^{3/2}$.

The exact and approximate solutions obtained by employing the same scheme were compared at the same points; however, the step was put equal to $h = 0.0125$. Calculations have shown that $|\varphi(t_k) - \varphi_n(t_k)| \le 0.00013$ for $t_k = 0.1, 0.2, \ldots, 1.0$ (see Figure 12.3).

For the same equation the case of the right-hand side being equal to unity was considered. From (12.1.18) it follows that in this case the results must be less accurate. Figure 12.4 shows that for $h = 0.1$ the calculated results (denoted by $\times\times$) differ substantially from the exact data in the neighborhood of zero; however, for $h = 0.00625$ (see open circles) the difference between the exact solution, $\varphi = 1/(\pi\sqrt{t})$, and the numerical results does not exceed 0.02.

We have also considered the two-dimensional equation

$$\int_0^x \int_0^y \frac{\varphi(\xi, \eta)\, d\xi\, d\eta}{\sqrt{x-\xi}\sqrt{y-\eta}} = xy, \tag{12.3.2}$$

whose exact solution is given by the formula

$$\varphi(x, y) = \frac{4}{\pi^2}\sqrt{xy}, \qquad x \in [0; 1],\ y \in [0; 1.5].$$

Numerical solutions were obtained for two cross sections: $x = 0.1$ and $x = 0.2$ (see Figures 12.5 and 12.6, respectively, where open circles correspond to $h = 0.1$, solid circles correspond to $h = 0.05$ and $\times\times$ correspond to $h = 0.025$). Different steps along the x and y directions were

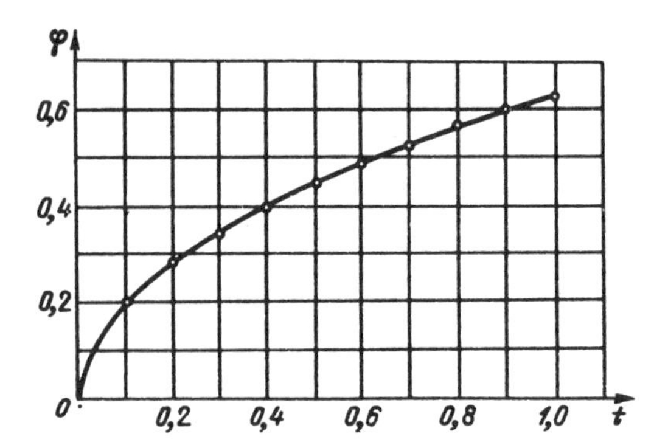

FIGURE 12.2. Comparison of the exact (solid line) and approximate (circles) solutions to Equation (12.3.1) at points $t_k = kh$, $h = 0.1$, $k = 1, 2, \ldots, 10$. The approximate solution was obtained by using mode (12.2.8).

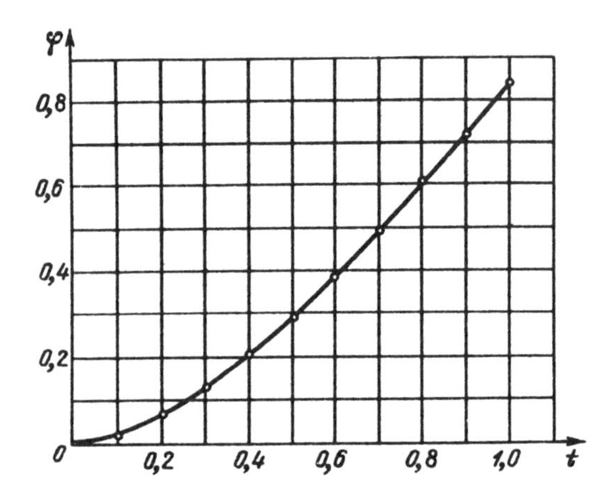

FIGURE 12.3. Comparison of the exact (solid line) and approximate (circles) solutions to Equation (12.3.1) at points $t_j = j \times 0.1$, $j = 1, \ldots, 10$, for $h = 0.00125$. The approximate solution was obtained by using mode (12.2.8).

used. Let us compare the calculated results with the exact solution at points $y_k = k \times 0.1$, $k = 1, 2, \ldots, 10$. For $x = 0.2$, we have $|\varphi(0.2, y_k) - \varphi_n(0.2, y_k)| \leq 0.008$ for $h = 0.05$ and $|\varphi(0.2, y_k) - \varphi_n(0.2, y_k)| \leq 0.003$ for $h = 0.025$.

12.4. NONLINEAR UNSTEADY PROBLEM FOR A THIN AIRFOIL

In a more accurate formulation the unsteady problem must be considered as a nonlinear one, because free vortices shedding from an airfoil do

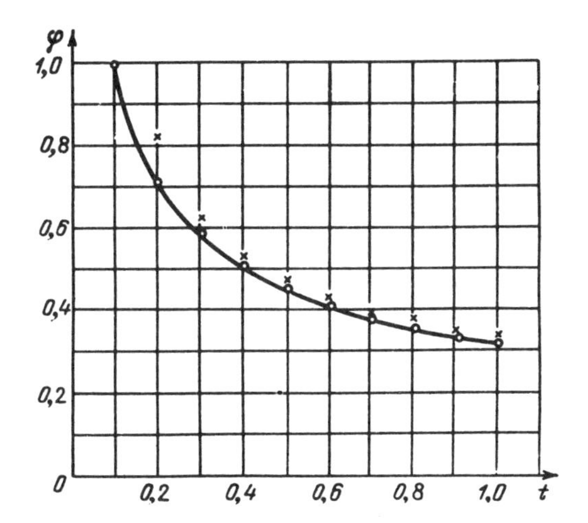

FIGURE 12.4. Comparison of the exact (solid line) and approximate solutions to Equation 12.3.1) with the right-hand side equal to unity, at points $t_j = 0.1 \times j$, $j = 1, 2, \ldots, 10$, for $h = 0.1$ ($\times\times\times$) and $h = 0.00625$ (circles). The approximate solution was obtained by using mode (12.2.8).

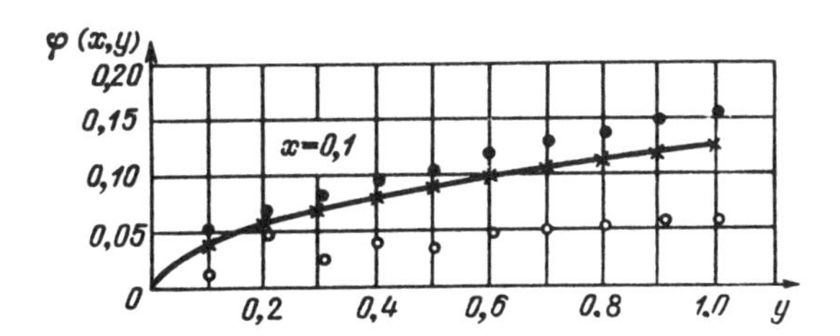

FIGURE 12.5. Comparison of the exact (solid line) and approximate solutions to Equation (12.3.2) at points $t_j = 0.1 \times j$, $j = 1, \ldots, 10$, for $x = 0.1$ and $h = 0.1$, 0.05, and 0.025 (denoted by ○ ○ ○, ●●●, and $\times\times\times$, respectively.

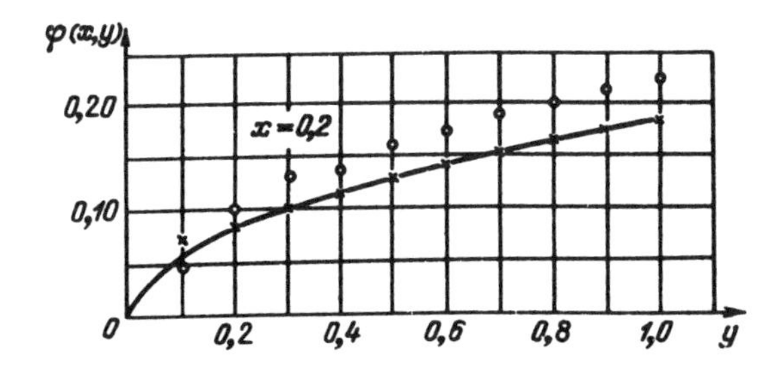

FIGURE 12.6. Comparison of the exact (solid line) and approximate solutions to Equations 12.3.2) at points $t_j = 0.1 \times j$, $j = 1, \ldots, 10$, for $x = 0.2$ and $h = 0.1$ and 0.025 (denoted by ○ ○ ○ and $\times\times\times$, respectively).

not move in its plane and their velocity differs from that of the uniform velocity $\mathbf{U}_0$ (Belotserkovsky and Nisht 1978).

Thus, we suppose that a sheet of free vortices sheds from the trailing edge of an airfoil, the velocities of the vortices coinciding with those of the particles occupying the same places. Let the instantaneous strength of the vortex sheet at point x of the airfoil be equal to $\gamma(x, t)$ and the strength of the free vortex sheet shed at the instant t be equal to $\delta(t)$. Suppose that the parametric equation of the curve where the sheet of free vortices is located at instant t is given by

$$x = x(t, \tau), \qquad y = y(t, \tau), \qquad 0 \leq \tau \leq t. \tag{12.4.1}$$

Because a new location of a free vortex is found by displacing it in the direction of the local velocity, we have the following system of equations for specifying the curve (12.4.1):

$$\frac{dx(t, \tau)}{dt} = V_x(x(t, \tau), y(t, \tau)), \qquad \frac{dy(t, \tau)}{dt} = V_y(x(t, \tau), y(t, \tau)), \tag{12.4.2}$$

where V_x and V_y are the flow velocity components at point $(x(t, \tau), y(t, \tau))$ due to the oncoming flow and the vortex sheet. Therefore, we get (see (9.3.14))

$$V_x = U_x + \int_{-1}^{1} \gamma(x, t)\, \omega_x(x(t, \tau), y(t, \tau), x, 0)\, dx$$
$$+ \int_0^t \delta(s)\, \omega_x(x(t, \tau), y(t, \tau), x(t, s), y(t, s)) \frac{dl}{ds}\, ds,$$
$$V_y = U_y + \int_{-1}^{1} \gamma(x, t)\, \omega_y(x(t, \tau), y(t, \tau), x, 0)\, dx$$
$$+ \int_0^t \delta(s)\, \omega_y(x(t, \tau), y(t, \tau), x(t, s), y(t, s)) \frac{dl}{ds}\, ds, \tag{12.4.3}$$

where

$$\omega_x(x_0, y_0, x, y) = \frac{1}{2\pi} \frac{y_0 - y}{(x_0 - x)^2 + (y_0 - y)^2},$$

$$\omega_y(x_0, y_0, x, y) = -\frac{1}{2\pi}\frac{x_0 - x}{(x_0 - x)^2 + (y_0 - y)^2},$$

$$dl = \sqrt{[x_s'(t,s)]^2 + [y_s'(t,s)]^2}\, ds, \qquad 0 \le \tau \le t.$$

For the system of differential equations (12.4.2) one has to find a special solution subject to the initial conditions

$$x(\tau, \tau) = 1, \qquad y(\tau, \tau) = 0. \tag{12.4.4}$$

Thus, we have to solve the Cauchy problem at the segment $[\tau, T]$, $0 \le \tau \le T$, subject to the initial conditions (12.4.4), where T is the instant prior to which the problem must be solved numerically.

If unsteady motion starts at a certain instant from the state of steady motion, then to find functions $\gamma(x,t)$ and $\delta(t)$ one has to use the no-penetration condition for the airfoil and the value of circulation at the latter. If an airfoil was initially at rest, then the total circulation of the whole of the vortex formation remains zero at any instant t, i.e.,

$$\int_{-1}^{1} \gamma(x,t)\, dx + \int_0^t \delta(s)\frac{dl}{ds}\, ds = 0. \tag{12.4.5}$$

The no-penetration condition generates another equation:

$$\int_{-1}^{1} \frac{\gamma(x,t)\, dx}{x_0 - x} + \int_0^t \frac{[x_0 - x(t,s)]\delta(s)(dl/ds)\, ds}{[x_0 - x(t,s)]^2 + [0 - y(t,s)]^2} = 2\pi U_y. \tag{12.4.6}$$

The system of Equations (12.4.2), (12.4.5), and (12.4.6) and the initial conditions (12.4.4) give a full solution to the nonlinear problem of unsteady flow past an airfoil with the vortex sheet shedding from the trailing edge only.

Let us show that the linear unsteady problem considered in the preceding section is a special case of the present problem. To do this one has to require that

$$y(t, \tau) \equiv 0, \tag{12.4.7}$$

i.e., the vortex sheet must travel along the OX axis. Hence,

$$V_x \equiv U_x,$$

$$V_y = U_y - \frac{1}{2\pi}\int_{-1}^{1} \frac{\gamma(x,t)\, dx}{x(t,\tau) - x} - \frac{1}{2\pi}\int_0^t \frac{\delta(s)(dl/ds)\, ds}{x(t,\tau) - x(t,s)} \equiv 0. \tag{12.4.8}$$

The latter identity must be fulfilled by the wake no-penetration condition, because in this case V_y is the normal velocity component at a point of the wake.

Then, by supposing that the angle of attack is small and the oncoming flow velocity is equal to unity, one gets

$$\frac{dx(t,\tau)}{dt} = U_x = 1, \qquad \frac{dy(t,\tau)}{dt} \equiv 0. \tag{12.4.9}$$

Hence, taking into account initial conditions (12.4.4), we arrive at

$$x(t,\tau) = 1 + (t - \tau), \qquad y(t,\tau) \equiv 0. \tag{12.4.10}$$

Thus, the system of Equations (12.4.2) that describes the form of the wake is solved, and the problem reduces to solving the system of Equations (12.4.5) and (12.4.6). Because in this case

$$dl = ds, \tag{12.4.11}$$

the system of equations coincides with System (12.1.9).

This nonlinear unsteady problem is solved numerically by the method of discrete vortices (Belotserkovsky and Nisht 1978) in the following way. The problem is considered at discrete instants t_r, $t_{r+1} - t_r = \Delta t$, $r = 1, 2, \ldots$. Similarly to the linear problem considered previously, we choose at the airfoil points x_i, $i = 1, \ldots, n$, where bound discrete vortices r_{ir} as positioned, and reference points x_{0_j}, $j = 1, \ldots, n$. At each instant $r = 1, 2, \ldots$ a discrete free vortex Λ_r sheds from the airfoil. At the instant r the free vortex is located at point $x_n + h$ of the axis OX, and at the next instant, $r + 1$, it is displaced (in the direction of the velocity vector) by the distance equal to the product of the flow velocity at the preceding point by Δt. In other words, at the instant r the coordinates of a vortex shed at the instant ν are given by

$$x_{r+1,\nu} = x_{r\nu} + \left[U_{xr} + \sum_{i=1}^{n} \Gamma_{ir}\, \omega_x(x_{r\nu}, y_{r\nu}, x_i, 0) \right.$$

$$\left. + \sum_{s=1}^{r} \Lambda_s\, \omega_x(x_{r\nu}, y_{r\nu}, x_{rs}, y_{rs}) \right] \Delta t,$$

$$y_{r_{r+1},\nu} = y_{r\nu} + \left[U_{yr} + \sum_{l=1}^{n} \Gamma_{ir}\, \omega_y(x_{r\nu}, y_{r\nu}, x_l, 0) \right.$$

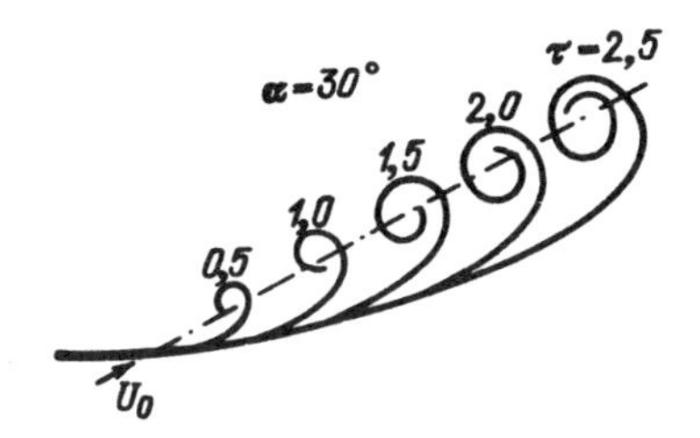

FIGURE 12.7. Computation of the starting Prandtl vortex.

$$+ \sum_{s=1}^{r} \Lambda_s \omega_y(x_{r\nu}, y_{r\nu}, x_{rs}, y_{rs}) \Bigg] \Delta t,$$

$$x_{\nu\nu} = x_n + h,\ y_{\nu\nu} = 0,\ r = 1,2,\ldots,\quad \nu = 1,\ldots,r. \quad (12.4.12)$$

For calculating circulations of discrete vortices we use n equations derived by applying the no-penetration condition at points x_{0_j}, $j = 1,\ldots,n$, as well as the conditions of constancy of the sum of all circulations of discrete vortices:

$$\sum_{i=1}^{n} \Gamma_{ir} \omega_y(x_{0j}, 0, x_i, 0) + \sum_{s=1}^{r} \Lambda_s \omega(x_{0j}, 0, x_{rs}, y_{rs})$$

$$= -U_{yr}, \quad j = 1,\ldots,n,\ r = 1,2,\ldots,$$

$$\sum_{i=1}^{n} \Gamma_{ir} + \sum_{s=1}^{r} \Lambda_s = 0, \quad r = 1,2,\ldots. \quad (12.4.13)$$

Because $\Lambda_1,\ldots,\Lambda_{r-1}$ are calculated at the preceding steps, only r_{ir}, $j = 1,\ldots,n$, and Λ_r are unknown at the instant r.

Thus, a numerical solution to the problem is obtained in the following way:

1. At the instant $r = 1$, Λ_1 is placed at point $M_{i,1} = (x_n + h, 0)$ and System (12.4.13) is solved. Then, the flow velocity is calculated at point $M_{1,1}$ and Formula (12.4.12) is used to calculate the position of the vortex Λ_1 at the instant $r = 2$, i.e., point $M_{2,1}$.
2. At the instant $r = 2$, Λ_2 is placed at point $M_{2,2} = M_{1,1}$ and Λ_1 is placed at point $M_{2,1}$, and System (12.4.13) is solved for $r = 2$. Next the flow velocities are calculated at points $M_{2,2}$ and $M_{2,1}$ and

Formulas (12.4.12) are used to calculate the coordinates of vortices Λ_1 and Λ_2 at the instant $r = 3$, i.e., points $M_{3,2}$ and $M_{3,1}$ are found. Then, the second stage is iterated until the instant T is attained.

Figure 12.7 demonstrates some calculated results. However, the problem of convergence of approximate solutions of unsteady problems remains unsolved (in the sense of strictly mathematical verification).

Here we note only that due to the choice of the position of vortex Λ_r at the instant r, System (12.4.13) is solvable in the same way as the corresponding system for the linear unsteady problem considered previously.

13

Aerodynamic Problems for Blunt Bodies

Of great practical importance is the problem of separated flow past blunt (high-drag) bodies whose contours contain sharp corners (certain buildings, monuments, towers, etc.). It is also important for designing various vehicles, such as automobiles, ships, aircraft, and so on.

13.1. MATHEMATICAL FORMULATION OF THE PROBLEM

The most general problem formulation at the level of selecting an adequate physical model describing flow past a lifting surface was presented in Section 9.1. Through Chapters 10–12 we moved step-by-step from a discrete vortex system approximating a vortex sheet that simulated a surface to the corresponding singular integral equations. However, this was done for models in which (especially in the case of three-dimensional flow past a wing) one can easily single out vortex sheet components affecting the formation of lift and moments of a lifting surface. However, for more complicated separated flow models constructed for finite-span wings of a complex plan form (Belotserkovsky and Nisht 1978), the vortex sheet cannot be modeled by horseshoe vortices (the more so for noncirculatory problems). Therefore, a vortex sheet was modeled by vortex segments augmented by rather complicated relationships between the latter. Simulation of flow past closed surfaces proved to be even more complicated. Although it is very difficult to substitute vortex segments for a vortex sheet modeling a closed surface, it was found that the sheet (both at the surface of a body and in the wake) could be readily modeled by a system of closed rectangular and triangular vortex frames or, in the case of plane flows, by pairs of discrete vortices (Aparinov et al. 1988; Belotserkovsky, Lifanov, and Mikhailov 1985, 1987). The strength of these vortex formations is equal to the density of a double layer potential distributed at the body and in the wake and generating the same potential as the disturbed flow (Sedov 1971–72).

Thus, let a body have the surface–contour σ modeled by a vortex sheet. Also, let k vortex sheets σ_p, $p = 1, 2, \ldots, k$, be shed from the surface–contour, which move with the fluid particles. The flow velocity induced by the vortex formations at point M at instant t will be denoted by $\mathbf{V}(M, t)$ and $\mathbf{V}_p(M, t)$, $p = 1, \ldots, k$, respectively, and the oncoming flow velocity by $\mathbf{U}_0(M, t)$. Then the no-penetration condition at point M_0 of the surface σ may be presented in the form

$$\mathbf{V}(M_0, t) \cdot \mathbf{n}_{M_0} + \sum_{p=1}^{k} \mathbf{V}_p(M_0, t)\mathbf{n}_{M_0} = -\mathbf{U}_0(M_0, t)\mathbf{n}_{M_0}. \quad (13.1.1)$$

Because the disturbed flow velocities will be presented as gradients of the corresponding double layer potentials in what follows, the condition that circulation along a closed contour embracing a body and the wake be equal to zero is fulfilled automatically. However, if the disturbed flow velocities are presented as flow velocities induced by vortex singularities residing at a lifting surface, then Equation (13.1.1) must be augmented by the zero-circulation condition for any contour embracing the surface and the wake.

Let $\mathbf{r}(M)$ be the radius-vector of point M. Then, the condition of point $M_p(\tau, t)$ of the vortex sheet σ_p that was shed from the surface σ at the instant τ, moving at the instant t along the fluid particle path, may be written in the form

$$r_t'\big(M_p(\tau, t)\big) = \mathbf{V}\big(M_p(\tau, t), t\big) + \sum_{m=1}^{k} \mathbf{V}_m\big(M_p(\tau, t), t\big)$$
$$+ \mathbf{U}_0\big(M_p(\tau, t), t\big) \quad (13.1.2)$$

subject to the initial condition

$$\mathbf{r}\big(M_p(\tau, \tau)\big) = \mathbf{r}(M_p),$$

where M_p is a point of the surface σ from which the particle was shed.

13.2. SYSTEM OF INTEGRODIFFERENTIAL EQUATIONS

The concrete form of the system of equations depends on the way flow velocities are calculated in the no-penetration condition (13.1.1) and hence, in (13.1.2).

Previously (see Chapters 9–12) a lifting surface was modeled by a vortex sheet.

Let us consider the most general two-dimensional flow when σ is a contour containing k angular points $M_p(x_p, y_p)$, $p = 1, \ldots, k$, from which vortex sheets are shedding; however, the contour may contain angular points from which no vortex sheets shed. The contour σ is assumed to be piecewise Lyapunov (Muskhelishvili 1952) and specified by the parametric equations

$$x = x(\theta), \qquad y = y(\theta).$$

If the contour is unclosed, then $\theta \in [-1, 1]$; if it is closed, then $\theta \in [0, 2\pi]$ and $x(0) = x(2\pi)$, $y(0) = y(2\pi)$.

Let the strength of the summary vortex layer at point $M(x(\theta), y(\theta))$ of the contour be equal to $\gamma(M, t) = \gamma(\theta, t)$ at instant t, and let the strength of a free vortex shed from the corner M_p at instant τ be equal to $\delta_p(\tau)$, $p = 1, \ldots, k$. Then, according to the Biot–Savart formula for a vortex filament, $\mathbf{V}(M_0, t)$ and $\mathbf{V}_p(M_0, t)$ entering Equation (13.1.1) are given by

$$\mathbf{V}(M_0, t) = \frac{1}{2\pi} \int_\sigma \frac{\mathbf{r}^*(M, M_0)}{|r(M, M_0)|^2} \gamma(M, t)\, d\sigma_M,$$

$$\mathbf{V}_p(M_0, t) = \frac{1}{2\pi} \int_{\sigma_p} \frac{\mathbf{r}^*\big(M_p(\tau, t), M_0\big)}{\big|\mathbf{r}\big(M_p(\tau, t), M_0\big|^2} \delta_p(\tau)\, \delta\sigma_{p,\, M_p(\tau, t)},$$

$$\mathbf{r}^* = (M, M_0) = (y_0 - y)i - (x_0 - x)j, \tag{13.2.1}$$

where the indices M and $M_p(\tau, t)$ for σ and σ_p signify that the length of the arc is taken care of by the coordinates of these points.

If contour σ is closed and the parameter σ is chosen in such a way that for an increasing θ the contour is passed counterclockwise, then

$$\mathbf{n}_{M_0} = \frac{\mathbf{r}^{*\prime}_{\theta_0}(M_0)}{\left|\mathbf{r}'_{\theta_0}(M_0)\right|} = \frac{y'(\theta_0)i - x'(\theta_0)j}{\sqrt{x'^2(\theta_0) + y'^2(\theta_0)}} \tag{13.2.2}$$

is an outward normal.

The condition that circulation around a material loop embracing both the body and the wake equal to zero has the form

$$\int_\sigma \gamma(M, t)\, d\sigma + \sum_{p=1}^{k} \int_{\sigma_p} \delta_p(\tau)\, d\sigma_{p,\, M_p(\tau, t)} = 0. \tag{13.2.3}$$

Because at instant t the parametric equation of the wake σ_p has the form

$$x = x_p(\tau, t), \qquad y = y_p(\tau, t), \qquad 0 \le \tau \le t, \tag{13.2.4}$$

then

$$d\sigma_{p,\, M(\tau, t)} = \sqrt{x'^2_{p,\tau} + y'^2_{p,\tau}(\tau, t)}\; dt.$$

In this case,

$$x_p(\tau, \tau) = x_p, \qquad y_p(\tau, \tau) = y_p.$$

Note that the set of points $\{x_p(\tau, s), y_p(\tau, s)\}$, $\tau \le s \le p$, describes, at instant t, the path of the particle shed from corner M_p at instant τ.

The formulas for $\mathbf{V}(M_p(\tau, t), t)$ and $\mathbf{V}_m(M_p(\tau, t), t)$ appearing in Equation (13.12) may be obtained from the corresponding formulas (13.2.1) by substituting $\mathbf{r}(M, M_p(\tau, t))$, $\mathbf{r}(M_m(s, t), M_p(\tau, t))$, $d\sigma_{m,\, M_m(s,t)}$ for $\mathbf{r}(M, M_0)$, $\mathbf{r}(M_p(\tau, t), M_0)$, $d\sigma_{p,\, M_p(\tau, t)}$, respectively.

The substitution of integral presentations (13.2.1) for velocities entering (13.1.1) and (13.1.2) results in Equation (13.1.1) becoming a singular integral equation and singular integrals appearing in (13.1.2).

In the special case when σ coincides with the segment $[-1, 1]$ of the OX axis, and a free vortex sheet sheds from point $x = 1$ only and moves linearly in the positive direction of the OX axis with average speed equal to unity, one arrives at Equations (12.4.8)–(12.4.11).

An important result was obtained by Poltavsky (1986), who showed that for Equation (12.1.9),

$$\lim_{x \to 1} \gamma(x, t) = \delta(t). \tag{13.2.5}$$

The latter equality confirms the validity of the Chaplygin–Joukowski hypothesis for a bound vortex sheet of an airfoil, according to which the strength of the vortex sheet vanishes when approaching the trailing edge (where the sheet sheds).

Hence, at the trailing edge of a thin airfoil the whole of the vortex sheet consists of a free vortex sheet shedding off the edge. This remark allows us to place a free discrete vortex shedding from an airfoil at the trailing edge and to assume that it continues moving along the local flow velocity vector. This is of special importance for analyzing flow past a finite-thickness airfoil containing angular points, when the question arises where should a free discrete vortex shedding from a corner be placed. Now it is clear that, in analogy to the linear unsteady problem (if the Chaplygin–Joukowski hypothesis is applied to the bound vortex layer in the framework of the nonlinear problem), the first discrete vortex shedding from a fixed corner must be placed at the corner itself and then allowed to move along the local velocity vector. This principle is used for constructing discrete vortex formations used in Belotserkovsky, Lifanov, and Mikhailov (1985) for obtaining numerical solutions to the problem of separated flow past a contour containing angular points.

When considering three-dimensional problems it was found that integral equations should be written with respect to the double layer potential jump. The same procedure will also be applied to two-dimensional problems. In this case, one gets the following equation for the velocity $\mathbf{V}(M_0, t)$ induced by a body at point M_0 at instant t:

$$\mathbf{V}(M_0,t) = \nabla_{M_0}\int_\sigma \left(\frac{\partial}{\partial n_M}B(M_0,M)\right)g(M,t)\,d\sigma_M. \qquad (13.2.6)$$

Here $B(M_0, M) = (2\pi)^{-1}\ln r_{MM_0}^{-1}$ for the two-dimensional case and $B(M_0, M) = (4\pi)^{-1} r_{MM_0}^{-1}$ for the three-dimensional case; $\nabla_M f(M) = \nabla_M f(x, y, z) = f_x'\mathbf{i} + f_y'\mathbf{j} + f_z'\mathbf{k}$. For the velocity $\mathbf{V}_p(M_0, t)$ induced at point M_0 at instant t, one gets

$$\mathbf{V}_p(M_0,t) = \nabla_{M_0}\int_{\sigma_p} \left(\frac{\partial}{\partial n_{M_p(\tau,t)}}B(M_0,M_p(\tau,t))\right)g_p(\tau)\,d\sigma_{p,\,M_p(\tau,t)}. \qquad (13.2.7)$$

The relationships for $\mathbf{V}(M_p(\tau, t), t)$ and $\mathbf{V}_m(M_p(\tau, t), t)$ entering Equation (13.1.2) may be obtained from the corresponding formulas (13.2.6) and (13.2.7) by using the same substitutions as used before.

Note 13.2.1. Because the oncoming flow is potential, the product $\mathbf{U}_0(M_0, t) \cdot \mathbf{n}_{M_0}$ is a normal derivative of a harmonic function, and, hence, if σ is a closed contour–surface, then the following equality is valid:

$$\int_\sigma \mathbf{U}_0(M_0,t)\cdot\mathbf{n}_{M_0}\,d\sigma_{M_0} \equiv 0. \qquad (13.2.8)$$

Note 13.2.2. If a plane contour σ is unclosed, then for $M_0 \in \sigma$ the integrals entering (13.2.1) have a singularity of the form $\cot(\theta_0 - \theta)^{-1}$, and those entering (13.2.6) have the form $(\theta_0 - \theta)^{-2}$. However, if σ is a closed contour, then the preceding integrals have singularities of the form $(\theta_0 - \theta)/2$ and $\sin^{-2}[(\theta_0 - \theta)/2]$, respectively.

13.3. SMOOTH FLOW PAST A BODY: VIRTUAL INERTIA

As a rule, a vortex sheet forms both at the surface of a body and in the wake downstream of it. Therefore, hydrodynamic loads must be calculated taking into account vortex formations developing in the wake. However, for unsteady motions of a body, aerodynamic loads must also be known if the vortex wake behind a body may be ignored, as, for example, in the case of the noncirculatory flow mode. The same mode is used for calculating virtual inertia that is, in effect, an augmentation of the inertial properties of a body (such as mass, inertial/moments, etc.) and enter the expressions for forces and moments exerted by outer flow upon a body. According to the mode, outer flow is assumed to be fully potential and it is supposed that no wake forms downstream of a body. In this case differential equation (13.1.2) is ignored, and the problem reduces to finding the flow potential satisfying the Laplace equation outside a body and the no-penetration condition for the body surface.

Generally, the no-penetration condition (13.1.1) for a solid body traveling through an incompressible fluid with translational velocity $\mathbf{U}_0$ and angular velocity $\boldsymbol{\Omega}$ may be presented in the form

$$\frac{\partial \Phi}{\partial n_{M_0}} = \mathbf{V}(M_0) \cdot \mathbf{n}_{M_0} = (\mathbf{U}_0 + \boldsymbol{\Omega} \times \mathbf{r}_{M_0}) \cdot \mathbf{n}_{M_0}, \qquad (13.3.1)$$

where $\mathbf{r}_{M_0}$ is the radius-vector of point M_0 at which the no-penetration condition is met and Φ is the disturbed velocity potential.

In what follows the subscript 0 is omitted when referring to point M_0 and its coordinates.

By presenting vectors $\mathbf{U}_0$ and $\boldsymbol{\Omega}$ in the coordinate form

$$\mathbf{U}_0 = U_{0x} i + U_{0y} j + U_{0z} k,$$

$$\boldsymbol{\Omega} = \Omega_{xi} + \Omega_{yj} + \Omega_z k,$$

boundary conditions (13.3.1) may be written in the form

$$\frac{\partial \Phi}{\partial n_M} = U_{0x} \cos \widehat{nx} + U_{0y} \cos \widehat{ny} + U_{0z} \cos \widehat{nz} + \Omega_x \big(y \cos \widehat{nz} - z \cos \widehat{ny} \big)$$

$$+ \Omega_y (z \cos \widehat{nx} - x \cos \widehat{nz}) + \Omega_z \big(x \cos \widehat{ny} - y \cos \widehat{nx} \big). \qquad (13.3.2)$$

Because the Laplace equation is linear, the potential may be presented in the form:

$$\Phi = U_{0x}\Phi_1 + U_{0y}\Phi_2 + U_{0z}\Phi_3 + \Omega_x\Phi_4 + \Omega_y\Phi_5 + \Omega_z\Phi_6$$

$$= U_1\Phi_1 + U_2\Phi_2 + U_3\Phi_3 + U_4\Phi_4 + U_5\Phi_5 + U_6\Phi_6 = \sum_{i=1}^{6} U_i\Phi_i.$$

Then the expressions for forces and moments acting upon a body become (Belotserkovsky 1967)

$$x = \sum_{i=1}^{6} \frac{dU_i}{dt}\lambda_{i1} + \sum_{i=1}^{6} U_5U_i\lambda_{i3} - \sum_{i=1}^{6} U_6U_i\lambda_{i2},$$

$$y = -\sum_{i=1}^{6} \frac{dU_i}{dt}\lambda_{i2} - \sum_{i=1}^{6} U_6U_i\lambda_{i1} + \sum_{i=1}^{6} U_4U_i\lambda_{i3},$$

$$z = -\sum_{i=1}^{6} \frac{dU_i}{dt}\lambda_{i3} - \sum_{i=1}^{6} U_4U_i\lambda_{i2} + \sum_{i=1}^{6} U_5U_i\lambda_{i1},$$

$$M_x = -\sum_{i=1}^{6} \frac{dU_i}{dt}\lambda_{i4} - \sum_{i=1}^{6} U_5U_i\lambda_{i6} + \sum_{i=1}^{6} U_6U_i\lambda_{i5}$$

$$- \sum_{i=1}^{6} U_2U_i\lambda_{i3} + \sum_{i=1}^{6} U_3U_i\lambda_{i2},$$

$$M_y = -\sum_{i=1}^{6} \frac{dU_i}{dt}\lambda_{i5} - \sum_{i=1}^{6} U_6U_i\lambda_{i4} + \sum_{i=1}^{6} U_4U_i\lambda_{i6}$$

$$- \sum_{i=1}^{6} U_3U_i\lambda_{i1} + \sum_{i=1}^{6} U_1U_i\lambda_{i3},$$

$$M_z = -\sum_{i=1}^{6} \frac{dU_i}{dt}\lambda_{i6} - \sum_{i=1}^{6} U_4U_i\lambda_{i5} + \sum_{i=1}^{6} U_5U_i\lambda_{i4}$$

$$- \sum_{i=1}^{6} U_1U_i\lambda_{i2} + \sum_{i=1}^{6} U_2U_i\lambda_{i1}, \qquad (13.3.3)$$

where

$$\lambda_{ij} = -\rho \int_{\sigma} \Phi_i \frac{\partial \Phi_j}{\partial n_M} \, d\sigma_M \qquad (13.3.4)$$

are the so-called *virtual inertia* coefficients (integration is implemented over the surface of a body, i.e., the outer side of σ).

It should be noted that in the case of a body moving through water, virtual inertia is often of the same order of magnitude as the mass and the inertia moments of the body itself.

Using the theory of functions of a complex variable, many authors have calculated virtual inertia coefficients of various bodies (Riemann and Kreps 1947). However, until now all attempts have failed to construct a unified technique for calculating them for arbitrary bodies participating in arbitrary motions. Nevertheless, this can be done with the help of the method of discrete vortex pairs (for two-dimensional flows) or by employing the method of closed quadrangular–triangular vortex frames (for three-dimensional flows), as discussed in the following text.

Thus, coefficients λ_{ij} are calculated by means of finding the flow potential Φ_i, and $\partial \Phi_j / \partial n_M$ are defined by boundary conditions (13.3.2) for the corresponding mode of motion:

$$\frac{\partial \Phi_1}{\partial n} = \cos \widehat{nx}, \qquad \frac{\partial \Phi_2}{\partial n} = \cos \widehat{ny}, \qquad \frac{\partial \Phi_3}{\partial n} = \cos nz,$$

$$\frac{\partial \Phi_4}{\partial n} = y \cos \widehat{nz} - z \cos \widehat{ny}, \qquad \frac{\partial \Phi_5}{\partial n} = z \cos \widehat{nx} - x \cos \widehat{nz},$$

$$\frac{\partial \Phi_6}{\partial n} = x \cos \widehat{ny} - y \cos \widehat{nx}. \qquad (13.3.5)$$

Let us denote

$$\frac{\partial \Phi_j}{\partial n} = f_j, \qquad j = 1, \ldots, 6. \qquad (13.3.6)$$

Then the problem of finding virtual inertia reduces to solving the six integral equations

$$\mathbf{V}_j(M_0) \cdot \mathbf{n}_{M_0} = f_j, \qquad j = 1, \ldots, 6, \, M_0 \in \sigma, \qquad (13.3.7)$$

where $\mathbf{V}_j(M_0)$ is specified by Formula (13.2.6), i.e., we arrive at the solution of integral equations with respect to the double layer potential jump.

In the two-dimensional case the latter equation has the form

$$\frac{1}{2\pi}\int_{\sigma}\frac{r_{MM_0}^2(\mathbf{n}_M,\mathbf{n}_{M_0}) - 2(\mathbf{r}_{MM_0},\mathbf{n}_M)(\mathbf{r}_{MM_0},\mathbf{n}_{M_0})}{r_{MM_0}^4}g(M)\,d\sigma_M = f(M_0),$$

$$M_0 \in \sigma. \quad (13.3.8)$$

If σ is the interval $[-1,1]$ of the OX axis lying in the plane OXY, then the equation has the form

$$\frac{1}{2\pi}\int_{-1}^{1} g(x)\frac{dx}{(x_0-x)^2} = f(x_0), \qquad x_0 \in (-1,1), \quad (13.3.9)$$

and if σ is a unit-radius circle centered at the origin of coordinates, then

$$\frac{1}{8\pi}\int_0^{2\pi} g(\theta)\frac{d\theta}{\sin^2((\theta_0-\theta)/2)} = f(\theta_0), \qquad \theta_0 \in [0,2\pi]. \quad (13.3.10)$$

In the three-dimensional case Equation (13.3.7) has the form

$$\frac{1}{4\pi}\int_{\sigma}\frac{r_{MM_0}^2(\mathbf{n}_M,\mathbf{n}_{M_0}) - 3(\mathbf{r}_{MM_0},\mathbf{n}_M)(\mathbf{r}_{MM_0},\mathbf{n}_{M_0})}{r_{MM_0}^5}g(M)\,d\sigma_M = f(M_0),$$

$$M_0 \in \sigma. \quad (13.3.11)$$

If σ is the quadrant $[-1,1]\times[-1,1]$ of the plane OXY belonging to the space $OXYZ$, then

$$\frac{1}{4\pi}\int_{-1}^{1}\int_{-1}^{1} g(x,y)\frac{dx\,dy}{\left[(x_0-x)^2+(y_0-y)^2\right]^{3/2}} = f(x_0,y_0),$$

$$x_0, y_0 \in (-1,1). \quad (13.3.12)$$

13.4. NUMERICAL CALCULATION OF VIRTUAL INERTIA COEFFICIENTS: SOME CALCULATED DATA

Let us present the main idea of numerical calculation of virtual inertia in the two-dimensional case. Let a closed contour σ be specified in the parametric form $x = x(\theta)$, $y = y(\theta)$, $\theta \in [0, 2\pi]$, where $x(\theta)$ and $y(\theta)$ are periodic functions with the period 2π. Let us next choose points $\theta_1, \dots, \theta_n$ interpreted as points of a circle dividing the latter into equal arcs, and points $\theta_{01}, \ldots, \theta_{0n}$, which are the middles of arcs $\widetilde{\theta_1\theta_2}, \ldots, \widetilde{\theta_n\theta_1}$. Points $M_{0k} = (x(\theta_{0k}), y(\theta_{0k}))$, $k = 1, \ldots, n$, are assumed to be the reference points, and by the following system of linear algebraic equations is substituted for Equation (13.3.8):

$$\sum_{k=1}^{n} \hat{w}_{km} g_k = f_m, \qquad m = 1, \ldots, n, \tag{13.4.1}$$

where

$$\hat{w}_{km} = \frac{1}{2\pi} \int_{\sigma_k} \frac{r^2_{MM_{0m}}(\mathbf{n}_M, \mathbf{n}_{M_{0m}}) - 2(\mathbf{r}_{MM_{0m}}, \mathbf{n}_M)(\mathbf{r}_{MM_{0m}}, \mathbf{n}_{M_{0m}})}{r^4_{MM_{0m}}} d\sigma_M,$$

$$g_k = g(M_{0k}), \qquad f_m = f(M_{0m}).$$

It is known (Sedov 1971–72) that a double potential jump at a surface of a body is determined to an accuracy of a constant. Hence, System (13.4.1) is ill-conditioned, and in order to single out a unique solution it suffices to specify a value of the potential jump at a point of the surface of a body:

$$g(M_{0q}) = g_q = 0. \tag{13.4.2}$$

In fact, to solve aerodynamic problems, not the function $g(M)$ itself, but its derivative is needed. Hence, the function may be ascribed an arbitrary value at a point of the surface.

System (13.4.1) subject to Condition (13.4.1) has the form

$$\sum_{\substack{k=1 \\ k \neq q}}^{n} \hat{w}_{km} g_k = f_m, \qquad m = 1, 2, \ldots, n. \tag{13.4.3}$$

This system has more equations than unknowns and has no solution. In order to obtain a well-posed problem, one has to introduce a regularizing variable γ_{0n} (Lifanov 1980a). This results in the system

$$\gamma_{0n} + \sum_{\substack{k=1 \\ k \neq q}}^{n} \hat{w}_{km} g_k = f_m, \qquad m = 1, \ldots, n. \tag{13.4.4}$$

The matrix of the system may be shown to be well-conditioned for any n (Dvorak 1986).

Because the value of the potential at the outer side of contour σ is calculated by employing the formula (Tikhonov and Samarskii 1966)

$$\Phi(M_0) = \frac{1}{2\pi}\int_\sigma \frac{\cos \widehat{n_M r_{MM_0}}}{r_{MM_0}} \cdot g(M)\, d\sigma_M - \frac{g(M_0)}{2},$$

the following formula must be used for calculating the value of the potential at point M_{0m} of the contour:

$$\Phi_r(M_{0m}) = \frac{1}{2\pi}\sum_{\substack{k=1\\k\neq m}}^{n} \frac{\cos \widehat{n_{M_{0k}} r_{M_{0k}M_{om}}}}{r_{M_{0k}m_{0m}}} g_k\, \Delta\sigma_k - \frac{g_m}{2}. \qquad (13.4.5)$$

Note 13.4.1. According to Sedov (1971–72), the velocity field due to the potential of a constant-strength double layer positioned at a given surface σ coincides with the velocity field induced by a vortex filament of the same strength placed at the boundary of surface σ (if the latter is unclosed). However, if σ is an unclosed curve, then the corresponding velocity field coincides with the velocity field induced by a pair of discrete vortices of the same strength, which is equal to the double layer potential strength. Therefore, for a two-dimensional problem one has

$$\hat{w}_{km} = w_{km} - w_{k-1m}, \qquad (13.4.6)$$

where w_{km} is the normal velocity component at point M_{0m} induced by a unit-strength vortex placed at point $M_k(x(\theta_k), y(\theta_k))$, $k, m = 1, \ldots, n$. Therefore, this method for calculating aerodynamic characteristics with the help of system (13.4.4) is called the *method of discrete vortex pairs* (see Figure 13.1).

Note 13.4.2. If the contour σ contains angular points, then points M_k are chosen in such a way that the angular points are among them; in other words, discrete vortices are placed at the angular points.

Let us consider the three-dimensional problem. In accordance with Note 13.4.1, we start by considering closed quadrangular–triangular vortex frames. Therefore, surface σ is modeled by a system of such frames, and reference points M_{0m} are placed at the centers of the frames (see Figure 13.2). Then, by fulfilling the no-penetration condition at the points, we arrive at the system of linear algebraic equations of the form (13.4.1), where w_{km} is the normal velocity component at point M_{0m} induced by the kth vortex frame of unit strength. This system is also ill-conditioned, and the condition of the form (13.4.2) singles out its unique solution again.

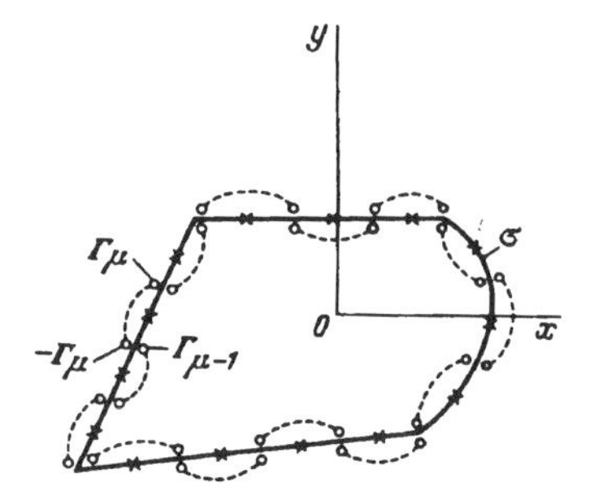

FIGURE 13.1. Simulation of a closed contour by the method of vortex pairs.

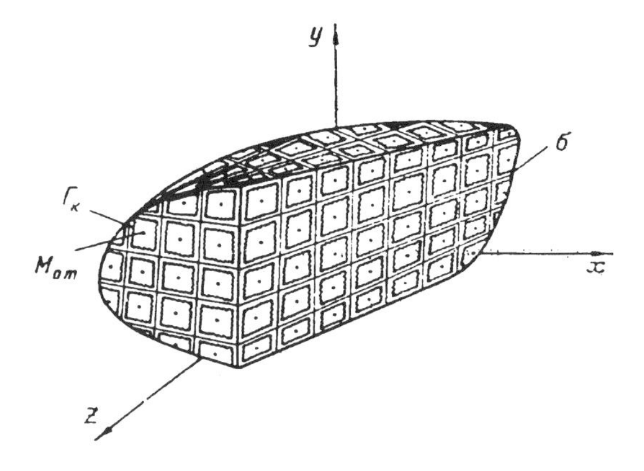

FIGURE 13.2. Simulation of a surface by closed vortex frames with the reference points positioned at the centers of the frames.

Using the latter condition, we arrive at a system of the form (13.4.3), which has no solution because it has more equations than unknowns. Then, introduction of the regularizing variable γ_{0n} allows us to obtain a system of the form (13.4.4), which is well-conditioned and in which

$$\widehat{w_{km}} = \frac{1}{4\pi}\int_{\sigma_k} \frac{r^2_{MM_{0m}}(\mathbf{n}_M, \mathbf{n}_{M_{0m}}) - 3(\mathbf{r}_{MM_{0m}}, \mathbf{n}_M)(\mathbf{r}_{MM_{0m}}, \mathbf{n}_{M_{0m}})}{r^5_{MM_{0m}}}\, d\sigma_M. \tag{13.4.7}$$

Note that $\widehat{w_{km}}$ is equal to the normal velocity component at point M_{0m}, induced by the vortex filament of unit strength located at the boundary of surface σ_k. It should be stressed that all the filaments must be similarly oriented.

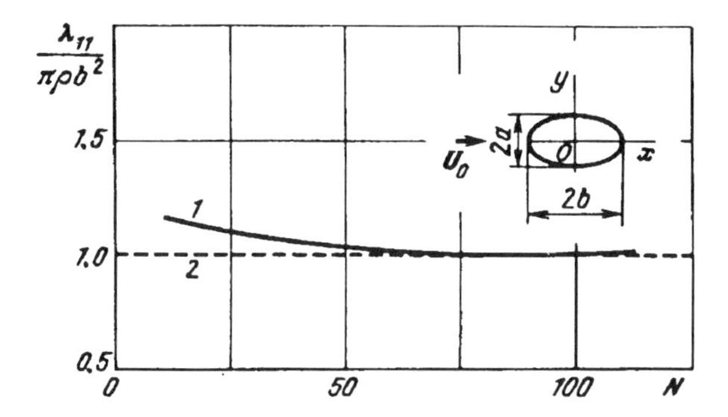

FIGURE 13.3. Convergence of calculated virtual inertia coefficients of an ellipse as a function of the number of discrete vortices. The solid and the dashed lines correspond to calculated results and the exact solution (Riemann and Kreps 1947), respectively.

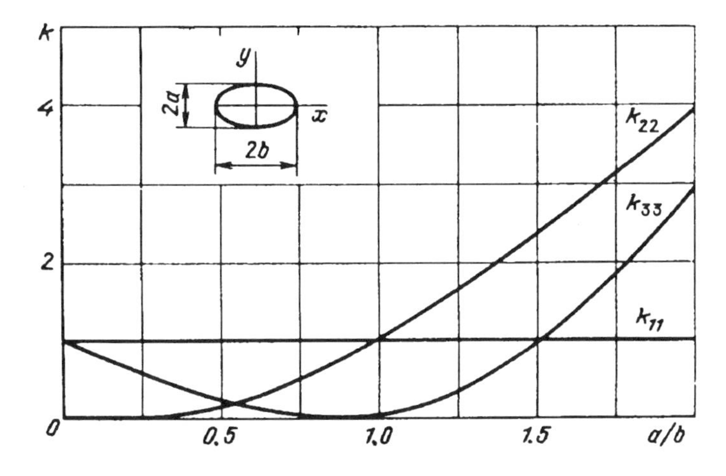

FIGURE 13.4. Virtual inertia coefficients for ellipses.

In order to demonstrate efficiency of the present method, it was used to calculate virtual inertia coefficients of ellipsoids, rectangles, and parallelepipeds (see Figures 13.3–13.7).

13.5. NUMERICAL SCHEME FOR CALCULATING SEPARATED FLOWS

In separated flows vortex sheets develop not only at a body surface, but also in the wake (see Sections 9.1 and 13.1). The latter sheet may also be modeled by free discrete vortices.

Let us start by considering two-dimensional flows. In this case, angular points of contour σ coincide with the flow separation points: it is quite

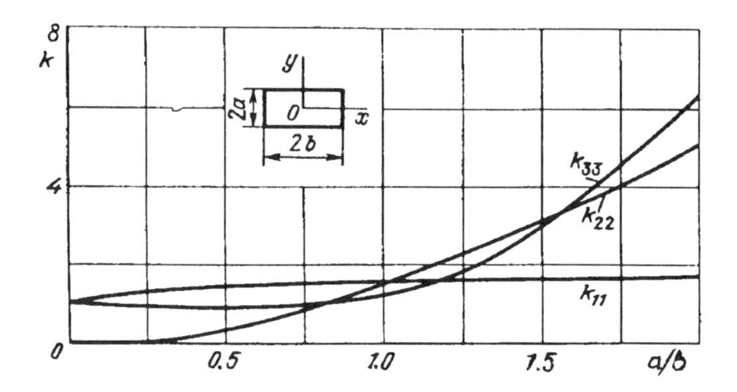

FIGURE 13.5. Virtual inertia coefficients for rectangles.

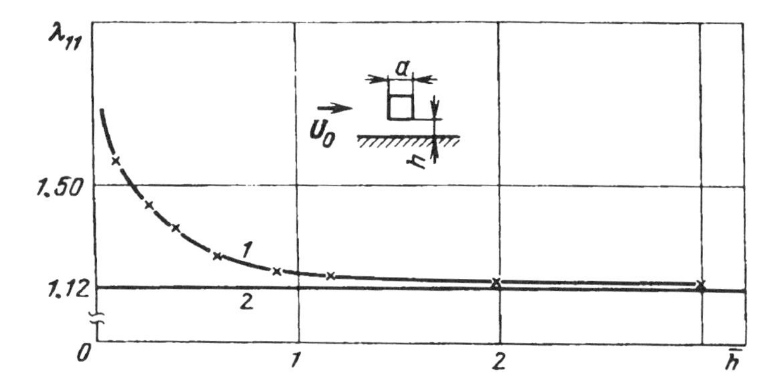

FIGURE 13.6. The effect of a screen on the virtual inertia of a square. Calculated results and the exact solution are shown by crosses and the solid line, respectively (Riemann and Kreps 1947).

natural for an ideal fluid, because otherwise the flow velocity disturbances at the points tend to infinity. The vortex sheet at a body surface and in the medium may be modeled by pairs of discrete vortices, as shown in the work of Belotserkovsky, Lifanov, and Mikhailov (1985); however, most often the simulation is carried with the help of discrete vortices. The approach is described here in detail.

Figure 13.8 presents the spatial distribution of discrete vortices. At the initial instant they are positioned only at the contour itself; the angular points of the latter are among the points occupied by discrete vortices (these are the discrete vortices assumed to be free, in accordance with Section 13.2). At the next instant, the free vortices are displaced into the

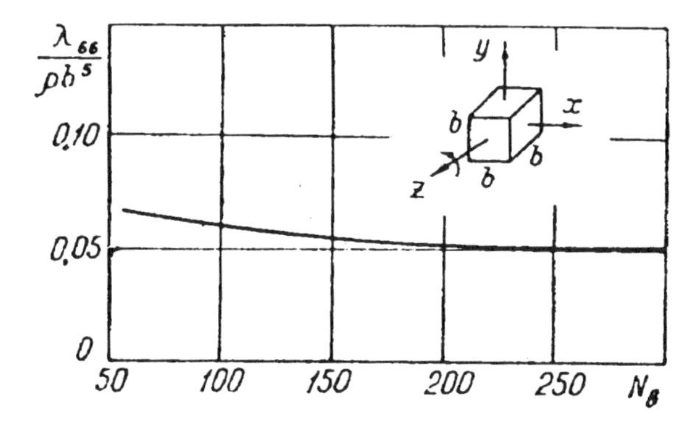

FIGURE 13.7. Calculated values of the virtual inertia for a cube, λ_{66}, depending on the number of vortex frames at the surface of the cube.

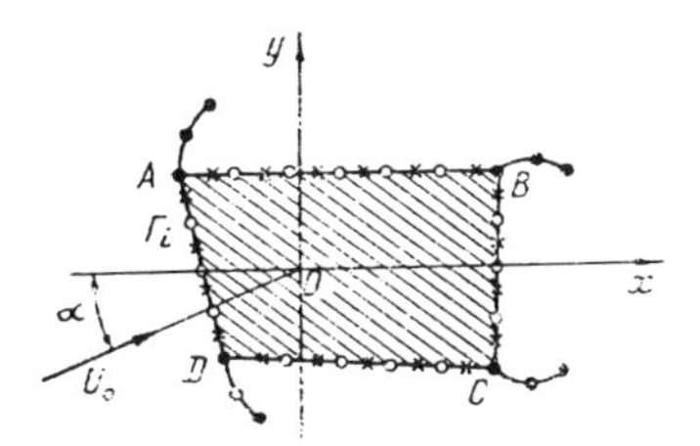

FIGURE 13.8. Distributions of discrete vortices (●●●) and reference points (× × ×) in the case of the unsteady problem for a contour containing angular points.

flow along the local velocity vector **v** (the displacement being equal to $\mathbf{v}\,\Delta\tau$, where τ is the time step) without changing their circulation, because the medium is ideal. At the second calculation instant, one has to deal with discrete vortices of unknown circulation located at the contour and free vortices of known circulation located near angular points. Subsequently, the free vortices travel with the fluid particles, newly born free vortices shedding from each of the corners at each reference instant. The Lord Kelvin (Thomson) condition (according to which circulation around a closed contour embracing both a body and the wake stays constant) is met at each reference instant. Thus, at each reference instant one has to consider a system of linear algebraic equations with respect to circulations of discrete vortices located at the contour, for which the number of equations exceeds the number of unknowns by 1, i.e., a system of equations generally has no solutions. By introducing a regularizing variable, one may obtain a system of linear algebraic equations possessing a unique solution. Thus, for the scheme shown in Figure 13.8 we get the system of

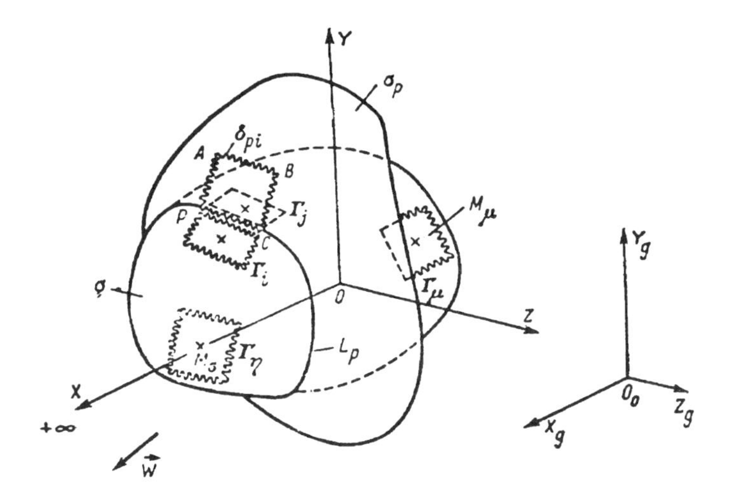

FIGURE 13.9. Distributions of closed vortex frames at the surface of a body and in the wake in the case of three-dimensional unsteady problems.

equations

$$\gamma_{0n} + \sum_{k=1}^{r} w_{k,m} = -\sum_{R=1}^{4}\sum_{s=1}^{r-1} \delta_{R,s} w_{s,m}^{R} + 2\pi f_m,$$

$$m = 1, \ldots, p,\ r = 1, 2, \ldots,$$

$$\sum_{k=1}^{n} \Gamma_k^r = \sum_{R=1}^{4}\sum_{s=1}^{r-1} \delta_{R,s}, \tag{13.5.1}$$

which corresponds to the rth reference instant.

The positions the free discrete vortices occupy at the next instant are determined by the formulas

$$x_{R,r+1} = x_{R,r} + v_x\,\Delta\tau,$$

$$y_{R,r+1} = y_{R,r} + v_y\,\Delta\tau. \tag{13.5.2}$$

For separated three-dimensional flows the discrete vortex sheet at the body and in the wake is represented by closed vortex frames (see Figure 13.9). At the initial instant, $\tau = 0$, the vortex wake is absent, and one has to solve the problem of noncirculatory flow past a body. Hence, the system

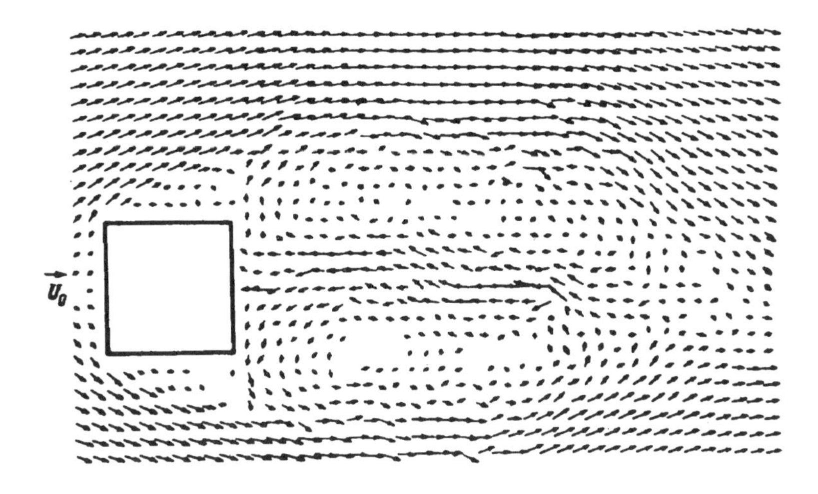

FIGURE 13.10. Symmetric unsteady separated flow past a square (the initial period of the flow).

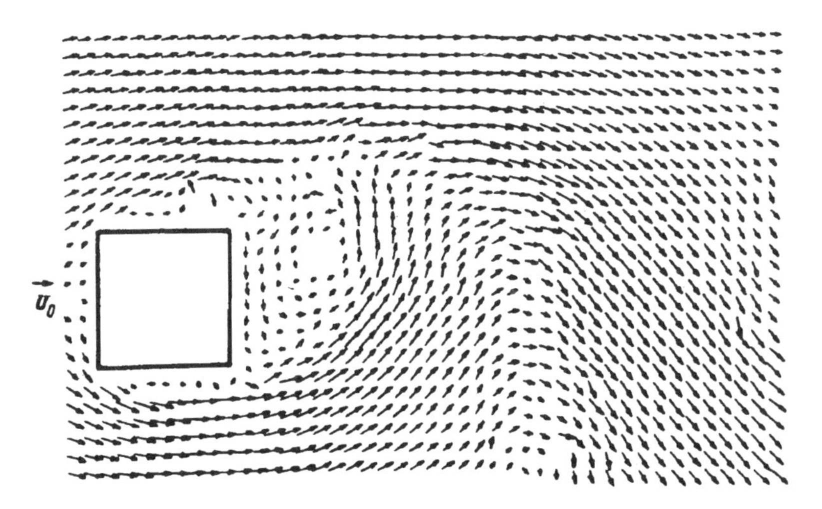

FIGURE 13.11. Asymmetric unsteady separated flow past a square.

of linear algebraic equations for calculating the unknown circulations of the vortex frames has the form (13.4.4). If a vortex sheet sheds from the angular line L_p of surface σ, then the side of a vortex frame lying at the body surface and bordering on L_p must lie at L_p. The vortex frames bordering on the line L_p on its opposite sides must have a common side

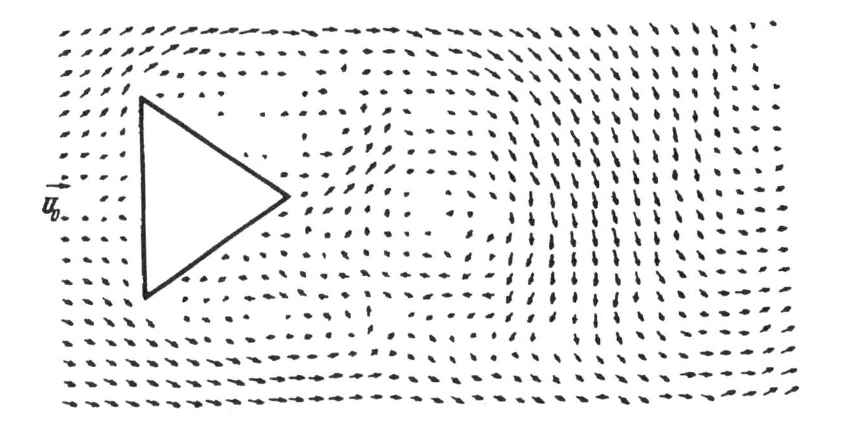

FIGURE 13.12. Flow field in the case of unsteady flow past a triangle.

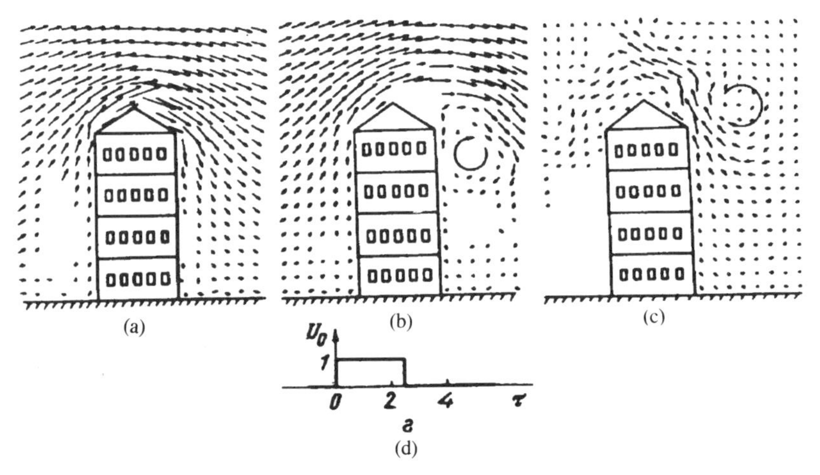

FIGURE 13.13. Flow field in the case of unsteady flow past a building. (a), (b), and (c) correspond to $\tau = 0.5, 2.5$, and 5.0, respectively; (d) shows the flow velocity variation in a gust.

(see Figure 13.9). For $\tau > 0$ a free vortex frame of circulation δ_{pi}^{k} equal to the difference of circulations of the corresponding vortex frames at the surface σ sheds into the flow from this common segment:

$$\delta_{pi}^{k} = \Gamma_{m}^{k-1} - \Gamma_{j}^{k-1} \tag{13.5.3}$$

Circulations of the free vortex frames do not vary as time passes, and each

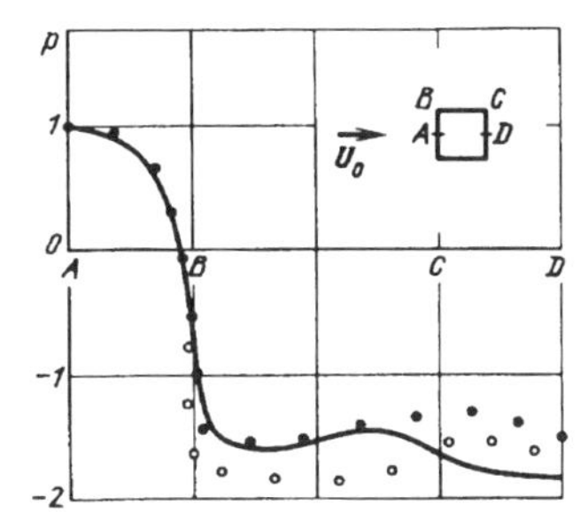

FIGURE 13.14. Pressure distribution at the surface of a square. The solid line shows calculated results, ●●● shows experimental data of Sluchanovskaya (1973) and ○ ○ ○ shows experimental data of Gorlin (1970).

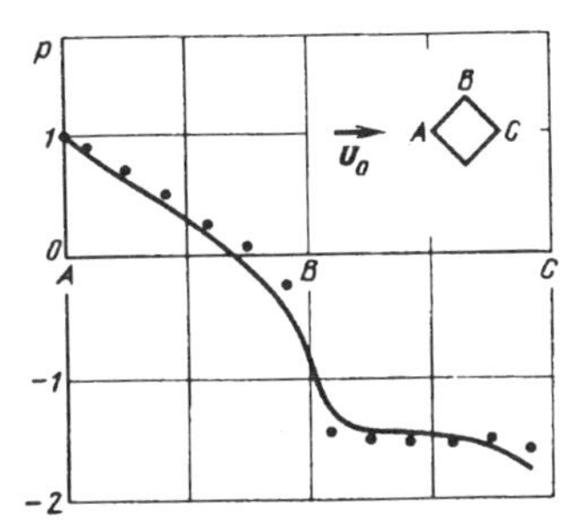

FIGURE 13.15. Pressure distribution at the surface of a rhombus. The solid line shows calculated results and ○ ○ ○ shows experimental data of Gorlin (1970).

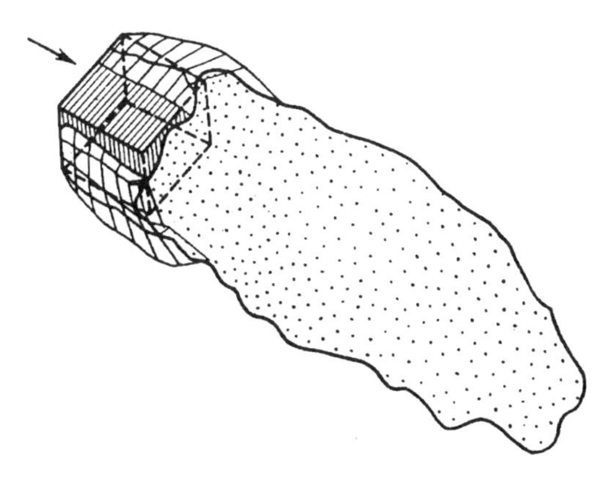

FIGURE 13.16. Unsteady flow past a cube. The form of the vortex sheet for $\tau = 6$.

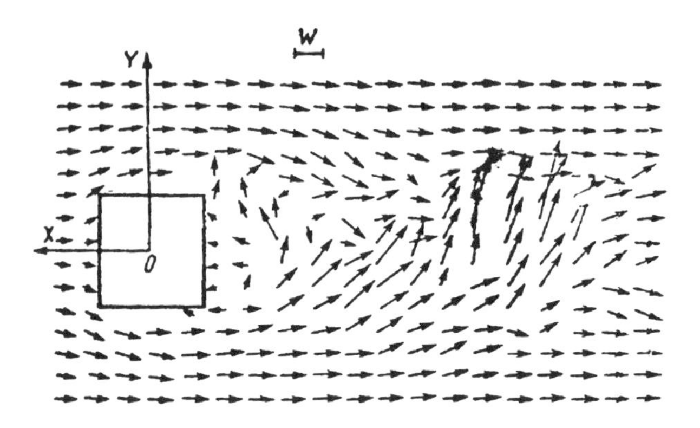

FIGURE 13.17. The vector flow velocity field in the plane OXY for $\tau = 6$.

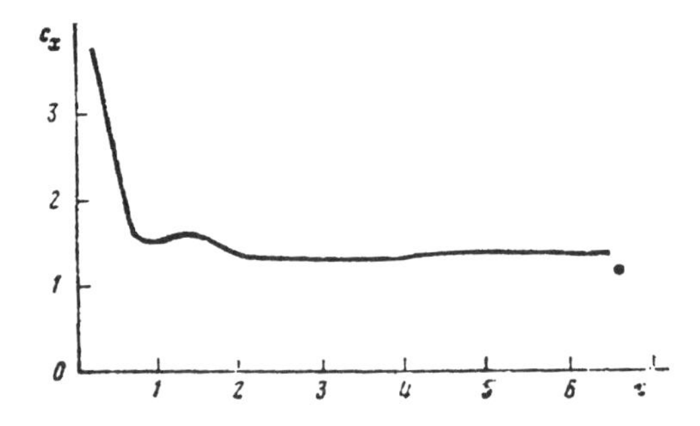

FIGURE 13.18. The drag coefficient C_x for a cube versus τ.

of the vortices displaces along the local flow velocity vector (see (13.5.2)). The preceding numerical technique for calculating separated unsteady flows was used to compile programs and to calculate two-dimensional flows past a square, a triangle, and building-like figures (Figures 13.10–13.15) and three-dimensional flows past a cube (Figures 13.16–13.18).

14 Some Questions of Regularization in the Method of Discrete Vortices and Numerical Solution of Singular Integral Equations

14.1. ILL-POSEDNESS OF EQUATIONS INCORPORATING SINGULAR INTEGRALS

While considering singular integrals, one often comes across situations when, due to measurement or calculation errors, the density of an integral is only known approximately. The operation of calculation of a regular integral is known to be stable in all commonly used metrics; in other words, small variations in a regular integral correspond to small variation in its density (Tikhonov and Arsenin 1979).

In Luzin (1951) and Khvedelidze (1957) it was shown that singular integrals, and hence equations incorporating the latter, are stable in the metrics L_2 with a corresponding weight at a given curve. On the other hand, Luzin (1951) constructed an example of such a function $f(\theta)$, $\int_{-\pi}^{\pi} f(\theta)\,d\theta = 0$, which is continuous on $[-\pi, \pi]$ and has a limited variation, whereas the function

$$\gamma(\theta_0) = \frac{-1}{2\pi}\int_0^{2\pi} \cot\frac{\theta_0 - \theta}{2} f(\theta)\,d\theta \tag{14.1.1}$$

suffers infinite discontinuities at a set that is dense all over $[-\pi, \pi]$.

This example shows that singular integrals are unstable in uniform metrics, i.e., if $f(\theta)$ is a continuous function close to zero in a uniform metric, then $\gamma(\theta)$ may differ from zero by a however large quantity. Singular integral (14.1.1) proves to be unstable also in the case when the function $f_1(\theta) - f_2(\theta)$ is smooth and has small values in a uniform metric.

The following theorem is true.

Theorem 14.1.1. *For any positive numbers ϵ and M there exist such periodic functions $f_1(\theta)$ and $f_2(\theta)$, infinitely differentiable on $[0, 2\pi]$, that the corresponding functions $\gamma_1(\theta)$ and $\gamma_2(\theta)$ (see Formula (14.1.1) are also infinitely differentiable, and the inequalities*

$$\|f_1 - f_2\| < \epsilon, \qquad \|\gamma_1 - \gamma_2\|_C > M. \tag{14.1.2}$$

hold.

Proof. In the theory of trigonometric series it is shown (Luzin 1951) that if function $f(\theta)$ belongs to the space L_2 on the segment $[-\pi, \pi]$, then it may be represented by the Fourier series

$$f(\theta) \sim \sum_{n=1}^{\infty} (a_n \cos n\theta + b_n \sin n\theta), \qquad \int_{-\pi}^{\pi} f(\theta)\, d\theta = 0, \tag{14.1.3}$$

and the conjugated function $\gamma(\theta)$ given by the formula

$$\gamma(\theta) \sim \sum_{n=1}^{\infty} (-b_n \cos n\theta + a_n \sin n\theta), \tag{14.1.4}$$

also belongs to L_2 on $[-\pi, \pi]$, and the equalities

$$f(\theta) = \frac{1}{2\pi} \int_0^{2\pi} \cot\frac{\theta - \theta_0}{2} \gamma(\theta_0)\, d\theta_0 \tag{14.1.5}$$

and (14.1.1) hold.

The function constructed by Luzin (1951) is such that the corresponding Fourier series converges to it uniformly. Let us denote by $f_n(\theta)$ $[\gamma_n(\theta)]$ the nth truncated sum of the Fourier series for function $f(\theta)$ $[\gamma(\theta)]$. Function $\gamma_n(\theta)$ is conjugated with respect to function $f_n(\theta)$, and hence the two functions satisfy Relationship (14.1.5). Because function $f_n(\theta)$ converges uniformly to function $f(\theta)$, a number $n_1(\epsilon)$ may be found for a given number $\epsilon > 0$, such that for all m, $n > n_1(\epsilon)$ the inequality

$$|f_m(\theta) - f_n(\theta)| < \epsilon \tag{14.1.6}$$

will be valid.

On the other hand, function $\gamma(\theta)$ is a limit of the sequence of functions $\gamma_n(\theta)$. Let point θ_1 be the point where function $\gamma(\theta)$ suffers an infinite discontinuity. Hence, at the point $|\gamma_n(\theta_1)| \to \infty$ for $n \to \infty$. Thus, if we consider numbers m and n satisfying the latter inequality, and fix n while m tends to infinity, then $|\gamma_n(\theta_1) - \gamma_m(\theta_1)| \to \infty$ for $m \to \infty$, i.e., for a given number $M > 0$ there exists $m(M)$ such that

$$|\gamma_n(\theta_1) - \gamma_{m(M)}(\theta)| > M.$$

Thus the theorem is proved, because functions $f_n(\theta)$, $f_{m(M)}(\theta)$, $\gamma_n(\theta)$, and $\gamma_{m(M)}(\theta)$ are infinitely differentiable. ■

As far as a singular integral over a circle is expressible via an integral with Hilbert kernel, it is unstable in a uniform metric.

In Gakhov (1977), Ivanov (1968), and Muskhelishvili (1952) it is shown that a singular integral over a segment or a system of nonintersecting segments may be reduced to a singular integral over a circle. Hence, the integrals are unstable in a uniform metric.

Because solutions to singular integral equations are expressed through singular integrals, the former are also unstable in a uniform metric. In practice, as a rule, one has to solve numerically singular integral equations whose right-hand sides and regular kernels are known approximately in a uniform metric. Therefore, the methods of numerical solution of these equations must be stable in a uniform metric because in practice one has to compare calculated results mainly in this metric). Such solutions may be constructed by the method of regularization developed by Tikhonov (Tikhonov and Arsenin 1979). Maslov (1967) has shown that a certain modification of the regularization proposed by Tikhonov for limited operators is applicable to singular integral operators that are unlimited operators in metric C. However, a reasonable use of singularities present in kernels of singular integral equations allows us to employ the simplest method of regularizing a numerical solution without constructing the regularizing operator explicitly. Originally, the idea of such regularization was used for integral equations of the first kind with a logarithmic singularity in works by Tikhonov, Dmitriev, and Zakharov (Dmitriev and Zakharov 1967, Tikhonov and Dmitriev 1968), who called it natural regularization or self-regularization. Later it was proved (Arsenin et al. 1985,

Lifanov 1980a, Matveev 1982b) that the method of discrete vortices, treated as a method of solving singular integral equations numerically, and its generalizations described in the preceding chapters, are also a method of natural regularization. Because the preceding regularizing methods of numerical solution are based on using various quadrature formulas for estimating singular integrals, in the following text the quadrature formulas will be shown to possess the regularizing property in both uniform and integral metrics.

14.2. REGULARIZATION OF SINGULAR INTEGRAL CALCULATION

In this section the following problem is considered (Matveev 1982b). Let a singular integral (one of those considered before) be calculated with the help of one of the previously mentioned quadrature formulas with (1) its density at the grid points known exactly and (2) the density known with a certain error whose value is known in a uniform metric. What will be the difference between the calculated values? By what means may the difference be made smaller by increasing the number of points?

Consider a singular integral on a segment in the case of using equally spaced grid points. The following theorem is true.

Theorem 14.2.1 (Matveev 1982b). *Let the density $\varphi(t)$ of singular integral* (5.1.1) *over segment $[a, b]$ be known at the equally spaced grid points $E = \{t_i = a + ih,\ h = (b - a)/(n + 1),\ i = 1, \ldots, n\}$ with an error ϵ_i such that $|\epsilon_i| \leq \epsilon$, $i = 1, \ldots, n$. Let $\tilde{S}_n(t_{0j})$, $t_{0j} = t_j + h/2$, $j = 0, 1, \ldots, n$, be the value of the quadrature sum for the integral calculated by using Formula* (1.3.6) *at the grid points $E_0 = \{t_{0j},\ j = 0, 1, \ldots, n\}$ when the values of $\varphi(t_i)$ are substituted by $\varphi(t_i) + \epsilon_i$. Then the inequality*

$$\left|S_n(t_{0j}) - \tilde{S}_n(t_{0j})\right| \leq O(\epsilon \ln n), \tag{14.2.1}$$

holds where $j = 0, 1, \ldots, n$ and $O(\epsilon \ln n) = 2\epsilon(3 + \ln n)$.

The proof of the theorem follows from Formula (1.3.2).

Note that the latter estimate depends only on the relative position of grids E and E_0 for $n \to \infty$, and hence, the grids may be translated along segment $[a, b]$ (see notes concerning the choice of grids in Section 1.3).

Note 14.2.1. Estimate (14.2.1) cannot be perfected as far as the order of magnitude is concerned.

In fact, let $n = 2m$ and $\epsilon_i = -\epsilon$, $i = 1, \ldots, m$; $\epsilon_i = \epsilon$, $i = m + 1, \ldots, 2m$. Then

$$\sum_{i=1}^{n} \frac{\epsilon_i h}{t_i - t_{0m}} = 2\epsilon \sum_{i=1}^{m} \frac{h}{t_{0m} - t_i} \geq 2\epsilon \sum_{k=0}^{m-1} \frac{2}{1 + 2k}$$

$$\geq 2\epsilon \ln(n + 1). \tag{14.2.2}$$

If $n = 2m + 1$, then by putting $\epsilon_i = \epsilon$, $i = m + 1, \ldots, 2m + 1$, one gets an estimate of the form (14.2.2) again.

Note 14.2.2. By denoting the value of a singular integral at points t_{0j} by $I(t_{0j})$, one gets

$$\left| I(t_{0j}) - \tilde{S}_n(t_{0j}) \right| \leq \left| I(t_{0j}) - S_n(t_{0j}) \right| + \left| S_n(t_{0j}) - \tilde{S}_n(t_{0j}) \right|$$

$$\leq \theta(t_{0j}) + O(\epsilon \ln n). \tag{14.2.3}$$

In the latter estimate the first summand, $\theta(t_{0j})$, depends on the form of the function $\varphi(t)$, and hence, this circumstance must be taken into consideration when choosing grids E and E_0. Thus, if $\varphi(t)$ has an integrable singularity at point $q \in (a, b)$ (as in the problem of an airfoil with a flap), then E and E_0 must be chosen in the following way:

1. Point q lies midway between the nearest points belonging to E and E_0.
2. Point q coincides with one of the points E if $\varphi(t)$ suffers a discontinuity of the first kind at the point; however, in the latter case one must take $[\varphi(t_{i_q} - 0) + \varphi(t_{i_q} + 0)]/2$ at point $t_{i_q} = q$.

The behavior of quantity $\theta(t_{0j})$ was analyzed in the preceding chapters. The second summand $O(\epsilon \ln n)$ depends only on the form and relative position of the grids E and E_0.

Next we consider a singular integral over a circle. In this case the following theorem is true.

Theorem 14.2.2. *Let us denote by $\tilde{S}_n(t_{0j})$ the quadrature sum for an integral over a circle at point t_{0j}, calculated with the help of Formula* (1.2.9), *where $\varphi(t_i) + \epsilon_i$ is substituted for $\varphi(t_i)$. Then estimate* (14.2.1), *where*

$$O(\epsilon \ln n) = 2\pi\epsilon(3 + \ln n) \tag{14.2.4}$$

is valid for $|S_n(t_{0j}) - \tilde{S}_n(t_{0j})|$, *where* $S_n(t_{0j})$ *is specified by the same formula.*

The proof of the theorem follows from refinement of Formula (1.3.2). It suffices to note that in Formula (1.2.8) we have $C = \pi$, because $\sin x \geq 2x/\pi$ for $0 \leq x \leq \pi/2$.

Note 14.2.3. An analogous statement is true for the quadrature sum for an integral with Hilbert kernel, obtained from Formula (1.5.4) or (2.1.19). In fact, in this case the inequality

$$\sum_{k=1}^{n} \left| \cot \frac{\theta_k - \theta_{0j}}{2} \right| \frac{2\pi}{n} \leq 2\pi(3 + \ln n) \tag{14.2.5}$$

holds.

Note 14.2.4. Theorem 14.2.2 is valid also for the interpolation quadrature formula for a singular integral over a circle. In fact, we have

$$\sum_{k=0}^{2n} \left| \frac{1}{t_k - t_{0m}} \right| \frac{|2\pi i t_k|}{2n+1} \leq \sum_{k=0}^{2n} \frac{\pi/(2n+1)}{|\sin(\theta_k - \theta_{0m})/2|} \leq 2\pi(3 + \ln n). \tag{14.2.6}$$

Next consider interpolation quadrature formulas for a singular integral over segment $[-1, 1]$. Thus, let

$$\varphi(t) = \omega(t)\psi(t),$$

$$\omega(t) = (1-t)^{\alpha}(1+t)^{\beta}, \qquad \psi(t) \in H. \tag{14.2.7}$$

Let $\alpha = \beta = -1/2$. Then, for $m = 1, \ldots, n-1$,

$$\sum_{k=1}^{n} \frac{\pi/n}{\left| \cos \frac{2k-1}{2n}\pi - \cos \frac{2m}{2n}\pi \right|}$$

$$= \frac{1}{2} \sum_{k=1}^{n} \frac{\pi/n}{\left| \sin \frac{2(k-m)-1}{4n} \right| \left| \sin \frac{2(k+m)-1}{4n}\pi \right|}$$

$$\leq \frac{\pi}{4} \sum_{k=1}^{n} \frac{\pi/n}{\frac{|2(k-m)-1|\pi}{4n} \left| \sin \frac{2(k+m)-1}{4n}\pi \right|}$$

$$\leq D_{-1/2,-1/2}(m, n), \tag{14.2.8}$$

where the quantity $D_{-1/2,-1/2}(m,n)$ satisfies the relationships:

1. For $t_{0m} \in [-1+\delta, 1-\delta]$,

$$D_{-1/2,-1/2}(m,n) = O_{1,\delta}(\ln n). \tag{14.2.9}$$

2. For $t_{0m} \in (-1, -1+\delta)$ or $t_{0m} \in (1-\delta, 1)$,

$$D_{-1/2,-1/2}(m,n) = \frac{1}{\sqrt{1-t_{0m}^2}} O_{2,\delta}(\ln n). \tag{14.2.10}$$

Let now $\alpha = \beta = 1/2$. Then, for $m = 0, 1, \ldots, n+1$,

$$\sum_{k=1}^{n} \frac{\dfrac{\pi}{n+1} \sin^2 \dfrac{k}{n+1}\pi}{\left| \cos \dfrac{2k}{2(n+1)}\pi - \cos \dfrac{2m-1}{2(n+1)}\pi \right|}$$

$$= \frac{1}{2} \sum_{k=1}^{n} \frac{\dfrac{\pi}{n+1} \sin^2 \dfrac{k}{n+1}\pi}{\left| \sin \dfrac{2(k-m)+1}{4(n+1)}\pi \right| \left| \sin \dfrac{2(k+m)-1}{4(n+1)}\pi \right|}$$

$$\leq \frac{\pi}{4} \sum_{k=1}^{n} \frac{\dfrac{\pi}{n+1} \sin^2 \dfrac{k\pi}{n+1}}{\dfrac{|2(k-m)+1|}{4(n+1)}\pi \left| \sin \dfrac{2(k+m)-1}{4(n+1)}\pi \right|}$$

$$\leq D_{1/2,1/2}(m,n), \tag{14.2.11}$$

where the quantity $D_{1/2,1/2}(m,n)$ satisfies Relationship (14.2.9) for any t_{0m}; in other words, the quantity of the order of $\ln n$ entering this relationship is independent of δ.

Finally, consider the case when $\alpha = 1/2$ and $\beta = -1/2$. Then,

$$\sum_{k=1}^{n} \frac{\dfrac{4\pi}{2n+1} \sin^2 \dfrac{k}{2n+1}\pi}{\left| \cos \dfrac{2k}{2n+1}\pi - \cos \dfrac{2m-1}{2n+1} \right|}$$

$$\leq \frac{\pi}{4} \sum_{k=1}^{n} \frac{\dfrac{4\pi}{2n+1} \sin^2 \dfrac{k}{2n+1} \pi}{\dfrac{|2(k-m)+1|}{4\pi+2} \pi \left| \sin \dfrac{2(k+m)-1}{4n+2} \pi \right|}$$

$$\leq D_{1/2,\,-1/2}(m, n), \qquad (14.2.12)$$

where the quantity $D_{1/2,\,-1/2}(m, n)$, $m = 1, \ldots, n$, satisfies Relationship (14.2.9) for all $t_{0m} \in (-1, +\delta, 1)$ and Relationship (14.2.10) for all $t_{0m} \in [-1, -1 + \delta]$.

Thus, the validity of the following theorem ensues from Formula (14.2.3) and the results obtained in Chapter 1 and this section.

Theorem 14.2.3. *All the quadrature formulas constructed previously by using two grids possess the property of regularization in a uniform metric. The number of points n forming the grid E is a regularizing parameter.*

Let us explain the idea of this theorem. Let functions $\varphi(t)$ and $\psi(t)$ be specified at points of grid E with an error ϵ in a uniform metric; in other words, let the function $\tilde{\varphi}(t)$ $[\tilde{\psi}(t)]$ be specified. Let $\theta(t_{0m}, n, \epsilon)$ be the absolute value of the difference of a singular integral at point t_{0m} and the quadrature sum calculated by using function $\tilde{\varphi}(t)$ $[\tilde{\psi}(t)]$.

Then, for any number $\Delta > 0$ there exists such a number $\epsilon(\Delta) > 0$ and such a function $n(\epsilon) \to \infty$ for $\epsilon \to 0$ that for any positive number $\epsilon < \epsilon(\Delta)$, the quantity $\theta(t_{0m}, n(\epsilon), \epsilon)$ does not exceed the number Δ at the corresponding set.

Thus, for a singular integral on segment $[-1, 1]$ when $\varphi(t) \in H^*$ on $[-1, 1]$, the preceding set is the segment $[-1 + \delta, 1 - \delta]$. For a singular integral at a circle when $\varphi(t) \in H$, the set is the whole of the circle, etc. (see the results obtained in Sections 1.2 and 1.3 and in this section).

Note 14.2.5. Quadrature formulas of the form (1.3.17), constructed by using either the grid E or the grid E_0, also possess regularizing properties.

14.3. METHOD OF DISCRETE VORTICES AND REGULARIZATION OF NUMERICAL SOLUTIONS TO SINGULAR INTEGRAL EQUATIONS

As far as singular integral equations are unstable in a uniform metric, the question arises of whether the preceding methods of approximate solution of the equations are stable in the metric. If the approximate methods were developed on the basis of some quadrature formulas for singular integrals, it is clear that the quadrature formulas must be stable with respect to small variations in the integral density in a uniform metric. Upon passing from singular integral equations to Fredholm integral equations of the second kind, one may use any quadrature formulas possessing regularizing properties, because a method of approximate solution of Fredholm equations of the second kind developed on the basis of the formulas is stable. It is of interest to analyze stability of direct methods of numerical solution of singular integral equations, developed on the basis of quadrature formulas possessing regularizing properties. It turns out that the use of such arbitrary quadratures may result in developing unstable methods of approximate solution of singular integral equations. For example, by using the quadrature formula of the form (1.3.17) on the set $E = \{t_i = a + ih,\ h = (b - a)/(n + 1),\ i = 1, \dots, n\}$, we deduce that the system of linear algebraic equations

$$\sum_{\substack{k=1 \\ k \neq j}}^{n} \frac{\varphi_n(t_k)h}{t_j - t_k} = f(t_j), \qquad j = 1, \dots, n, \tag{14.3.1}$$

contains n unknowns and n equations. The system approximates the characteristic singular integral equation (5.1.1) on $[a, b]$, but its determinant vanishes for any odd n.

In Ivanov (1968) it is shown that a similar situation may take place for a singular integral equation of the second kind on a circle.

By using two grids for constructing a numerical method of solving singular integral equations, one may develop well-conditioned systems of linear algebraic equations providing a method that is stable in a uniform metric. Originally, the idea of using a pair of grids was formulated in Belotserkovsky (1955) while constructing the method of discrete vortices for solving aerodynamic problems.

In this section we will show that the methods of solving singular integral equations, described previously and based on the ideas of the method of discrete vortices, are in fact methods of regularization for the equations.

1. Let us consider the problem of stability of the preceding methods with respect to small variations in right-hand sides. We will start by analyzing Equation (5.1.1) and presenting Systems (5.1.3)–(5.1.5) in matrix form:

$$A\varphi_n = f_n, \tag{14.3.2}$$

where A is a matrix of coefficients appearing in the preceding systems of equations, and φ_n and f_n are the columns consisting of unknowns and right-hand sides, respectively. Next consider the system

$$A\tilde{\varphi}_n = f_n + \epsilon_n, \tag{14.3.3}$$

where ϵ_n is a column of random errors, such that each of the coordinates is smaller in the absolute value than the number $\epsilon > 0$.

The following theorem is true.

Theorem 14.3.1. *For a difference of solutions to Systems* (14.3.2) *and* (14.3.3), *the inequality*

$$|\varphi_n(t_k) - \tilde{\varphi}_n(t_k)| \leq \mathfrak{Z}_n^{(1)}(t_k)\epsilon \ln n \tag{14.3.4}$$

holds, where $\mathfrak{Z}_n^{(1)}(t_k)$ *has the following properties*:

1. *For all points* $t_k \in [-1+\delta, 1-\delta]$, *where* δ *is an arbitrary small positive number*,

$$\mathfrak{Z}_n^{(1)}(t_k) \leq D_{1,\delta}. \tag{14.3.5}$$

2. *For all points* $t_k \in [-1, 1]$,

$$\sum_{k=1}^{n} \mathfrak{Z}_n^{(1)}(t_k)h \leq B_1. \tag{14.3.6}$$

(*here constants* $D_{1,\delta}$ *and* B_1 *are independent of* n).

Proof. System (14.3.3) has the solution

$$\tilde{\varphi}_n(t_k) = A^{-1}f_n + A^{-1}\epsilon_n = \varphi_n(t_k) + A^{-1}\epsilon_n, \tag{14.3.7}$$

where elements of the inverse matrix A^{-1} are specified by Formulas (5.1.13). By this formula and Equation (14.3.7) the validity of Inequality (14.3.4) ensues from Equations (5.1.14), (5.1.15), and (5.1.23)–(5.1.26), because

$$\frac{1}{h} I_{\kappa,k}^{(n)} \sum_{j=1}^{n} \frac{1}{h} I_{\kappa,0j}^{(n)} \frac{h}{|t_k - t_{0j}|} \leq \mathfrak{Z}_n^{(1)}(t_k) \ln n. \tag{14.3.8}$$

Theorem 14.3.1 is also true for the characteristic equation at a circle (6.1.1) if System (6.1.2) is substituted for Equation (14.3.2). However, in this case, $\mathfrak{Z}_n^{(1)}(t_k) \leq D_2$ for all t_k at the circle where the constants D_2 is independent of n. This conclusion follows from Formulas (6.1.6), (6.1.10), (6.1.11), and (14.2.4).

Theorem 14.3.1 is valid also for characteristic equation (6.2.1) with the Hilbert kernel. In the latter case either System (6.2.6) or System (6.2.19) must be substituted for Equation (14.3.2), and the matrix A^{-1} is given by

Formula (6.2.13) or (6.2.23), respectively. Then one has to employ Formula (14.2.5) according to which $\mathfrak{Z}_n^{(1)}(\theta_k) \le D_3$ for $\theta_k \in [0, 2\pi]$, where D_3 is independent of n. ■

Next we consider Equation (5.1.1) for the case when $[a,b] = [-1,1]$ and unequally spaced grid points are used. Let us write the systems of linear algebraic equations (5.2.2), (5.2.15), and (5.2.16) in the form

$$A\psi_n = f_n, \tag{14.3.9}$$

and the system incorporating errors on the right-hand sides in the form

$$A\tilde{\psi}_n = f_n + \epsilon_n. \tag{14.3.10}$$

Then the following theorem is true.

Theorem 14.3.2. *For the difference of solutions to Systems* (14.3.9) *and* (14.3.10), *the inequality*

$$\left|\psi_n(t_k) - \tilde{\psi}_n(t_k)\right| \le \mathfrak{Z}_n^{(2)}(t_k)\varepsilon \ln n \tag{14.3.11}$$

holds where the quantity $\mathfrak{Z}_n^{(2)}(t_n)$ *possesses the following properties*:

1. *If* $\kappa = 1$, *then*

$$\mathfrak{Z}_n^{(2)}(t_k) \le D_4 \tag{14.3.12}$$

for all $t_k \in [-1,1]$;

2. *If* $\kappa = -1$, *then*

$$\mathfrak{Z}_n^{(2)}(t_k) \le D_{5,\delta}, \qquad t_k \in [-1+\delta, -\delta], \tag{14.3.13}$$

$$\mathfrak{Z}_n^{(2)}(t_k) \le \frac{D_{6,\delta}}{\sqrt{1-t_k^2}}, \qquad t_k \in [-1, -1+\delta] \cup [1-\delta, 1]. \tag{14.3.14}$$

3. *If* $\kappa = 0$, *then the quantity* $\mathfrak{Z}_n^{(2)}(t_k)$ *satisfies Inequality* (14.3.13) *for points* $t_k \in [-1, 1-\delta]$ *and Inequality* (14.3.14) *for points* $t_k \in [1-\delta, 1]$;
4. *For all* κ,

$$\sum_{k=1}^{n} \mathfrak{Z}_n^{(2)}(t_k)a_k \le B_2. \tag{14.3.15}$$

Here D_4, D_5, D_6, and B_2 are constants independent of both n and k.

The validity of Theorem 14.3.2 follows from the results obtained in Section 5.2 and Formulas (14.2.8), (14.2.11), and (14.2.12).

Note 14.3.1. Estimates (14.3.4) and (14.3.11) cannot be refined. In the simplest way this may be demonstrated for an equation with the Hilbert kernel. Let us choose points θ_k in such a way that the point π is one of them for $k = k_\pi$. The errors will be selected as follows. If $\theta_{0j} < \pi$, then $\epsilon_{n,j} = -\epsilon$; if $\theta_{0j} > \pi$, then $\epsilon_{n,j} = \epsilon$. Then the following inequality holds at point π:

$$\left| \sum_{j=1}^{n} \cot \frac{\pi - \theta_{0j}}{2} \epsilon_{n,j} \frac{2\pi}{n} \right| \geq B_3 \varepsilon \ln n, \tag{14.3.16}$$

where the constant $B_3 > 0$ is independent of n.

Analogous examples may be considered for other cases also.

2. Let us continue by considering the problem of stability of the numerical methods under consideration with respect to small variations in the elements of matrices of systems of linear algebraic equations.

Let us consider the following system of linear algebraic equations:

$$A_1 \tilde{\tilde{\varphi}}_n = f_n + \epsilon_n, \tag{14.3.17}$$

where matrix A_1 may be obtained from matrix A (see System (14.3.2) as follows.

If uniform grids E and E_0 are used in a singular integral equation on a segment, then t_1 and t_{0j} are substituted by

$$\tilde{t}_i = t_i + \epsilon_{1,n,i} h, \qquad |\epsilon_{1,n,i}| < \epsilon,$$

$$\tilde{t}_{0j} = t_{0j} + \epsilon_{1,n,0j} h, \qquad |\epsilon_{1,n,0j}| < \epsilon, \tag{14.3.18}$$

respectively, where the number ϵ must be sufficiently small (say, $\epsilon \leq 1/8$), so that points $\tilde{t}_1$ and $\tilde{t}_{0j}$ are not coincident, and h is substituted by

$$\tilde{h} = h + \epsilon^0 h, \qquad |\epsilon^0| < \epsilon. \tag{14.3.19}$$

If an equation with the Hilbert kernel is considered, then, in analogy to Formulas (14.3.18) and (14.3.19), we put $\tilde{\theta}_k = \theta_k + \epsilon_{1,n,k} h$ and $\tilde{\theta}_{0j} = \theta_{0j} + \epsilon_{1,n,0j} h$, $h = 2\pi/n$, *and for an equation over a circle*, $\tilde{t}_k = \exp(i\tilde{\theta}_k)$ *and* $\tilde{t}_{0j} = \exp(i\tilde{\theta}_{0j})$.

The following theorem is true.

Theorem 14.3.3. *For a difference of solutions to Systems* (14.3.2) *and* (14.3.17), *the inequality*

$$\left|\varphi_n(t_k) - \tilde{\tilde{\varphi}}_n(t_k)\right| \le \mathfrak{Z}_n^{(3)}(t_k)(\epsilon + \eta_{n,\epsilon} + \eta_{n,\epsilon}^2)\ln n,$$

$$\eta_{n,\epsilon} = \epsilon \frac{1 - (\epsilon \ln n)^n}{1 - \epsilon \ln n} \ln n \tag{14.3.20}$$

holds, where the quantity $\mathfrak{Z}_n^{(3)}(t_k)$ *is analogous to the quantity* $\mathfrak{Z}_n^{(1)}(t_k)$.

Proof. System (14.3.17) is solvable, and

$$\tilde{\tilde{\varphi}}_n(t_k) = A_1^{-1} f_n + A_1^{-1} \epsilon_n, \tag{14.3.21}$$

where matrix A_1^{-1} may be obtained from matrix A^{-1} by replacing the factors $I_{\kappa,k}^{(n)}$ and $I_{\kappa,0j}^{(n)}$, $\kappa = 0, 1, -1$, by the factors $\tilde{I}_{\kappa,k}^{(n)}$ and $\tilde{I}_{\kappa,0j}^{(n)}$ obtained from the original ones by replacing t_k and t_{0j} by $\tilde{t}_k$ and $\tilde{t}_{0j}$, respectively. By the choice of points $\tilde{t}_k$ and $\tilde{t}_{0j}$, one gets

$$\tilde{I}_{\kappa,k}^{(n)} = I_{\kappa,k}^{(n)}(1 + \beta_{\kappa,k}) \prod_{\substack{m=1 \\ m \ne k}}^{n} (1 + \alpha_{k,m}),$$

$$\tilde{I}_{\kappa,0j}^{(n)} = I_{\kappa,0j}^{(n)}(1 + \beta_{\kappa,0j}) \prod_{\substack{m=1 \\ m \ne j}}^{n} (1 + \alpha_{0j,m}),$$

$$|\beta_{\kappa,k}||\beta_{\kappa,0j}| \le \beta_4 \epsilon, \qquad |\alpha_{k,m}| \le B_5 \frac{\epsilon}{|k - m|},$$

$$|\alpha_{0j,m}| \le B_6 \frac{\epsilon}{|j - m|}, \tag{14.3.22}$$

where the constants B_4, B_5, and B_6 are independent of n. Let us prove Formulas (14.3.22) for $\tilde{I}_{0,k}^{(n)}$ and $\tilde{I}_{0,0j}^{(n)}$ in the case of equally spaced grid points. We have [see (5.1.12)]:

$$\tilde{I}_{0k} = (\tilde{t}_{0k} - \tilde{t}_k) \prod_{\substack{m=1 \\ m \ne k}}^{n} \frac{\tilde{t}_{0m} - \tilde{t}_k}{\tilde{t}_m - \tilde{t}_k}, \qquad \frac{\tilde{t}_{0m} - \tilde{t}_m}{\tilde{t}_m - \tilde{t}_k} = \frac{t_{0m} - t_k}{t_m - t_k}(1 + \alpha_{k,m}), \tag{14.3.23}$$

and hence, Formula (14.3.22) is valid.

Thus,

$$\tilde{I}_{\kappa,k} = I_{\kappa,k}(1+\xi_{\kappa,k}),$$

$$\tilde{I}_{\kappa,0j} = I_{\kappa,0j}(1+\xi_{\kappa,0j}), \qquad |\xi_{\kappa,k}|, |\xi_{\kappa,0j}| \le B_7 \eta_{n,\epsilon}, \tag{14.3.24}$$

where B_7 is independent of n.

The validity of Theorem 14.3.3 follows from Formulas (14.3.21)–(14.3.24) and Theorem 14.2.1. ■

Note 14.3.1. If $\epsilon \ln n < 1$, then the factor preceding $\mathfrak{Z}_n^{(3)}(t_k)$ in Inequality (14.3.20) is of the same order of magnitude as $\epsilon \ln^2 n$.

If $[a, b] = [-1, 1]$ in Equation (5.1.1) and unequally spaced grid points are under study, then one has to consider the system of linear algebraic equations

$$A_1 \tilde{\tilde{\psi}}_n = f_n + \epsilon_n, \tag{14.3.25}$$

where matrix A_1 is obtained from matrix A entering System (14.3.9) as follows. If we denote $t_k = \cos\theta_{1,k}$, $t_{0j} = \cos\theta_{1,0j}$, and $a_k = a(\theta_k)$, then $\tilde{t}_k = \cos(\theta_{1,k} + \epsilon_{1,n,k} h_1)$, $\tilde{t}_{0j} = \cos(\theta_{1,0j} + \epsilon_{1,n,0j} h_1)$, $h_1 = \pi/n$ and $\tilde{a}_k = a(\theta_k + \epsilon_{1,n,k} h_1)$ must be substituted for $\tilde{t}_k$ and $\tilde{t}_{0j}$. Next, matrix A $\tilde{t}_k$, t_{0j}, and $\tilde{a}_k$ are to be substituted for t_k, t_{0j}, and a_k entering matrix A, respectively. Then, an inequality of the form (14.3.20) is also valid for the absolute value of the difference $\psi_n - \tilde{\tilde{\psi}}_n$, where the quantity $\mathfrak{Z}_n^{(3)}(t_k)$ has the same properties as quantities $D_{1/2,1/2}(m,n)$, $D_{-1/2,1/2}(m,n)$, and $D_{-1/2,-1/2}(m,n)$ for $\kappa = 1, -1$, and 0, respectively, and the solution is limited at point $t = 1$.

Let us consider the differences between solutions $\varphi(t_k)$ to Equations (5.1.1) and (6.1.1) and $\varphi_k(\theta_k)$ to Equation (6.2.1) and the corresponding solutions to systems of linear algebraic equations of the form (14.3.3) and (14.3.17); in other words, we will analyze the quantities $\theta^{(1)}(\epsilon, n, t_k)$ and $\theta^{(2)}(\epsilon, n, t_k)$, where $\theta^{(1)} = |\varphi(t_k) - \tilde{\varphi}_n(t_k)|$ and $\theta^{(2)} = |\varphi(t_k) - \tilde{\tilde{\varphi}}_n(t_k)|$.

The following theorem follows from Theorems 14.3.1 and 14.3.3.

Theorem 14.3.4. *Let equally spaced grid points be chosen for Equation* (5.1.1), *and let function f belong to the class H on* $[a, b]$. *Then for any* $\Delta > 0$ *there exist such a number* $\epsilon(\Delta) > 0$ *and such a function* $n(\epsilon) \to \infty$ *for* $\epsilon \to 0$ *that for any positive number* $\epsilon < \epsilon(\Delta)$, *the following inequalities are valid*:

a. *For all the points* $t_k \in [a+\delta, b-\delta]$, *where* $\delta > 0$ *is a fixed small number*,

$$\theta^{(i)}(\epsilon, n(\epsilon), t_k) \le \Delta, \qquad i = 1, 2. \tag{14.3.26}$$

b. *For all the points* $t_k \in [a, b]$,

$$\sum_{k=1}^{n} \theta^{(i)}(\epsilon, n(\epsilon), t_k)h \le \Delta, \qquad i = 1, 2. \qquad (14.3.27)$$

To prove this theorem it suffices to use the inequalities

$$|\varphi(t_k) - \tilde{\varphi}_n(t_k)| \le |\varphi(t_k) - \varphi_n(t_k)| + |\varphi_n(t_k) - \tilde{\varphi}_n(t_k)|,$$

$$\left|\varphi(t_k) - \tilde{\tilde{\varphi}}_n(t_k)\right| \le |\varphi(t_k) - \varphi_n(t_k)| + \left|\varphi_n(t_k) - \tilde{\tilde{\varphi}}_n(t_k)\right| \quad (14.3.28)$$

and Theorems 5.1.1, 14.3.1, and 14.3.3.

For all t_k, Inequality (14.3.26) holds for equations on a circle as well as for equations with the Hilbert kernel.

If $[a, b] = [-1, 1]$ in Equation (5.1.1) and the grid points are unequally spaced, then by $\theta^{(i)}(\epsilon, n, t_k)$, $i = 1, 2$, one has to denote absolute values of the differences $\psi(t_k) - \tilde{\psi}_n(t_k)$ and $\psi(t_k) - \tilde{\tilde{\psi}}_n(t_k)$ for which Theorem 14.3.4 also holds.

In accordance with the latter theorem, the preceding algorithm for obtaining approximate solutions to Eqs. (5.1.1), (6.1.1), and (6.2.1) is a regularizing one. The regularizing parameter n is equal to the number of equations of a system of linear algebraic equations substituting for the original singular integral equation.

Numerical methods for full singular integral equations of the first kind on a segment–circle and with the Hilbert kernel possess an analogous regularizing property.

14.4. REGULARIZATION IN THE CASE OF UNSTEADY AERODYNAMIC PROBLEMS

In the preceding section we have shown that the method of discrete vortices is in fact a method of regularization of numerical solutions to ill-posed (in a uniform metric) problems of steady two-dimensional flow of an ideal incompressible fluid past an airfoil.

Presently, there are no analogous mathematical proofs for unsteady problems of flow past an airfoil. Moreover, in order to ensure convergence of a numerical solution in the case of a nonlinear unsteady problem, one has to take into consideration the following specific feature of the problem (Belotserkovsky and Nisht 1978). From Formula (9.3.14) it follows that flow velocities induced at a point approaching a vortex filament tend to infinity. One can come across a similar situation in the case of a nonlinear unsteady problem, when the solution may start oscillating at a segment. In order to suppress the oscillations one has to artificially limit flow velocities in the neighborhood of the vortex axis.

However, the following simple and most efficient method of regularization may be proposed. Let ϵ be the shortest distance between reference points and discrete vortices at an airfoil (both the points and vortices may be unequally spaced). For a chosen vortex system one cannot be sure that the flow velocity field will be determined correctly within the neighborhood of the vortex axis. Therefore, if in the process of solution the distance between free vortices becomes less than ϵ, the induced flow velocities must be limited. According to (9.2.1), disturbed flow velocities vanish at the axis of a vortex filament. Hence, the flow velocity field within a circle centered at the vortex axis, whose radius is equal to ϵ, should be either determined by means of linear interpolation between the values of the velocity at the circle and at the vortex axis or just put equal to zero.

Calculations show that due to discreteness of the scheme with respect to both coordinates and time, free vortices sometimes "jump across" the airfoil surface. However, this deficiency of the method was removed by introducing a condition according to which such a vortex returns to its original position at the next step.

Part IV: Some Problems of the Theory of Elasticity, Electrodynamics, and Mathematical Physics

15

Singular Integral Equations of the Theory of Elasticity

15.1. TWO-DIMENSIONAL PROBLEMS OF THE THEORY OF ELASTICITY

Following M. M. Soldatov (Belotserkovsky, Lifanov, and Soldatov 1983), we will demonstrate how two-dimensional problems may be reduced to singular integral equations.

The basic relationships of the theory of elasticity are given by the Kolosov–Muskhelishvili formulas (Muskhelishvili 1966, Parton and Perlin 1984), which determine stresses through two functions of a complex variable, φ and ψ, analytic in the domain D of the plane OXY:

$$\sigma_x + \sigma_y = 4\,\mathrm{Re}[\varphi'(z)],$$

$$\sigma_y - \sigma_x + 2i\tau_{xy} = 2[\bar{z}\varphi''(z) + \psi'(z)]. \qquad (15.1.1)$$

In the absence of bulk (mass) forces, the functions satisfy the homogeneous biharmonic equation of the problem; thus, it suffices to meet the boundary condition at contour L bounding the domain D:

$$\kappa_k \varphi(t) - t\varphi'(t) - \overline{\psi(t)} = f_k(t), \qquad t \in L, k = 1, 2. \quad (15.1.2)$$

If the latter condition is written in displacements, then $k = 1$, and

$$\kappa_1 = \kappa = \begin{cases} 3 - 4\nu & \text{for plane strain,} \\ \dfrac{3 - \nu}{1 + \nu} & \text{for plane stress,} \end{cases}$$

$$f_1 = 2\mu(u + iv), \qquad (15.1.3)$$

where ν is the Poisson coefficient, μ is the shear modulus, and u and v are displacements along the axes OX and OY, respectively.

However, if Condition (15.1.2) is given in stresses, then $k = 2$, and

$$\kappa_2 = -1, \qquad f_2 = -i\int_{s_0}^{s} (\sigma_{x\rho} + i\sigma_{y\rho})\, ds + C_2,$$

$$\sigma_{x\rho} = \sigma_x l + \tau_{xy} m, \qquad \sigma_{y\rho} = \tau_{xy} l + \sigma_{ym}, \tag{15.1.4}$$

where σ_x, σ_y, and τ_{xy} are the normal and tangential stresses in the coordinate system OXY; $l = \cos(n, x) = dy/ds$ and $m = \cos(n, y) = dx/ds$ are the direction cosines, and C_2 is a complex constant determined by the condition to be presented in the following text.

For a singly connected domain, contour L in (15.1.2) is passed around counterclockwise; in the case of a multiply connected domain, contour L is passed around in such a way that domain D stays to the left. For functions $\varphi(t)$, $\psi(t)$, and $\varphi'(t)$ and complex conjugated functions $\overline{\psi(t)}$ and $\overline{\varphi'(t)}$, limiting values are taken at points located inside and outside D in the cases of solving the internal and the external problem, respectively.

Analytic functions $\varphi(z)$ and $\psi(z)$ are sought in the form

$$\varphi(z) = \frac{1}{2\pi i}\int_L \frac{\omega(t)}{t - z}\, dt, \qquad \psi(z) = \frac{1}{2\pi i}\int_L \frac{\bar{C}\,\overline{\omega(t)} - \bar{t}\omega'(t)}{t - z}\, dt,$$

$$z \in D, \tag{15.1.5}$$

where the auxiliary function $\omega(t)$ is defined by the integral equation

$$\frac{\kappa_k - C}{2}\omega(t) + \frac{\kappa_k}{2\pi i}\int_L \omega(\tau)\, d\left(\ln\frac{\bar{\tau} - \bar{t}}{\tau - t}\right) + \frac{\kappa_k + C}{2\pi i}$$

$$\times \int_L \omega(\tau)\, d(\ln(\tau - t)) + \frac{1}{2\pi i}\int_L \omega(\tau)\, d\left(\frac{\tau - t}{\bar{\tau} - \bar{t}}\right) = f_k(t),$$

$$k = 1, 2, \tag{15.1.6}$$

obtained from (15.1.2). Here C is a complex parameter determining the type of the resulting equation.

For $C = -\kappa_k$, $k = 1, 2$, one arrives at the Fredholm equation of the second kind (Muskhelishvili 1966):

$$\kappa_k \omega(t) + \frac{\kappa_k}{2\pi i}\int_L \omega(\tau)\, d\left(\ln\frac{\bar{\tau}-\bar{t}}{\tau - t}\right) + \frac{1}{2\pi i}\int_L \overline{\omega(\tau)}\, d\left(\frac{\tau - t}{\bar{\tau}-\bar{t}}\right) = f_k(t). \tag{15.1.7}$$

Due to the presence of an eigenfunction, the latter equation is hardly solvable by the method of formal quadratures, because the determinant of the corresponding system of linear algebraic equations is equal to zero, and calculated values of the function we seek are unstable (Parton and Perlin 1982, 1984).

One comes across similar complications when reducing the problem to other regular equations of the Muskhelishvili and Sherman–Lauricella types [see Parton and Perlin (1984) where some methods of removing the drawback are considered, such as fixation of ω at some points and exclusion of the corresponding equations, and the use of errors of quadrature formulas for refining the structure of algebraic equations].

If $C = 0$, then Equation (15.1.6) is a degenerate singular integral equation of the second kind. A nondegenerate equation of the second kind may be obtained, for instance, for $C = i$. However, here we analyze the reduction of Equation (15.1.6) to a singular integral equation of the first kind, which is achievable for $C = \kappa_k$. Then, Equation (15.1.6) reduces to

$$\frac{\kappa_k}{\pi i}\int_L \omega(\tau)\mathrm{Re}\left(\frac{d\tau}{\tau - t}\right) + \frac{1}{2\pi i}\int_L \overline{\omega(\tau)}\, d\left(\frac{\tau - t}{\bar{\tau}-\bar{t}}\right) = f_k(t), \qquad k = 1, 2. \tag{15.1.8}$$

As we see, this equation contains Hilbert's kernel. Let D be a singly connected domain whose contour L is a Lyapunov curve, i.e., is given by the same parametric equation, $x = x(\theta)$, $y = y(\theta)$, as in (10.4.1). Then we have

$$\mathrm{Re}\left(\frac{d\tau}{\tau - t}\right) = \frac{x'(x - x_0) + y'(y - y_0)}{(x - x_0)^2 + (y - y_0)^2}\, d\theta = \cot\frac{\theta - \theta_0}{2} K_1(\theta, \theta_0)\, d\theta,$$

$$\tau = x + iy, \qquad t = x_0 + iy_0, \qquad x_0 = x(\theta_0), \qquad y = y(\theta_0),$$

$$\theta, \theta_0 \in [0, 2\pi]. \tag{15.1.9}$$

Note that the function $K_1(\theta, \theta_0)$ is periodic in both θ and θ_0 with the period 2π, $K_1(\theta_0, \theta_0) = 0.5$, and belongs to the class H on $[0, 2\pi]$ (see Section 10.4).

Let us analyze the properties of Equation (15.1.8) for the case when L is a circle and the equation has the form

$$\kappa_k \int_0^{2\pi} \omega(\theta) \cot \frac{\theta - \theta_0}{2} \, d\theta - i \int_0^{2\pi} \overline{\omega(\theta)} \, e^{i(\theta + \theta_0)} \, d\theta = 2\pi i f_k(\theta_0),$$

$$k = 1, 2. \quad (15.1.10)$$

If an internal problem formulated in stresses is self-balanced, then both the force and the moment vector due to external loadings must vanish. Therefore, Equation (15.1.10) is to be supplemented by conditions imposed on the right-hand side. The condition that the vector of external forces equal to zero is fulfilled automatically if $f_2(\theta)$ is a unique function. However, to be unique, the function f_2 must be periodic on L; then

$$f_2(0) = f_2(2\pi). \quad (15.1.11)$$

From Equation (15.1.10) it follows that the latter condition does not impose any restrictions on function ω. According to Parton and Perlin (1982), the condition that the vector of the moment due to external forces equal to zero,

$$\mathrm{Re} \int_L \overline{f_2(t)} \, dt = 0, \quad (15.1.12)$$

results, in the case of a circle, in the relationship

$$-iR \int_0^{2\pi} \left[\overline{\omega(\theta)} \, e^{i\theta} + \omega(\theta) \, e^{-i\theta} \right] d\theta = \int_L \overline{f_2(t)} \, dt, \qquad t = Re^{i\theta}$$

(for $\kappa_k = \kappa_2 = -1$). Here the left-hand side is a fully imaginary expression for any ω, and that is why Equation (15.1.12) does not impose any restrictions on ω.

Continuing the analysis of the problem on a circle, we conclude that the homogeneous equation (15.1.10) has (for $f_k \equiv 0$) eigenfunctions whose form is determined by presenting $\omega(\theta)$ in the form of, for instance, the Fourier series. For $k = 1, 2$ the eigenfunctions of Equations (15.1.8) and (15.1.10) are given by the complex constant

$$\omega = a + ib \quad (15.1.13)$$

and for $k = 2$ are given by the function

$$\omega(\theta) = ia_1 \, e^{i\theta} \quad (15.1.14)$$

[for the homogeneous equation (15.1.10), whereas for Equation (15.1.8) they are given by the function $\omega(\tau) = ia_2\tau$, where a_1 and a_2 are arbitrary constants].

Thus, Equations (15.1.8) and (15.1.10) must be solved simultaneously, subject to certain auxiliary conditions "cutting off" the eigenfunctions (15.1.13) and (15.1.14). The conditions may be presented in the form

$$\int_L \frac{\omega(\tau)\, d\tau}{\tau} = 0, \qquad k = 1,2,$$

$$\int_L \left[\frac{\omega(\tau)}{\tau^2}\, d\tau + \frac{\overline{\omega(\tau)}}{\bar{\tau}^2}\, d\bar{\tau} \right] = 0, \qquad k = 2. \tag{15.1.15}$$

D. I. Sherman was the first to introduce the left-hand sides of the latter equations into integral equation (15.1.7) to ensure uniqueness of its solution (Parton and Perlin 1982). For a circle Conditions (15.1.15) acquire the form

$$\int_0^{2\pi} \omega(\theta)\, d\theta = 0, \qquad k = 1,2,$$

$$\operatorname{Im} \int_0^{2\pi} \overline{\omega(\theta)}\, e^{i\theta}\, d\theta = 0, \qquad k = 2, \tag{15.1.16}$$

and hence ω in the form (15.1.13) does not satisfy the first of the latter two equations, whereas in the form (15.1.14) it does not satisfy the second one.

From Equation (15.1.10) it follows that

$$\int_0^{2\pi} f_k(\theta)\, d\theta = 0, \tag{15.1.17}$$

and hence the integrals of both the real and the imaginary part of function f_k are equal to zero. If the problem is solved in stresses, then Condition (15.1.17) determines the complex constant C_2 in (15.1.4). However, if the problem is solved in strains, then the condition is fulfilled automatically for a function f_1 that is analytic within the domain D and continuous on L. For a circle, Condition (15.1.12) results in the relationship

$$\int_0^{2\pi} (f_R \sin\theta - f_I \cos\theta)\, d\theta = 0, \qquad f_2 = f_R + if_I. \tag{15.1.18}$$

For an arbitrary smooth closed contour L, integration of (15.1.10) with respect to θ results in the relationship

$$\int_L f_k(\theta)\, d\theta = 0, \tag{15.1.19}$$

analogous to Condition (15.1.17) on a circle.

Thus, the second major problem for a region bounded by a closed Lyapunov contour L is reducible to integral equation (15.1.8) and Conditions (15.1.11), (15.1.12), (15.1.15), and (15.1.19), allowing us to determine the function ω. In the case when L is a circle, the conditions must be replaced by (15.1.10), (15.1.11), and (15.1.16)–(15.1.18). These equations are singular and have the Hilbert kernel.

Let us demonstrate the application of the numerical method developed in Section 6.2 to solving the two-dimensional problem of the theory of elasticity by considering Equation (15.1.10) subject to Conditions (15.1.11) and (15.1.16)–(15.1.18) for $k = 2$, i.e., $\kappa_k = -1$. We start by singling out the real and the imaginary parts of Equation (15.1.10) and Condition (15.1.16). Then we arrive at the system of equations

$$\int_0^{2\pi} \omega_R(\theta)\left[\cot\frac{\theta - \theta_0}{2} - \sin(\theta + \theta_0)\right] d\theta + \int_0^{2\pi} \omega_I(\theta)\cos(\theta + \theta_0)\, d\theta$$
$$= 2\pi f_I(\theta_0),$$

$$\int_0^{2\pi} \omega_R(\theta)\cos(\theta + \theta_0)\, d\theta + \int_0^{2\pi} \omega_I(\theta)\left[\cot\frac{\theta - \theta_0}{2} + \sin(\theta + \theta_0)\right] d\theta$$
$$= 2\pi f_R(\theta_0),$$

$$\int_0^{2\pi} \omega_R(\theta)\, d\theta = 0, \qquad \int_0^{2\pi} \omega_I(\theta)\, d\theta = 0,$$

$$\int_0^{2\pi} [\omega_R(\theta)\sin\theta - \omega_I(\theta)\cos\theta]\, d\theta = 0. \tag{15.1.20}$$

where the subscripts R and I denote the real and the imaginary parts of the corresponding function, respectively.

Suppose functions $f_R(\theta)$ and $f_I(\theta)$ belong to the class H on $[0, 2\pi]$. Choose points θ_i and θ_{0i}, $i = 1, \ldots, n$, as was done in Theorem 6.2.1. Replace System (15.1.10) composed of integral equations by the following system of linear algebraic equations:

$$\sum_{i=1}^{n} \omega_{n,R}(\theta_i)\left[\cot\frac{\theta_i - \theta_{0j}}{2} - \sin(\theta_i + \theta_{0j})\right]\frac{2\pi}{n}$$

$$+ \sum_{i=1}^{n} \omega_{n,I}(\theta_i)\cos(\theta_i - \theta_{0j})\frac{2\pi}{n} + \beta_1 + \beta_3 \cos\theta_{0j} = 2\pi f_I(\theta_{0j}),$$

$$j = 1, \ldots, n,$$

$$\sum_{i=1}^{n} \omega_{n,R}(\theta_i)\cos(\theta_i + \theta_{0j})\frac{2\pi}{n}$$

$$+ \sum_{i=1}^{n} \omega_{n,I}(\theta_i)\left[\cot\frac{\theta_i - \theta_{0j}}{2} + \sin(\theta_i + \theta_{0j})\right]\frac{2\pi}{n}$$

$$+\beta_2 + \beta_3 \sin\theta_{0j}$$

$$= -2\pi f_R(\theta_{0j}), \qquad j = 1, \ldots, n,$$

$$\sum_{i=1}^{n} \omega_{n,R}(\theta_i)\frac{2\pi}{n} = 0, \qquad \sum_{i=1}^{n} \omega_{n,I}(\theta_i)\frac{2\pi}{n} = 0,$$

$$\sum_{i=1}^{n} \left[\omega_{n,R}(\theta_i)\sin\theta_i - \omega_{n,I}(\theta_j)\cos\theta_i\right]\frac{2\pi}{n} = 0. \qquad (15.1.21)$$

Here β_1, β_2, and β_3 are regularizing factors that make the system well-conditioned. Note that for $n \to \infty$ all three factors tend to zero, when and only when all Conditions (15.1.17) and (15.1.18) are met. In fact, by summing up first the first n equations of System (15.1.21) and then the second group of n equations, one gets

$$\beta_1 + \alpha_1 \beta_3 = \alpha_2, \qquad \beta_2 + \alpha_3 \beta_3 = \alpha_4, \qquad (15.1.22)$$

where $\alpha_i \to 0$ for $n \to \infty$, $i = 1, 2, 3, 4$. Then, by multiplying the first n equations by $\cos\theta_{0j}$, $j = 1, \ldots, n$, respectively, and the second group of n equations by $\sin\theta_{0j}$, $j = 1, \ldots, n$, and summing all the $2n$ equations, one gets (taking into account Condition (15.1.18)

$$\alpha_5 \beta_1 + \alpha_6 \beta_2 + (1 + \alpha_7)\beta_3 = \alpha_8, \qquad (15.1.23)$$

where $\alpha_i \to 0$ for $n \to \infty$, $i = 5, 6, 7, 8$.

The statement made previously for β_1, β_2, and β_3 follows from Systems (15.1.22) and (15.1.23).

Note that the regularizing factors β_1, β_2, and β_3 may be introduced in an alternative manner; however, in any case, they must tend to zero for $n \to \infty$, and the system must remain well-conditioned. For instance, one may take for the first n equations β_1, $\theta_{0j}\beta_2$, and $\theta_{0j}^2\beta_3$, and for the second group of n equations β_1, $(2\pi + \theta_{0j})\beta_2$, and $(2\pi + \theta_{0j})^2\beta_3$.

If a circle is loaded by uniformly distributed concentrated forces, then the functions $f_R(\theta)$ and $f_I(\theta)$ suffer discontinuities of the first kind at the points where the forces are applied. In this case θ_i and θ_{0j} are chosen in such a way that the discontinuity points coincide with the reference points θ_{0j}, and $f_R(\theta)$ and $f_I(\theta)$ are represented by the arithmetic mean of one-sided limits.

If a domain under consideration has symmetry axes, then the system of integral equations (15.1.20) may be transformed into a system of singular integral equations of the first kind on a segment, and either all or some of the conditions imposed on ω will be satisfied. If the resulting system of equations is solved numerically, then one has to use the results obtained in Chapter 5, with the points of the sets E and E_0 chosen for each of the equations with respect to the function in which an equation is singular.

Thus, generally, the sets of points θ_i and θ_{0j} chosen for ω_R and ω_I differ. Figures 15.1 and 15.2 present calculated results for System (15.1.21) for a circle subjected to different loadings.

In all cases the calculated data are stable and thoroughly convergent for the order of the system varying from 30 to 110. This conclusion is also substantiated by comparing with exact solutions.

The problem is similarly solved for any singly–multiply connected domain whose contour is composed of a finite number of closed Lyapunov curves. To solve the problem one has only to specify the contour in parametric form.

15.2. CONTACT PROBLEM OF INDENTATION OF A UNIFORMLY MOVING PUNCH INTO AN ELASTIC HALF-PLANE WITH HEAT GENERATION

Following Saakyan (Lifanov and Saakyan 1982, Saakyan 1978) we will show how the problems formulated in the headings of Sections 15.2 and 15.3 may be reduced to singular integral equations.

Let a rigid punch whose base has an arbitrary smooth configuration move over the boundary of an elastic half-plane with velocity V_0 less than or equal to the Rayleigh wave speed. At the same time, the punch is being indented into the half-plane by force P (see Figure 15.3).

It is supposed that there is dry friction between the punch and the half-plane; in other words, the tangential stresses within the contact zone

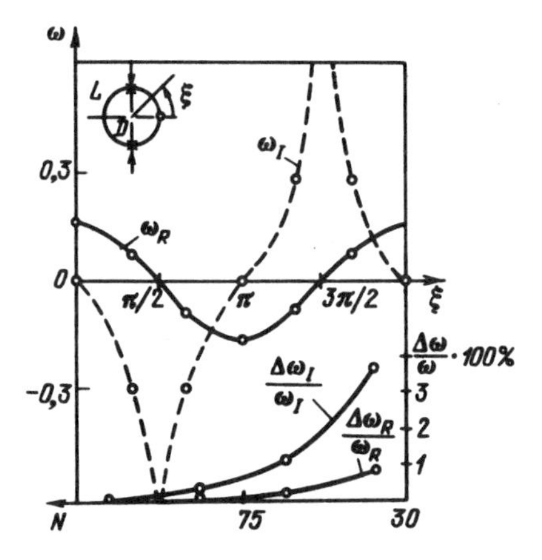

FIGURE 15.1. Approximate values of functions w_R and w_I for the case of a circle loaded with two concentrated balanced compressing forces.

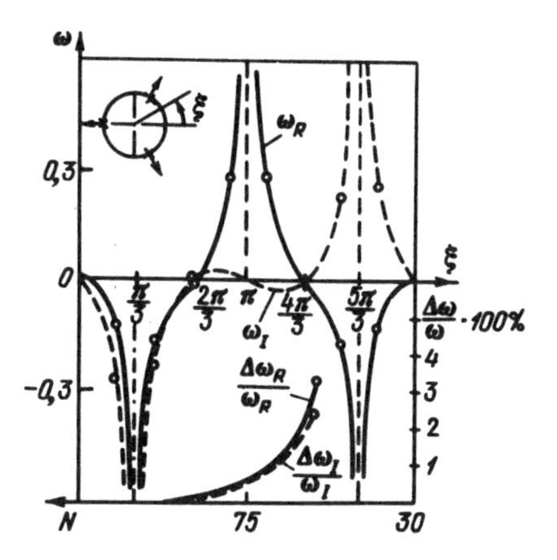

FIGURE 15.2. Approximate values of functions w_R and w_I for the case of three concentrated balanced extending forces.

are proportional to the normal pressure. Hence, heat is generated within the contact zone whose quantity is proportional to the speed of relative motion, the friction coefficient, and the normal contact pressure (Aleksandrov et al. 1969). The portions of the punch and the half-plane that are not in contact are supposed to be thermally insulated.

Let us choose a stationary system of coordinates $O_1X_1Y_1$ and a moving system of coordinates OXY attached to the punch. Obviously, the two systems are related by the formulas $X = X_1 - V_0t$ and $y = y_1$, and in the latter system of coordinates the quantities being sought are independent of

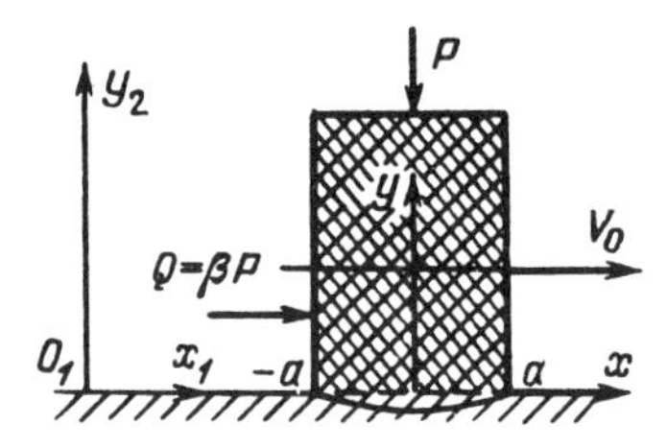

FIGURE 15.3. Motion of a rigid punch over an elastic strip in the presence of indenting force P.

time; in other words, a steady-state situation will be analyzed. Heat generation within the contact zone results in the appearance of heat fluxes $Q_1(x)$ and $Q_2(x)$ directed into the half-plane and the punch, respectively, and related to the contact pressure $p(x)$ by the relationship

$$Q_1(x) + Q_2(x) = \beta V_0 p(x), \tag{15.2.1}$$

where β is the friction coefficient. According to the Fourier heat conduction law, one has

$$Q_k(x) = (-1)^{k+1}\lambda_k \frac{\partial T_k(x,y)}{\partial y}, \qquad k = 1,2,$$

where λ_1 and λ_2 are the heat conductivity coefficients, and $T_1(x, y)$ and $T_2(x, y)$ are the temperatures of points belonging to the half-plane and the punch, respectively.

We will suppose that deformations of the elastic half-plane do not affect the temperature field. Then the problem splits into the following two problems: the problem of calculating the temperature field and the two-dimensional problem of elastodynamics in the presence of a temperature field. Let us start by considering the former problem.

First we construct the influence function for the problem; in other words, we construct a solution to the heat-conduction problem for an elastic half-plane, with a point source of unit heat flux moving along a thermally insulated boundary at a constant speed V_0. Then we arrive at the heat equation

$$\nabla^2 T(x_1, y_1, t) - \frac{\rho c_\epsilon}{\lambda_1}\frac{\partial T(x_1, y_1, t)}{\partial t} = 0 \tag{15.2.2}$$

subject to the boundary conditions

$$\frac{\partial T(x_1, y_1, t)}{\partial y_1} = -\frac{1}{\lambda_1}\delta(x_1 - V_0 t - \xi) \tag{15.2.3}$$

for $y_1 = 0$, and $T(x_1, y_1, t) < \infty$ for $y_1 \to -\infty$.

Here $T(x_1, y_1, t)$ is the temperature of points belonging to the half-plane, and ρ, c_ϵ, and λ_1 are the density, specific heat capacity, and heat conductivity of the material of the half-plane, respectively; $\delta(x)$ is the Dirac function, and ξ is the coordinate of a heat source at the instant $t = 0$.

In the movable system of coordinates Equations (15.2.2) and (15.2.3) become

$$\nabla^2 T(x, y) + \frac{\rho c_\epsilon V_0}{\lambda_1}\frac{\partial T(x, y)}{\partial x} = 0, \tag{15.2.4}$$

$$\left.\frac{\partial T(x, y)}{\partial y}\right|_{y=0} = -\frac{1}{\lambda_1}\delta(x - \xi) \tag{15.2.5}$$

and $T(x, y) < \infty$ for $y \to -\infty$.

The following complex Fourier transform with respect to the variable x may be applied to the latter boundary problem:

$$\frac{d^2 \overline{T}(y, \alpha)}{dy^2} - (\alpha^2 + i\alpha\kappa)\overline{T}(y, \alpha) = 0, \tag{15.2.6}$$

$$\frac{d\overline{T}(y, \alpha)}{dy} = -\frac{e^{i\alpha\xi}}{\lambda_1} \quad \text{for } y = 0,$$

$$\overline{T}(y, \alpha) < \infty \quad \text{for } y \to -\infty, \tag{15.2.7}$$

where α is a complex parameter of the transform, $\kappa = \rho c_\epsilon v_0/\lambda_1$, and $\overline{T}(y,\alpha)$ is the Fourier temperature transformant given by the formula

$$\overline{T}(y, \alpha) = \int_{-\infty}^{\infty} T(x, y) e^{ix\alpha}\, dx.$$

By solving Equation (15.2.6) subject to the boundary conditions (15.2.7) one gets

$$\overline{T}(y, \alpha) = -\frac{e^{i\alpha\xi}}{\lambda_1 \eta(\alpha)} e^{y\eta(\alpha)}, \tag{15.2.8}$$

where $\eta(\alpha) = \sqrt{\alpha^2 + i\alpha\kappa}$. The function $\eta(\alpha)$ branches at points $\alpha = 0$ and $\alpha = -i\kappa$ in the plane of the complex variable $\alpha = \sigma + i\tau$. In order to isolate a single-valued branch of the function, the plane α must be cut along the lines connecting the branching points with the infinitely distant point and lying in the upper and the lower half-plane, respectively (see Figure 15.4). This cut allows us to choose a single-valued branch of the root, meeting the condition $\eta(\alpha) \to |\alpha|$ for $\alpha \to \pm\infty$ along the real axis.

Let us next construct the influence function for the two-dimensional problem of elastodynamics in the presence of a temperature field generated by a moving concentrated source of strength S. The motion is accompanied by those of concentrated normal P and tangential Q forces.

Then we arrive at the Lamé equations written for the stationary coordinate system $O_1X_1Y_1$ in the presence of a temperature field:

$$\mu\nabla^2 u + (\lambda + \mu)\frac{\partial\theta}{\partial x_1} = \rho\frac{\partial^2 u}{\partial t^2} + \gamma\frac{\partial T}{\partial x_1},$$

$$\mu\nabla^2 v + (\lambda + \mu)\frac{\partial\theta}{\partial y_1} = \rho\frac{\partial^2 v}{\partial t^2} + \gamma\frac{\partial T}{\partial y_1}, \qquad (15.2.9)$$

where $u(x_1, y_1, t)$ and $v(x_1, y_1, t)$ are the displacement components, $\theta(x_1, y_1, t) = \partial u/\partial x_1 + \partial v/\partial y_1$ is the volumetric expansion, ρ is density, λ and μ are the Lamé coefficients, $\gamma = (3\lambda + 2\mu)\alpha_t$, α_t is the linear expansion coefficient, and $T(x_1, y_1, t)$ is the temperature determined beforehand.

By using the Duhamel–Neumann relationships

$$\sigma_y = \lambda\theta + 2\mu\frac{\partial v}{\partial y_1} - \gamma T, \qquad \tau_{xy} = \mu\left(\frac{\partial u}{\partial y_1} + \frac{\partial v}{\partial x_1}\right),$$

the boundary conditions may be presented in the form

$$\lambda\theta + 2\mu\frac{\partial v}{\partial y_1} - \gamma T = -P\delta(x_1 - V_0 t - \xi),$$

$$\mu\left(\frac{\partial u}{\partial y_1} + \frac{\partial v}{\partial x_1}\right) = Q\delta(x_1 - V_0 T - \xi) \quad \text{for } y = 0;$$

$$\sigma_y, \sigma_x, \tau_{xy} \to 0 \quad \text{for } y \to -\infty. \qquad (15.2.10)$$

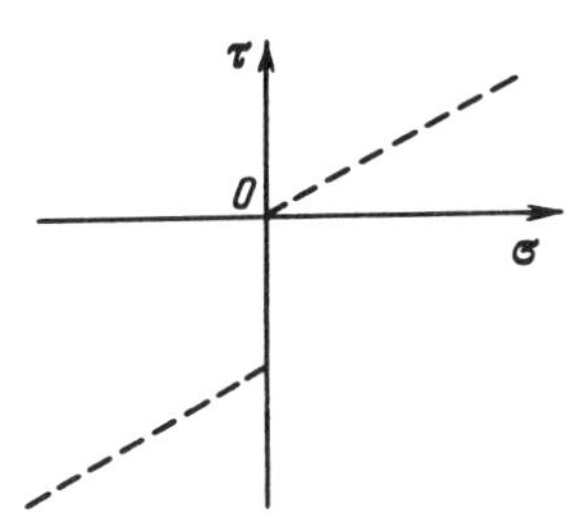

FIGURE 15.4. The complex variable plane with cuts (see the dashed lines) singling out the unique branch of the function $\eta(\alpha) = \sqrt{\alpha^2 + i\alpha k}$.

Upon passing again to the moving system of coordinates, we get for the latter two equations,

$$\mu \nabla^2 u + (\lambda + \mu)\frac{\partial \theta}{\partial x} = \rho V_0^2 \frac{\partial^2 u}{\partial x^2} + \gamma \frac{\partial T}{\partial x},$$

$$\mu \nabla^2 v(\lambda + \mu)\frac{\partial \theta}{\partial y} = \rho V_0^2 \frac{\partial^2 v}{\partial x^2} + \gamma \frac{\partial T}{\partial y}; \qquad (15.2.11)$$

$$\lambda\theta + 2\mu\frac{\partial v}{\partial y} - \gamma T = -P\delta(x - \xi) \quad \text{for } y = 0,$$

$$\mu\left(\frac{\partial u}{\partial y} + \frac{\partial v}{\partial x}\right) = Q\delta(x - \xi). \qquad (15.2.12)$$

By differentiating the first of Equations (15.2.11) with respect to x, the second with respect to y, and summing the results, one gets the following equation in $\theta(x, y)$:

$$(\lambda + 2\mu)\nabla^2\theta = \rho V_0^2 \frac{\partial^2 \theta}{\partial x^2} + \gamma \nabla^2 T. \qquad (15.2.13)$$

Let us subject Equations (15.2.11)–(15.2.13) to the generalized Fourier transform given by the formula

$$\overline{\varphi}(y, \sigma) = \int_{-\infty}^{\infty} \varphi(x, y) e^{i\sigma x}\, dx.$$

Then we get a system of ordinary differential equations in the variable y with respect to the functions $\bar{u}(y, \sigma)$, $\bar{v}(y, \sigma)$, and $\bar{\theta}(y, \sigma)$, subject to boundary conditions obtained from (15.2.12). Let us find some special solutions to the system, meeting the latter boundary conditions (taking into account (15.2.10)). To do this we first observe that if at the real axis the function $\eta(\sigma)$ assumes the values chosen for the single-valued branch of $\eta(\alpha)$ beforehand, then the generalized Fourier temperature transform coincides with the complex Fourier transform. However, in order to solve a contact problem one has to consider displacements of the half-plane's boundary points only, i.e., to consider $\bar{u}(0, \sigma)$ and $\bar{v}(0, \sigma)$ only. By subjecting $\bar{u}(0, \sigma)$ and $\bar{v}(0, \sigma)$ to the inverse Fourier transform, one obtains the influence functions $u(x, 0)$ and $v(x, 0)$ for the two-dimensional problem of elastodynamics formulated for a half-plane in the case of uniformly moving concentrated forces and a heat source. Next we use the values of the integrals

$$\frac{1}{2\pi}\int_{-\infty}^{\infty}\frac{ie^{i\sigma\xi}}{\sigma}e^{-i\sigma x}\,d\sigma = \frac{1}{2}\operatorname{sign}(x-\xi),$$

$$\frac{1}{2\pi}\int_{-\infty}^{\infty}\frac{e^{i\sigma\xi}}{|\sigma|}e^{-i\sigma x}\,d\sigma = \frac{1}{\pi}\ln\frac{1}{|x-\xi|} + C,$$

known from the theory of generalized functions (Brychkov and Prudnikov 1977) where C is a certain constant. In the theory of generalized functions, the latter is equal to the Euler constant; in the two-dimensional problem of elastodynamics for a half-plane it is equal to infinity, because the system of forces applied to the half-plane is unbalanced. Then the influence function finally becomes

$$u(x,0) = -\pi\nu_0\nu_1\frac{P}{2\mu}\operatorname{sign}(x-\xi) + \nu_1\frac{Q}{\mu}\ln\frac{1}{|x-\xi|}$$

$$+ C_1 + \frac{k_2\gamma S}{\mu\lambda_1\nu_3}R_1(\xi - x),$$

$$v(x,0) = -\pi\nu_0\nu_1\frac{Q}{2\mu}\operatorname{sign}(x-\xi) - \nu_1\frac{P}{\mu}\ln\frac{1}{|x-\xi|}$$

$$+ c_2 + \nu_1\nu_2\frac{\gamma S}{\lambda_1\mu}R_2(\xi - x), \tag{15.2.14}$$

where

$$R_1(\xi - x) = \int_{-\infty}^{\infty} \left[\frac{k_1}{\eta(\sigma)} - \frac{1}{|\sigma|} \right] \frac{ie^{i\sigma(\xi - x)}}{\sigma + i\kappa c_1^2/V_0^2}\, d\sigma,$$

$$R_2(\xi - x) = \int_{-\infty}^{\infty} \left[\frac{k_1}{\eta(\sigma)} - \frac{1}{|\sigma|} \right] \frac{e^{i\sigma(\xi - x)}\operatorname{sign}\sigma}{\sigma + i\kappa c_1^2/V_0^2}\, d\sigma,$$

$$\nu_0 = \frac{(1 + k_2^2) - 2k_1k_2}{k_1(1 - k_2^2)}, \qquad \nu_1 = \frac{k_1(1 - k_2^2)}{\pi\left[4k_1k_2 - (1 + k_2^2)^2\right]},$$

$$\nu_2 = \frac{1 + k_2^2}{2k_1(1 - k_2^2)},$$

$$\nu_3 = \pi\left[4k_1k_2 - (1 + k_2^2)^2, \qquad k_i = \sqrt{1 - V_0^2/c_i^2}, \qquad i = 1,2,$$

$$c_1 = \sqrt{(\lambda + 2\mu)/\rho}, \qquad c_2 = \sqrt{\mu/\rho}.$$

Next we consider the contact problem. Evidently, the effect of a punch on a half-plane is equivalent to applying unknown normal $p(\xi)$ and tangential $q(\xi)$ contact stresses and a heat source $Q_1(\xi)$ to the boundary of the half-plane, all of them distributed over the segment $[-a, a]$. By the principle of superposition, the displacements of boundary points of the half-space may be obtained by integrating Equations (15.2.14) with respect to ξ and replacing P, Q, and S by $p(\xi)$, $q(\xi)$, and $Q_1(\xi)$, respectively. Then one gets

$$u(x,0) = -\pi\nu_0\nu_1 \frac{1}{2\mu} \int_{-a}^{a} p(\xi)\operatorname{sign}(x - \xi)\, d\xi$$
$$+ \frac{\nu_1}{\mu} \int_{-a}^{a} \ln \frac{1}{|x - \xi|} q(\xi)\, d\xi + \frac{k_2\gamma}{\mu\lambda_1\nu_3}$$
$$\times \int_{-a}^{b} R_1(\xi - x) Q_1(\xi)\, d\xi + c_3,$$

$$v(x,0) = -\frac{\pi\nu_0\nu_1}{2\mu}\int_{-a}^{a} q(\xi)\operatorname{sign}(x-\xi)\,d\xi - \nu_1\frac{1}{\mu}\int_{-a}^{a}\ln\frac{1}{|x-\xi|}p\xi\,d\xi$$

$$+ \nu_1\nu_2\frac{\gamma}{\lambda_1\mu}\int_{-a}^{a} R_2(\xi - x)Q_1(\xi)\,d\xi + c_4, \qquad (15.2.15)$$

where c_3 and c_4 are certain infinite constants. The contact conditions are given by equalities

$$v(x,0) = f(x) - d, \qquad T_1(x,0) = T_2(x,0), \qquad |x| < a, \quad (15.2.16)$$

where $f(x)$ describes the base of a punch, d is the punch immersion, and T_1 and T_2 are the temperatures of the half-plane and the punch, respectively. Suppose the punch is much longer than the contact zone, and hence it may be replaced by a half-plane as far as its temperature is concerned. Then the temperature of the punch's boundary points is given by

$$T_2(x,0) = \frac{1}{\pi\lambda_2}\int_{-a}^{a}\ln|x-\xi|Q_2(\xi)\,d\xi + c_5, \qquad (15.2.17)$$

where c_5 is also an infinite constant.

For the temperature of the boundary points one gets from (15.2.8),

$$T_1(x,0) = \frac{1}{\pi\lambda_1}\int_{-a}^{a}\ln|x-s|Q_1(s)\,ds - \frac{1}{\lambda_1}$$

$$\times\int_{-a}^{a} R(s-x)Q_1(s)\,ds + c_6,$$

$$R(s-x) = \frac{1}{2\pi}\int_{-\infty}^{\infty}\left[-\frac{1}{\sqrt{\alpha^2 + i\alpha\kappa}} - \frac{1}{|\alpha|}\right]e^{i\alpha(s-x)}\,d\alpha. \quad (15.2.18)$$

If one considers a harmonic (in time) temperature distribution within the punch and the half-plane and singles out the wave propagating into infinity, then by equating the temperatures of the punch and the half-plane within the contact zone (see Equations 15.2.17) and (15.2.18), respectively) in the limit of the wave's frequency tending to zero, one arrives at the heat contact equation

$$\frac{1}{\pi\lambda_2}\int_{-a}^{a}\ln|x-s|Q_2(s)\,ds = \frac{1}{\pi\lambda_1}\int_{-a}^{a}\ln|x-s|Q_1(s)\,ds$$

$$-\frac{1}{\lambda_1}\int_{-a}^{a} R(s-x)Q_1(s)\,ds \quad (15.2.19)$$

subject to the condition that the infinite constants c_5 and c_6 are self-eliminating:

$$\frac{1}{\lambda_1}\int_{-a}^{a} Q_1(s)\,ds = \frac{1}{\lambda_2}\int_{-a}^{a} Q_2(s)\,ds. \qquad (15.2.20)$$

Excluding the quantity $Q_2(s)$ appearing in Equation (15.2.19), using relationship Equation (15.2.1), and differentating the equation with respect to x, one gets

$$\frac{\lambda_1+\lambda_2}{\pi\lambda_1\lambda_2}\int_{-a}^{a}\frac{Q_1(s)\,ds}{s-x} - \frac{\beta V_0}{\pi\lambda_2}\int_{-a}^{a}\frac{p(s)\,ds}{s-x}$$

$$+\frac{1}{\lambda_1}\int_{-a}^{a}\frac{\partial R(s-x)}{\partial x}Q_1(s)\,ds = 0. \qquad (15.2.21)$$

Then by substituting the expression for the displacement $v(x,0)$ for $q(x)=\beta p(x)$ into the first contact condition (15.2.16) and differentiating with respect to x, we obtain

$$\frac{\nu_1}{\mu}\int_{-a}^{a}\frac{p(s)}{s-x}\,ds + \frac{\beta}{\mu}\pi\nu_0\nu_1 p(x)$$

$$-\nu_1\nu_2\frac{\gamma}{\lambda_1\mu}\int_{-a}^{a}\frac{\partial R_2(s-x)}{\partial x}Q_1(s)\,ds = -f'(x). \quad (15.2.22)$$

The latter equation should be supplemented with the equation of the punch equilibrium:

$$\int_{-a}^{a} p(x)\,dx = P. \qquad (15.2.23)$$

Thus, we have derived a system of singular integral equations of the first and second kind, Equations (15.2.21) and (15.2.22), which, subject to Conditions (15.2.20) and (15.2.23), has a unique solution.

Let us introduce the nondimensional quantities entering the latter system of equations in the following manner:

$$\xi = \frac{x}{a}, \qquad \psi(\xi) = \frac{a}{P}p(x), \qquad \hat{Q}_i(\xi) = \frac{\gamma a^2}{P\lambda_1}Q_i(x),$$

$$i = 1,2,$$

$$\varphi(\xi) = \frac{\mu a}{P} f'(x), \qquad \lambda = \frac{\lambda_1}{\lambda_2}, \qquad \zeta = \frac{\gamma\beta V_0 a}{\lambda_2},$$

$$\hat{\kappa} = \kappa a = \frac{\rho c_\epsilon V_0 a}{\lambda_1}.$$

Then by introducing for the sake of convenience a new function $\chi(\xi) = (1+\lambda)\hat{Q}_1(\xi) - \zeta\hat{P}(\xi)$, we arrive finally at the following system of singular integral equations:

$$\int_{-1}^{1} \frac{\chi(s)\,ds}{s-\zeta} + \frac{\pi}{1+\lambda}\int_{-1}^{1} \frac{\partial R(s-\xi)}{\partial \xi} \times [\chi(s) + \zeta\psi(s)]\,ds = 0,$$

$$\int_{-1}^{1} \frac{\psi(s)\,ds}{s-\xi} + \pi\beta\nu_0\psi(s) - \frac{\nu_2}{1+\lambda}\int_{-1}^{1} \frac{\partial R_2(s-\xi)}{\partial \xi} \times [\chi(s) + \zeta\psi(s)]\,ds = -\frac{1}{\nu_1}\varphi(\xi), \quad (15.2.24)$$

subject to the conditions

$$\int_{-1}^{1} \chi(s)\,ds = 0, \qquad \int_{-1}^{1} \psi(s)\,ds = 1. \quad (15.2.25)$$

The first equation in (15.2.24) is a singular integral equation of the first kind with respect to the function $\chi(s)$, and the second equation is a singular integral equation of the second kind with respect to the function $\psi(s)$. Because in both functions the index 1 solution is sought, we deduce (by using the results of Sections 5.2 and 7.1) that Equations (15.2.24) and Conditions (15.2.25) must be replaced by the following system of linear algebraic equations:

$$\frac{\pi}{n}\sum_{k=1}^{n}\left[\frac{1}{\tau_k - \xi_i} + \frac{\pi}{1+\lambda}\frac{\partial R(\tau_k - \xi_i)}{\partial \xi_i}\right]\chi_k^*$$

$$+ \frac{\pi\zeta}{1+\lambda}\sum_{p=1}^{m} a_p \frac{\partial R(t_p - \zeta_i)}{\partial \xi_i}\psi_p^* = 0, \qquad i = 1,\ldots,n-1,$$

$$\frac{\pi\nu_2}{(1+\lambda)n}\sum_{k=1}^{n}\frac{\partial R_2(\tau_k - s_j)}{\partial s_j}\chi_k^*$$

$$-\sum_{p=1}^{m} \partial_p \left[\frac{1}{t_p - s_j} - \frac{\zeta \nu_2}{1+\lambda} \frac{\partial R_2(t_p - s_j)}{\partial s_j} \right] \psi_p^* = \frac{\varphi(s_j)}{\nu_1},$$

$$j = 1, \ldots, m-1,$$

$$\sum_{k=1}^{n} \chi_k^* \frac{\pi}{n} = 0, \qquad \sum_{p=1}^{m} \psi_p^* a_p = 1, \tag{15.2.26}$$

where τ_k and ξ_i are the same as in System (5.2.3) and t_p, s_j, and a_p are the same as in System (7.1.9). The number α is given by the formula

$$\alpha = -\frac{1}{\pi} \arctan(\beta \nu_0).$$

Thus, the contact problem of motion of a punch has been reduced to that of solving a system of linear algebraic equations.

System (15.2.26) was solved numerically for a variety of speeds of motion of a punch and the following values of the parameters:

$$\lambda = 35, \qquad \zeta = 642 V_0/c_2, \qquad c_2^2/c_1^2 = 0.275,$$

$$\beta = 0.27, \qquad \kappa = 1.807 \times 10^{-5} V_0/c_2.$$

The contour of the base of the punch was described by the function $f(x) = 0.1x^2$. Figure 15.5 presents distributions of heat fluxes directed into the half-plane and into the punch (see the solid and dashed curves, respectively).

Contact pressures calculated at the points corresponding to the roots of the polynomial $P_{10}^{(\alpha, -1-\alpha)}(x)$ were compared with the values of the function

$$\frac{a}{p} P(x) = \omega(x) \frac{-\sin \pi\alpha}{5\pi\theta_1} \left[5\theta_1 - 2\alpha(1+\alpha) - (1+2\alpha)x - x^2 \right],$$

which gives an accurate analytic solution for the contact pressure for a similar problem without heat generation, obtained by the method of orthogonal polynomials. The comparison presented in Figure 15.5 shows that the heat generation does not practically affect the contact stresses distribution (for $V_0/c_2 = 0.2$ the difference is less than 0.03%).

Figures 15.6 and 15.7 show the contact pressure distributions and their regular portions for different speeds of motion of the punch. We see that an increase in the speed of motion results in a decrease in pressure in the

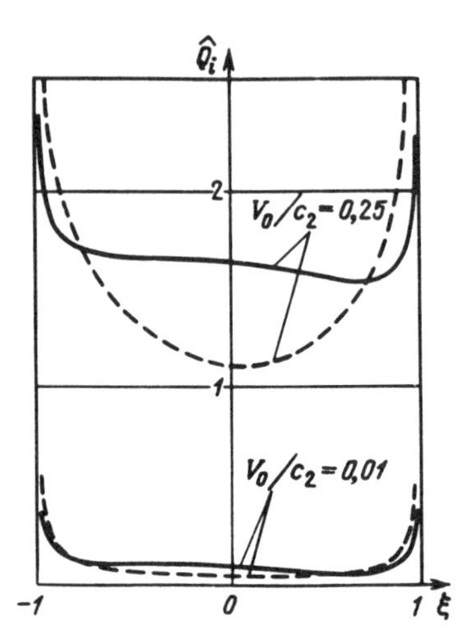

FIGURE 15.5. Distributions of heat fluxes directed into the half-plane (solid line) and into the punch (dashed line) for $V_0/c_2 = 0.25$.

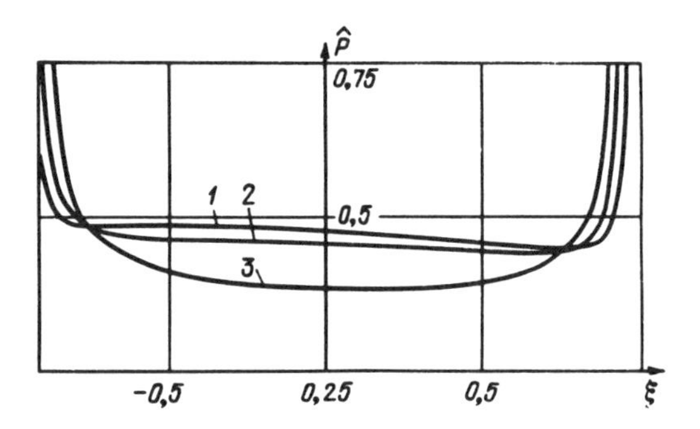

FIGURE 15.6. Distributions of contact pressures and their regular portions for various speeds of the punch. 1 indicates $V_0/c_2 = 0.44 \times 10^{-5}$, 2 indicates $V_0/c_2 = 0.5$, and 3 indicates $V_0/c_2 = 0.8$.

central part of the contact zone. At the same time, the stress concentration coefficients at the ends of the contact zone decrease.

15.3. ON THE INDENTATION OF A PAIR OF UNIFORMLY MOVING PUNCHES INTO AN ELASTIC STRIP

Let us consider a pair of punches moving along the edges of an elastic strip whose thickness is equal to $2d$ (see Figure 15.8). The forces P and Q are applied in such a way that the summary moment acting on a punch is equal to zero.

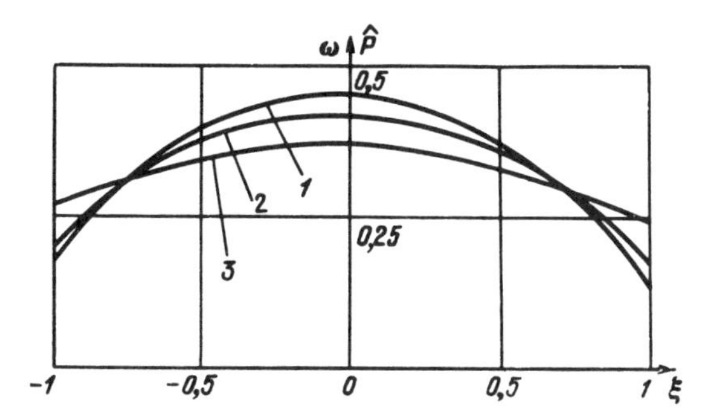

FIGURE 15.7. Distributions of contact pressures and their regular portions for various speeds of the punch. 1 indicates $V_0/c_2 = 0.44 \times 10^{-5}$, 2 indicates $V_0/c_2 = 0.5$, and 3 indicates $V_0/c_2 = 0.8$.

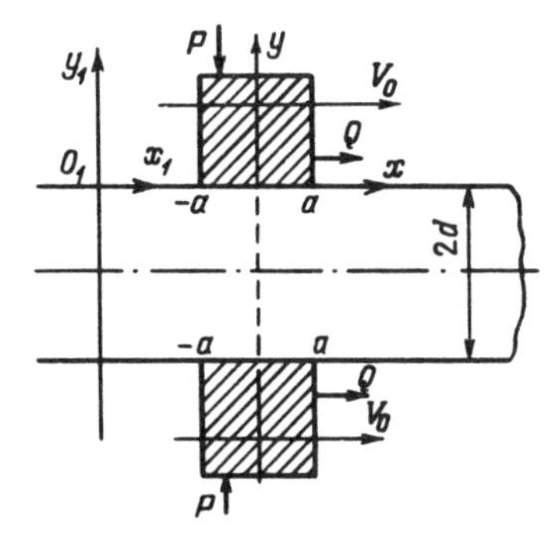

FIGURE 15.8. Motion of rigid punches over the boundary of an elastic strip in the presence of indenting forces P.

It is supposed that the tangential stress $q(x)$ acting at the contact surface between a punch and the strip is subject to the Coulomb law $q(x) = \beta p(x)$, where β is the friction coefficient.

Let us derive the governing equations. We will start by choosing a stationary system of coordinates $O_1X_1Y_1$ and a moving system of coordinates OXY attached to a punch and determined by the formulas $x = x_1 - V_0t$ and $y = y_1$. As far as the strip is symmetric with respect to the line $y = -d$, we will consider its upper half only.

Let us use the results presented by Saakyan (1978), who considered the problem of concentrated forces and heat sources moving along the edges of an elastic strip at speed V_0. By using the principle of superposition in the absence of heat sources, one may immediately derive expressions for the vertical component of the boundary point displacements (in the presence of distributed normal and tangential stresses):

$$v(x,0) = \frac{1}{\mu}\int_{-a}^{a} K_1(x-s)q(s)\,ds - \frac{1}{\mu}\int_{-a}^{a} K_2(x-s)p(s)\,ds, \quad (15.3.1)$$

where

$$K_1(u) = \frac{i}{2\pi}\int_{-\infty}^{\infty}\Big[2k_1k_2\sinh(\sigma k_1 d)\cosh(\sigma k_2 d)$$

$$-(1+k_2^2)\sinh(\sigma k_2 d)\cosh(\sigma k_1 d)\Big]\frac{e^{-i\sigma u}}{\sigma\Delta}\,d\sigma,$$

$$K_2(u) = \frac{k_1(1-k_2)}{2\pi}\int_{-\infty}^{\infty}\frac{\sinh(\sigma k_1 d)\sinh(\sigma k_2 d)}{\sigma\Delta}e^{-i\sigma u}\,d\sigma,$$

$$\Delta = 4k_1k_2\sinh(\sigma k_1 d)\cosh(\sigma k_2 d)$$

$$-(1+k_2^2)\sinh(\sigma k_2 d)\cosh(\sigma k_1 d),$$

$$k_i = \sqrt{1-\frac{V_0^2}{c_i^2}}, \quad i=1,2, \quad c_1^2 = \frac{\lambda+2\mu}{\rho}, \quad c_2^2 = \frac{\mu}{\rho},$$

λ and μ are the Lamé coefficients and ρ is the density of the strip material.

Within the contact zone we have the usual contact condition (Shtaerman 1949)

$$v(x,0) = f(x) - \delta, \qquad -a < x < a, \qquad (15.3.2)$$

where $f(x)$ is a function describing the punch base and δ is the measure of identation. Although $f(x) \equiv 0$ for the punches shown in Figure 15.8, we preserve the function $f(x)$ in what follows.

The functions $K_1(u)$ and $K_2(u)$ may be presented in the form

$$K_1(u) = \frac{1}{2}\,\frac{2k_1k_2-(1+k_2^2)}{4k_1k_2-(1+k_2^2)^2}\operatorname{sign} u + R_1(u),$$

$$K_2(u) = \frac{k_1(1-k_2^2)}{\pi\left[4k_1k_2-(1+k_2^2)^2\right]}\ln\frac{1}{|u|} + R_2(u) + c_0, \quad (15.3.3)$$

where

$$R_1(u) = \frac{i}{2\pi}\int_{-\infty}^{\infty}\left[\frac{2k_1k_2\tanh(\sigma k_1 d) - (1+k_2^2)\tanh(\sigma k_2 d)}{4k_1k_2\tanh(\sigma k_1 d) - (1+k_2^2)\tanh(\sigma k_2 d)}\right.$$

$$\left. - \frac{2k_1k_2 - (1+k_2^2)}{4k_1k_2 - (1+k_2^2)^2}\right]\frac{e^{-i\sigma u}}{\sigma}\, d\sigma,$$

$$R_2(u) = \frac{k_1(1-k_2^2)}{2\pi}\int_{-\infty}^{\infty}\left[\frac{\tanh(\sigma k_1 d)\tanh(\sigma k_2 d)}{4k_1k_2\tanh(\sigma k_1 d) - (1+k_2^2)^2\tanh(\sigma k_2 d)}\right.$$

$$\left. - \frac{\operatorname{sign}\sigma}{4k_1k_2 - (1+k_2^2)^2}\right]\frac{e^{-i\sigma u}}{\sigma}\, d\sigma$$

are regular functions in the neighborhood of zero and c_0 is an infinite constant. It can be easily shown that all derivatives of the functions are also regular in the neighborhood of zero.

By substituting (15.3.3) into (15.3.1) and (15.3.1) into the contact condition (15.3.2), and taking into account that $q(x) = \beta p(x)$, one gets

$$\int_{-a}^{a}\left[\ln\frac{1}{|x-s|} + \frac{\pi\beta}{2}\theta_0 \operatorname{sign}(x-s) - \frac{1}{\theta_1}R(x-s)\right]p(s)\, ds$$

$$= \frac{\mu}{\theta_1}(\delta - f(x) - c_0 P),$$

where

$$\theta_0 = \frac{1 + k_2^2 - 2k_1k_2}{4k_1k_2 - (1+k_2^2)^2}, \qquad \theta_1 = \frac{k_1(1-k_2^2)}{\pi\left[4k_1k_2 - (1+k_2^2)^2\right]},$$

$$R(u) = R_2(u) - \beta R_1(u), \qquad P = \int_{-u}^{a} p(x)\, dx. \tag{15.3.4}$$

After differentiating the latter equation with respect to x, one obtains

$$\int_{-a}^{a}\left[\frac{1}{s-x} - \frac{1}{\theta_1}\frac{\partial R(x-s)}{\partial x}\right]p(s)\, ds + \pi\beta\theta_0 p(x) = -\frac{\mu}{\theta_1}f'(x). \tag{15.3.5}$$

Upon passing to nondimensional quantities $x = a\xi$, $d^* = d/a$, $\bar{p}(\xi) = ap(x)/P$, and $g = P/(\mu a)$, Equation (15.3.5) and Condition (15.3.4) become

$$\int_{-1}^{1} \frac{\bar{p}(s)}{s - \xi} ds + \pi\beta\theta_0 \bar{p}(\xi) + \frac{1}{\theta_1} \int_{-1}^{1} N(\xi - s)\bar{p}(s)\, ds = -\frac{f'(\xi)}{g\theta_1} \tag{15.3.6}$$

$$\int_{-1}^{1} \bar{p}(s)\, ds = 1. \tag{15.3.7}$$

where

$$N(\xi - s) = -\frac{ik_1(1 - k_2^2)}{2\pi} \int_{-\infty}^{\infty} \times \left[\frac{\tanh(\sigma k_1 d^*)\tanh(\sigma k_2 d^*)}{4k_1 k_2 \tanh(\sigma k_1 d^*) - (1 - k_2^2)^2 \tanh(\sigma k_2 d^*)} - \frac{\operatorname{sign}\sigma}{4k_1 k_2 - (1 + k_2^2)^2} \right] e^{-i\sigma(\xi - s)}\, d\sigma - \frac{\beta}{2\pi} \times \int_{-\infty}^{\infty} \left[\frac{2k_1 k_2 \tanh(\sigma k_1 d^*) - (1 + k_2^2)\tanh(\sigma k_2 d^*)}{4k_1 k_2 \tanh(\sigma k_1 d^*) - (1 + k_2^2)^2 \tanh(\sigma k_2 d^*)} - \frac{2k_1 k_2 - (1 + k_2^2)}{4k_1 k_2 - (1 + k_2^2)^2} \right] e^{-i\sigma(\xi - s)}\, d\sigma.$$

From Gakhov (1977) and Muskhelishvili (1952) it is known that a solution to singular integral equation (15.3.6) may incorporate a singularity at the ends of the interval $(-1, 1)$ (see (7.1.2)):

$$\bar{p}(\xi) = \omega(\xi)\varphi(\xi) = (1 - \xi)^{\alpha}(1 + \xi)^{-1-\alpha}\varphi(\xi), \tag{15.3.8}$$

where $\varphi(\xi)$ is a limited function and the number α is given by the relationships

$$\beta\theta_0 + \cot \pi\alpha = 0, \qquad 0 > \alpha > -1. \tag{15.3.9}$$

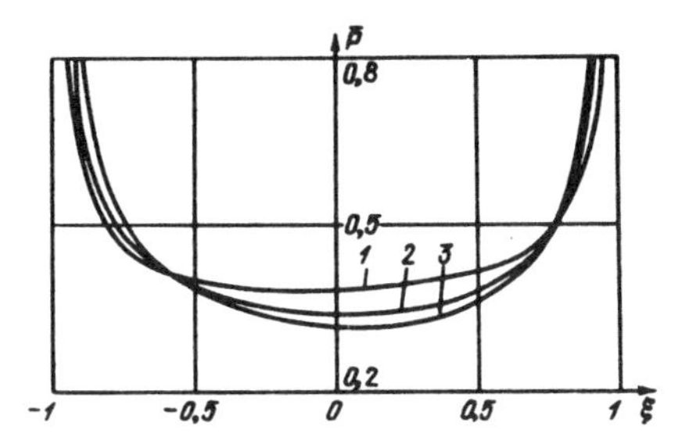

FIGURE 15.9. Contact pressure distribution for $V_0/c_2 = 0.5$: 1 indicates $d/a = 1.5$, 2 indicates $d/a = 3$, and 3 indicates d/a 24.

Singular integral equation (15.3.6) may be solved numerically as shown in Section 7.1; in other words, we pass to a system of linear algebraic equations with respect to the values of the function $\varphi(\xi)$ at the corresponding grid points:

$$\sum_{i=1}^{n} a_i \varphi_i \left[\frac{1}{t_i - s_j} + \frac{1}{\theta_1} N(s_j - t_i) \right] = -\frac{f'(s_j)}{g\theta_1}, \qquad j = 1, \dots, n-1,$$

$$\sum_{i=1}^{n} a_i \varphi_j = 1, \qquad j = n. \qquad (15.3.10)$$

Here t_i, $i = 1, \dots, n$, are the roots of the Jacobi polynomial $P_n^{(\alpha, -1-\alpha)}(\xi)$ and s_j, $j = 1, \dots, n-1$, are the roots of the polynomial $P_{n-1}^{(-\alpha, 1+\alpha)}(\xi)$, $\varphi_i = \varphi(t_i)$. In this case, a_i may be written as

$$a_i = \frac{1}{n \sin \pi\alpha} \frac{P_{n-1}^{(-\alpha, 1+\alpha)}(t_i)}{P_{n-1}^{(1+\alpha, -\alpha)}(t_i)}, \qquad i = 1, \dots, n,$$

because according to Bateman and Erdèlyi (1953) and Erdogan, Gupta, and Cook (1973),

$$\frac{d}{dx}\left[P_n^{(\alpha, \beta)}(x) \right] = \frac{1}{2}(n + \alpha + \beta + 1) P_{n-1}^{(\alpha+1, \beta+1)}(x). \quad (15.3.11)$$

Numerical calculations were performed for the following values of the constants: the number of reference points $n = 10$, the Poisson coefficient for the strip material $\nu = 0.31$, the friction coefficient $\beta = 0.27$; the values of V_0/c_2 (where c_2 is the Rayleigh wave speed) and the ratio of the strip width to the length of the contact zone, d/a, were varied. Figures 15.9 and

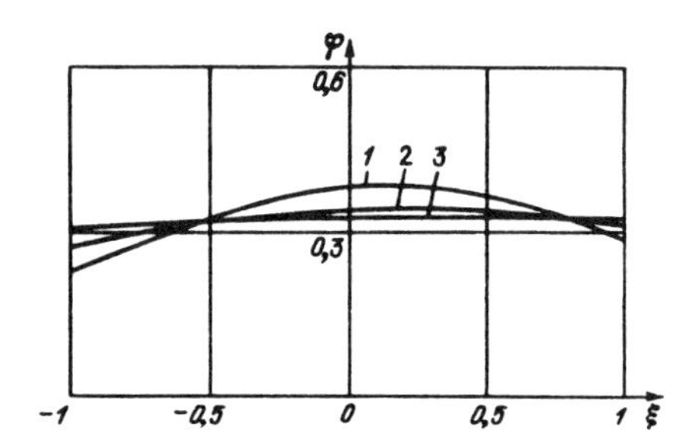

FIGURE 15.10. Values of the function $\varphi(\xi)$, for $V_0/c_2 = 0.5$: 1 indicates $d/a = 1.5$, 2 indicates $d/a = 3$, and 3 indicates d/a 24.

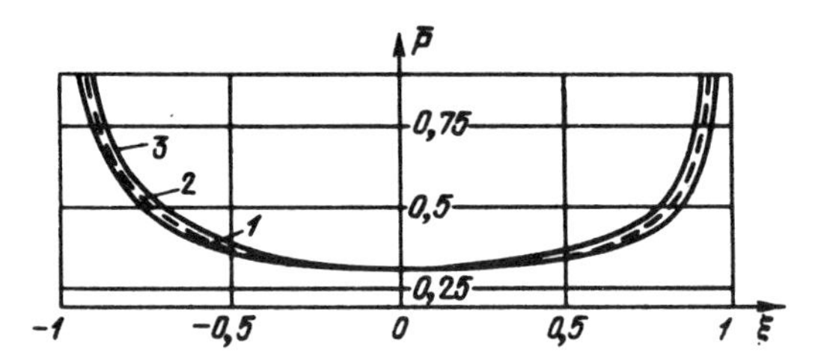

FIGURE 15.11. Contact pressure distribution for $d/a = 100$: 1 indicates $V_0/c_2 = 0.01$, 2 indicates $V_0/c_2 = 0.2$, and 3 indicates $V_0/c_2 = 0.9$.

15.10 show the contact pressure and the function $\varphi(\xi)$ versus ξ. Here $\varphi(+1)$ and $\varphi(-1)$ are the stress concentration coefficients whose values increase with an increase in the width of the strip. Additionally, we note that for $d/a = 24$ the plot of function $\varphi(\xi)$ virtually coincides with the straight line corresponding to the solution of an analogous problem for an elastic half-plane. Hence, for an elastic strip whose width exceeds the length of the contact zone by a factor of 25 or more, the interaction of the punches is negligibly small.

It should be noted that exceeding the limiting value of d/a causes the influence of the second punch to become negligibly small. Also note that d/a depends, though only slightly, on the speed of motion of a punch: the higher the speed, the larger the critical value of d/a. Figure 15.11 shows the contact pressure versus ξ.

16

Numerical Method of Discrete Singularities* in Boundary Value Problems of Mathematical Physics

In this chapter, following Gandel (1982, 1983), we show how a wide class of boundary problems of mathematical physics may be reduced to singular integral equations and solved by employing the numerical method of discrete singularities developed previously.

16.1. DUAL EQUATIONS FOR SOLVING MIXED BOUNDARY VALUE PROBLEMS

Let us consider dual equations of the form

$$A_0 + \sum_{n=1}^{\infty} (A_n \cos ny + B_n \sin ny) = 0, \qquad y \in CE,$$

$$bA_0 + \sum_{n=1}^{\infty} (1 - \epsilon_n)(nA_n \cos ny + nB_n \sin ny) = f(y), \qquad y \in E, \tag{16.1.1}$$

* The method of discrete singularities in the theory of elasticity and electrodynamics is a numerical method similar in its physical and mathematical structure to the method of discrete vortices in aerodynamics.

$$E = \bigcup_{k=1}^{m} (\alpha_k, \beta_k), \qquad CE = [-\pi, \pi] \setminus E,$$

$$-\pi < \alpha_1 < \beta_1 < \cdots < \alpha_m \ll \beta_m < \pi. \tag{16.1.2}$$

Here the smooth function $f(y)$, $y \in \bar{E} = \bigcup_{k=1}^{m} [\alpha_i, \beta_k]$, the constant b, and sequence ϵ_n, $n \in N \equiv \{1, 2, \ldots\}$, are given, and ϵ_n tends to zero for $n \to \infty$ no slower than $O(n^{-2})$. The coefficients A_0, A_n, and B_n, $n \in N$, must be found.

The possibility of using the method of discrete singularities for numerical solution of the preceding equations was considered in Gandel (1982, 1983), where some mixed boundary problems for the Laplace and Helmholtz equations were reduced to dual equations under analysis.

Let us start by considering some of the problems and writing the corresponding dual equations.

1. The simplest example is the following mixed boundary-value problem for the Laplace equation inside a circle:

$$\Delta u = 0, \qquad r < R, \tag{16.1.3}$$

$$u|_{r=R} = 0, \qquad \varphi \in CE, \tag{16.1.4}$$

$$\left.\frac{\partial u}{\partial r}\right|_{r=R} = f(\varphi), \qquad \varphi \in E, \tag{16.1.5}$$

where $f(\varphi)$, $\varphi \in \bar{E}$, is a given smooth function. The solution $u = u(r, \varphi)$ will be sought in the class of functions that are twice continuously differentiable inside the circle and remain continuous up to its boundary.

The harmonic function is presented in the form

$$u(r, \varphi) = A_0 + \sum_{n=1}^{\infty} \left(\frac{r}{R}\right)^n (A_n \cos n\varphi + B_n \sin n\varphi),$$

and the unknown coefficients A_0, A_n, and B_n, $n \in N$, will be found from the dual equation of the type (16.1.1) and (16.1.2):

$$A_0 + \sum_{n=1}^{\infty} (A_n \cos n\varphi + B_n \sin n\varphi) = 0, \qquad \varphi \in CE,$$

$$\sum_{n=1}^{\infty} n(A_n \cos n\varphi + B_n \sin n\varphi) = Rf(\varphi), \qquad \varphi \in E, b = 0, \epsilon_n \equiv 0$$

derived from boundary conditions (16.1.4) and (16.1.5).

2. Let us consider the problem of a stationary continuum limited by two infinitely long coaxial cylinders at whose external surface and a portion of internal surface composed of a finite number of longitudinal strips temperature is maintained constant, and at the remaining portion of the internal surface heat flux is specified.

   Let the axis of the cylinder coincide with the OZ axis of the Cartesian system of coordinates, and let the cross section by the plane OXY be a ring whose internal and external radii are R_1 and R_2, respectively. In the cross section we introduce polar coordinates r and φ. Next the internal cylindrical surface is divided into two sets of longitudinal strips $\{r = R_1,\ \varphi \in CE,\ z \in \tilde{R}\}$ and $\{r = R_1,\ \varphi \in E,\ z \in \tilde{R}\}$, where $\tilde{R}$ is the set of real numbers.

   For the sake of simplicity we will consider the case when the temperature is independent of z. Let the temperature field between the cylinders be described by $u = u(r, \varphi)$, $R_1 < r < R_2$. Thus, we arrive at the boundary-value problem

$$\Delta u = 0, \qquad R_1 < r < R_2, \tag{16.1.6}$$

$$u|_{r=R_2} = 0, \tag{16.1.7}$$

$$u|_{r=R_1} = 0, \qquad \varphi \in CE, \tag{16.1.8}$$

$$u|_{r=R_1} = f(\varphi), \qquad \varphi \in E. \tag{16.1.9}$$

Here f is a smooth function, and $u = u(r, \varphi)$ is sought in the class of functions that are twice continuously differentiable inside the ring and remain continuous up to the boundary.

A function that is harmonic inside the ring and meets Condition (16.1.7) has the form

$$u(r, \varphi) = A_0 \frac{\ln R_2/r}{\ln R_2/R_1} + \sum_{n=1}^{\infty} \frac{(R_2/r)^n - (r/R_2)^n}{(R_2/R_1)^n - (R_1/R_2)^n} \times (A_n \cos n\varphi + B_n \sin n\varphi),$$

and the unknown coefficients A_0, A_n, and B_n, $n \in N$, may be found from the dual equation of the type (16.1.1) and (16.1.2):

$$\frac{A_0}{\ln R_2/R_1} + \sum_{n=1}^{\infty} \frac{1 + (R_1/R_2)^{2n}}{1 - (R_1/R_2)^{2n}} (nA_n \cos n\varphi + nB_n \sin n\varphi) = Rf(\varphi),$$

$$\varphi \in E.$$

$$A_0 + \sum_{n=1}^{\infty} (A_n \cos n\varphi + B_{i1} \sin n\varphi) = 0,$$

$$\varphi \in CE,$$

$$b = \left(\ln \frac{R_2}{R_1}\right)^{-1}, \qquad \epsilon_n = \frac{2(R_1/R_2)^{2n}}{(R_1/R_2)^{2n} - 1},$$

derived from boundary conditions (16.1.8) and (16.1.9).

3. Next we consider a three-dimensional mixed boundary-value problem for the Laplace equation, namely, the problem of stationary temperature distribution within a finite cylinder at whose side surface composed of a finite number of longitudinal strips and the two bases, temperature is maintained constant, whereas at the remaining side surface heat flux is specified.

We will seek a function $u = u(r, \varphi z)$ that is twice continuously differentiable inside the cylinder, remains continuous up to its boundary, and meets the conditions

$$\Delta u = 0, \qquad r < R, 0 < z < H, \tag{16.1.10}$$

$$u|_{z=0} = u|_{z=H} = 0, \qquad r < R, \tag{16.1.11}$$

$$u|_{r=R} = 0, \qquad 0 < z < H, \varphi \in CE, \tag{16.1.12}$$

$$\left.\frac{\partial u}{\partial r}\right|_{r=R} = \frac{1}{R} f(\varphi, z), \qquad 0 < z < H, \varphi \in E. \tag{16.1.13}$$

A solution to the Laplace equation (16.1.10) meeting Condition (16.1.11) will be sought in the form

$$u(r, \varphi, z) = \sum_{k=1}^{\infty} u_k(r, \varphi) \sin \frac{\pi k z}{H}, \qquad r < R, 0 < z < H,$$

where $u_k(r, \varphi)$ satisfies both the Helmholtz equation

$$\Delta u_k - \left(\frac{\pi k}{H}\right)^2 u_k = 0, \qquad r < R, \tag{16.1.14}$$

and the boundary conditions corresponding to (16.1.12) and (16.1.13):

$$u_k|_{r=R} = 0, \qquad \varphi \in CE, \tag{16.1.15}$$

$$\left.\frac{\partial u_k}{\partial r}\right|_{r=R} = \frac{1}{R} f_k(\varphi), \qquad \varphi \in E, \tag{16.1.16}$$

where the Fourier series coefficients

$$f_k(\varphi) = \frac{2}{H}\int_0^H f(\varphi, \xi)\sin\frac{\pi k \xi}{H}\, d\xi, \qquad k \in N.$$

The function $u_k = u_k(r, \varphi)$ is sought in the form

$$u_k(r,\varphi) = u_{k0}^{(1)}(r) + \sum_{n=1}^{\infty} \left(u_{kn}^{(1)}(r)\cos n\varphi + u_{kn}^{(2)}(r)\sin n\varphi\right).$$

By (16.1.14) the unknown functions $u_{kn}^{(i)}(r)$, $k, n \in N$, $i = 1, 2$, and $n = 0$, $i = 1$, are solutions to the modified Bessel equation

$$\omega'' + \frac{1}{r}\omega' - \left(\left(\frac{\pi k}{H}\right)^2 + \frac{n^2}{r^2}\right)\omega = 0, \qquad |\omega(+0)| < \infty.$$

Hence, $u_{kn}^{(i)}(r)$ coincide, to an accuracy of a constant factor, with modified Bessel functions

$$I_n(x) = \sum_{s=0}^{\infty} \frac{(x/2)^{2s+n}}{s!(n+s)!}, \quad \text{where } x = \frac{\pi k}{H} r.$$

Thus, the sought after functions $u_k(r, \varphi)$, $k \in N$, may be presented in the form

$$u_k(r,\varphi) = \frac{I_0(\pi k r/H)}{I_0(\pi k R/H)} A_{k0} + \sum_{n=1}^{\infty} \frac{I_n(\pi k r/H)}{I_n(\pi k R/H)}$$

$$\times (A_{kn}\cos n\varphi + B_{kn}\sin n\varphi),$$

and by boundary conditions (16.1.15) and (16.1.16), the unknown coefficients A_{k0}, A_{kn}, and B_{kn}, $n \in N$, must be found from the dual equation

$$A_{k0} + \sum_{n=1}^{\infty} (A_{kn} \cos n\varphi + B_{kn} \sin n\varphi) = 0, \qquad \varphi \in CE,$$

$$\frac{(\pi kR/H) I_0'(\pi kR/H)}{I_0(\pi kR/H)} A_{k0} + \sum_{n=1}^{\infty} \frac{(\pi kR/H) I_n'(\pi kR/H)}{n I_n(\pi kR/H)}$$

$$(nA_{kn} \cos n\varphi + nB_{kn} \sin n\varphi) = f_k(\varphi), \qquad \varphi \in E.$$

Let us show that this is a dual equation of the type (16.1.1) and (16.1.2). In fact, from the recurrent relationship for modified Bessel functions (Bateman and Erdèlyi 1953),

$$\frac{dI_n(x)}{dx} \equiv I_n'(x) = I_{n+1}(x) + \frac{n}{x} I_n(x),$$

one gets

$$\frac{(\pi kR/H) I_n'(\pi kR/H)}{n I_n(\pi kR/H)} = 1 - \epsilon_n,$$

$$\text{where } \epsilon_n = \frac{(\pi kR/H) I_{n+1}(\pi kR/H)}{-n I_n(\pi kR/H)},$$

and from the asymptotic equality $I_n(x) \sim (x/2)^n/n!$ for a fixed $x \neq 0$ and $n \to \infty$, it follows that $\epsilon_n = O(n^{-2})$.

4. A number of mixed boundary-value problems for a plane layer may be reduced to the dual equation (16.1.1) and (16.1.2) also, if the sought after solution is a periodic function of one of the Cartesian coordinates. The simplest example of the problem is the two-dimensional problem of a stationary temperature distribution within a uniform layer between a pair of parallel planes when a given temperature is maintained at one of the planes as well as at a periodically repeated system of strips of the other boundary plane, and heat flux is specified at the remaining portion of the boundary. For determining the temperature field $u = u(x, z)$, $-\infty < x < +\infty$, $0 \leq z \leq H$, one has the following boundary-value problem:

$$\Delta u = 0, \quad u(x + 2\pi, z) = u(x, z), \quad x \in R, 0 < z < H, \tag{16.1.17}$$

$$u|_{z=H} = 0, \qquad -\pi \leq x \leq \pi, \tag{16.1.18}$$

$$u|_{z=0} = 0, \qquad x \in CE, \tag{16.1.19}$$

$$-\left.\frac{\partial u}{\partial z}\right|_{z=0} = f(x), \qquad x \in E. \tag{16.1.20}$$

The harmonic function (16.1.17), whose period is 2π in x and that meets Condition (16.1.18), has the form

$$u(x,z) = \left(1 - \frac{z}{H}\right)A_0 + \sum_{n=1}^{\infty} \frac{\sinh n(H-z)}{\sinh nH}(A_n \cos nx + B_n \sin nx).$$

From boundary conditions (16.1.19) and (16.1.20) for $z = 0$, one gets the following dual equation of the type (16.1.1) and (16.1.2), which may be used for determining the unknown coefficients A_0, A_n, and B_n, $n \in N$:

$$\dot{A}_0 + \sum_{n=1}^{\infty} (A_n \cos nx + B_n \sin nx) = 0, \qquad x \in CE,$$

$$\frac{1}{H}A_0 + \sum_{n=1}^{\infty} \coth nH(nA_n \cos nx + nB_n \sin nx) = f(x), \qquad x \in E,$$

where $b = 1/H$, $\epsilon_n = 1 - \coth(nH) = O[\exp(-2Hn)]$.

5. Finally, we will consider a more complicated boundary-value problem reducible to a system of two dual equations of the type under consideration. Specifically, we will consider the mixed boundary-value problem for a stationary equation describing a layer limited by a "double lattice." Let the layer be located between the planes $z = \pm H/2$ of the Cartesian system of coordinates and let the "lattices" be composed of periodically repeated systems of strips whose edges are parallel to the axis OY and that are positioned at the upper and the lower boundary planes, respectively. For the sake of simplicity we will consider the plane problem for the Laplace equation. The mathematical formulation of the problem is as follows.

   We seek 2π-periodic function in x,

$$u = u(x,z), \qquad u(x + 2\pi, z) = u(x,z), \qquad x \in R, |z| \leq H/2,$$

that is continuous up to the boundary and satisfies the conditions

$$\Delta u = 0, \qquad |z| < H/2, \tag{16.1.21}$$

$$\frac{\partial u}{\partial z}\left(x, \frac{H}{2}\right) = f_1(\dot{x}), \qquad x \in E_1, \tag{16.1.22}$$

$$u\left(x, \frac{H}{2}\right) = 0, \qquad x \in CE_1, \qquad (16.1.23)$$

$$\frac{\partial u}{\partial z}\left(x, -\frac{H}{2}\right) = f_2(x), \qquad x \in E_2, \qquad (16.1.24)$$

$$u(x, -H/2) = 0, \qquad x \in CE_2, \qquad (16.1.25)$$

where $E_i = \bigcup_{k=1}^{m_i}(\alpha_{ik}, \beta_{ik}), \ -\pi < \alpha_{i1} < \beta_{i1} < \cdots < \alpha_{im_i} < \pi, \ i = 1, 2$; $f_i(x), \ x \in \bar{E}_i$, are given smooth functions. A solution to the problem is sought in the form

$$u(x, z) = a_0 + b_0 z + \sum_{n=1}^{\infty} \{(a_n \cosh nz + b_n \sinh nz)\cos nx$$

$$+ (c_n \cosh nz + d_n \sinh nz)\sin nx\}, \qquad |z| \leq H/2.$$

The first dual equation is derived subject to Conditions (16.1.22) and (16.1.23) for $z = H/2$:

$$a_0 + \frac{b_0 H}{2} + \sum_{n=1}^{\infty} [(A_n + B_n)\cos nx + (C_n + D_n)\sin nx] = 0,$$

$$x \in CE_1, \qquad (16.1.26)$$

$$b_0 + \sum_{n=1}^{\infty} [n(A_n + B_n)\cos nx + n(C_n + D_n)\sin nx]$$

$$- \sum_{n=1}^{\infty} \{(\epsilon_n' A_n - n\epsilon_n'' B_n)\cos nx + (\epsilon_n' nC_n - \epsilon_n'' nD_n)\sin nx\}$$

$$= f_1(\dot{x}), \qquad x \in E_1; \qquad (16.1.27)$$

the second one is obtained from Conditions (16.1.24) and (16.1.25) for $z = -H/2$:

$$a_0 - \frac{b_0 H}{2} + \sum_{n=1}^{\infty} [(A_n - B_n)\cos nx + (C_n - D_n)\sin nx] = 0,$$

$$x \in CE_2, \qquad (16.1.28)$$

$$b_0 - \sum_{n=1}^{\infty} [n(A_n - B_n)\cos nx + n(C_n - D_n)\sin nx]$$

$$+ \sum_{n=1}^{\infty} \{(\epsilon_n' nA_n + \epsilon_n'' nB_n)\cos nx + (\epsilon_n' nC_n + \epsilon_n'' nD_n)\sin nx\}$$

$$= f_2(x), \qquad x \in E_2, \tag{16.1.29}$$

where

$$A_n = a_n \cosh nH/2, \quad B_n = b_n \sinh nH/2, \quad C_n = C_n \cosh nH/2,$$

$$D_n = d_n \sinh nH/2, \quad \epsilon_n' = 1 - \tanh nH/2, \quad \epsilon_n'' = \coth nH/2 - 1.$$

From these examples it follows that in order to obtain solutions to the preceding boundary-value problems of mathematical physics in the form of Fourier series, one has to find their coefficients from the dual equations of the type (16.1.1) and (16.1.2).

However, often there is no need to calculate all the coefficients, because only one or some of the first coefficients are of physical importance as, for instance, in the case of studying wave diffraction (see paragraphs 3 and 4). On the other hand, in the case of the heat conduction problems considered before, it suffices to know the function being sought (temperature) only at the portions of the boundary where it is not specified (but the heat flux is given).

It is clearly desirable to find the values independently, without calculating the coefficients of the series.

Thus, for applications a method of solving dual equations is needed that would be sufficiently versatile and would allow us to answer questions of physical importance without carrying out extremely complicated calculations. A method for solving dual equations of the type under consideration, based on reduction to the Riemann–Hilbert problem, was developed in connection with solving problems of electromagnetic wave diffraction on a lattice (Agranovich, Marchenko, and Shestopalov 1962; Shestopalov 1971).

Another numerical method of solving dual equations of the form (16.1.1) and (16.1.2) as well as of their continual analog—the corresponding dual integral equations—based on the reduction to a singular integral equation on a system of segments (Gandel 1982, 1983) and its subsequent solution by the method of discrete singularities (Lifanov and Matveev 1983b) is discussed in the next section. The method is quite simple and, independently of the number of segments composing the major set E (see (16.1.2), results in solving similar systems of linear algebraic equations.

It should also be noted that all the quantities discussed in the preceding text may be readily expressed through the solution to a singular integral equation.

16.2. METHOD FOR SOLVING DUAL EQUATIONS

Let us continue by reducing the dual equation (16.1.1) and (16.1.2) to a singular integral equation on a system of intervals. Denote

$$U(y) \equiv A_0 + \sum_{n=1}^{\infty} (A_n \cos ny + B_n \sin ny), \qquad y \in [-\pi, \pi], \quad (16.2.1)$$

$$F(y) \equiv \frac{dU(y)}{dy} = \sum_{n=1}^{\infty} (-nA_n \sin ny + nB_n \cos ny),$$

$$y \in [-\pi, \pi]. \quad (16.2.2)$$

For the problems of mathematical physics under consideration $U(y)$, $y \in [-\pi, \pi]$, is a continuous function, and

$$F(y)|_{y \in E} = O(q^{-1/2}), \quad (16.2.3)$$

where q is the distance from point y to the boundary ∂E of set E.

By Equation (16.1.1),

$$F(y) = 0, \qquad y \in CE, \quad (16.2.4)$$

and

$$\int_{\alpha_k}^{\beta_k} F(y)\, dy = 0, \qquad k = 1, 2, \dots, m. \quad (16.2.5)$$

Let us add the relationship

$$A_0 + \sum_{n=1}^{\infty} (-1)^n A_n = 0, \quad (16.2.6)$$

obtained from Equation (16.1.1) for $y = \pi$. Then the latter equation will be replaced by Equations (16.2.4)–(16.2.6).

Taking into account Equation (16.2.4), one gets from Equation (16.2.2) for $n \in N$,

$$A_n = -\frac{1}{\pi n} \int_E F(y) \sin ny\, dy, \quad (16.2.7)$$

$$B_n = \frac{1}{\pi n} \int_E F(y) \cos ny\, dy, \quad (16.2.8)$$

Also taking into account Equation (16.2.7) and using the known expansion of function $g(x) = x/2$, $x \in (-\pi, \pi)$, into the Fourier series, we get from (16.2.6),

$$A_0 = -\frac{1}{2\pi}\int_E F(y)y\,dy. \tag{16.2.9}$$

Thus, all the unknown coefficients are expressed by function $F(y)$, $y \in E$, which is to be found. Note that function $U(x)$ is also directly expressible by $F(y)$:

$$U(x) = \int_{-\pi}^{\pi} F(y)\,dy, \qquad x \in [-\pi, \pi],$$

and by (16.2.4) and (16.2.5), $U(x)$ is continuous.

Next we apply the Hilbert transform for periodic functions to function $F(y)$, $y \in [-\pi, \pi]$:

$$(\Gamma F)(x) \equiv \frac{1}{2\pi}\int_{-\pi}^{\pi} F(y)\cot\frac{y-x}{2}\,dy,$$

which transfers $\cos ny$ into $-\sin nx$ and $\sin ny$ into $\cos nx$. Taking into account (16.2.4), one gets from (16.2.2),

$$\sum_{n=1}^{\infty}(nA_n\cos nx + nB_n\sin nx) = -\frac{1}{2\pi}\int_E F(y)\cot\frac{y-x}{2}\,dy. \tag{16.2.10}$$

Let us substitute the sum $\sum_{n=1}^{\infty}(nA_n\cos nx + nB_n\sin nx)$ appearing in (16.1.2) by (16.2.10), and let us substitute all the remaining required coefficients entering the other summands on the left-hand side of (16.2.2) by their presentations through $F(y)$, $y \in E$ (see (16.2.7)–(16.2.9)). After some evident manipulation, we conclude that function $F(y)$, $y \in E$, satisfies the singular integral equation

$$\frac{1}{\pi}\int_E \frac{F(y)\,dy}{y-x} + \frac{1}{\pi}\int_E\left\{K(y-x) + \frac{by}{2}\right\}F(y)\,dy = -f(x), \quad x \in E, \tag{16.2.11}$$

where

$$K(x) = \frac{1}{2}\cot\frac{x}{2} - \frac{1}{x} - \sum_{n=1}^{\infty}\epsilon_n\sin nx.$$

Let us describe the class of functions in which a solution to Equation (16.2.11) must be sought.

Let us denote the construction of function $F(y)$ in interval (α_k, β_k) by $F_k(y)$; in other words, $F_k(y) = F(y)|_{y \in (\alpha_k, \beta_k)}$. By (16.2.3),

$$F_k(y) = \frac{U_k(y)}{\sqrt{(\beta_k - y)(y - \alpha_k)}}, \qquad y \in (\alpha_k, \beta_k),\ k = 1, 2, \ldots, m, \tag{16.2.12}$$

where $U_k(y)$, $y \in [\alpha_k, \beta_k]$, are Hölder-continuous functions, nonzero at the ends of the interval. Also, m additional conditions (16.2.5) must be met:

$$\int_{\alpha_k}^{\beta_k} \frac{U_k(y)\, dy}{\sqrt{(\beta_k - y)(y - \alpha_k)}} = 0, \qquad k = 1, 2, \ldots, m.$$

The characteristic equation for (16.2.11) in the said class of functions, subject to the additional conditions (16.2.5), has a unique solution. Approximate values of the functions $u_k(y)$, $k = 1, 2, \ldots, m$, may be readily obtained by using the numerical method proposed in Lifanov and Matveev (1983) and previously described in Section 5.4. As applied to the case under consideration, the major result may be formulated as follows. Let

$$y_i^{(n_k)} = \frac{\beta_k - \alpha_k}{2} \cos \frac{2i - 1}{2n_k} \pi + \frac{\beta_k + \alpha_k}{2}, \qquad i = 1, \ldots, n_k, \tag{16.2.13}$$

$$x_j^{(n_k)} = \frac{\beta_k - \alpha_k}{2} \cos \frac{j}{n_k} \pi + \frac{\beta_k + \alpha_k}{2}, \qquad j = 1, \ldots, n_k - 1,$$

$$k = 1, \ldots, m, \tag{16.2.14}$$

and $K(x), f(x) \in H_r(\alpha)$. If the singular integral equation (16.2.11) subject to additional conditions (16.2.5) is uniquely solvable in the class of functions determined by Conditions (16.2.12), then at a sufficiently large

$$N_1 = \min_{1 \le k \le m} n_k,$$

the system of linear algebraic equations [with respect to unknowns

$u_{k,n_k}(y^{(n_k)_i})]$

$$\sum_{k=1}^{m}\sum_{i=1}^{n_k}\left\{\frac{1}{y_i^{(n_k)}-x_j^{(n_p)}}+K\left(y_i^{(n_k)}-x_j^{(n_p)}\right)+\frac{b}{2}y_i^{(n_k)}\right\}$$

$$\times\, u_{k,n_k}(y_i^{(n_k)})\frac{1}{n_k}=-f\left(x_j^{(n_p)}\right),\qquad j=1,\ldots,n_p-1,$$

$$\sum_{i=1}^{n_k}u_{k,n_k}(y_i^{(n_k)})\frac{1}{n_k}=0,\qquad j=n_p,\, p=1,2,\ldots,m, \quad (16.2.15)$$

has a unique solution, and

$$\left|u_{k,n_k}(y_i^{(n_k)})-u_k(y_i^{(n_k)})\right|\le O\left(\frac{\ln n}{n_k^{r+\alpha}}\right),\qquad k=1,2,\ldots,m, \quad (16.2.16)$$

for $N_1\to\infty$.

Having obtained the values of $u_{k,n_k}(y_i^{(n_k)})$, $i=1,\ldots,n_k$, $k=1,\ldots,m$, one may find approximate values of the sought after coefficients A_0, A_n, and B_n, $n\in N$, by using Formulas (16.2.7)–(16.2.9), as well as approximate values of the functions $F_k(y)$, $y\in(\alpha_k,\beta_k)$, composing an approximate solution to Equation (16.2.11), in the following way.

By using the values of $u_{k,n_k}(y_i^{(n_k)})$, $i=1,\ldots,n_k$, we construct an interpolation polynomial of the (n_k-1)th degree [denoted by $P_{k,n_k-1}(y)$, $y\in(\alpha_k,\beta_k)$] in such a way that

$$P_{k,n_k-1}(y_i^{(n_k)})=u_{k,n_k}(y_i^{(n_k)}),\qquad i=1,\ldots,n_k,\, k=1,\ldots,m.$$

Then approximate values of the functions $F_k(y)$, $y\in(\alpha_k,\beta_k)$, are given by the formulas

$$F_{k,n_k}(y)=\frac{P_{k,n_k-1}(y)}{\sqrt{(\beta_k-y)(y-\alpha_k)}},\qquad \alpha_k<y<\beta_k,\, k=1,\ldots,m. \quad (16.2.17)$$

Following Longhanns and Selbermann (1981), one arrives at the error estimate

$$\|F_k-F_{k,n_k}\|_{L^2\omega_k}=O\left(\frac{\ln n}{n_k^{r+\alpha}}\right),\qquad \omega_k=\sqrt{(\beta_k-y)(y-\alpha_k)}\,. \quad (16.2.18)$$

Then by using Formulas (16.2.7)–(16.2.9), one obtains the following approximate values of the coefficients A_q, B_q, $q \in N$, and A_0:

$$\tilde{A}_q = -\frac{1}{q}\sum_{k=1}^{m}\frac{1}{n_k}\sum_{i=1}^{n_k} u_{k,n_k}(y_i^{(n_k)})\sin qy_i^{(n_k)},$$

$$\tilde{B}_q = \frac{1}{q}\sum_{k=1}^{m}\frac{1}{n_k}\sum_{i=1}^{n_k} u_{k,n_k}(y_i^{(n_k)})\cos qy_i^{(n_k)},$$

$$\tilde{A}_0 = -\frac{1}{2}\sum_{k=1}^{n}\frac{1}{n_k}\sum_{i=1}^{n_k} u_{k,n_k}(y_i^{(n_k)})\,y_i^{(n_k)}.$$

Taking into account (16.2.16), the latter formulas permit us to conclude that

$$|A_q - \tilde{A}_q| \le \frac{1}{q}O\left(\frac{\ln n}{N_1^{r+\alpha}}\right), \tag{16.2.19}$$

where it is supposed that $n_k/N_1 \le D^* < +\infty$, $k = 1,2,\ldots,m$, $N_1 \to \infty$.

Next we write a system of singular integral equations whose solution is obtained by solving the system of dual equations (16.1.26), (16.1.27), (16.1.28), and (16.1.29). Acting similarly to what was done when reducing the dual equation (16.1.1) and (16.1.2) to a singular integral equation, we introduce two functions

$$F_1(x) \equiv \sum_{n=1}^{\infty} - n[(A_n + B_n)\sin nx - (C_n + D_n)\cos nx], \qquad |x| \le \pi, \tag{16.2.20}$$

$$F_2(x) \equiv \sum_{n=1}^{\infty} - n[(A_n - B_n)\sin nx - (C_n - D_n)\cos nx]. \tag{16.2.21}$$

From (16 .1.26) and (16.1.28), one gets

$$F_i(x) = 0, \qquad x \in CE_i,\ i = 1,2, \tag{16.2.22}$$

together with $m_1 + m_2$ relationships

$$\int_{\alpha_{ik}}^{\beta_{ik}} F_i(y)\,dy = 0, \qquad k = 1,\ldots,m_i,\ i = 1,2. \tag{16.2.23}$$

All the sought after coefficients a_0, b_0, A_n, B_n, C_n, and D_n, $n \in N$, may be expressed through the functions $F_i(y)$, $y \in E_i$, $i = 1, 2$, which are found from a system of singular integral equations.

After some manipulation, we arrive at the final result. The function $F_i(x)$, $i = 1, 2$, is sought in the class of functions presentable in the form

$$F_i(x) = \frac{\Phi_i(x)}{\sqrt{\prod_{k=1}^{m_i} |(x - \alpha_{ik})(\beta_{ik} - x)|}},$$

where $\Phi_i(x) \in H$ and does not vanish at the boundary ∂E_i of the set E_i. The sought after functions $F_i(x)$, $i = 1, 2$, meet Conditions (16.2.23) and the system of singular integral equations

$$\frac{1}{\pi}\int_{E_1} \frac{F_1(y)\,dy}{y - x} + \frac{1}{\pi}\int_{E_1} K_1(x, y)F_1(y)\,dy$$

$$- \frac{1}{\pi}\int_{E_2} K_2(x, y)F_2(y\,dy = -f_1(x), \qquad x \in E_1, \quad (16.2.24)$$

$$\frac{1}{\pi}\int_{E_2} \frac{F_2(y)\,dy}{y - x} + \frac{1}{\pi}\int_{E_2} K_1(x, y)F_2(y)\,dy$$

$$- \frac{1}{\pi}\int_{E_1} K_2(x, y)F_1(y)\,dy = f_2(x), \qquad x \in E_2, \quad (16.2.25)$$

where

$$K_1(x, y) = \frac{1}{2}\cot\frac{y - x}{2} - \frac{1}{y - x} + \frac{1}{2}\sum_{n=1}^{\infty} \epsilon_{1n} \sin n(y - x) + \frac{y}{2H},$$

$$K_2(x, y) = \frac{1}{2}\sum_{n=1}^{\infty} \epsilon_{2n} \sin n(y - x) + \frac{y}{2H},$$

$$\epsilon_{1n} \equiv \epsilon_n'' - \epsilon_n' = \frac{4e^{-2nH}}{1 - e^{-2nH}}, \qquad \epsilon_{2n} \equiv \epsilon_n'' + \epsilon_n' = \frac{4e^{-nH}}{1 - e^{-nH}}.$$

From the general theory of singular integral equations it follows that in the class of functions under consideration, characteristic equations have unique solutions if the corresponding additional conditions (16.2.23) are met.

As long as a singular integral over only one of the sets E_i, $i = 1, 2$, is present in each of the equations composing a system, an approximate solution to the boundary problem discussed in the preceding section (paragraph 5) may be obtained by the method of discrete singularities. Accordingly, the system of linear algebraic equations is composed of systems of such equations for each of Equations (16.2.24) and (16.2.25) (see (16.2.15)).

By solving the system of n linear algebraic equations with n unknowns thus obtained, we first find approximate values of the functions $u_{sk}(y)$, then those of $F_{sk}(y) = u_{sk}(y)[(\beta_{sk} - y)(y - \alpha_{sk})]^{-1/2}$ and F_s, $s = 1, 2$, where $F_{sk}(y)$ is the *restriction of function* F_s on the interval $(\alpha_{sk}, \beta_{sk})$. Now we can find approximate values of both coefficients of the series for function $u(x, z)$ and the values of the function $u(x, z)$ itself at the portions of the boundary where it was not specified beforehand:

$$u\left(x, \frac{H}{2}\right) = \int_{-\pi}^{x} F_1(y)\, dy, \qquad u\left(x, -\frac{H}{2}\right) = \int_{-\pi}^{x} F_2(y)\, dy,$$

$$x \in [-\pi, \pi].$$

Note concerning a continuum analog of the dual equations (16.1.1) and (16.1.2): Let us consider the dual integral equations (Gandel 1983)

$$\frac{1}{\sqrt{2\pi}} \int_{-\infty}^{\infty} Q(\lambda) e^{i\lambda x}\, d\lambda = 0, \qquad x \in CE, \tag{A}$$

$$\frac{1}{\sqrt{2\pi}} \int_{-\infty}^{\infty} |\lambda| Q(\lambda)(1 + \epsilon(\lambda)) e^{i\lambda x}\, d\lambda = f(x), \qquad x \in E, \tag{B}$$

where $E = \bigcup_{k=1}^{m} (a_k, b_k)$, $-\infty < a_1 < b_1 < \cdots < a_m < b_m < +\infty$, $CE = R \setminus E$; $f(x)$, $x \in \bar{E}$, and $\epsilon(\lambda)$, $\lambda \in R$, are given functions, and $\epsilon(\lambda) = O(\lambda^{-2})$ for $\lambda \to \infty$, while $Q(\lambda)$, $\lambda \in R$, is an unknown function.

One comes across these equations when considering boundary problems of electrostatics–electrodynamics in the case when the boundary is an isolated lattice composed of a finite number of plane strips. Usually, to solve such a problem it suffices to calculate the values of the integral

$$\frac{1}{\sqrt{2\pi}} \int_{-\infty}^{\infty} Q(\lambda) e^{i\lambda x}\, d\lambda \equiv \tilde{U}(x)$$

on the left-hand side of the first equation for $x \in E$.

Acting in a formal way, we denote

$$F(x) \equiv \frac{d\tilde{U}(x)}{dx} = \frac{i}{\sqrt{2\pi}} \int_{-\infty}^{\infty} \lambda Q(\lambda) e^{i\lambda x}\, d\lambda, \qquad x \in R,$$

and derive a singular integral equation for function $F(x)$, $x \in E$, that will be sought in the class of functions presentable in the form

$$F(x) = \frac{\Phi(x)}{\sqrt{\prod_{k=1}^{m} |(b_k - x)(x - a_k)|}}, \qquad x \in E,$$

where $\Phi(x)$, $x \in \overline{E}$, is a Hölder-continuous function nonzero at the boundary of the set E, and, by Equation (A),

$$F(x) = 0, \qquad x \in CE, \tag{C}$$

$$\int_{a_k}^{b_k} F(x)\, dx = 0, \qquad k = 1, \ldots, m. \tag{D}$$

By applying to the function $F(x)$, $x \in R$, the Hilbert transform

$$(HF)(x) = \frac{1}{\pi} \int_{-\infty}^{\infty} \frac{F(y)\, dy}{-y + x}$$

and taking into account (C) and the fact that

$$H\colon \exp(i\lambda y) \mapsto i[(|\lambda|/\lambda)\exp(i\lambda x)],$$

we get

$$\frac{1}{\sqrt{2\pi}} \int_{-\infty}^{\infty} |\lambda| Q(\lambda) e^{i\lambda x}\, d\lambda = \frac{1}{\pi} \int_E \frac{F(y)\, dy}{y - x}, \qquad x \in R.$$

Then, using the presentation for $F(x)$, $x \in R$, and taking into account (C), one obtains

$$\lambda Q(\lambda) = \frac{1}{i\sqrt{2\pi}} \int_E F(y) e^{-i\lambda y}\, dy, \qquad \lambda \in R.$$

By using the latter two relationships and Equation (B), one may derive the singular integral equation

$$\frac{1}{\pi} \int_E \frac{F(y)\, dy}{y - x} + \int_E K(y - x) F(y)\, dy = f(x), \qquad x \in E,$$

which function $F(y)$, $y \in E$, satisfies. Here

$$K(z) = \frac{1}{2\pi i} \int_{-\infty}^{\infty} \frac{|\lambda|}{\lambda} \epsilon(\lambda) e^{-i\lambda z} \, d\lambda$$

is a kernel coinciding with the Fourier transform of function $\epsilon(\lambda)$sign λ, $\lambda \in R$, to an accuracy of the factor $1/(i\sqrt{2\pi})$.

It should be remembered that the required function is sought in the class indicated previously and meets Conditions (D), thus ensuring unique solvability of the corresponding characteristic equation.

After obtaining the function $F(y)$, $y \in E$, the sought after function $\tilde{U}(x)$ may be calculated by using the formula $\tilde{U}(x) = \int_{a_1}^{x} F(y)\, dy$, $x \in R$.

16.3. DIFFRACTION OF A SCALAR WAVE ON A PLANE LATTICE: DIRICHLET AND NEUMANN PROBLEMS FOR THE HELMHOLTZ EQUATION

Let us consider two important examples of boundary problems of mathematical physics resulting in dual equations of the type (16.1.1) and (16.1.2).

Specifically, we will consider three-dimensional Dirichlet and Neumann problems for the Helmholtz equation in the case when the boundary is formed by a plane lattice. Both the problems of diffraction of a plane monochromatic wave on a plane ideally conducting lattice (numerical solution of these problems is discussed in the following section) and the problems of diffraction of acoustic waves on "soft" and "rigid" lattices reduce to the stated two boundary problems.

Both the Dirichlet and Neumann problems considered in this section reduce to a mixed boundary problem for the Helmholtz equation in a half-space, whose dual equations have the form (16.1.1).

The method described in the preceding section was used to solve model problems numerically; the solutions agreed favorably with accurate solutions.

16.3.1. Formulation of the Dirichlet Problem and Derivation of Dual Equations

Let a lattice formed of periodically repeated (the period being equal to 21) systems of strips whose edges are parallel to the OX' axis be considered in the plane $OX'Y'$ of the Cartesian system of coordinates

$$\{(x', y', z'): z' = 0, \quad -\infty < x' < +\infty, y' \in CE'\},$$

where $E = \bigcup_{j=1}^{m} E'_j$, $E'_j = (\alpha'_j, \beta'_j)$, $CE' = [-l, l] \setminus E'$, $-l < \alpha'_1 < \beta'_1 < \cdots < \alpha'_m < \beta'_m < l$. Let a plane monochromatic wave $\mathring{u} = i\exp(-ikz')$, where $k = \omega/c$ is the wave number and c is the speed of light, fall onto a lattice from the half-space $z' > 0$ perpendicularly to the former.

Let us denote by u_0 the sum of the incident and reflected waves in the upper half-space for the case when the first boundary condition $u_0|_{z'=0} = 0$ is met throughout the plane:

$$u_0 = ie^{-ikz'} - ie^{ikz'} = 2\sin kz', \qquad z' \geq 0. \tag{16.3.1}$$

If the total field vanishes at the lattice, then it may be sought in the form

$$u_{\text{total}} = \begin{cases} u_0 + u^+, & z > 0, \\ u^-, & z < 0, \end{cases}$$

where the functions $u^+(y', z')$ and $u^-(y', z')$ are $2l$-periodic with respect to y',

$$u^{\pm}(y' + 2l, z') = u^{\pm}(y', z'), \qquad -\infty < y' < \infty,\ z' \gtrless 0,$$

and meet both the Helmholtz equation

$$\Delta u^{\pm} + k^2 u = 0, \qquad z' \gtrless 0, \tag{16.3.2}$$

and the matching conditions

$$u^-|_{z'=0} = u^+|_{z'=0}, \qquad y' \in E', \tag{16.3.3}$$

$$\left.\frac{\partial u^-}{\partial z'}\right|_{z'=0} = 2k + \left.\frac{\partial u^+}{\partial z'}\right|_{z'=0}, \qquad y' \in E', \tag{16.3.4}$$

which together with (16.3.2) ensure validity of the Helmholtz equation for the total field $\Delta u_{\text{total}} + k^2 u_{\text{total}} = 0$ throughout the space outside the lattice. At the lattice itself the total field vanishes:

$$u^-|_{z'=0} = (u_0 + u^+)|_{z'=0} \equiv u^+|_{z'=0} = 0, \qquad y' \in CE'. \tag{16.3.5}$$

Additionally, in order to ensure that a solution to the boundary problem under consideration is unique, both radiation conditions at infinity and Meixner conditions at the ribs of the lattice must be met (Hönl, Maue, and Westphal 1964; Shestopalov et al. 1973).

Solutions to the Helmholtz equations in the upper ($z' > 0$) and the lower ($z' < 0$) half-space meeting the periodicity conditions have the form

$$u^{\pm}(y', z') = A_0^{\pm} e^{\pm ikz'} + \sum_{n=1}^{\infty} e^{\pm \gamma_n z'} \left(A_n^{\pm} \cos \frac{\pi n y'}{l} + B^{\pm} \sin \frac{\pi n y'}{l} \right),$$

$$z' \gtrless 0,$$

where $\gamma_n = \sqrt{(\pi n/l)^2 - k^2}$ and the radiation conditions will be fulfilled if the branch of the radical is chosen in such a way that

$$\operatorname{Im} \gamma_n < 0 \quad \text{for } n < kl/\pi,$$

$$\gamma_n > 0 \quad \text{for } n > kl/\pi. \tag{16.3.6}$$

From (16.3.4) and (16.3.5) it is seen that the functions $u^{-}(y',0)$ and $u^{+}(y',0)$ coincide for all $y' \in [-l, 1]$; hence,

$$A_n^- = A_n^+ \equiv A_n, \qquad n = 0, 1, 2, \ldots,$$

$$B_n^- = B_n^+ \equiv B_n, \qquad n = 1, 2, \ldots .$$

Thus we have

$$u^{\pm} = A_0 e^{\pm ikz'} + \sum_{n=1}^{\infty} e^{\pm \gamma_n z'} \left(A_n \cos \frac{\pi n y'}{l} + B_n \sin \frac{\pi n y'}{l} \right), \qquad z' \gtrless 0, \tag{16.3.7}$$

and after using boundary condition (16.3.5) at the lattice, we arrive at the equation

$$A_0 + \sum_{n=1}^{\infty} \left(A_n \cos \frac{\pi n y'}{l} + B_n \sin \frac{\pi n y'}{l} \right) = 0, \qquad y' \in CE'. \tag{16.3.8}$$

Then from Condition (16.3.4), the other equation for determining unknown coefficients in the presentation (16.3.7) of a solution to the problem under consideration may be obtained:

$$-ikA_0 + \sum_{n=1}^{\infty} \gamma_n \left(A_n \cos \frac{\pi n y'}{l} + B_n \sin \frac{\pi n y'}{l} \right) = k, \qquad y' \in E'. \tag{16.3.9}$$

Let us introduce the nondimensional quantities

$$y = \pi y'/l, \qquad \alpha_j = \pi\alpha_j'/l, \qquad \beta_j = \pi\beta_j'/l, \qquad \kappa = lk/\pi = 2l/\lambda, \tag{16.3.10}$$

where λ is the incident field wavelength. Then,

$$\gamma_n = \frac{\pi n}{l}\sqrt{1 - \left(\frac{\kappa}{n}\right)^2}, \qquad n \in N,$$

and the branch of the radical is selected in order to meet Conditions (16.3.6). Let us also introduce

$$E = \bigcup_{j=1}^{m} E_j, \qquad E_j = (\alpha_j, \beta_j), \qquad CE = [-\pi, \pi] \setminus E,$$

$$-\pi < \alpha_1 < \beta_1 < \cdots < \alpha_m < \beta_m < \pi.$$

Then, Equations (16.3.8) and (16.3.9) become

$$A_0 + \sum_{n=1}^{\infty} (A_n \cos ny + B_n \sin ny) = 0, \qquad y \in CE, \tag{16.3.11}$$

$$-i\kappa A_0 + \sum_{n=1}^{\infty} n\sqrt{1 - \left(\frac{\kappa}{n}\right)^2}(A_n \cos ny + B_n \sin ny) = \kappa, \qquad y \in E. \tag{16.3.12}$$

These equations correspond to Equations (16.1.1) and (16.1.2) for

$$b = -i\kappa, \qquad f(y) \equiv \kappa, \qquad \epsilon_n = 1 - \sqrt{1 - (\kappa/n)^2},$$

where $\epsilon_n = O(\kappa^2/(2n^2))$.

As was shown in the preceding section, all the sought after coefficients A_n, B_n, $n \in N$, and A_0 are expressed by Formulas (16.2.7)–(16.2.9) through solution $F(y)$, $y \in E$, to the singular integral equation (16.2.11), where one should put

$$b = -i\kappa, \qquad f(x) = \kappa, \qquad \epsilon_n = 1 - \sqrt{1 - (\kappa/n)^2}, \qquad n \in N,$$

and the choice of the branch of the radical is determined by Conditions (16.2.5).

The Meixner conditions determine behavior of both the wave field and its first derivatives in the neighborhood of a strip edge. These conditions generate presentations (16.2.12) for construction of function $F(y)$ in intervals E_j.

16.3.2. Model Problems

Consider the function

$$g(x) = \begin{cases} 0, & x \in [-\pi, -1], \\ -\sqrt{1-x^2}, & x \in (-1, 1), \\ 0, & x \in [1, \pi], \end{cases}$$

which after being expanded into the Fourier series becomes

$$g(x) = -\frac{1}{4} - \sum_{n=1}^{\infty} \frac{\Phi_1(n)}{n} \cos nx, \qquad x \in [-\pi, \pi], \quad (16.3.13)$$

because according to Bateman and Erdèlyi (1953),

$$\frac{1}{\pi} \int_{-1}^{1} \sqrt{1-t^2} \cos nt \, dt = \frac{\Phi_1(n)}{n}, \qquad n = 0, 1, \ldots,$$

where $\Phi_1(z)$ is the Bessel function defined by the integral.

By differentiating (16.3.13) and using the Hilbert transform, one gets

$$\sum_{n=1}^{\infty} \Phi_1(n) \cos nx = \frac{1}{2\pi} \int_{-1}^{1} \frac{t}{\sqrt{1-t^2}} \cot \frac{t-x}{2} \, dt,$$

and because for $x \in (-1, 1)$,

$$\frac{1}{\pi} \int_{-1}^{1} \frac{t}{\sqrt{1-t^2}} \frac{dt}{t-x} = 1;$$

hence we have

$$\sum_{n=1}^{\infty} \Phi_1(n) \cos nx = 1 + \frac{1}{\pi} \int_{-1}^{1} \frac{t}{\sqrt{1-t^2}} \left[\frac{1}{2} \cot \frac{t-x}{2} - \frac{1}{t-x} \right] dt \quad (16.3.14)$$

(Bateman and Erdèlyi 1953).

Thus, from (16.3.13) it follows that

$$\frac{1}{4} + \sum_{n=1}^{\infty} \frac{\Phi_1(n)}{n} \cos nx = 0, \qquad x \in [-\pi, \pi] \setminus (-1, 1), \quad (16.3.15)$$

and employing (16.3.14), one gets

$$\frac{b}{4} + \sum_{n=1}^{\infty} (1 - \epsilon_n)\Phi_1(n) \cos nx = f(x), \qquad x \in (-1, 1), \quad (16.3.16)$$

where

$$f(x) \equiv \frac{4+b}{4} + \frac{1}{\pi}\int_{-1}^{1}\left[\frac{1}{2}\cot\frac{t-x}{2} - \frac{1}{t-x}\right]\frac{t\,dt}{\sqrt{1-t^2}}$$

$$- \sum_{n=1}^{\infty} \epsilon_n \Phi_1(n) \cos nx, \qquad x \in [-1, 1]. \qquad (16.3.17)$$

Finally, we arrive at the following result. The sequence

$$A_0 = 1/4, \qquad A_n = \Phi_1(n)/n, \qquad B_n = 0, \qquad n \in N,$$

gives a solution to the dual equation (16.1.1) and (16.1.2) with the right-hand side $f(x)$.

Note also that in the case under consideration a solution to the corresponding singular integral equation (16.2.11) is given by the function

$$F(x) = g'(x)|_{x \in (-1,1)} = \frac{x}{\sqrt{1-x^2}}. \qquad (16.3.18)$$

In what follows we will briefly discuss the model problem on a system of intervals $E = E_1 \cup E_2$ (for $m = 2$).

It is obvious that the function $F(y) = F_1(y) + F_2(y)$, where

$$F_k(y) = \begin{cases} \dfrac{C_k(y - (\alpha_k + \beta_k/2))}{\sqrt{(\beta_k - y)(y - \alpha_k)}}, & x \in E_k, \\ 0, & x \notin E_k, \end{cases} \qquad (16.3.19)$$

$k = 1, 2$, and C_1, C_2 are given constants, is a solution to the integral equation (16.2.11) with right-hand side

$$f(x) = -\frac{1}{2\pi}\int_E F(y)\cot\frac{y-x}{2}\,dy - \frac{b}{2\pi}\int_E yf(y)\,dy + \frac{1}{\pi}\int_E F(y)\sum_{n=1}^{\infty}\epsilon_n \sin n(y-x)\,dy, \qquad x \in E, \quad (16.3.20)$$

and meets Relationships (16.2.5).

By substituting $F(y)|_{y \in E_k} = F_k(y)$ (see (16.3.20)) by the corresponding expressions (16.3.19), we calculate $f_k(x) \equiv f(x)|_{x \in E_k}$, $k = 1, 2$:

$$f_k(x) = -\frac{1}{\pi}\int_{E_k}\frac{F_k(y)\,dy}{y-x} - \frac{1}{\pi}\int_{E_l}\frac{F_l(y)\,dy}{y-x} - \frac{1}{\pi}\int_E F(y)\left\{\frac{1}{2}\cot\frac{y-x}{2} - \frac{1}{y-x}\right\}dy + \frac{1}{\pi}\int_E F(y)\left[\sum_{n=1}^{\infty}\epsilon_n \sin n(y-x) - \frac{by}{2}\right]dy, \qquad x \in E_k,$$

where k, l take on the values of 1, 2 and $k \neq l$.

The first two integrals may be expressed explicitly:

$$\frac{1}{\pi}\int_{E_k}\frac{F_k(y)\,dy}{y-x} = C_k, \qquad x \in E_k,$$

$$\frac{1}{\pi}\int_{E_l}\frac{F_l(y)\,dy}{y-x} = C_l\left(1 - \frac{|\alpha_l + \beta_l - 2x|}{\sqrt{x^2 - (\alpha_l + \beta_l)x + \alpha_l\beta_l}}\right), \qquad x \in E_k.$$

All the remaining integrals were calculated approximately by using Gaussian quadratures.

For one of the intervals, $E = (-1, 1)$, the model problem was solved numerically by employing a high-speed supercomputer*. Table 16.1 presents approximate values of $u_n(y_i^{(n)})$, $i = 1, \ldots, n$, along with the accurate values $u(y_i^{(n)}) \equiv y_i^{(n)}$ calculated for the number of grid points $n = 20$. The two sets of data differ (in absolute values) by less than 2×10^{-5}.

* This is probably the first mention in press about the Russian supercomputer on which General Belotserkovsky has worked (G. Ch.).

TABLE 16.1

i	$u_n(y_i^{(n)}$	$u(y_i^{(n)})$	i	$u_n(y_i^{(n)})$	$u(y_i^{(n)})$
1	0.99693	0.99692	11	−0.07846	−0.07846
2	0.97235	0.97237	12	−0.23345	−0.23344
3	0.92389	0.92388	13	−0.38269	−0.38268
4	0.85266	0.85264	14	−0.52251	−0.52250
5	0.76042	0.76041	15	−0.64945	−0.64945
6	0.62945	0.64945	16	−0.76041	−0.76041
7	0.52249	0.52250	17	−0.85264	−0.85264
8	0.38269	0.38268	18	−0.92389	−0.92388
9	0.23344	0.23345	19	−0.97236	−0.97237
10	0.07846	0.07846	20	−0.99692	−0.99692

Then we have calculated approximate values of the sought after coefficients A_0, A_q, and B_q, $q = 1, 2, 3$, by using the formulas

$$\tilde{A}_q = -\frac{1}{nq} \sum_{i=1}^{n} u_n(y_i^{(n)}) \sin qy_i^{(n)},$$

$$\tilde{B}_q = \frac{1}{nq} \sum_{i=1}^{\infty} u_n(y_i^{(n)}) \cos qy_i^{(n)}, \qquad q \in N,$$

$$\tilde{A}_0 = -\frac{1}{2n} \sum_{i=1}^{n} u_n(y_i^{(n)}) y_i^{(n)}.$$

As before, approximate values of the integrals at the segment $(-1, 1)$ were calculated by employing the weight function $(1 - y^2)^{-1/2}$ and Gaussian quadrature formulas.

Table 16.2 shows approximate values of coefficients $\tilde{A}_0$, $\tilde{A}_1$, $\tilde{A}_2$, and $\tilde{A}_3$ and numerical values obtained by using the formulas of the accurate solution to the model problem $A_0 = 1/4$ and $A_q = \Phi_1(q)/q$, $q = 1, 2, 3$, and employing the Bessel functions tables (Abramowitz and Stegun 1964). The difference between the approximate and accurate values of the coefficients does not exceed 12×10^{-6}. To the same accuracy, coefficients $\tilde{B}_1$, $\tilde{B}_2$, and $\tilde{B}_3$ were equal to zero.

The model problem for two intervals was solved numerically for $C_1 = 2$, $C_2 = 0.5$, $\alpha_1 = -3$, $\beta_1 = -2$, $\alpha_2 = -1$, $\beta_2 = 0.5$, $n_1 = 4$, and $n_2 = 6$. Table 16.3 permits comparison of approximate values of $u_{k,n_k}(y_i^{(n_k)})$, $i = 1, \ldots, n_k$, $k = 1, 2$ with the accurate ones $u_k(y_i^{(n_k)}) = C_k(y_i^{(n_k)} - (\beta_k + \alpha_k)/2)$. The absolute error does not exceed 2×10^{-6} throughout the domain.

TABLE 16.2

q	$\tilde{A}_q$	A_q	q	$\tilde{A}_q$	A_q
0	0.250008	0.25	2	0.288364	0.2883624
1	0.440062	0.4400506	3	0.113010	0.1130197

TABLE 16.3

$k = 1$			$k = 2$		
i	$u_{k,n_k}(y_i^{(n_k)})$	$u_k(y_i^{(n_k)})$	i	$u_{k},n_k(y_i^{(n_k)})$	$u_k(y_i^{(n_k)})$
1	0.923879	0.923879	1	0.362221	0.362222
2	0.382683	0.382684	2	0.265164	0.265165
3	−0.382683	−0.382683	3	0.097056	0.097057
4	−0.923879	−0.923879	4	−0.097058	−0.097057
			5	−0.265164	−0.265165
			6	−0.362220	−0.362222

16.3.3. Neumann Problem

Let a plane monochromatic wave $\mathring{v} = i\exp(-ikz')$ fall onto the lattice $\{(x', y', z')\colon\ -\infty < x' < \infty,\ y' \in E' + 2nl,\ n \in Z,\ z' = 0\}$ from the half-space $z' > 0$ perpendicularly to the lattice. By V_0 we denote the sum of the incident and reflected waves in the upper half-plane in the case where the second boundary condition

$$\left.\frac{\partial V_0}{\partial z'}\right|_{z'=0} = 0, \qquad V_0 = 2i\cos kz',\ z' \geq 0, \tag{16.3.21}$$

is met at the whole of the plane $OX'Y'$.

If the normal derivative of the total field vanishes at the lattice, then the field is sought in the form

$$V_{\text{total}} = \begin{cases} V_0 + V^+, & z' > 0, \\ V^-, & z' < 0. \end{cases}$$

The functions $V^+(y', z')$ and $V^-(y', z')$ are $2l$-periodic in y' and satisfy the Helmholtz equation for $z' \gtrless 0$, respectively, as well as the matching conditions

$$V^-|_{z'=0} = 2i + V^+|_{z'=0}, \qquad y' \in CE', \tag{16.3.22}$$

$$\left.\frac{\partial V^-}{\partial z'}\right|_{z'=0} = \left.\frac{\partial V^+}{\partial z'}\right|_{z'=0}, \qquad y' \in CE'. \tag{16.3.23}$$

At the lattice the normal derivative of the total field vanishes:

$$\left.\frac{\partial V^-}{\partial z'}\right|_{z'=0} = \left.\left(\frac{\partial V_0}{\partial z'} + \frac{\partial V^+}{\partial z'}\right)\right|_{z'=0} \equiv \left.\frac{\partial V^+}{\partial z'}\right|_{z'=0} = 0, \qquad y' \in E'. \tag{16.3.24}$$

In order to ensure uniqueness of a solution to the boundary-value problem under consideration, both the radiation and Meixner conditions must be met (Hönl, Maue, and Westphal 1964).

A solution to the second boundary-value problem is sought in the form

$$V^{\pm} = A_0^{\pm} e^{\pm ikz'} + \sum_{n=1}^{\infty} e^{\pm \gamma_n z'} \left(A_n^{\pm} \cos \frac{\pi n y'}{l} + B_n^{\pm} \sin \frac{\pi n y'}{l} \right), \quad z' \gtrless 0,$$

where γ_n is defined in Section 16.3.1.

From (16.3.22) and (16.3.23), it follows that

$$A_n^- = -A_n^+ \equiv A_n, \qquad n = 0, 1, 2, \ldots,$$

$$B_n^- = -B_n^+ \equiv B_n, \qquad n = 1, 2, \ldots .$$

By introducing the nondimensional quantities (16.3.9) and using Conditions (16.3.22) and (1.28), one arrives at the dual equation

$$A_0 + \sum_{n=1}^{\infty} (A_n \cos ny + B_n \sin ny) = i, \qquad y \in CE,$$

$$-i\kappa A_0 + \sum_{n=1}^{\infty} (1 - \epsilon_n)(nA_n \cos ny + nB_n \sin ny) = 0, \qquad y \in E,$$

or by putting

$$A_0 = A_0' + i, \tag{16.3.25}$$

one obtains finally a dual equation of the form (16.1.1) and (16.1.2):

$$A_0' + \sum_{n=1}^{\infty} (A_n \cos ny + B_n \sin ny) = 0, \qquad y \in CE,$$

$$bA_0' + \sum_{n=1}^{\infty} (1 - \epsilon_n)(nA_n \cos ny + nB_n \sin ny) = 0, \qquad y \in E,$$

where $b = -i\kappa$, $\epsilon_n = 1 - \sqrt{1 - (\kappa/n)^2}$, $n \in N$. Thus, the Neumann problem has been reduced to the dual equation previously obtained for the Dirichlet problem.

As shown in Section 16.2, unknown coefficients A_n and B_n, $n \in N$, appearing in the representations of the fields V^- and V^+, are defined by Formulas (16.3.7) and (16.3.8) via function $F(y)$, $y \in E$, which satisfies the singular integral equation of the first kind (16.2.11) with the right-hand side equal to $-f(x) = \kappa$, $x \in \overline{E}$, $b = -i\kappa$, and $\epsilon_n = 1 - \sqrt{1 - (\kappa/n)^2}$, $n \in N$, and the choice of the branch of the radical is determined by Conditions (16.2.5) and (16.2.12). The coefficient A_0 entering (16.3.25) is calculated from the formula

$$A_0 = i - \frac{1}{2\pi} \int_E F(y)\, dy,$$

where A_0' was found by using Formula (16.2.9).

16.4. APPLICATION OF THE METHOD OF DISCRETE SINGULARITIES TO NUMERICAL SOLUTION OF PROBLEMS OF ELECTROMAGNETIC WAVE DIFFRACTION ON LATTICES

Much research has been done on wave diffraction on lattices. However, of special importance is the work by Agranovich, Marchenko, and Shestopalov (1962), where a method for solving the problem of diffraction of a plane monochromatic electromagnetic wave on a plane ideally conducting lattice was proposed. The method, based on reduction to the Riemann–Hilbert problem, was widely used for solving various problems of wave diffraction on periodic structures. The results are presented in monographs (Shestopalov 1971, Shestopalov et al. 1973) where the interested reader will find extensive lists of references.

The problems of diffraction of plane monochromatic electromagnetic waves on a plane ideally conducting lattice are reduced to the Dirichlet and Neumann problems considered in detail in the preceding section.

Some problems were solved by the method of discrete singularities for the purpose of comparison with calculated results obtained by the method of reduction to the Riemann–Hilbert problem (Shestopalov 1971).

The approach proposed herewith ensures a simple and uniform method for solving the problems of wave diffraction on lattices irrespective of the number of strips per period.

Let a lattice be placed in the plane OXY, which is composed of periodically repeated (with period $2l$) groups of infinitely thin, ideally conducting strips parallel to the OX axis. Let, further, a plane monochromatic wave fall onto the lattice from the upper half-space ($z > 0$):

$$\mathbf{E}^{(\text{fall})} = \mathbf{E}_0^{(\text{fall})} e^{-ikz}, \qquad \mathbf{H}^{(\text{fall})} = \mathbf{H}_0^{(\text{fall})} e^{-ikz},$$

where $k = \omega/c$ is the wave number and the time dependence is determined by the factor $\exp(-i\omega t)$.

First we will consider the problem of reflection of such a plane wave from an infinitely thin, ideally conducting screen placed in the plane OXY. The field in the upper half-space will be presented in the form of a sum of an incident and a reflected wave:

$$\mathbf{E}^{(\text{ref})} = \mathbf{E}_0^{(\text{ref})} e^{ikz}, \qquad \mathbf{H}^{(\text{ref})} = \mathbf{H}_0^{(\text{ref})} e^{ikz}.$$

A solution to the Maxwell equations will be sought in the form of a sum of the initial wave (zero for $z < 0$ and a superposition of the incident and the reflected waves for $z > 0$) and the scattered wave:

$$\mathbf{E} = \mathbf{E}^{(\text{inc})} + \mathbf{E}^{(\text{scat})}, \qquad \mathbf{H} = \mathbf{H}^{(\text{inc})} + \mathbf{H}^{(\text{scat})}.$$

The field must be continuous outside the strips; the components of the electric vector parallel to the plane OXY must vanish at the strips, and hence, the magnetic vector components normal to the plane the strips lie in must vanish too. Additionally, both the radiation conditions at infinity and the Meixner conditions at the ribs must be met.

It is obvious that in the framework of the problem under consideration, all the planes perpendicular to the OX axis are physically equivalent, and hence, the scattered field is independent of x. The Maxwell equations split into a pair of independent systems of equations for the components E_x, H_y, H_z and H_x, E_y, E_z (Hönl, Maue, and Westphal 1964).

The requirement that the electric vector components parallel to the plane be equal to zero results in the boundary conditions $E_x = 0$ and $\partial H_x / \partial z = 0$ (at the strips). Thus, one may consider separately the cases of E and H polarization for which the corresponding vector is polarized parallel to the OX axis (Belotserkovsky and Nisht 1978).

Because the problem is linear, the amplitudes of the components $E_x^{(\text{inc})}$ and $H_x^{(\text{inc})}$ will be assumed to be equal to i. Then, from the boundary conditions for $z = 0$ one can easily obtain the amplitudes of the components $E_x^{(\text{ref})}$ and $H_x^{(\text{ref})}$ of the wave reflected from the screen:

$$E_x^{(\text{ref})} = -ie^{ikz}, \qquad H_x^{(\text{ref})} = ie^{ikz}.$$

Hence,

$$E_x^{(\mathrm{inc})} = \begin{cases} 2\sin kz, & z > 0, \\ 0, & z < 0, \end{cases}$$

$$H_x^{(\mathrm{inc})} = \begin{cases} 2i\cos kz, & z > 0, \\ 0, & z < 0. \end{cases}$$

a. Let us start by considering the case of E polarization. The problem reduces to finding the function $u \equiv E_x$ presentable in the upper and lower half-spaces in the form

$$u = \begin{cases} 2\sin kz + u^{+}, & z > 0, \\ u^{-}, & z < 0, \end{cases}$$

and satisfying the Helmholtz equation outside the lattice, the first boundary condition at the lattice, the radiation conditions at infinity, and the Meixner conditions at the ribs.

This is the Dirichlet problem. As shown in the preceding section (see Section 16.3.1), it reduces to solving singular integral equation (16.2.11) subject to additional conditions (16.2.5) and (16.2.12). A numerical solution is found by the method of discrete singularities discussed in Section 16.2.

After finding u, all the other components of the total field are determined by the formulas (Hönl, Maue, and Westphal 1964)

$$E_y = E_z = H_z = 0, \qquad H_y = \frac{1}{ik}\frac{\partial u}{\partial z}, \qquad H_z = -\frac{1}{ik}\frac{\partial u}{\partial y}.$$

b. In the case of the H polarization, we arrive at the Neumann problem for the function $V = H_x$ sought in the upper and the lower half-spaces in the form

$$V = \begin{cases} 2i\cos kz + V^{+}, & z > 0, \\ V^{-}, & z < 0. \end{cases}$$

The problem was considered in the preceding section (see Section 16.3.3), where it was shown to be solvable by the method of discrete singularities.

The remaining components of the total field are defined by the formulas (Hönl, Maue, and Westphal 1964)

$$E_x = H_y = H_z = 0, \qquad E_y = -\frac{1}{ik}\frac{\partial V}{\partial z}, \qquad E_z = \frac{1}{ik}\frac{\partial V}{\partial y}.$$

TABLE 16.4

κ	$\lvert A_0 \rvert$	κ	$\lvert A_0 \rvert$	$\lvert A_1 \rvert$	$\lvert A_2 \rvert$
0.1	0.0694	1.2	0.6086	0.9108	
0.2	0.1394	1.4	0.5538	0.8320	
0.3	0.2106	1.6	0.5256	0.7942	
0.4	0.2838	1.8	0.5060	0.7721	
0.5	0.3598	2	0.4769	0.7576	
0.6	0.4401	2.2	0.4642	0.7009	0.3527
0.7	0.5267	2.4	0.4696	0.6801	0.3426
0.8	0.6232	2.6	0.4826	0.6665	0.3361
0.9	0.7383	2.8	0.5050	0.6589	0.3314
1	0.9476	3	0.6506	0.6506	0.3264

TABLE 16.5

n	$\lvert A_0 \rvert$	$\lvert A_1 \rvert$
4	0.6056	0.9123
6	0.6085	0.9110
8	0.6085	0.9108

The problem of diffraction of an E-polarized plane monochromatic wave falling downward onto a simple (one strip l wide per period) lattice was solved for 20 values of the parameter $\kappa \in (0; 3]$. The number of grid points used in the method of discrete singularities was chosen equal to 10.

The values of A_0 and A_q, and then of $|A_0|$ and $|A_q|$, $q = 1, 2$, were calculated for nonattenuating diffraction harmonics. The results are presented in Table 16.4. The calculated values of B_1 and B_2 were equal to zero.

A comparison of the numerical results with those presented in the monograph (Shestopalov 1971) shows that calculation of amplitudes of diffraction spectra by the method of discrete singularities produces quite satisfactory results and is very economical.

For $\kappa = 1, 2$ the problem was solved with the number of grid points equal to 4, 6, and 8 (see Table 16.5).

17
Reduction of Some Boundary Value Problems of Mathematical Physics to Singular Integral Equations

In this chapter it is shown how some boundary value problems of mathematical physics may be reduced to singular integral equations by employing the theory of potential.

17.1. DIFFERENTIATION OF INTEGRAL EQUATIONS OF THE FIRST KIND WITH A LOGARITHMIC SINGULARITY

Let a problem under consideration be reduced somehow to the solution of one of the following integral equations:

$$\frac{1}{\pi}\int_0^{2\pi} \ln\left|\sin\frac{t_0 - t}{2}\right| \gamma(t)\,dt + \int_0^{2\pi} K(t_0, t)\gamma(t)\,dt = f(t_0), \qquad t_0 \epsilon [0, 2\pi], \tag{17.1.1}$$

$$\frac{1}{\pi}\int_{-1}^{1} \ln|t_0 - t|\gamma(t)\,dt + \int_{-1}^{1} K(t_0, t)\gamma(t)\,dt = f(t_0), \qquad t_0 \epsilon (-1, 1), \tag{17.12}$$

where the functions $K(t_0, t)$ and $f(t_0)$ are such that their derivatives with respect to t_0 and t belong to the class $H(\alpha)$ on the corresponding segment, and both the functions and their derivatives are periodic (with the period equal to 2π) in Equation (17.1.1). As a rule, the equations have unique solutions: the former, in the class H on $[0, 2\pi]$; the latter, in the class of functions of the form

$$\gamma(t) = \frac{\psi(t)}{\sqrt{1 - t^2}}, \tag{17.1.3}$$

where function $\psi(t)$ belongs to the class H on $[-1, 1]$.

Let us differentiate each of Equations (17.1.1) and (17.1.2) with respect to t_0. As shown in Gakhov (1977) and Muskhelishvili (1952), in the presence of a logarithmic singularity, the signs before a derivative and an integral may be changed to

$$\frac{1}{2\pi}\int_0^{2\pi}\cot\frac{t_0 - t}{2}\gamma(t)\,dt + \int_0^{2\pi}K'_{t_0}(t_0, t)\gamma(t)\,dt = f'(t_0), \tag{17.1.4}$$

$$\frac{1}{\pi}\int_{-1}^{1}\frac{\gamma(t)\,dt}{t_0 - t} + \int_{-1}^{1}K'_{t_0}(t_0, t)\gamma(t)\,dt = f'(t_0). \tag{17.1.5}$$

As a rule, the latter two equations have, in the class of functions, solutions to an accuracy of a constant, whereas for the former one the condition

$$\int_0^{2\pi}\left[f'(t_0) - \int_0^{2\pi}K'_{t_0}(t_0, t)\gamma(t)\,dt\right]dt_0 = 0 \tag{17.1.6}$$

must also be met. The latter requirement is fulfilled if $K(t_0, t) \equiv 0$. In order to single out a unique solution of Equations (17.1.4) and (17.1.5) it suffices to specify an integral characteristic of the solution. Therefore, instead of Equation (2.1.1), one has to consider the system

$$\frac{1}{2\pi}\int_0^{2\pi}\cot\frac{t_0 - t}{2}\gamma(t)\,dt + \int_0^{2\pi}K'_{t_0}(t_0, t)\gamma(t)\,dt = f'(t_0), \quad t_0 \epsilon[0, 2\pi],$$

$$\frac{1}{\pi}\int_0^{2\pi}\ln\left|\frac{a - t}{2}\right|\gamma(t)\,dt + \int_0^{2\pi}K(a, t)\gamma(t)\,dt = f(a), \quad a \ \epsilon[0, 2\pi],$$

(17.1.7)

where a is a fixed point of the segment $[0, 2\pi]$. Equation (17.1.2) must be substituted by the system

$$\frac{1}{\pi}\int_{-1}^{1}\frac{\gamma(t)\,dt}{t_0 - t} + \int_{-1}^{1} K'_{t_0}(t_0, t)\gamma(t)\,dt = f'(t_0), \qquad t_0 \epsilon(-1,1),$$

$$\frac{1}{\pi}\int_{-1}^{1}\ln|a - t|\gamma(t)\,dt + \int_{-1}^{1} K(a, t)\gamma(t)\,dt = f(a), \qquad a \epsilon(-1,1). \tag{17.1.8}$$

In a similar way one may reduce the following equation with a logarithmic singularity

$$B\int_{-b}^{b} j(t) H_0^{(2)}\left(b\left|\sin\frac{t_0 - t}{2}\right|\right) dt = f(t_0), \qquad t_0 \epsilon[-b, b] \tag{17.1.9}$$

(where $0 \leq b \leq \pi$) to a singular integral equation of the form (17.1.7) or (17.1.8). Equation (17.1.9) is derived when solving various plane problems of radio wave diffraction (Zakharov and Pimenov 1982, Nazarchuk 1989).

17.2. DIRICHLET AND NEUMANN PROBLEMS FOR THE LAPLACE EQUATION

It is well known that numerous applied problems of fluid mechanics reduce to the Dirichlet or Neumann problems for the Laplace equation (Bitsadze 1981, Vladimirov 1976, Zakharov and Pimenov 1982, Abramov and Matveev 1987, Nazarchuk 1989, Tikhonov and Samarskii 1966). Thus, the problems of aerodynamics considered in the preceding chapters of this book are variants of the Neumann problem for the Laplace equation.

Let the Laplace equation

$$\Delta\phi = 0 \tag{17.2.1}$$

be given on the plane *OXY* within a limited closed domain D with a piecewise smooth boundary σ (Muskhelishvili 1952).

It is required to find a solution to the Dirichlet problem subject to the condition

$$\phi/\sigma = f(M_0), \qquad M_0 \in \sigma, \tag{17.2.2}$$

where $f(M_0)$, $f'_x(M_0)$, and $f'_y(M_0)$, $M_0 \in \sigma$, belong to the class H_0 on L, i.e., belong to the class H on all smooth portions of σ.

The problem thus formulated has a unique solution (Tikhonov and Samarskii 1966) that will be sought in the form of a simple layer potential on σ, i.e., in the form

$$\phi(M_0) = \frac{1}{2\pi}\int_\sigma \mu(M)\ln\frac{1}{r_{MM_0}}\,d\sigma_M, \tag{17.2.3}$$

where M_0 is an arbitrary point of the domain D. Because the simple layer potential is continuous on D, the boundary condition (17.2.2) may be written in the form

$$\frac{1}{2\pi}\int_\sigma \mu(M)\ln\frac{1}{r_{MM_0}}\,d\sigma_M = f(M_0), \qquad M_0 \in \sigma. \tag{17.2.4}$$

As long as σ is a closed contour, differentiation permits us to reduce Equation (17.2.4) to a singular integral equation of the first kind with Hilbert kernel (see Section 17.1).

Let us now find a solution to the Neumann problem, i.e., consider the case when the condition

$$\left.\frac{\partial\phi}{\partial n_{M_0}}\right|_\sigma = f(M_0), \qquad M_0 \in \sigma, \tag{17.2.5}$$

is specified, where n_{M_0} is a unit vector of the outward normal of the curve σ at point M_0, and function $f(M_0)$ may have the same form as in Equation (17.2.2).

A solution to the problem will be sought in the form of a double layer potential

$$\phi(M_0) = \frac{1}{2\pi}\int_\sigma g(M)\frac{\partial}{\partial n_M}\ln\frac{1}{r_{MM_0}}\,d\sigma_M, \tag{17.2.6}$$

though traditionally it is sought in the form of a simple layer potential (Bitsadze 1981, Vladimirov 1976, Tikhonov and Samarskii 1966). Because the normal derivative of a double layer potential is continuous in the domain D, boundary condition (17.2.5) acquires the form

$$\frac{1}{2\pi}\int_\sigma g(M)\frac{\partial}{\partial n_{M_0}}\frac{\partial}{\partial n_M}\ln\frac{1}{r_{MM_0}}\,d\sigma_M = f(M_0), \qquad M_0 \in \sigma, \tag{17.2.7}$$

or the form (13.3.8).

Note that Equation (13.3.8) furnishes simultaneously a solution to the external Neumann problem [subject to the condition that circulation due the gradient of function $\Phi(M_0)$ along any path encompassing domain D is zero] and the solution at infinity vanishes. The latter equation, as well as Equation (13.3.8), may be used for solving the Neumann problem when σ is an unclosed contour and domain D is an external one with respect to the latter. In this case Equation (17.2.7) (or Equation (13.3.8)) has a unique solution. However, if σ is a closed contour, then Equation (17.2.7) (or Equation (13.3.8)) has a solution accurate to a constant (Lifanov 1988), and its right-hand side must satisfy the equality

$$\int_\sigma f(M)\, d\sigma = 0. \tag{17.2.8}$$

If σ is an open curve and the second derivatives of the functions presenting it in the parametric form belong to the class $H(\alpha)$ on $[-1, 1]$, then, similarly to what was done in Section 10.4 for an airfoil with an unclosed contour, one can show that Equation (17.2.7) transforms to the equation

$$\frac{1}{2\pi}\int_{-1}^{1}\frac{g(t)\,dt}{(t_0 - t)^2} + \int_{-1}^{1} K(t_0, t) g(t)\, dt = f(t_0), \tag{17.2.9}$$

whose solution has the form $g(t) = \sqrt{1 - t^2}\,\psi(t)$, where $\psi'(t)$ is a function belonging to the class H on $[-1, 1]$.

If σ is a closed contour, then it transforms to the equation

$$\frac{1}{8\pi}\int_{0}^{2\pi}\frac{g(t)\,dt}{\sin^2((t_0 - t)/2)} + \int_{0}^{2\pi} K(t_0, t) g(t)\, dt = f(t_0), \tag{17.2.10}$$

whose solution is periodic and has a derivative that belongs to the class H on $[0, 2\pi]$.

17.3. MIXED BOUNDARY VALUE PROBLEMS

In electrostatics one comes across problems for the Laplace equation subject to the boundary conditions

$$\phi/\sigma_1 = f_1(M_0), \qquad \left.\frac{\partial \phi}{\partial n_{M_0}}\right|_{\sigma_2} = f_2(M_0), \tag{17.3.1}$$

where σ_1 and σ_2 are curves crossing each other at the end points only, and $\sigma_1 \cup \sigma_2 = \sigma$.

A solution to the problem is sought in the form

$$\phi(M_0) = \frac{1}{2\pi}\int_{\sigma_1} \mu(M)\ln\frac{1}{r_{MM_0}}\,d\sigma_M + \frac{1}{2\pi}\int_{\sigma_2} g(M)\frac{\partial}{\partial n_M}\ln\frac{1}{r_{MM_0}}\,d\sigma_M. \tag{17.3.2}$$

Then boundary conditions (17.3.1) result in a system of two integral equations with respect to two functions $\mu(M)$, $M \in \sigma_1$ and $g(M)$, $M \in \sigma_2$:

$$\frac{1}{2\pi}\int_{\sigma_1} \mu(M)\ln\frac{1}{r_{MM_0}}\,d\sigma_M + \frac{1}{2\pi}\int_{\sigma_2} g(M)\frac{\partial}{\partial n_M}\ln\frac{1}{r_{MM_0}}\,d\sigma_M$$

$$= f_1(M_0), \qquad M_0 \in \sigma_1,$$

$$\frac{1}{2\pi}\int_{\sigma_1} \mu(M)\frac{\partial}{\partial n_{M_0}}\ln\frac{1}{r_{MM_0}}\,d\sigma_M + \frac{1}{2\pi}\int_{\sigma_2} g(M)\frac{\partial}{\partial n_{M_0}}\frac{\partial}{\partial n_M}\ln\frac{1}{r_{MM_0}}\,d\sigma_M$$

$$= f_2(M_0), \qquad M_0 \in \sigma_2, \tag{17.3.3}$$

The first equation of the latter system has a logarithmic singularity on σ_1 and hence must be differentiated with respect to a parameter. As a result, one gets a diagonal system of two singular integral equations. A unique solution to the system is singled out with the help of an integral relationship for functions $\mu(M)$ and $g(M)$, obtained by fixing an arbitrary value of the parameter in the first equation (17.3.3). Thus, one is looking for a solution $(\mu(M), g(M))$ to System (17.3.3), where function $\mu(M)$ tends to infinity at the end points of the curve σ_1 as a reciprocal of the square root of the distance from the end points, and function $g(M)$ tends to zero at the end points of curve σ_2 in the same manner.

Conclusion

The advent and wide use of high-speed supercomputers in practically all areas of human activities* is one of the most remarkable manifestations of the scientific and technological revolution. The importance of quantitative description and mathematical modeling is constantly growing. A new method of research based on numerical experiment has emerged and is developing aggressively; it has promoted establishing close relations between the physical, mathematical, and computerized approaches to studying natural phenomena and on developing a universal language of communication for all three. The most practical proved to be discrete methods of description that are most natural, result in simpler mathematical logic, and are adequate to computer languages. Fundamental problems of mathematics and applied problems of mechanics, electrodynamics, mathematical physics, and the like tend to be brought closer together, to interweave, and to affect each other favorably.

Numerical experiments become a virtual laboratory of a mathematician. First, they permit an *a priori* check of the facts forming the foundation of a theory. Second, realistic computer calculations help develop, verify, and perfect numerical methods. Presently, the methods are required to be much more strict and versatile, because they form the production basis of modern science. The methods must not only be correct, but also highly efficient, stable, universal, and logically simple.

This book is dedicated to detailed discussion and verification of a method for solving numerically (with the help of a computer) both one- and multidimensional singular integral equations characteristic of various

*This remark by General Belotserkovsky about supercomputers was not in the original book published in 1985. At that time there were no supercomputers in Russia. It is amazing how quickly they have penetrated all areas of human activity in Russia (G. Ch.).

areas of mathematical physics and applications. The methods of using the equations were illustrated by considering problems of aerodynamics, elastodynamics, and electrodynamics. Under present conditions, the transformations to which the original ideas of the method described in this book were subjected should be considered natural.

1. Heuristic analysis and numerical experiments used for perfecting the method by applying it to special test problems of the theory of airfoils possessing exact solutions.
2. Generalization of the method, its extension onto wings of an arbitrary plan form, and development of the method at a new level with the help of logical considerations and numerical experiments.
3. Rigorous mathematical verification by means of constructing special quadrature formulas for both singular integrals and numerical solution of the equations.
4. Generalization of the method onto a wider class of singular integral equations and its applications in a variety of new areas.

Viability of the method is mostly due to its logic, versatility of the employed approach, and stability of calculated results.

In order to further promote development of the body of mathematics under consideration as well as to meet everyday needs of a number of applied sciences, we would like to draw the reader's attention to the following problems:

1. Many applied problems reduce to solving integral equations and their systems. Because an effective method for solving singular integral equations has been developed and widely tested, it is worthwhile to seek ways to derive equations belonging to the type, some of which were presented in the foregoing text.
2. Construct formulas of the type of quadrature formulas of the method of discrete vortices, for the singular integral

$$I(t_0) = \int_L \frac{\gamma(t)\,dt}{t_0 - t}, \tag{C.1}$$

which would ensure uniform convergence of a fixed accuracy for the cases when:

   a. L is a segment/circle, and $\gamma(t)$ suffers discontinuities of the first kind at a number of internal points.
   b. L has points of bifurcation (say, L is T- or +-like in shape),
   c. At some of the points $\gamma(t)$ suffers power-law integrable discontinuities.

3. Construct quadrature formulas of the type used in the method of discrete vortices for the integral of the theory of a finite-span wing,

$$A = \iint_\sigma \frac{\gamma(x,z)}{(z_0 - z)^2}\left(1 + \frac{x_0 - x}{\sqrt{(x_0 - x)^2 + (z_0 - z)^2}}\right) dx\,dz, \quad \text{(C.2)}$$

which would ensure uniform convergence at reference points, depending on differential properties of the regular part of function $\gamma(x, z)$. For example, if σ is the rectangle $[-b, b] \times [-l, l]$, let us suppose that $\gamma(x, z)$ has the form

$$\sqrt{l^2 - z^2}\sqrt{\frac{b - x}{b + x}}\, u(x, z),$$

where function $u(x, z)$ [the regular part of function $\gamma(x, z)$] has all the derivatives of the rth order that belong to the class $H(\alpha)$ on σ.

4. Consider Problem 3 for the two-dimensional singular integral

$$B = \iint_\sigma \frac{f(x_0, y_0, \theta)u(x, y)\,dx\,dy}{(x_0 - x)^2 + (y - y_0)^2}, \quad \text{(C.3)}$$

where θ is the polar angle of point (x, y) with respect to point (x_0, y_0), and $f(x_0, y_0, \theta)$ is a characteristic of the integral B. Consider the same problem for the integral

$$G = \iint_\sigma \frac{Q(x, y)\,dx\,dy}{\left[(x_0 - x)^2 + (y_0 - y)^2\right]^{3/2}}. \quad \text{(C.4)}$$

5. Construct numerical methods of the type of the method of discrete vortices for the singular integral equation

$$a\gamma(t_0) + \frac{b}{\pi}\int_L \frac{\gamma(t)\,dt}{t_0 - t}\int_L K(t_0, t)\gamma(t)\,dt = f(t_0) \quad \text{(C.5)}$$

(L is one of the curves mentioned previously), ensuring uniform convergence in cases when:

a. $f(t_0)$ suffers one or a number of discontinuities of the first kind or power integrable discontinuities.

b. Both kernel $K(t_0, t)$ and function $f(t_0)$ suffer discontinuities at the same points t_0,

c. a and b are complex numbers.

6. Construct for the Abel equation

$$\int_a^t \frac{\varphi(\tau)\,d\tau}{(t-\tau)^\alpha} + \int_a^t K(t,\tau)\varphi(\tau)\,d\tau = f(t) \qquad \text{(C.6)}$$

a numerical method analogous to the one constructed in Section 12.2, whose convergence rate would depend on differential properties of $f(t)$ and $K(t,\tau)$.

7. For equations

$$\iint_\sigma \frac{\gamma(x,z)}{(z_0-z)^2}\left(1+\frac{x_0-x}{\sqrt{(x_0-x)^2+(z_0-z)^2}}\right)dx\,dz$$

$$= f(x_0, z_0), \qquad \text{(C.7)}$$

$$\iint_\sigma \frac{G(x,z)}{\left[(x_0-x)^2+(z_0-z)^2\right]^{3/2}}\,dx\,dz = f(x_0,z_0): \qquad \text{(C.8)}$$

a. Prove the existence and uniqueness theorems (under corresponding additional conditions),
b. Construct and substantiate numerical methods ensuring uniform convergence over the entire region σ,
c. Do the same in the presence of a summand with a regular kernel in Equations (C.7) and (C.8).

Perhaps you will succeed in finding such a system of orthogonal (on the rectangle $[-b,b]\times[-l,l]$ of the plane OXY) polynomials $P_{n,m}(x,z)$ of degree n in x and m in z, and with the weight $w(x,z) = (\sqrt{b^2-x^2}\sqrt{l^2-z^2})^{-1}$. Using such a system and the presentation

$$G(x,z) = w(x,z)u(x,z), \qquad \text{(C.9)}$$

where $u(x,z)$ is a function of the class H on the rectangle, it will be possible to construct an interpolation polynomial $u_n(x,z)$,

$$u_n(x_i,z_j) = u(x_i,z_j), \qquad i=1,\dots,n,\ j=1,\dots,m, \qquad \text{(C.10)}$$

by using roots (x_i, z_i) of polynomial $P_{n,m}(x,z)$. Then, with the help of the polynomial, it will be possible to construct the system of algebraic equations

$$\sum_{i=1}^{n} \sum_{j=1}^{m} u_n(x_i, z_j) A_{i,j}^{k,m} = f(x_{0k}, z_{0m}), \qquad k = 1, \dots, n, m = 1, \dots, m. \tag{C.11}$$

Here $A_{i,j}^{k,m}$ are coefficients of the quadrature formula for the integral

$$\int_{-b}^{b} \int_{-l}^{l} \frac{w(x, z) u(x, z)}{\left[(x_0 - x)^2 + (z_0 - z)^2\right]^{3/2}} \, dx \, dz, \tag{C.12}$$

constructed with the help of polynomial $u_n(x, z)$ at point (x_{0k}, x_{0m}). It should be noted that points (x_{0k}, z_{0m}) must correspond to the roots of the polynomial–function

$$Q_{n,m}(x_0, z_0) = \int_{-b}^{b} \int_{-l}^{l} \frac{w(x, z) P_{n,m}(x, z)}{\left[(x_0 - x)^2 + (z_0 - z)^2\right]^{3/2}} \, dx \, dz. \tag{C.13}$$

On the other hand, the use of two-dimensional splines also seems to be attractive. Both theoretical analysis and numerical experiments in this area would be of interest.

8. Mathematical verification of the method of discrete vortices in the framework of the nonlinear unsteady problem (see Section 12.4).

In addition to the development of the preceding approaches, numerical experiments have inflicted more radical changes upon numerical methods, because derivation of integral equations for solving nonlinear problems of the theory of wings (and even more so for the study of separated flows past various lifting surfaces) proved to be not only quite laborious but also unnecessary (Belotserkovsky 1968, Belotserkovsky and Nisht 1978). The experience of using discrete models considered in this book permitted us to describe various phenomena by using one and the same language at all stages of research—starting from the construction of a physical scheme and finishing with computer simulation. A vortex proved to be not only a clear-cut physical concept, but also a convenient mathematical abstraction ensuring, in particular, stability of calculations.

Figure C.1 shows an example of calculated separated flow past a plane that is perpendicular to the oncoming flow and travels with speed u_0. As usual, the plate is replaced by a system of discrete vortices, and the

FIGURE C.1. Construction of the Kármán vortex street by the method of discrete vortices.

Chaplygin–Joukowski condition that the flow velocity finite at both edges is used. Because circulations of bound vortices vary in time, the flow past the plate is accompanied by shedding free vortices. This ensures conservation of circulation in accordance with the theorems of hydrodynamics and allows us to meet the Chaplygin–Joukowski conditions at both edges of the plate. The shape of the vortex wake is determined in the process of calculation subject to the conditions that the free vortices be frozen into the fluid and their circulation be independent of time. As a result, we managed to simulate not only integral but local effects, including the loss of stability of vortex surfaces and the formation of clusters composing the Kármán vortex street. The method also proved to be most efficient for solving three-dimensional problems of aerohydrodynamics.

The analysis of the problems called upon somewhat different mathematical formulations as well as required us to extend notions related to both the solution procedure and peculiarities of organization of numerical calculations. The interested reader will find a detailed description of the approach in Belotserkovsky and Nisht (1978).

In connection with the foregoing comments, let us point out the following problems whose actuality becomes ever more evident. First, the mentioned ideas must be actively transferred into other areas of mathematical physics and applied science. The first steps in this direction, made in elastodynamics, have already brought promising results. Second, rigorous verification of new approaches to solving boundary problems should be undertaken; the approaches should be fully based on discrete presentations. Third, it is necessary to analyze and put into practice the possibilities of creating optimal software for high-speed supercomputers. The problem acquires even greater importance in view of the development of a unified mathematical methodology that permits us to solve various problems of different physical nature by organizing conveyerized and co-current calculations.

In conclusion, we would like to draw the reader's attention to some new processes under way in science. Numerical experiment is a qualitatively new method of study possessing a number of most unusual features; moreover, it incorporates what is presently called "artificial intelligence," because after being fully developed, a model finds its way from contradictory situations without employing some special algorithms.

It is common knowledge that no discrete distribution of point vortices allows simulation of a stable Kármán vortex street. However, we were able to do this in the framework of our model of separated flow past a plate by forming finite vortex clusters (Belotserkovsky and Nisht 1978).

As known, flow velocities induced at the ends of a thin vortex surface tend to infinity. Nevertheless, a model describing the formation of an initial vortex wake downstream of a plate resulted in constructing a vortex spiral—a model of the initial Prandtl vortex.

The so-called "effect of beneficial separation" is widely used in modern aviation. It is realized by creating favorable conditions for flow separation at the leading edges of small-aspect-ratio triangular wings or at a "bulb" of a swept-back wing. To induce flow separation, it suffices to sharpen the leading edges. Subsonic flow separates from a sharp edge, because otherwise its velocity would tend to infinity. Under real conditions a separation zone forms, which may be modeled by a vortex surface. The method of discrete vortices permits development of mathematical models of such flows (Belotserkovsky and Nisht 1978), which incorporated a special algorithm for describing the roll-up of vortex sheets into vortex cores. However, in our calculations bow vortex cores were formed automatically (Belotserkovsky and Nisht 1978). Note that the presence of the cores results in increasing lift of a wing and extending the working range of the angles of attack.

The selection of stable vortex structures takes place in the process of either analyzing flow development (unsteady problems) or carrying out iterations (steady problems). Sometimes the results are quite startling, as in the case of a steady jet outflowing through a square orifice into a space filled with a fluid at rest. The pressure at the boundary of the jet must be equal to that in the space. High flow velocities developing in the corner points of the square-like orifice result in a local decrease of pressure and indentation of the jet surface. Because the outflowing fluid is supposed to be incompressible, the continuity equation results in conservation of the cross-sectional area of the jet, which acquires a star-like configuration.

The accumulated experience of mathematical simulation permits us to state that without numerical experiments one cannot comprehend either mathematics or physics of a complicated phenomenon.

Only after carrying out a numerical experiment can a mathematician be sure that the very essence of a problem and its solution are understood. Applied scientists may be sure that they thoroughly understand a more or less complicated phenomenon only if they are able to construct its mathematical model for calculation on a computer.

One of the realistic ways of increasing research efficiency in a number of novel areas, which was prepared by the long period of development of the method of discrete vortices, is the use of the latter's achievements on the basis of the method of discrete singularities.

Along with the development and practical use of the two methods, it is worthwhile to develop specialized software and architecture of multiprocessor computers oriented onto the methods.

The authors hope that this book will help to solve these problems.

References

Abramov, B. D. and Matveev, A. F. 1987. On the Reduction of Boundary Value Problems of the Theory of Neutron Transfer to Singular Integral Equations, Preprint No. 46, Institute for Theoretical and Experimental Physics, Moscow (in Russian).

Abramowitz, M. and Stegun, I. A., Eds. 1964. *Handbook of Mathematical Functions with Formulas, Graphs and Mathematical Tables*. National Bureau of Standards, Washington, D. C.

Afendikova, N. G., Lifanov, I. K., and Matveev, A. F. 1986. About Numerical Solution of Singular Integral Equations with Cauchy and Hilbert Kernels, Preprint No. 73, Institute for Theoretical and Experimental Physics, Moscow (in Russian).

Afendikova, N. G., Lifanov, I. K., and Matveev, A. F. 1987. *Differential Equations*, 23, 1392–1402.

Agranovich, Z. S., Marchenko, V. A., and Shestopalov, V. P. 1962. *J. Tech. Phys.*, 32(4), 381–394.

Aleksandrov, V. M., et al. 1969. Determination of thermoelastic contact pressures in bearings with a polymer covering, in *Contact Problems and Their Engineering Applications*. Galin, L. A., Ed., NIIMASH, Moscow (in Russian).

Aparinov, V. A., Belotserkovsky, S. M., Lifanov, I. K., and Mikhailov, A. A. 1988. *J. Numer. Math. Math. Phys.*, 28, 1558–1566.

Arsenin, V. Ya., Belotserkovsky, C. M., Lifanov, I. K., and Matveev, A. F. 1985. *Differential Equations*, 21, 455–464.

Ashley, H. and Landahl, M. 1967. *Aerodynamics of Wings and Bodies*. Addison-Wesley, Reading, MA.

Bateman, H. and Erdèlyi, A. 1953. *Higher Transcendental Functions*. McGraw-Hill, New York.

Bearman, P. W. and Obasaju, E. D. 1982. *J. Fluid Mechanics*, vol. 119.

Belokopytova, L. V. and Filshtinsky, L. A. 1979. *J. Appl. Math. Mech.*, 43, 138–143.

Belotserkovsky, S. M. 1955a. *J. Appl. Math. Mech.*, 19(2), 159–164.

Belotserkovsky, S. M. 1955b. *J. Appl. Math. Mech.*, 19(4), 410–420.

Belotserkovsky, S. M. 1955c. Studies in Aerodynamics of Modern Lifting Surfaces, Doctoral thesis (in Russian).

Belotserkovsky, S. M. 1967. *The Theory of Thin Wings in Subsonic Flow*. Plenum Press, New York.

Belotserkovsky, S. M. 1968. *Fluid Dynamics*, 3(4).

Belotserkovsky, S. M. 1977. *Annu. Rev. Fluid Mech.*, 9, 469–494.

Belotserkovsky, S. M. 1983. About the methodology of developing, verifying and applying mathematical models in aviation, in *Problems of Cybernetics, Problems of Developing and Applying Mathematical Models*. Moscow (in Russian).

Belotserkovsky, S. M. and Lifanov, I. K. 1981. *Differential Equations*, 17, 1539–1547.

Belotserkovsky, S. M. and Nisht, M. I. 1978. *Separation and Nonseparation of Thin Streamlined Wings in Ideal Fluids*. Nauka, Moscow.

Belotserkovsky, S. M. and Skripach, B. K. 1975. *Aerodynamic Derivatives of an Aircraft and a Wing at Subsonic Flow Velocities*. Nauka, Moscow (in Russian).

Belotserkovsky, S. M., Lifanov, I. K., and Mikhailov, A. A. 1985. Modeling on computer separation streamline profiles with corner points. *Dokl. Akad. Nauk SSSR*, 285, 1348–1352.

Belotserkovsky, S. M., Lifanov, I. K., and Mikhailov, A. A. 1987. *Uchyonye Zapiski ZAGI*, 18, 1–10.

Belotserkovsky, S. M., Lifanov, I. K., and Nisht, M. I. 1978. Discrete-vortex method in aerodynamic problems and multi-dimensional singular integral equations theory, in *Proc. 4th Int. Conf. on Numerical Methods in Fluid Dynamics*, Tbilisi, pp. 30–34.

Belotserkovsky, S. M., Lifanov, I. K., and Soldatov, M. M. 1983. *J. Appl. Math. Mech.* 47, 781–789.

Belotserkovsky, S. M., Nisht, M. I., Ponomarev, A. G., and Rysev, O. V. 1987. *Computer Investigations of Parachutes and Deltaplanes. Mashinostroenie, Moscow (in Russian).*

Belotserkovsky, S. M., Skripach, B. K., and Tabachnikov, V. G. 1971. *A Wing in Nonstationary Gas Flow*. Nauka, Moscow (in Russian).

Berezin, I. S. and Zhidkov, N. P. 1962. *Methods of Calculations*, vols. 1, 2. Nauka, Moscow (in Russian).

Bisplinghoff, R. L., Ashley, H., and Halfman, R. L. 1955. *Aeroelasticity*. Addison-Wesley, Reading, MA.

Bitsadze, A. V. 1981. *Some Classes of Partial Differential Equations*. Nauka, Moscow (in Russian).

Boikov, I. V. 1972. *J. Numer. Math. Math. Phys.*, 12, 1381–1390.

Brychkov, Yu. A. and Prudnikov, A. P. 1977. *Integral Transforms for Generalized Functions*. Nauka, Moscow (in Russian).

Chaplygin, S. A. 1976. Pressure exerted by plane-parallel flow on obstacles, in *Selected Works*. Nauka, Moscow (in Russian).

Chaplygin, S. A. 1976. Results of theoretical studies on the motion of aeroplanes, in *Selected Works*. Nauka, Moscow (in Russian).

Dmitriev, V. I. and Zakharov, E. V. 1967. *Izv. Akad. Nauk SSSR Physics of Earth*, 5.

Dmovska, R. and Kostrov, B. V. 1973. *Arch. Mech. Stosow.*, 25, 421–440.

Dvorak, A. V. 1986. *Trudy Akad. Zhukovskogo*, 1313, 441–453 (in Russian).

Dzhishkariani, A. V. 1979. *J. Numer. Math. Math. Phys.*, 19, 1149–1161.

Dzhishkariani, A. V. 1981. *J. Numer. Math. Math. Phys.*, 21, 355–362.

Dzhvarsheishvili, A. G. 1978. *Izv. VUZov. Mathematics*, 6; 63–72 (in Russian).

Efremov, I. I. 1966. On an approximate solution to the singular integral equation of the theory of wings, in *Dynamics of Systems of Solid and Liquid Bodies*. Kiev (in Russian).

Elliott, D. 1980. Technical Report, No. 144, Math. Dept., Univ. of Tasmania.

Erdogan, F. E. and Gupta, G. D. 1972. *Quart. J. Appl. Math.*, 7, 525–534.

Erdogan, F. E., Gupta, G. D., and Cook, T. S. 1973. The numerical solutions of singular integral equations, in *Methods of Analysis and Solutions of Crack Problems*. Broek, A., Ed., Noordhoff, Leyden.

Falkner, V. M. 1947. Reports and Memoranda No. 2591.

Fan Van Hap 1969. *Vestnik MGU. Series I. Mathematics, Mechanics*, 3 (in Russian).

Fikhtengoltz, G. M. 1959. *A Course in Differential and Integral Calculuses*, vol. 2. Fizmatgiz, Moscow (in Russian).

Gabdulkhaev, B. G. 1968. *Dokl. Akad. Nauk SSSR*, 179, 260–263.

Gabdulkhaev, B. G. 1975–76. *Izv. VUZov. Mathematics*, 7, 30–41 (1975); *Izv. VUZov. Mathematics*, 1, 30–41 (1976) (in Russian).

Gabdulkhaev, B. G. and Dushnov, P. N. 1973. *Izv. VUZov. Mathematics*, 7, 12–24 (in Russian).
Gabedava, G. V. 1974. *Trudy Mat. Inst. Akad. Nauk USSR*, 44, 52–56 (in Russian).
Gagua, I. B. 1960. A special case of application of multiple Cauchy integrals, in *Studies in Modern Problems of the Theory of Complex Variables*. Fizmatgiz, Moscow (in Russian).
Gakhov, F. D. 1958. *Boundary Value Problems*. Fizmatgiz, Moscow (in Russian).
Gakhov, F. D. 1977. *Boundary Value Problems*. Nauka, Moscow (in Russian).
Galin, L. A., Ed. 1976. *Development of the Theory of Contact Problems in the USSR*. Nauka, Moscow (in Russian).
Gandel, Yu. V. 1982. *Theory of Functions, Functional Analysis and Applications*, 38, 15–18.
Gandel, Yu. V. 1983. *Theory of Functions, Functional Analysis and Applications*, 40, 33–36.
Golubev, V. V. 1949. *Lectures on the Theory of Wings*. GITTL, Moscow (in Russian).
Gorlin, S. M. 1970. *Experimental Aerodynamics*. Vysshaya Shkola, Moscow (in Russian).
Goursat, E. 1934. *Cours d'Anasyse Mathématique*. vol. III. Gauthier-Villars, Paris (in French).
Hadamard, J. 1978. *The Cauchy Problem for Linear Partial Hyperbolic Equations*. Nauka, Moscow (in Russian).
Hardy, G. H. 1949. *Divergent Series*. Oxford University Press.
Hönl, H., Maue, A. W., and Westpfahl, K. 1964. *Theory of Diffraction*. Mir, Moscow (in Russian).
Ivanov, V. V. 1968. *The Theory of Approximate Methods and Its Applications to Numerical Solution of Singular Integral Equations*. Naukova Dumka, Kiev (in Russian).
Kakichev, V. A. 1959. *Uchyonye Zapiski Shakht. Ped. Inst.*, 2(6), 25–90 (in Russian).
Kakichev, V. A. 1967. *Izv. VUZov. Mathematics*, 7, 54–64 (in Russian).
Kalandiya, A. I. 1973. *Mathematical Methods of the Two-Dimensional Theory of Elasticity*. Nauka, Moscow (in Russian).
Kantorovich, L. V. and Krylov, V. I. 1952. *Approximate Methods of the Higher Analysis*. GITTL, Moscow (in Russian).
Karpenko, L. N. 1971. *Visnik Kiivs'kogo Universitetu. Mathematics and Technology*, 13, 74–79 (in Ukrainian).
Keldysh, M. V. and Sedov, L. I. 1937. *Dokl. Akad. Nauk SSSR*, 16, 7–10.
Khapaev, M. M. 1982. *Differential Equations*, 18, 498–505 (in Russian).
Khudyakov, G. E. 1973. *Proc. Inst. Mechanics, Moscow Univ.*, 24, 61–67 (in Russian).
Khvedelidze, B. V. 1957. *Trudy Tbilisskogo Matematicheskogo Inst. Akad. Nauk USSR*, 23, 3–158 (in Russian).
Kogan, Kh. M. 1967. *Differential Equations*, 3, 278–293.
Kolesnikov, G. A. 1957. Method for calculating circulation around small-aspect-ratio wings, in *Theoretical Works in Aerodynamics*. Moscow (in Russian).
Korneichuk, A. A. 1964. Quadrature formulas for singular integrals, in *Numerical Methods for Solving Differential and Integral Equations and Quadrature Formulas*. Nauka, Moscow (in Russian).
Kotovsky, V. N., Nisht, M. I., and Fyodorov, R. M. 1980. *Dokl. Akad. Nauk SSSR*, 252, 1341–1345.
Krenk, S. 1975. *Quart. J. Appl. Math.* 33, 225–232.
Krikunov, Yu. I. 1962. Differentiation of singular integrals with the Cauchy kernel and a boundary property of holomorphic functions, in *Boundary Value Problems of the Theory of Functions of a Complex Variable*. Kazan University Press, Kazan (in Russian).
Kupradze, V. D. 1950. *Boundary Value Problems of the Theory of Oscillations and Integral Equations*. GITTL, Leningrad (in Russian).
Lavrent'ev, M. A. 1932. *Trudy ZAGI*, 118, 3–56 (in Russian).
Lavrent'ev, M. A. and Shabat, B. V. 1973. *Methods of Functions of a Complex Variable*. Nauka, Moscow (in Russian).
Lifanov, I. K. 1978a. *Dokl. Akad. Nauk SSSR*, 239, 265–268.
Lifanov, I. K. 1978b. *Dokl. Akad. Nauk SSSR*, 243, 22–25.

Lifanov, I. K. 1979a. *Dokl. Akad. Nauk SSSR*, 249, 1306–1309.

Lifanov, I. K. 1979b. *J. Appl. Math. Mech.*, 43, 184–188.

Lifanov, I. K. 1979c. Topology of curves and numerical solution of singular integral equations of the first kind, in *Proc. 4th Tiraspolskii Symp. General Topology and Its Applications* (in Russian).

Lifanov, I. K. 1980a. *Dokl. Akad. Nauk SSSR*, 255, 1046–1050.

Lifanov, I. K. 1980b. *Izv. VUZov. Mathematics*, 6, 44–51 (in Russian).

Lifanov, I. K. 1980c. *Siberian Math. J.*, 21, 46–60 (in Russian).

Lifanov, I. K. 1981a. About approximate calculation of multi-dimensional singular integral, in *Proc. Semin. Vecua Inst. Applied Mathematics*, No. 15, 13–16 (in Russian).

Lifanov, I. K. 1981b. *Differential Equations*, vol. 17.

Lifanov, I. K. 1988. *Differential Equations*, vol. 24, 110–115.

Lifanov, I. K. and Matveev, A. F. 1983a. Approximate Solution to a Singular Integral Equation on a Segment with Variable Coefficients, Preprint No. 185, Institute for Theoretical and Experimental Physics, Moscow (in Russian).

Lifanov, I. K. and Matveev, A. F. 1983b. *Theory of Functions, Functional Analysis and Applications*, 30, 104–110.

Lifanov, I. K. and Polonskii, Ya. E. 1975. *J. Appl. Math. Mech.*, 39, 742–746.

Lifanov, I. K. and Saakyan, A. V. 1982. *J. Appl. Math. Mech.*, 46, 494–501.

Longhanns, P. and Selbermann, B. 1981. *Math. Nachr.*, 103, 199–244.

Luzin, N. N. 1951. *Integrals and Trigonometric Series*. GITTL, Moscow (in Russian).

Maslov, V. P. 1967. *Dokl. Akad. Nauk SSSR*, 176, 1012–1014.

Matveev, A. F. and Molyakov, N. M. 1988. Method for Numerical Solution of the Problem of Stationary Flow Past a Slightly Curved Airfoil, Preprint No. 88, Institute for Theoretical and Experimental Physics, Moscow (in Russian).

Matveev, A. F. 1982a. Approximate Solution of Some Singular Integro-Differential Equations, Institute for Theoretical and Experimental Physics, Preprint No. 83, Moscow (in Russian).

Matveev, A. F. 1982b. On Self-Regularization of the Problem of Calculating Singular Integrals with Cauchy and Hilbert Kernels in the C-Metrics, Preprint No. 165, Institute for Theoretical and Experimental Physics, Moscow (in Russian).

Matveev, A. F. 1988. On the Construction of Approximate Solutions to Singular Integral Equations of the Second Kind, Preprint No. 88-35, Institute for Theoretical and Experimental Physics, Moscow (in Russian).

Mikhlin, S. G. 1948. *Achievements of Mathematical Sciences*, 3(3), 29–112 (in Russian).

Mikhlin, S. G. 1962. *Multi-Dimensional Singular Integrals and Integral Equations*. Fizmatgiz, Moscow (in Russian).

Mitra, Ed. 1977. *Numerical Methods in Electrodynamics*. Mir, Moscow.

Moiseev, N. N. 1979. *Mathematical Experiment*. Nauka, Moscow (in Russian).

Mokin, Yu. I. 1978. *Mat. Sbornik*, 106(148), 234–264 (in Russian).

Multhoff, H. 1938. *Luftfahrtforschung*, 15, 153–169.

Muskhelishvili, N. N. 1952. *Singular Integral Equations*. Noordhoff, Groningen.

Muskhelishvili, N. N. 1966. *Some Fundamental Problems of Mathematical Theory of Elasticity*. Nauka, Moscow (in Russian).

Nazarchuk, Z. T. 1989. *Numerical Analysis of Wave Diffraction on Cylindrical Structures*. Naukova Dumka, Kiev (in Russian).

Nekrasov, A. I. 1947. *The Theory of Nonstationary Flow Past a Wing*. Akad. Nauk SSSR Press, Moscow (in Russian).

Panasyuk, V. V., Savruk, M. P., and Datsyshin, A. P. 1976. *Stress Distribution in the Neighborhood of Cracks in Plates and Shells*. Naukova Dumka, Kiev (in Russian).

Parton, V. Z. and Perlin, P. I. 1982. *Integral Equations in Elasticity*. Mir, Moscow.

Parton, V. Z. and Perlin, P. I. 1984, *Mathematical Methods of the Theory of Elasticity*, vols. 1, 2. Mir, Moscow.

Petrovsky, I. G. 1981. *Lectures on the Theory of Integral Equations*. Nauka, Moscow (in Russian).
Poltavsky, L. N. 1986. *Trudy Akad. Zhukovskogo*, 1313, 419–423 (in Russian).
Polyakhov, N. N. 1973. *Vestnik LGU. Mathematics, Mechanics, Astronomy*, 7, 115–121 (in Russian).
Privalov, I. I. 1935. *Integral Equations*. ONTI, Moscow (in Russian).
Proskuryakov, I. V. 1967. *Collection of Problems in Linear Algebra*. Nauka, Moscow (in Russian).
Prudnikov, A. P., Brychkov, Yu. A., and Marichev, O. I. 1983. *Integrals and Series. Special Functions*. Nauka, Moscow (in Russian).
Pykhteev, G. N. 1972. *J. Numer. Math. Math. Phys.*, vol. 12.
Riemann, I. S. and Kreps, R. L. 1947. *Trudy ZAGI*, issue 635 (in Russian).
Saakyan, A. V. 1978. *Dokl. Akad. Nauk ArmSSR*, 67, 78–85 (in Russian).
Safronov, I. D. 1956. *Dokl. Akad. Nauk SSSR*, 111, 37–39.
Samarskii, A. A. and Andreev, V. B. 1976. *Difference Methods for Elliptic Equations*. Nauka, Moscow (in Russian).
Sanikidze, J. G. 1970. *Ukrainian Math. J.*, 22, 106–114 (in Russian).
Savruk, M. P. 1981. *Two-Dimensional Problems of the Theory of Elasticity for Bodies with Cracks*. Naukova Dumka, Kiev (in Russian).
Sedov, L. I. 1971–72. *A Course in Continuum Mechanics*, vols. 1–4. Wolters-Noordhoff, Groningen.
Shabat, B. V. 1976. *Introduction to Complex Analysis*, parts I, II. Nauka, Moscow (in Russian).
Sheshko, M. A. 1976. *Izv. VUZov. Mathematics*, no. 12 (in Russian).
Shestopalov, V. P. 1971. *The Method of the Riemann–Hilbert Problem in the Theory of Diffraction and Propagation of Electromagnetic Waves*. Kharkov University Press, Kharkov (in Russian).
Shestopalov, V. P., Litvinenko, L. N., Maslov, S. A., and Sologub, V. G. 1973. *Diffraction of Waves on Lattices*. Kharkov University Press, Kharkov (in Russian).
Shipilov, S. D. 1986. *Trudy Akad. Zhukovskogo*, 1313, 476–487 (in Russian).
Shtaerman, I. Ya. 1949. *Contact Problem of the Theory of Elasticity*. GITTL, Moscow (in Russian).
Sluchanovskaya, Z. P. 1973. Pressure distributions at the surfaces of rectangular, trihedral and semi-circular cylinders and their aerodynamic coefficients, in *Collection of Works*, No. 24. Institute of Mechanics, Moscow State University, Moscow (in Russian).
Spence, D. A. 1958. *The Aeronautical Quarterly*, 9, 287–299.
Stark, I. 1971. *AIAA J.*, 9(9), 244–245.
Thamasphyros, G. J. and Theocaris, P. S. 1977. *BTI*, 17, 458–464.
Tikhonov, A. N. and Arsenin, V. Ya. 1979. *Methods for Solving Incorrect Problems*. Nauka, Moscow (in Russian).
Tikhonov, A. N. and Dmitriev, V. I. 1968. Methods for calculating current distribution in a system of linear vibrators and the direction diagrams of the system, in *Numerical Methods and Programming*, issue 10. Moscow University Press, Moscow (in Russian).
Tikhonov, A. N. and Samarskii, A. A. 1966. *Equations of Mathematical Physics*. Nauka, Moscow (in Russian).
Tumashev, G. G. and Il'insky, N. B. 1967. *Izv. VUZov. Mathematics*, 7, 100–103 (in Russian).
Vladimirov, V. S. 1976. *Equations of Mathematical Physics*. Nauka, Moscow (in Russian).
Voevodin, V. V. 1977. *Numerical Fundamentals of Algebra*. Nauka, Moscow (in Russian).
Volokhin, V. A. 1981. *Izv. VUZov. Mathematics*, 1, 11–14 (in Russian).
Weighardt, K. 1939. *Z. Angew. Math. Mech.*, 19, 257–270.
Weissinger, J. 1947. Technical Memo 1120, NACA.
Zakharov, E. V. and Pimenov, Yu. V. 1982. *Numerical Analysis of Radio Waves Diffraction*. Radio i Svyaz, Moscow (in Russian).

Index

INDEX

A

B

C

D

E

O

P

Q

R

S

T

U

V

W

Z